MW00721355

3rd Edition

Construction Methods and Management

S.W. NUNNALLY

Consulting Engineer

Professor Emeritus
North Carolina State University

REGENTS/PRENTICE HALL
Englewood Cliffs, N. J. 07632

Library of Congress Cataloging-in-Publication Data

Nunnally, S. W.
Construction methods and management / S.W. Nunnally. -- 3rd ed.
p. cm.
Includes bibliographical references and index.
ISBN 0-13-175274-X
1. Building. 2. Construction industry--Management. I. Title.
TH145.N86 1993
624--dc20 91-42585
CIP

Acquisition Editor: *Robert Koehler*
Editorial/production supervision and
interior design: *Penelope Linskey*
Cover design: *Diane Conner*
Prepress Buyer: *Ilene Levy*
Manufacturing Buyer: *Ed O'Dougherty*

Printed in the United States of America
10 9 8 7 6 5 4 3 2 1

ISBN 0-13-175274-X

Prentice-Hall International (UK) Limited, *London*
Prentice-Hall of Australia Pty. Limited, *Sydney*
Prentice-Hall Canada Inc., *Toronto*
Prentice-Hall Hispanoamericana, S. A., *Mexico*
Prentice-Hall of India Private Limited, *New Delhi*
Prentice-Hall of Japan, Inc., *Tokyo*
Simon & Schuster Asia Pte. Ltd., *Singapore*
Editora Prentice-Hall do Brasil, Ltda., *Rio de Janeiro*

To Joan, Steve, Jan, and John

CONTENTS

8 PAVING 172

Part Two Building Construction 196

9 FOUNDATIONS 196

10 CONCRETE CONSTRUCTION 226

11 CONCRETE FORM DESIGN 261

12 WOOD CONSTRUCTION 286

Preface

Construction is a dynamic process that is becoming increasingly complex. Today's construction manager is faced with unprecedented challenges in planning and constructing the public and private facilities required to satisfy society's needs. Among the many new or vastly increased problems facing the industry are increasingly larger and more complex projects, resulting in greater contractor risk, ever-increasing governmental regulation, high cost of borrowed funds, continued inflation and escalation of costs, the need for energy conservation, and a greater demand for renovation and reconstruction services. This situation has produced an increasing demand for technically competent, innovative construction managers. As a result of this demand, the number of construction management and construction engineering programs offered by colleges and universities has grown rapidly. This book is intended to serve the needs of such programs by providing a comprehensive introduction to the methods and management of construction.

This third edition incorporates new and revised material to reflect recent developments in the construction industry. Major additions include the topics of aggregate production, linear scheduling, repair and rehabilitation of transportation facilities, robots in construction, and trenchless excavation. Other new or expanded topics include asphalt plant production, building codes and regulations, concrete formwork design, grouting, hydraulic excavators/shovels, impact rippers, pile foundations, quality control, and soil improvement techniques, in addition to updated text and references.

It is hoped that the material presented is comprehensive enough to serve as the basic text for a variety of construction courses. For an introductory course, upper-division college-level students should be able to cover much of the material in one semester. For more in-depth coverage, the material should be split between two or more courses. Topics may, of course, be omitted or augmented as appropriate to the nature of the course and the desires of the instructor. It is strongly recommended that study of the text be supplemented by visits to construction projects and/or audiovisual material.

It is the author's conviction that all architecture and civil engineering students should be required to complete a basic course in construction as part of their professional studies. Because construction is the product of all design, no design can be a good one unless it can be readily and safely constructed. An appreciation of construction procedures will not only produce a better designer but will also be invaluable to the many graduates of such programs who end up directly involved in the construction industry. This book should serve as a suitable text for such a course.

It would not be possible to produce a book of this type without the assistance of many individuals and organizations. The assistance of construction industry associations and construction equipment manufacturers in providing information and photographs and in permitting reproduction of certain elements of their material is gratefully acknowledged. Where possible, appropriate credit has been provided. I would also like to express my appreciation to my former students, as well as to D.T. Iseley, M.K. Kurtz, N.O. Schmidt, and H.E. Wahls for their helpful comments and suggestions.

Comments from readers regarding errors and suggestions for improvement are solicited.

S.W. Nunnally

1
INTRODUCTION

1-1
THE CONSTRUCTION INDUSTRY

The construction industry is one of the largest industries in the United States, historically accounting for about 10% of the nation's gross national product and employing some 5 million workers. Because construction is an exciting, dynamic process which often provides high income for workers and contractors, it provides an appealing career opportunity. However, the seasonal and sporadic nature of construction work often serves to significantly reduce the annual income of many workers. Construction contracting is also a very competitive business with a high rate of bankruptcy. Therefore, it is essential that construction professionals have a knowledge of the business and management aspects of construction as well as a knowledge of its technical aspects. The topics discussed throughout this book provide a basic understanding of the methods and management of construction.

Companies and individuals engaged in the business of construction are commonly referred to as *construction contractors* (or simply *contractors*) because they operate under a contract arrangement with the owner. Construction contractors may be classified as general contractors or specialty contractors. *General contractors* engage in a wide range of construction activities and execute most major construction projects. When they enter into a contract with an owner to provide complete construction services, they are called *prime contractors. Specialty contractors* limit their activities to one or more construction specialties, such as electrical, plumbing, heating

Figure 1-1 Construction of St. Louis Gateway Arch. (Courtesy of American Institute of Steel Construction)

and ventilating, or earthmoving. Specialty contractors are often employed by a prime contractor to accomplish some specific phase of a construction project. Since the specialty contractors are operating under subcontracts between themselves and the prime contractor, the specialty contractors are referred to as *subcontractors.* Thus, the terms "subcontractor" and "prime contractor" are defined by the contract arrangement involved, not by the work classification of the contractors themselves. For example, a specialty contractor employed by an owner to carry out a particular project might employ a general contractor to execute some phase of the project. In this situation, the specialty contractor becomes the prime contractor for the project and the general contractor becomes a subcontractor.

While the number of construction contractors in the United States has been estimated to exceed 800,000, some 60% of these firms employ three or fewer workers. In one survey, contractors employing 100 or more workers made up less than 1 percent of the nation's construction firms but accounted for about 30 percent of the value of work performed. The trend in recent years has been for the large construction firms to capture an increasing share of the total U.S. construction market.

The major divisions of the construction industry consist of building construction (also called "vertical construction") and heavy construction (also called "horizontal construction"). *Building construction* (Figure 1-2), as the name implies,

Figure 1-2 Modern building construction project. (Courtesy of Brick Association of North Carolina)

involves the construction of buildings. This category may be subdivided into residential and nonresidential building construction. While building construction accounts for a majority of the total U.S. new construction market, many of the largest and most spectacular projects fall in the heavy construction area. *Heavy construction* (Figure 1-3) includes highways, airports, railroads, bridges, canals, harbors, dams, and other major public works. Some other divisions of the construction industry include industrial construction, process plant construction, marine construction, and utility construction.

1-2 THE CONSTRUCTION PROCESS

Project Development and Contract Procedures

The major steps in the construction contracting process include bid solicitation, bid preparation, bid submission, contract award, and contract administration. These activities are described in Chapter 17. However, before the bidding process can take place, the owner must determine the requirements for the project and have the necessary plans, specifications, and other documents prepared. These activities make

Figure 1-3 Heavy Construction Project—Kennedy Space Center launch complex. (U.S. Air Force photograph)

up the project development phase of construction. For major projects, steps in the project development process include:

- Recognition of the need for the project.
- Determination of the technical and financial feasibility of the project.
- Preparation of detailed plans, specifications, and cost estimates for the project.
- Approval by regulatory agencies. This involves ascertaining compliance with zoning regulations, building codes, and environmental and other regulations.

For small projects, many of these steps may be accomplished on a very informal basis. However, for large or complex projects this process may require years to complete.

How Construction Is Accomplished

The principal methods by which facilities are constructed are illustrated in Figures 1-4 to 1-8. These include:

- Construction employing an owner construction force.
- Owner management of construction.
- Construction by a general contractor.
- Construction using a design/build (turnkey) contract.
- Construction utilizing a construction management contract.

Many large industrial organizations, as well as a number of governmental agencies, possess their own construction forces. Although these forces are utilized primarily for performing repair, maintenance, and alteration work, they are often

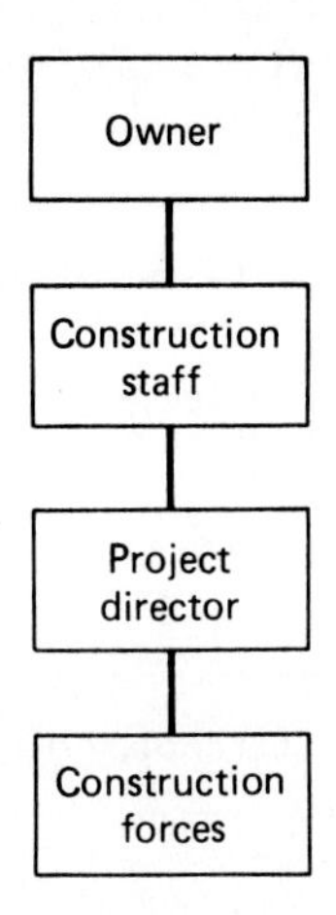

Figure 1-4 Construction employing owner construction forces.

capable of undertaking new construction projects (Figure 1-4). More frequently, owners utilize their construction staffs to manage their new construction (Figure 1-5). The work may be carried out by workers hired directly by the owner (force account), by specialty contractors, or by a combination of these two methods.

Construction by a general contractor operating under a prime contract is probably the most common method of having a facility constructed (Figure 1-6). However, two newer methods of obtaining construction services are finding increasing use: design/build (or turnkey) construction and construction utilizing a construction management contract. Under the *design/build* or *turnkey* construction concept

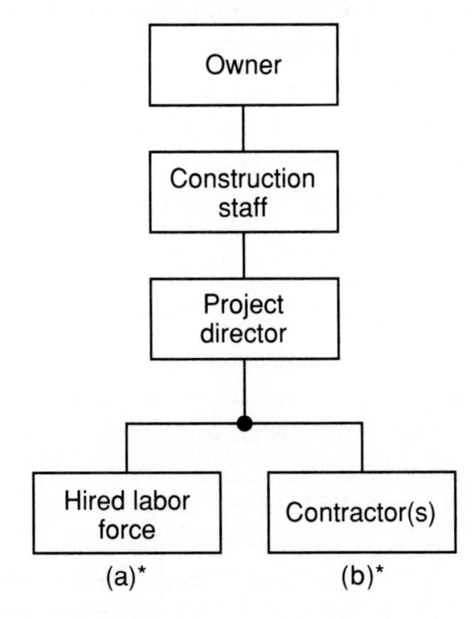

Figure 1-5 Owner-managed construction. [Either (a) or (b) or both may be employed.]

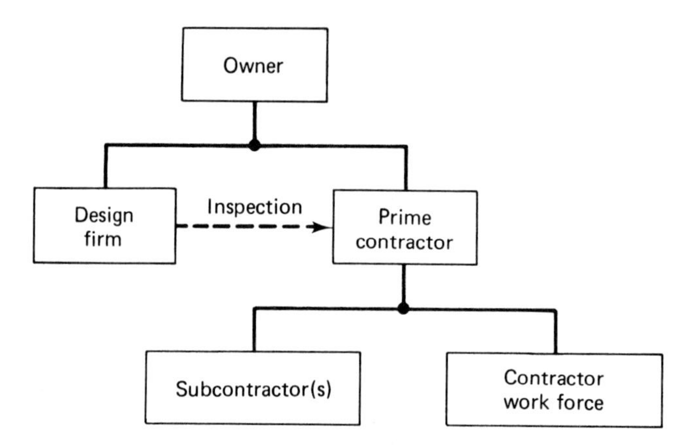

Figure 1-6 Construction by a general contractor.

(Figure 1-7), an owner contracts with a firm to both design and build a facility meeting certain specified (usually, performance-oriented) requirements. Such contracts are frequently utilized by construction firms that specialize in a particular type of construction and possess standard designs which they modify to suit the owner's needs. Since the same organization is both designing and building the facility, coordination problems are minimized and construction can begin before completion of final design. (Under conventional construction procedures it is also possible to begin construction before design has been completed. In this case, the construction contract is normally on a cost-reimbursement basis. This type of construction is referred to as *fast-track* construction.) The major disadvantages of the design/build concept are the difficulty of obtaining competition between suppliers and the complexity of evaluating their proposals.

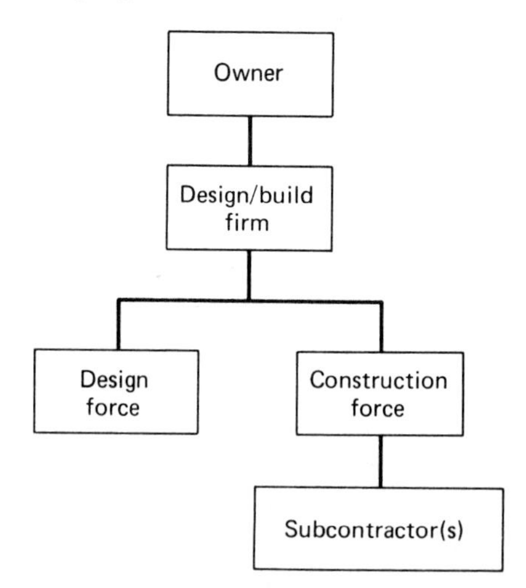

Figure 1-7 Construction employing a design/build firm.

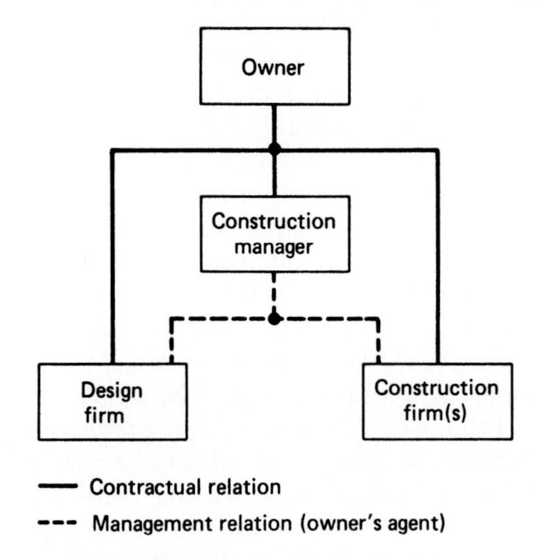

Figure 1-8 Construction utilizing a construction management contract.

Construction of a facility utilizing a *construction management contract* (Figure 1-8) is also somewhat different from the conventional construction procedure. Under the usual arrangement, also known as Agency Construction Management, a professional *construction manager* (CM) acts as the owner's agent to direct both the design and construction of a facility. Three separate contracts are awarded by the owner for design, construction, and construction management of the project. This arrangement offers potential savings in both time and cost compared to conventional procedures, as a result of the close coordination between design and construction. However, opponents of the method point out that the construction manager (CM) typically assumes little or no financial responsibility for the project and that the cost of his/her services may outweigh any savings resulting from improved coordination between design and construction. There is another, less common form of construction management contract known as Guaranteed Maximum Price Construction Management. Under this arrangement, the construction manager guarantees that the project cost will not exceed a specified amount. Under this procedure, which entails a certain amount of contractor risk, the construction contract is also normally held by the construction manager.

1-3 CODES AND REGULATIONS

Projects constructed in most areas of the United States must comply with a number of governmental regulations. These include building codes, zoning regulations, environmental regulations, and contractor licensing laws, among others.

Building Codes

Building codes, which are concerned primarily with public safety, provide minimum design and construction standards for structural and fire safety. In the United States, the Board of Fire Underwriters in 1905 published a Recommended National Building Code which provided minimum standards for fire protection and structural safety. This code, now known as the *Basic/National Building Code,* published by the Building Officials and Code Administrators International, was the only nationally recognized building code for a number of years. Other major building codes now include the *Uniform Building Code,* published by the International Conference of Building Officials, and the *Standard Building Code,* published by the Southern Building Code Congress International. These three model code groups cooperated in 1971 to publish the *CABO One and Two Family Dwelling Code* to provide a simplified standard for the construction of detached one- and two-family dwellings not more than three stories high. Most building codes include provisions for plumbing and mechanical systems. However, electrical work is commonly governed by the *National Electrical Code®,* published by the National Fire Protection Association under the auspices of the American National Standards Institute (ANSI).

Since the national model codes are purely advisory, a building code must be put into effect by local ordinance. While local building codes are usually based on one of the model codes, they often contain local modifications which are unnecessarily restrictive. Such restrictions, along with delays in updating local codes, result in increased building costs. Another problem associated with building codes at the local level is the quality of code administration. The lack of an adequate number of technically qualified building officials often leads to cursory inspections using a checklist approach and discourages contractors from utilizing new materials and procedures.

In most cases, a *building permit* must be obtained before construction of a building can begin. After a permit is issued, the local building department will inspect the project at designated points during construction. The scheduling of these inspections may pose problems for the contractor and often results in construction delays.

Zoning, Environmental, and Other Regulations

Zoning regulations, which control land use, limit the size, type, and density of structures that may be erected at a particular location. Some typical zoning classifications include commercial, residential (with specified density), industrial, office, recreational, and agricultural. Zoning classifications are normally designated by a combination of letters and numbers. As an example, the R-4 zoning classification might represent residential housing with a maximum density of 4 units per acre. To construct a facility not conforming to the current zoning, it would be necessary to obtain a change in zoning or an administrative exception.

Environmental regulations protect the public and environment by controlling such factors as water usage, vehicular traffic, precipitation runoff, waste disposal, and preservation of beaches and wetlands. Large projects, such as new highways and airports, waste disposal facilities, major shopping centers, large industrial plants, large housing developments, and athletic centers may require preparation and approval of an *Environmental Impact Statement* (EIS) describing and quantifying the effect the project will have on the environment. The preparation of an EIS is a complex, time-consuming, and expensive task which should be undertaken only with the assistance of a professional experienced in such matters. If municipal utility services are not available at the project site, additional permits may be required for water treatment plants, wells, sewage treatment facilities, and similar facilities.

The construction profession is also regulated by a number of governmental licensing and certification procedures. Communities having building departments usually require construction contractors to have their professional qualifications verified by licensing or certification. This may be done at the local level or by the state. State certification or licensing often requires satisfactory completion of a comprehensive written examination plus proof of financial capacity and verification of character. A business or occupational license is also normally required of all contractors. In addition, bonding is often required of construction contractors to further protect the public against financial loss.

1-4 STATE OF THE INDUSTRY

Construction Productivity

U.S. construction productivity (output per labor hour), which had shown an average annual increase of about 2% during the period after World War II until the mid-1960s, actually declined between 1965 and 1970. This decline continued until productivity had decreased some 20% between 1970 and 1980. During the same period, inflation in construction costs rose even faster than inflation in the rest of the economy.

Concerned about the effects of declining construction industry productivity on the U.S. economy, the Business Roundtable (an organization made up of the chief executive officers of some 200 major U.S. corporations) sponsored a detailed study of the U.S. construction industry. Completed in 1982, the resulting Construction Industry Cost Effectiveness (CICE) Study is probably the most comprehensive ever made of the U.S. construction industry. The study identified a number of construction industry problems and suggested improvements in the areas of project management, labor training and utilization, and governmental regulation (see references 2 and 5). It concluded that while much of the blame for industry problems should be shared by owners, contractors, labor, and government, many of the problems could be overcome by improved management of the construction effort by owners and contractors with the cooperation of the other parties. Some techniques for improving construction productivity and performance are discussed below and in Chapter 19.

Reducing Construction Costs

Some of the best opportunities for construction cost savings occur in the design process even before construction begins. Some design factors that can reduce construction costs include the use of modular dimensions, grouping plumbing and other equipment to minimize piping and conduit runs, incorporating prefabricated components and assemblies, utilizing economical materials (eliminating "gold plating"), and employing new technology. Injecting constructability considerations into the design process is one of the advantages claimed for the use of the construction management contract arrangement.

Some ways in which productivity can be increased and costs minimized during construction include:

- Good work planning.
- Careful selection and training of workers and managers.
- Efficient scheduling of labor, materials, and equipment.
- Proper organization of work.
- Use of laborsaving techniques such as prefabrication and preassembly.
- Minimizing rework through timely quality control.
- Preventing accidents through good safety procedures.

1-5 CONSTRUCTION MANAGEMENT

Elements of Construction Management

The term *construction management* may be confusing since it has several meanings. As explained earlier, it may refer to the contractual arrangement under which a firm supplies construction management services to an owner. However, in its more general use, it refers to the control of construction's basic resources of workers, material, equipment, money, and time. Thus, every construction supervisor from crew chief to company president is engaged in construction management.

While the principal objectives of every construction manager should be to complete the project on time and within budget, he or she has a number of other important responsibilities. These include safety, worker morale, public and professional relations, productivity improvement, innovation, and improvement of technology.

The scope of construction management is broad and includes such topics as construction contracts, construction methods and materials, production and cost estimating, progress and cost control, quality control, and safety. These are the problems to which succeeding portions of this book are addressed.

Quality Control

The terms *quality control* and *quality assurance* refer to the process of assuring that all elements of the constructed project meet the requirements established by the designer in the project plans and specifications. Regardless of the procedures established, the construction contractor is primarily responsible for quality. Inspections by an owner's representative or government agency provide little more than spot checks to verify that some particular aspect of the project meets minimum standards. Contractors should realize that the extra costs associated with rework are ultimately borne by the contractor, even on cost-type contracts. Poor quality control will result in the contractor gaining a reputation for poor work. The combined effect of increased cost and poor reputation often lead to construction company failure.

Safety and Health

Construction is inherently a dangerous process. Historically, the construction industry has had one of the highest accident rates among all industries. In the United States, concern over the frequency and extent of industrial accidents and health hazards led to the passage of the *Occupational Safety and Health Act of 1970* (OSHA), which established specific safety and health requirements for virtually all industries, including construction. As a result, management concern has tended to focus on OSHA regulations and penalties. However, the financial impact of a poor safety record is often more serious than are OSHA penalties. Safety and health programs and procedures are discussed in detail in Chapter 18.

Organization for Construction

There are probably as many different forms of construction company organization as there are construction firms. However, Figure 1-9 presents an organization chart that is reasonably representative of a general construction company that is medium to large in size.

Reasons for Construction Company Failure

Dun & Bradstreet and others have investigated the reasons for the high rate of bankruptcy in the construction industry. Some of the major factors they have identified include lack of capital, poor cost estimating, inadequate cost accounting, and lack of general management ability. All of these factors can be categorized as elements of

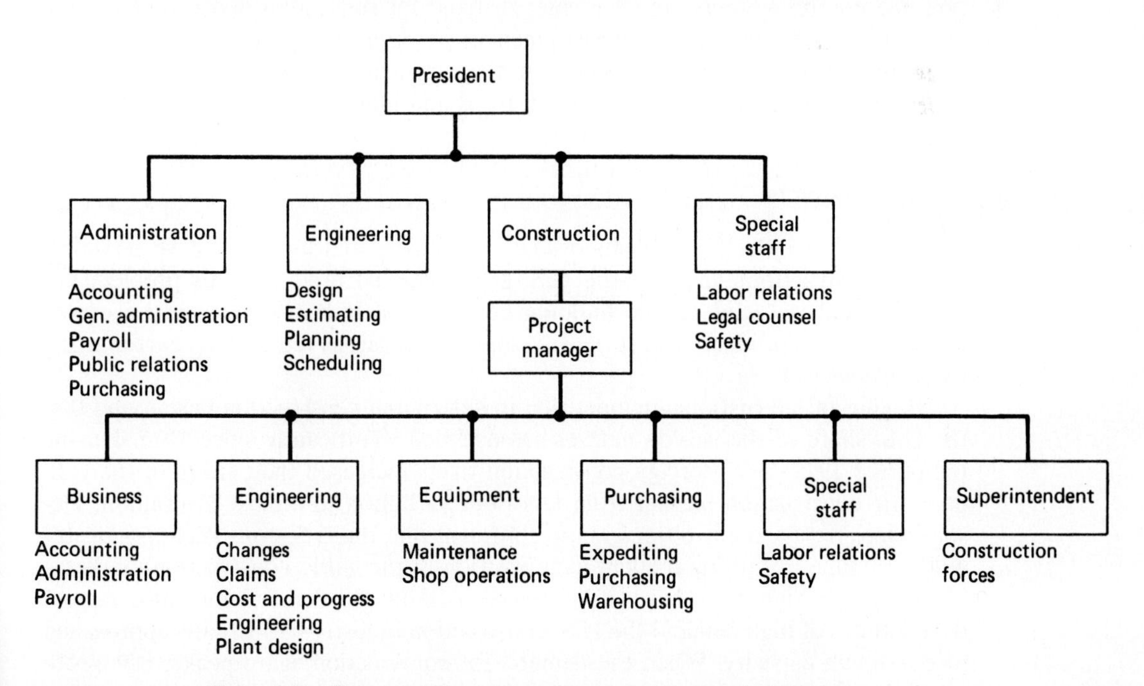

Figure 1-9 Representative construction company organization chart.

poor management. Such studies indicate that at least 90% of all construction company failures can be attributed to inadequate management.

Use of Computers

The wide availability and low cost of personal computers have placed these powerful tools at the disposal of every construction professional. Construction applications of computers are almost unlimited. They include word processing, cost estimating, financial planning, planning and scheduling, project accounting and management, and operations analysis and simulation, among others. These construction applications of computers are discussed in more detail in Chapter 19. Examples of construction applications of personal computers are presented in the end-of-chapter problems of each chapter.

1-6 CONSTRUCTION TRENDS AND PROSPECTS

Construction Trends

Major construction trends noted in recent years include increasingly larger and more complex projects, resulting in greater contractor risk, ever-increasing governmental regulation, high cost of borrowed funds, continuing inflation and escalation of costs, the need for energy conservation, greater use of prefabricated components and building systems, an increasing application of computers and management information and control systems, and a greater demand for renovation and reconstruction services. As a result of these developments, the larger well-managed construction firms are capturing an increasing share of the total construction market. All of these trends indicate an expanding need for trained and innovative construction management personnel.

Problems and Prospects

In recent years industry problems of low productivity and high cost have served to reduce construction's share of the U.S. gross national product. This problem has been particularly acute in the building construction industry because the use of larger and more productive earthmoving equipment has served to keep earthmoving costs relatively stable.

Studies of international competition in design and construction have found that the U.S. share of the world's market has declined significantly since 1975. During the period 1975–1985, foreign construction firms increased their share of the U.S. domestic construction market from less than $1 billion to almost $10 billion. Despite these trends, many observers are confident that the U.S. construction industry will, over time, regain its predominant position in the world construction market.

Although high costs have often served to limit the demand for construction, during times of high demand the U.S. construction industry has actually approached its maximum capacity. When the demand for construction again peaks, it is probable that new forms of construction organization and management as well as new construction methods will have to be developed to meet these demands. In any event, the U.S. construction industry will continue to provide many opportunities and rewards to the innovative, professionally competent, and conscientious construction professional.

In summary, the future of construction appears as dynamic as does its past. There are an abundance of problems, challenges, opportunities, and rewards waiting for those who choose to enter the construction industry. May the contents of this book provide you a firm foundation on which to build an exciting and rewarding career.

PROBLEMS

1. Explain the meaning of the term *horizontal construction.*
2. Evaluate the opportunities for increasing construction cost-effectiveness during the project development phase of the construction process.
3. Briefly explain the purpose of a building code.
4. Briefly explain the importance of the Business Roundtable CICE Study.
5. Explain the difference in meaning between the terms *construction management* and *construction employing a construction management contract.*
6. What is an Environmental Impact Statement (EIS)?
7. What major factors have been identified as responsible for construction company bankruptcy?
8. What is quality control, and whose responsibility is it?
9. Perform individual research to identify three commercially available personal computer (PC) programs suitable for use by a construction professional. Briefly describe the capabilities of each program and indicate its cost.
10. Describe three specific construction applications of a microcomputer or personal computer that you believe would be valuable to a construction professional.

REFERENCES

1. *Construction Dictionary,* 6th ed. Greater Phoenix, Arizona Chapter, National Association of Women in Construction, Phoenix, Ariz., 1985.
2. *Construction Industry Cost Effectiveness Project Report* (22 vols.). The Business Roundtable, New York, 1980–1982.
3. Frein, Joseph P. *Handbook of Construction Management and Organization,* 2nd ed. New York: Van Nostrand Reinhold, 1980.
4. Grow, Thomas A. *Construction: A Guide for the Profession.* Englewood Cliffs, N.J.: Prentice Hall, 1975.
5. More *Construction for the Money.* The Business Roundtable, New York, 1983.
6. Stein, J. Stewart. *Construction Glossary.* New York: Wiley, 1986.

PART ONE
EARTHMOVING AND HEAVY CONSTRUCTION

2
Earthmoving Materials and Operations

2-1
INTRODUCTION TO EARTHMOVING

The Earthmoving Process

Earthmoving (Figure 2-1) is the process of moving soil or rock from one location to another and processing it so that it meets construction requirements of location, elevation, density, moisture content, and so on. Activities involved in this process include excavating, loading, hauling, placing (dumping and spreading), compacting, grading, and finishing. The construction procedures and equipment involved in earthmoving are described in Chapters 3 to 6. Efficient management of the earthmoving process requires accurate estimating of work quantities and job conditions, proper selection of equipment, and competent job management.

Equipment Selection

The choice of equipment to be used on a construction project has a major influence on the efficiency and profitability of the construction operation. Although there are a number of factors that should be considered in selecting equipment for a project, the most important criterion is the ability of the equipment to perform the required work. Among those items of equipment capable of performing the job, the principal criterion for selection should be maximizing the profit or return on the investment produced by the equipment. Usually, but not always, profit is maximized when the

Figure 2-1 Earthmoving: an elevating scraper self-loads. (Courtesy of Clark Equipment Company)

lowest cost per unit of production is achieved. (Chapter 16 provides a discussion of construction economics.) Other factors that should be considered when selecting equipment for a project include possible future use of the equipment, its availability, the availability of parts and service, and the effect of equipment downtime on other construction equipment and operations.

After the equipment has been selected for a project, a plan must be developed for efficient utilization of the equipment. The final phase of the process is, of course, competent job management to assure compliance with the operating plan and to make adjustments for unexpected conditions.

Production of Earthmoving Equipment

The basic relationship for estimating the production of all earthmoving equipment is

$$\text{Production} = \text{Volume per cycle} \times \text{Cycles per hour} \tag{2-1}$$

The term "cycles per hour" must include any appropriate efficiency factors, so that it represents the number of cycles actually achieved (or expected to be achieved) per hour. The Construction Industry Manufacturers Association has developed standard production tables for shovels and draglines which may be used instead of Equation 2-1 for estimating the production of shovels and draglines. Manufacturers may also provide charts or tables for estimating the production of their equipment. The cost per unit of production may be calculated as follows:

$$\text{Cost per unit of production} = \frac{\text{Equipment cost per hour}}{\text{Equipment production per hour}} \quad (2\text{-}2)$$

Methods for determining the hourly cost of equipment operation are explained in Chapter 16.

There are two principal approaches to estimating job efficiency in determining the number of cycles per hour to be used in Equation 2-1. One method is to use the number of effective working minutes per hour to calculate the number of cycles achieved per hour. This is equivalent to using an efficiency factor equal to the number of working minutes per hour divided by 60. The other approach is to multiply the number of theoretical cycles per 60-min hour by a numerical efficiency factor. A table of efficiency factors based on a combination of job conditions and management conditions is presented in Table 2-1. Both methods are illustrated in the example problems.

2-2 EARTHMOVING MATERIALS

Soil and Rock

Soil and rock are the materials that make up the crust of the earth and are, therefore, the materials of interest to the constructor. In the remainder of this chapter, we will consider those characteristics of soil and rock that affect their construction use, including their volume-change characteristics, methods of classification, and field identification.

Table 2-1 Job efficiency factors for earthmoving operations

	*Management Conditions**			
*Job Conditions***	*Excellent*	*Good*	*Fair*	*Poor*
Excellent	0.84	0.81	0.76	0.70
Good	0.78	0.75	0.71	0.65
Fair	0.72	0.69	0.65	0.60
Poor	0.63	0.61	0.57	0.52

*Management conditions include:

Skill, training, and motivation of workers.

Selection, operation, and maintenance of equipment.

Planning, job layout, supervision, and coordination of work.

**Job conditions are the physical conditions of a job that affect the production rate (not including the type of material involved). They include:

Topography and work dimensions.

Surface and weather conditions.

Specification requirements for work methods or sequence.

General Soil Characteristics

Several terms relating to a soil's behavior in the construction environment should be understood. *Trafficability* is the ability of a soil to support the weight of vehicles under repeated traffic. In construction, trafficability controls the amount and type of traffic that can use unimproved access roads, as well as the operation of earthmoving equipment within the construction area. Trafficability is usually expressed qualitatively, although devices are available for quantitative measurement. Trafficability is primarily a function of soil type and moisture conditions. Drainage, stabilization of haul routes, or the use of low-ground-pressure construction equipment may be required when poor trafficability conditions exist. Soil drainage characteristics are important to trafficability and affect the ease with which soils may be dried out. *Loadability* is a measure of the difficulty in excavating and loading a soil. Loose granular soils are highly loadable, whereas compacted cohesive soils and rock have low loadability.

Unit soil weight is normally expressed in pounds per cubic yard or kilograms per cubic meter. Unit weight depends on soil type, moisture content, and degree of compaction. For a specific soil, there is a relationship between the soil's unit weight and its bearing capacity. Thus soil unit weight is commonly used as a measure of compaction, as described in Chapter 5. Soil unit weight is also a factor in determining the capacity of a haul unit, as explained in Chapter 4.

In their natural state, all soils contain some moisture. The moisture content of a soil is expressed as a percentage that represents the weight of water in the soil divided by the dry weight of the soil:

$$\text{Moisture content (\%)} = \frac{\text{Moist weight} - \text{Dry weight}}{\text{Dry weight}} \times 100 \qquad (2\text{-}3)$$

If, for example, a soil sample weighed 120 lb in the natural state and 100 lb after drying, the weight of water in the sample would be 20 lb and the soil moisture content would be 20%. Using Equation 2-3, this is calculated as follows:

$$\text{Moisture content} = \frac{120 - 100}{100} \times 100 = 20\%$$

2-3 SOIL VOLUME-CHANGE CHARACTERISTICS

Soil Conditions

There are three principal conditions or states in which earthmoving material may exist: bank, loose, and compacted. The meanings of these terms are as follows:

- *Bank:* Material in its natural state before disturbance. Often referred to as "in-place" or "in situ." A unit volume is identified as a *bank cubic yard* (BCY) or a *bank cubic meter* (Bm^3).

- *Loose:* Material that has been excavated or loaded. A unit volume is identified as a *loose cubic yard* (LCY) or *loose cubic meter* (Lm^3).
- *Compacted:* Material after compaction. A unit volume is identified as a *compacted cubic yard* (CCY) or *compacted cubic meter* (Cm^3).

Swell

A soil increases in volume when it is excavated because the soil grains are loosened during excavation and air fills the void spaces created. As a result, a unit volume of soil in the bank condition will occupy more than one unit volume after excavation. This phenomenon is called *swell.* Swell may be calculated as follows:

$$\text{Swell (\%)} = \left(\frac{\text{Weight/bank volume}}{\text{Weight/loose volume}} - 1\right) \times 100 \qquad (2\text{-}4)$$

Example 2-1

PROBLEM Find the swell of a soil that weighs 2800 lb/cu yd in its natural state and 2000 lb/cu yd after excavation.

SOLUTION

$$\text{Swell} = \left(\frac{2800}{2000} - 1\right) \times 100 = 40\% \qquad (2\text{-}4)$$

That is, 1 bank cubic yard of material will expand to 1.4 loose cubic yards after excavation.

Shrinkage

When a soil is compacted, some of the air is forced out of the soil's void spaces. As a result, the soil will occupy less volume than it did under either the bank or loose conditions. This phenomenon, which is the reverse of the swell phenomenon, is called *shrinkage.* The value of shrinkage may be determined as follows:

$$\text{Shrinkage (\%)} = \left(1 - \frac{\text{Weight/bank volume}}{\text{Weight/compacted volume}}\right) \times 100 \qquad (2\text{-}5)$$

Soil volume change due to excavation and compaction is illustrated in Figure 2-2. Note that both swell and shrinkage are calculated from the bank (or natural) condition.

Example 2-2

PROBLEM Find the shrinkage of a soil that weighs 2800 lb/cu yd in its natural state and 3500 lb/cu yd after compaction.

SOLUTION

$$\text{Shrinkage} = \left(1 - \frac{2800}{3500}\right) \times 100 = 20\% \qquad (2\text{-}5)$$

Hence 1 bank cubic yard of material will shrink to 0.8 compacted cubic yard as a result of compaction.

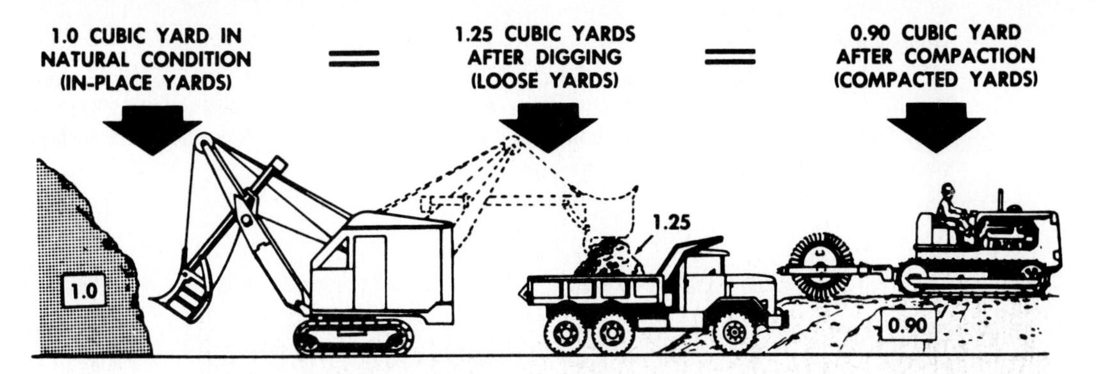

Figure 2-2 Typical soil volume change during earthmoving. (U.S. Department of the Army)

Load and Shrinkage Factors

In performing earthmoving calculations, it is important to convert all material volumes to a common unit of measure. Although the bank cubic yard (or meter) is most commonly used for this purpose, any of the three volume units may be used. A *pay yard* (or meter) is the volume unit specified as the basis for payment in an earthmoving contract. It may be any of the three volume units.

Because haul unit and spoil bank volume are commonly expressed in loose measure, it is convenient to have a conversion factor to simplify the conversion of loose volume to bank volume. The factor used for this purpose is called a *load factor.* A soil's load factor may be calculated by use of Equation 2-6 or 2-7. Loose volume is multiplied by the load factor to obtain bank volume.

$$\text{Load factor} = \frac{\text{Weight/loose unit volume}}{\text{Weight/bank unit volume}} \tag{2-6}$$

or

$$\text{Load factor} = \frac{1}{1 + \text{swell}} \tag{2-7}$$

A factor used for the conversion of bank volume to compacted volume is sometimes referred to as a *shrinkage factor.* The shrinkage factor may be calculated by use of Equation 2-8 or 2-9. Bank volume may be multiplied by the shrinkage factor to obtain compacted volume or compacted volume may be divided by the shrinkage factor to obtain bank volume.

$$\text{Shrinkage factor} = \frac{\text{Weight/bank unit volume}}{\text{Weight/compacted unit volume}} \tag{2-8}$$

or

$$\text{Shrinkage factor} = 1 - \text{shrinkage} \tag{2-9}$$

Example 2-3

PROBLEM A soil weighs 1960 lb/LCY (1163 kg/Lm^3), 2800 lb/BCY (1661 kg/Bm^3), and 3500 lb/CCY (2077 kg/Cm^3). (a) Find the load factor and shrinkage factor for the soil. (b) How many bank cubic yards (Bm^3) and compacted cubic yards (Cm^3) are contained in 1 million loose cubic yards (593,300 Lm^3) of this soil?

SOLUTION

$$\text{(a) Load factor} = \frac{1960}{2800} = 0.70 \qquad (2\text{-}6)$$

$$\left[= \frac{1163}{1661} = 0.70 \right]$$

$$\text{Shrinkage factor} = \frac{2800}{3500} = 0.80 \qquad (2\text{-}8)$$

$$\left[= \frac{1661}{2077} = 0.80 \right]$$

$$\text{(b) Bank volume} = 1{,}000{,}000 \times 0.70 = 700{,}000 \text{ BCY}$$

$$[= 593\,300 \times 0.70 = 415\,310 \text{ Bm}^3]$$

$$\text{Compacted volume} = 700{,}000 \times 0.80 = 560{,}000 \text{ CCY}$$

$$[= 415\,310 \times 0.80 = 332\,248 \text{ Cm}^3]$$

Typical values of unit weight, swell, shrinkage, load factor, and shrinkage factor for some common earthmoving materials are given in Table 2-2.

2-4 SPOIL BANKS

When planning and estimating earthwork, it is frequently necessary to determine the size of the pile of material that will be created by the material removed from the excavation. If the pile of material is long in relation to its width, it is referred to as a *spoil bank*. Spoil banks are characterized by a triangular cross section. If the material is dumped from a fixed position, a *spoil pile* is created which has a conical

Table 2-2 Typical soil weight and volume change characteristics*

	Unit Weight [lb/cu yd (kg/m^3)]						
	Loose	Bank	Compacted	Swell (%)	Shrinkage (%)	Load Factor	Shrinkage Factor
Clay	2310 (1370)	3000 (1780)	3750 (2225)	30	20	0.77	0.80
Common earth	2480 (1471)	3100 (1839)	3450 (2047)	25	10	0.80	0.90
Rock (blasted)	3060 (1815)	4600 (2729)	3550 (2106)	50	−30**	0.67	1.30**
Sand and gravel	2860 (1697)	3200 (1899)	3650 (2166)	12	12	0.89	0.88

*Exact values vary with grain size distribution, moisture, compaction, and other factors. Tests are required to determine exact values for a specific soil.

**Compacted rock is less dense than is in-place rock.

shape. To determine the dimensions of spoil banks or piles, it is first necessary to convert the volume of excavation from in-place conditions (BCY or Bm^3) to loose conditions (LCY or Lm^3). Bank or pile dimensions may then be calculated using Equations 2-10 to 2-13 if the soil's angle of repose is known.

A soil's *angle of repose* is the angle that the sides of a spoil bank or pile naturally form with the horizontal when the excavated soil is dumped onto the pile. The angle of repose (which represents the equilibrium position of the soil) varies with the soil's physical characteristics and its moisture content. Typical values of angle of repose for common soils are given in Table 2-3.

Triangular Spoil Bank

$$\text{Volume} = \text{Section area} \times \text{Length}$$

$$B = \left(\frac{4V}{L \times \tan R}\right)^{1/2} \tag{2-10}$$

$$H = \frac{B \times \tan R}{2} \tag{2-11}$$

where B = base width (ft or m)
H = pile height (ft or m)
L = pile length (ft or m)
R = angle of repose (deg)
V = pile volume (cu ft or m^3)

Conical Spoil Pile

$$\text{Volume} = \frac{1}{3} \times \text{Base area} \times \text{Height}$$

$$D = \left(\frac{7.64\ V}{\tan R}\right)^{1/3} \tag{2-12}$$

$$H = \frac{D}{2} \times \tan R \tag{2-13}$$

where D is the diameter of the pile base (ft or m).

Table 2-3 Typical values of angle of repose of excavated soil

Material	*Angle of Repose (deg)*
Clay	35
Common earth, dry	32
Common earth, moist	37
Gravel	35
Sand, dry	25
Sand, moist	37

Example 2-4

PROBLEM Find the base width and height of a triangular spoil bank containing 100 BCY (76.5 Bm^3) if the pile length is 30 ft (9.14 m), the soil's angle of repose is 37°, and its swell is 25%.

SOLUTION

$$\text{Loose volume} = 27 \times 100 \times 1.25 = 3375 \text{ cu ft}$$

$$[= 76.5 \times 1.25 = 95.6 \text{ m}^3]$$

$$\text{Base width} = \left(\frac{4 \times 3375}{30 \times \tan 37°}\right)^{1/2} = 24.4 \text{ ft} \qquad (2\text{-}10)$$

$$\left[= \left(\frac{4 \times 95.6}{9.14 \times \tan 37°}\right)^{1/2} = 7.45 \text{ m}\right]$$

$$\text{Height} = \frac{24.4}{2} \times \tan 37° = 9.2 \text{ ft} \qquad (2\text{-}11)$$

$$\left[= \frac{7.45}{2} \times \tan 37° = 2.80 \text{ m}\right]$$

Example 2-5

PROBLEM Find the base diameter and height of a conical spoil pile that will contain 100 BCY (76.5 Bm^3) of excavation if the soil's angle of repose is 32° and its swell is 12%.

SOLUTION

$$\text{Loose volume} = 27 \times 100 \times 1.12 = 3024 \text{ cu ft}$$

$$[= 76.5 \times 1.12 = 85.7 \text{ m}^3]$$

$$\text{Base diameter} = \left(\frac{7.64 \times 3024}{\tan 32°}\right)^{1/3} = 33.3 \text{ ft} \qquad (2\text{-}12)$$

$$\left[= \left(\frac{7.64 \times 85.7}{\tan 32°}\right)^{1/3} = 10.16 \text{ m}\right]$$

$$\text{Height} = \frac{33.3}{2} \times \tan 32° = 10.4 \text{ ft} \qquad (2\text{-}13)$$

$$\left[= \frac{10.16}{2} \times \tan 32° = 3.17 \text{ m}\right]$$

2-5 SOIL IDENTIFICATION AND CLASSIFICATION

Fundamental Soil Types

Soil is considered to be composed of five fundamental soil types: gravel, sand, silt, clay, and organic material. *Gravel* is composed of individual particles larger than about ¼ in. (0.6 cm) in diameter but smaller than 3 in. (7.6 cm) in diameter. Rock particles larger than 3 in. (7.6 cm) in diameter are called cobbles or boulders. *Sand* is material smaller than gravel but larger than the No. 200 sieve opening (0.07 mm). *Silt* particles pass the No. 200 sieve but are larger than 0.002 mm. *Clay* is composed of particles less than 0.002 mm in diameter. *Organic soils* contain partially decomposed vegetable matter.

Because a soil's characteristics are largely determined by the amount and type of each of the five basic soils present, these factors are used for the identification and classification procedures described in the remainder of this section.

Soil Classification Systems

There are two principal soil classification systems used for design and construction in the United States. These are the *Unified System* and the *AASHTO* [American Association of State Highway and Transportation Officials, formerly known as the American Association of State Highway Officials (AASHO)] *System.* In both systems soil particles 3 in. or larger in diameter are removed before performing classification tests.

The *liquid limit* (LL) of a soil is the water content (expressed in percentage of dry weight) at which the soil will just start to flow when subjected to a standard shaking test. The *plastic limit* (PL) of a soil is the moisture content in percent at which the soil just begins to crumble when rolled into a thread ⅛ in. (0.3 cm) in diameter. The *plasticity index* (PI) is the numerical difference between the liquid and plastic limits and represents the range in moisture content over which the soil remains plastic.

The Unified System assigns a two-letter symbol to identify each soil type. Field classification procedures are given in Table 2-4. Soils that have less than 50% by weight passing the No. 200 sieve are further classified as *coarse-grained soils,* whereas soils that have more than 50% by weight passing the No. 200 sieve are *fine-grained soils.* Gradation curves for well-graded and poorly graded sand and gravel are illustrated in Figure 2-3.

Under the AASHTO System, soils are classified as types A-1 through A-7, corresponding to their relative value as subgrade material. Classification procedures for the AASHTO system are given in Table 2-5.

Field Identification of Soil (Unified System)

When identifying soil in connection with construction operations, adequate time and laboratory facilities are frequently not available for complete soil classification. The use of the procedures described below together with Table 2-4 should permit a reasonably accurate soil classification to be made in a minimum of time.

All particles over 3 in. in diameter are first removed. The soil particles are then separated visually at the No. 200 sieve size: this corresponds to the smallest particles that can be seen by the naked eye. If more than 50% of the soil be weight is larger than the No. 200 sieve, it is a coarse-grained soil. The coarse particles are then divided into particles larger and smaller than ¼ in. (0.6 cm) in diameter. If over 50% of the coarse fraction (by weight) is larger than ¼ in. (0.6 cm) in diameter, the soil is classified as gravel; otherwise, it is sand. If less than 10% by weight of the total sample is smaller than the No. 200 sieve, the second letter is assigned based on grain

Table 2-4 Unified system of soil classification—field identification

Coarse-Grained Soils (Less Than 50% Pass No. 200 Sieve)

Symbol	Name	Percent of Coarse Fraction Less Than ¼ in.	Percent of Sample Smaller Than No. 200 Sieve	Comments
GW	Well-graded gravel	50 max.	< 10	Wide range of grain sizes with all intermediate sizes
GP	Poorly graded gravel	50 max.	< 10	Predominately one size or some sizes missing
SW	Well-graded sand	51 min.	< 10	Wide range of grain sizes with all intermediate sizes
SP	Poorly graded sand	51 min.	< 10	Predominately one size or some sizes missing
GM	Silty gravel	50 max.	≥ 10	Low-plasticity fines (see ML below)
GC	Clayey gravel	50 max.	≥ 10	Plastic fines (see CL below)
SM	Silty sand	51 min.	≥ 10	Low-plasticity fines (see ML below)
SC	Clayey sand	51 min.	≥ 10	Plastic fines (see CL below)

*Tests on Fraction Passing No. 40 Sieve (Approx. 1/64 in. or 0.4 mm)**

Fine-Grained Soils (50% or More Pass No. 200 Sieve)

Symbol	Name	Dry Strength	Shaking	Other
ML	Low-plasticity silt	Low	Medium to quick	
CL	Low-plasticity clay	Low to medium	None to slow	
OL	Low-plasticity organic	Low to medium	Slow	Color and odor
MH	High-plasticity silt	Medium to high	None to slow	
CH	High-plasticity clay	High	None	
OH	High-plasticity organic	Medium to high	None to slow	Color and odor
Pt	Peat	Identified by dull brown to black color, odor, spongy feel, and fibrous texture		

*Laboratory classification based on liquid limit and plasticity index values.

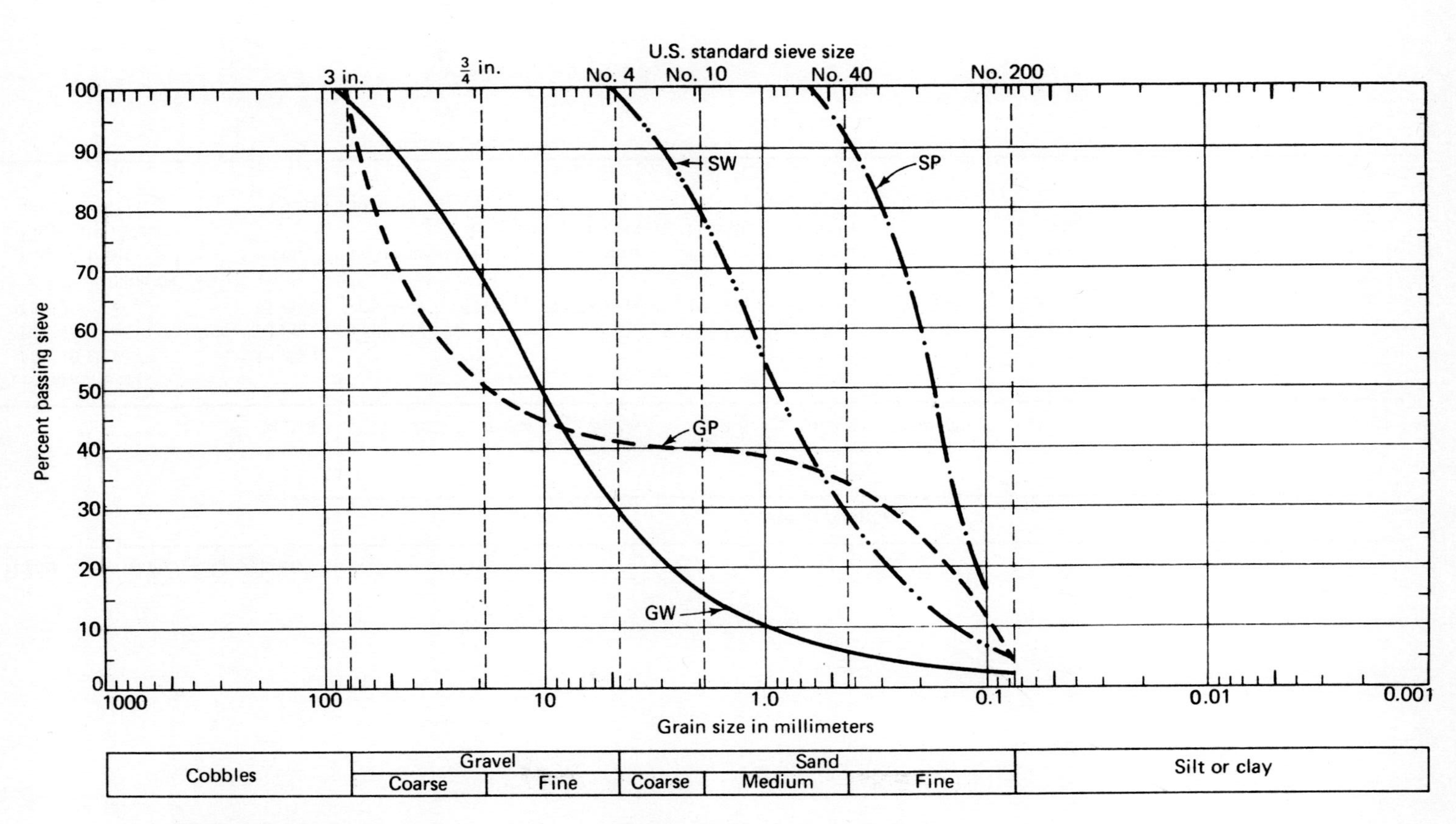

Figure 2-3 Typical gradation curves for coarse-grained soils. (U.S. Army Engineer School)

Table 2-5 AASHTO system of soil classification

	Group Number										
	A-1		*A-2*								
	A-1-a	*A-1-b*	*A-2-4*	*A-2-5*	*A-2-6*	*A-2-7*	*A-3*	*A-4*	*A-5*	*A-6*	*A-7*
Percent passing											
No. 10 sieve	50 max.										
No. 40 sieve	30 max.	50 max.					51 min.				
No. 200 sieve	15 max.	25 max.	35 max.	35 max.	35 max.	35 max.	10 max.	36 min.	36 min.	36 min.	36 min.
Fraction passing No. 40											
Liquid limit	6 max.	6 max.	40 max.	41 min.	40 max.	41 min.		40 max.	41 min.	40 max.	41 min.
Plasticity index			10 max.	10 max.	11 min.	11 min.		10 max.	10 max.	11 min.	11 min.
Typical material	Gravel and sand		Silty or clayey sand or gravel				Fine sand	Silt	Silt	Clay	Clay

size distribution. That is, it is either well graded (W) or poorly graded (P). If more than 10% of the sample is smaller than the No. 200 sieve, the second classification letter is based on the plasticity of the fines (L or H), as shown in the table.

If the sample is fine-grained (more than 50% by weight smaller than the No. 200 sieve), classification is based on dry strength and shaking tests of the material smaller than 1⁄64 in. (0.4 mm) in diameter.

DRY STRENGTH TEST

Mold a sample into a ball about the size of a golf ball to the consistency of putty, adding water as needed. Allow the sample to dry completely. Attempt to break the sample using the thumb and forefinger of both hands. If the sample cannot be broken, the soil is highly plastic. If the sample breaks, attempt to powder it by rubbing it between the thumb and forefinger of one hand. If the sample is difficult to break and powder, it has medium plasticity. Samples of low plasticity will break and powder easily.

SHAKING TEST

Form the material into a ball about 3⁄4 in. (19 mm) in diameter, adding water until the sample does not stick to the fingers as it is molded. Put the sample in the palm of the hand and shake vigorously. Observe the speed with which water comes to the surface of the sample to produce a shiny surface. A rapid reaction indicates a nonplastic silt.

Construction Characteristics of Soils

Some important construction characteristics of soils as classified under the Unified System are summarized in Table 2-6.

Table 2-6 Construction characteristics of soils (Unified System)

Soil Type	*Symbol*	*Drainage*	*Construction Workability*	*Suitability for Subgrade (No Frost Action)*	*Suitability for Surfacing*
Well-graded gravel	GW	Excellent	Excellent	Good	Good
Poorly graded gravel	GP	Excellent	Good	Good to excellent	Poor
Silty gravel	GM	Poor to fair	Good	Good to excellent	Fair
Clayey gravel	GC	Poor	Good	Good	Excellent
Well-graded sand	SW	Excellent	Excellent	Good	Good
Poorly graded sand	SP	Excellent	Fair	Fair to good	Poor
Silty sand	SM	Poor to fair	Fair	Fair to good	Fair
Clayey sand	SC	Poor	Good	Poor to fair	Excellent
Low-plasticity silt	ML	Poor to fair	Fair	Poor to fair	Poor
Low-plasticity clay	CL	Poor	Fair to good	Poor to fair	Fair
Low-plasticity organic	OL	Poor	Fair	Poor	Poor
High-plasticity silt	MH	Poor to fair	Poor	Poor	Poor
High-plasticity clay	CH	Very poor	Poor	Poor to fair	Poor
High-plasticity organic	OH	Very poor	Poor	Very poor to poor	Poor
Peat	Pt	Poor to fair	Unsuitable	Unsuitable	Unsuitable

PROBLEMS

1. Observations indicate that an excavator carries an average bucket load of 3.0 LCY (2.3 Lm^3) per cycle. The soil's load factor is 0.80. Cycle time averages 0.35 min. Job conditions are rated as fair, and management conditions are rated as good. Estimate the hourly production in BCY (Bm^3) for the excavator.
2. The hourly cost of a hydraulic shovel is $65 and of a truck is $35. If an equipment fleet consisting of one shovel and six trucks achieves a production of 300 BCY (229 Bm^3/h), what is the unit cost of loading and hauling?
3. A soil weighs 2400 lb/cu yd (1089 kg/m^3) loose, 3050 lb/cu yd (1383 kg/m^3) in-place, and 3600 lb/cu yd (1633 kg/m^3) compacted. Find the swell and shrinkage of this soil.
4. A soil weighs 2480 lb/cu yd (1471 kg/m^3) loose, 3100 lb/cu yd (1839 kg/m^3) in-place, and 3450 lb/cu yd (2047 kg/m^3) compacted. How many bank cubic yards (meters) of this material can a 10-cu yd (7.6-m^3)(heaped load) truck haul assuming no weight limitation?
5. An earth-fill dam requires a compacted volume of 3 million cubic yards (2.3×10^6 m^3). How many loose cubic yards (meters) of the soil of Problem 3 must be hauled to construct the dam?
6. Find the size of a conical spoil pile resulting from the excavation of 700 BCY (535 Bm^3) of gravel.
7. A rectangular ditch having a cross-sectional area of 24 sq ft (2.2 m^3) is being excavated in clay. The soil's angle of repose is 35° and its swell is 30%. Find the height and base width of the triangular spoil bank that will result from the trench excavation.
8. Using the AASHTO System, classify the soil whose test results are as follows:

   Passing No. 200 sieve = 34%
   Liquid limit = 30
   Plasticity index = 8

9. Write a computer program to calculate the height and base diameter (ft or m) of a conical spoil pile that will result from a rectangular excavation. Input should include the width, length, and depth of the excavation (ft or m), or the bank volume of the excavation, as well as the soil's angle of repose (deg) and swell (%). Solve Problem 6 using your program.
10. Write a computer program to determine the height and base width (ft or m) of the triangular spoil bank that will result from a rectangular trench excavation. Input should include the ditch width and depth (ft or m) as well as the soil's angle of repose (deg) and swell (%). Solve Problem 7 using your program.

REFERENCES

1. Ahlvin, Robert G., and Vernon A. Smoots. *Construction Guide for Soils and Foundations,* 2nd ed. New York: Wiley, 1988.
2. Bowles, Joseph E. *Engineering Properties of Soils and Their Measurements,* 3rd ed. New York: McGraw-Hill, 1985.

3. *Caterpillar Performance Handbook,* 21st ed. Caterpillar Inc., Peoria, Ill., 1990.
4. NUNNALLY, S.W. *Managing Construction Equipment.* Englewood Cliffs, N. J.: Prentice Hall, 1977.
5. *Production and Cost Estimating of Material Movement with Earthmoving Equipment.* Terex Corporation, Hudson, Ohio, 1981.
6. *Soils Manual for Design of Asphalt Pavement Structures* (MS-10), 2nd ed. The Asphalt Institute, Lexington, Ky., 1978.

3
Excavating and Lifting

3-1 INTRODUCTION

Excavating and Lifting Equipment

An *excavator* is defined as a power-driven digging machine. The major types of excavators used in earthmoving operations include hydraulic excavators and the members of the cable-operated crane-shovel family (shovels, draglines, hoes, and clamshells). Dozers, loaders, and scrapers can also serve as excavators. In this chapter we focus on hydraulic excavators and the members of the crane-shovel family used for excavating and lifting operations. Operations involving the dozer, loader, and scraper are described in Chapter 4. Special considerations involved in rock excavation are discussed in Chapter 6.

Hydraulic Excavators

The *hydraulic excavator,* illustrated in Figure 3-1 with a backhoe front end, is a hydraulically powered machine that has largely replaced the cable-operated backhoe and shovel of the crane-shovel family. Hydraulic excavators have a number of advantages over cable-operated excavators, including faster cycle time, higher bucket penetrating force, more precise digging, and easier operator control. In addition to backhoe and shovel front ends, there are a number of attachments available for hydraulic excavators. Among these are clamshells, augers, vibratory plate compactors, and hammers. Most of these attachments are designed to fit the backhoe front end.

Figure 3-1 Hydraulic excavator-backhoe. (Courtesy of Caterpillar Inc.)

Excavator Production

To utilize Equation 2-1 for estimating the production of an excavator, it is necessary to know the volume of material actually contained in one bucket load. The methods by which excavator bucket and dozer blade capacity are rated are given in Table 3-1. *Plate line capacity* is the bucket volume contained within the bucket when following the outline of the bucket sides. *Struck capacity* is the bucket capacity when the load is struck off flush with the bucket sides. *Water line capacity* assumes a level of material flush with the lowest edge of the bucket (i.e., the material level corresponds to the water level that would result if the bucket were filled with water). *Heaped volume* is the maximum volume that can be placed in the bucket without spillage based on a specified angle of repose for the material in the bucket.

Since bucket ratings for the cable shovel, dragline, and cable backhoe are based on struck volume, it is often assumed that the heaping of the buckets will compen-

Table 3-1 Bucket-capacity rating methods

Machine	*Rated Bucket Capacity*
Backhoe and shovel	
Cable	Struck volume
Hydraulic	Heaped volume at 1:1 angle of repose
Clamshell	Plate line or water line volume
Dragline	90% of struck volume
Loader	Heaped volume at 2:1 angle of repose

sate for the swell of the soil. That is, a 5-cu yd bucket would be assumed to actually hold 5 bank cu yd of material. A better estimate of the volume of material in one bucket load will be obtained if the nominal bucket volume is multiplied by a *bucket fill factor* or bucket efficiency factor. Suggested values of bucket fill factor for common soils are given in Table 3-2. The most accurate estimate of bucket load is obtained by multiplying the heaped bucket volume (loose measure) by the bucket fill factor. If desired, the bucket load may be converted to bank volume by multiplying its loose volume by the soil's load factor. This procedure is illustrated in Example 3-1.

Example 3-1

PROBLEM Estimate the actual bucket load in bank cubic yards for a loader bucket whose heaped capacity is 5 cu yd (3.82 m^3). The soil's bucket fill factor is 0.90 and its load factor is 0.80.

SOLUTION

$$\text{Bucket load} = 5 \times 0.90 = 4.5 \text{ LCY} \times 0.80 = 3.6 \text{ BCY}$$
$$[= 3.82 \times 0.90 = 3.44 \text{ Lm}^3 \times 0.80 = 2.75 \text{ Bm}^3]$$

The Crane-Shovel Family

In 1836, William S. Otis developed a machine that mechanically duplicated the motion of a man digging with a hand shovel. From this machine evolved the family of construction machines known as the *crane-shovel.* Members of this family include the mobile crane, shovel, dragline, backhoe, clamshell, and pile driver, shown in Figure 3-2.

The crane-shovel consists of three major assemblies: a carrier or mounting, a revolving superstructure containing the power and control units (also called the revolving deck or turntable), and a front-end attachment. Carriers available include crawler, truck, and wheel mountings, as shown in Figure 3-3. The crawler mounting provides excellent on-site mobility and its low ground pressure enables it to operate in areas of low trafficability. Crawler mountings are widely used for drainage and trenching work as well as for rock excavation. Truck and wheel mountings provide greater mobility between job sites but are less stable than crawler mountings and require better surfaces over which to operate. Truck mountings use a modified truck chassis as a carrier and thus have separate stations for operating the carrier and the revolving superstructure. Wheel mountings, on the other hand, use a single opera-

Table 3-2 Bucket fill factors for excavators

Material	*Bucket Fill Factor*
Common earth, loam	0.80–1.10
Sand and gravel	0.90–1.00
Hard clay	0.65–0.95
Wet clay	0.50–0.90
Rock, well blasted	0.70–0.90
Rock, poorly blasted	0.40–0.70

tor's station to control both the carrier and the crane-shovel mechanism. Truck mountings are capable of highway travel speeds of 50 mi/h or more, whereas wheel mountings are usually limited to 30 mi/h or less.

The name of a particular member of the crane-shovel family is determined by the front-end attachment used. Thus a crane-shovel with a shovel attachment is referred to simply as a shovel.

In this chapter we discuss the principles of operation, methods of employment, and techniques for estimating production of all members of the crane-shovel family except for the pile driver. Pile drivers and their operation are discussed in Chapter 9.

Trenchless Excavation

There is a growing demand for methods of installing utility systems below ground with minimum open excavation. Such construction is often called *trenchless excavation.* Trenchless excavation is much less disruptive to urban areas than are conventional trenching methods. While a number of different techniques are used for

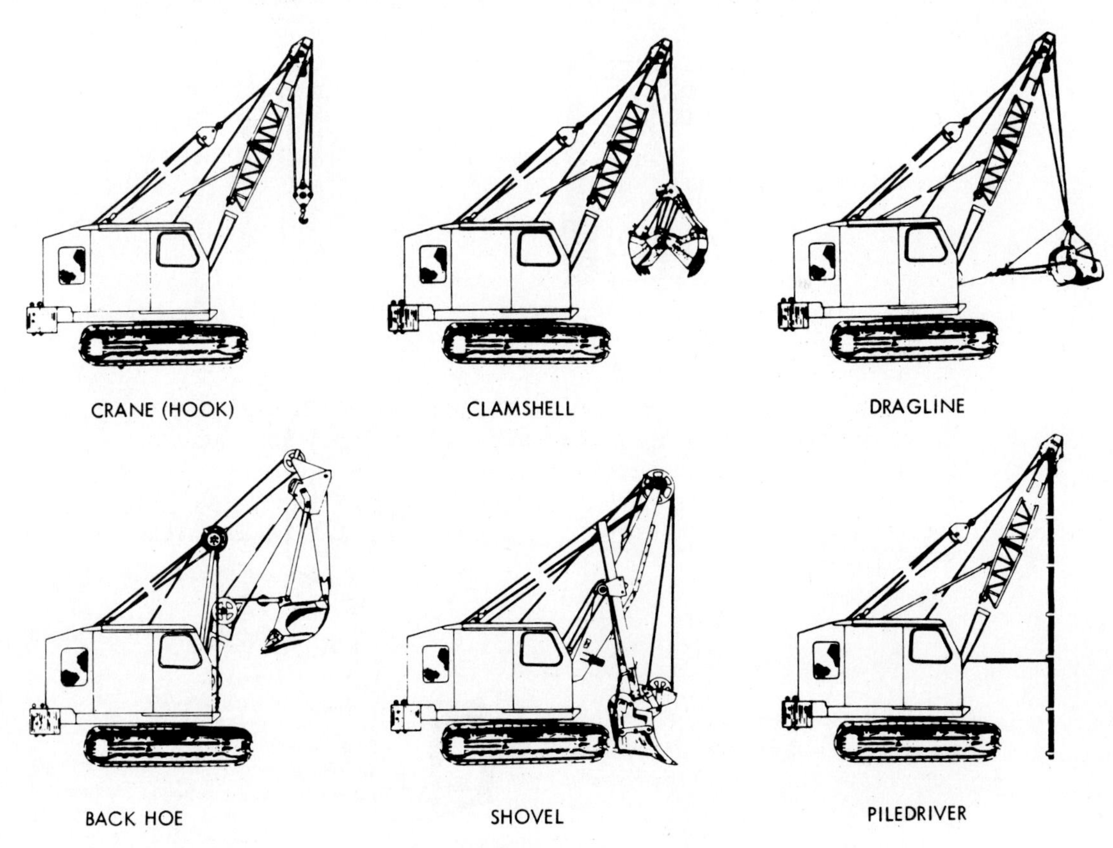

Figure 3-2 Members of the cable-operated crane-shovel family. (U.S. Department of the Army)

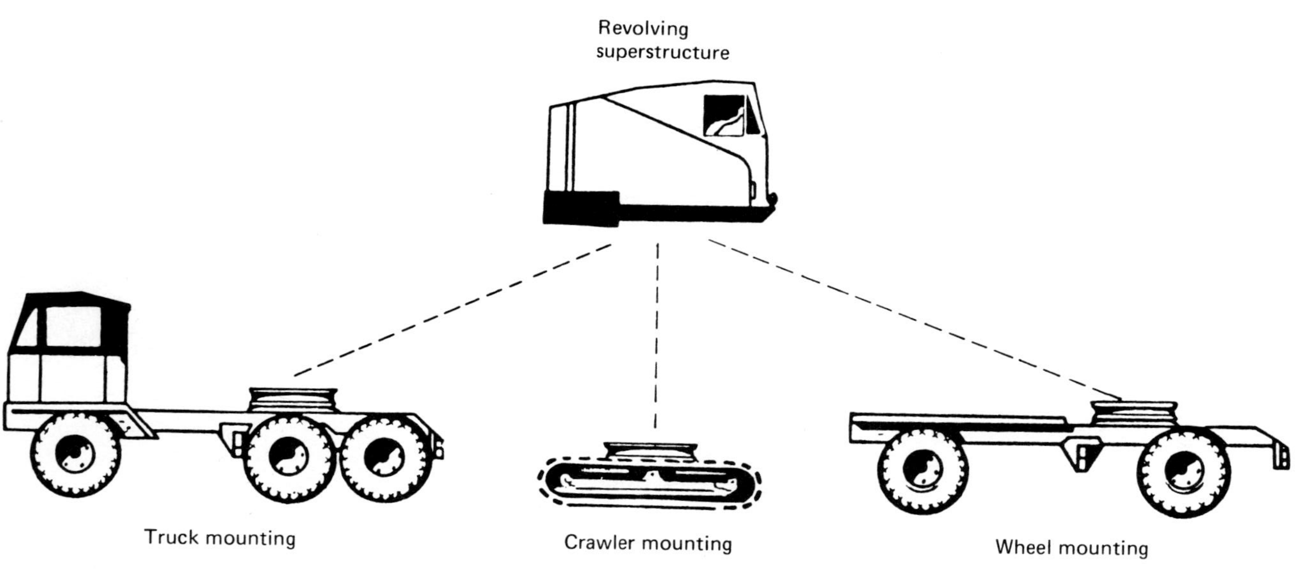

Figure 3-3 Crane-shovel mountings and revolving superstructure. (U.S. Department of the Army)

trenchless excavation, the principal categories include pipe jacking, horizontal earth boring, and utility tunneling.

The process of *pipe jacking* (Figure 3-4) involves forcing pipe horizontally through the soil. Working from a vertical shaft, a section of pipe is carefully aligned and advanced through the soil by hydraulic jacks braced against the shaft sides. As the pipe advances, spoil is removed through the inside of the pipe. After the pipe section has advanced far enough, the hydraulic rams are retracted and another section of pipe placed into position for installation. The excavation and spoil removal process can be manual or mechanical. The process requires workers to enter the pipe sections during the pipe jacking operation.

In *horizontal earth boring* a horizontal hole is created mechanically or hydraulically with the pipe to be installed serving as the casing for the hole. Some of the many installation methods include auger boring, rod pushing (thrust boring), rotational compaction boring, impact piercing, horizontal (directional) drilling, fluid boring, and microtunneling. Many of these techniques utilize lasers and television cameras for hole alignment and boring control. The use of a pneumatic piercing tool to create a borehole for a utility line is illustrated in Figure 3-5. After the bore has been completed, several methods are available to place pipe into the borehole. In one method, pipe is pulled through the bore using the tool's air hose or a steel cable pulled by the air hose. Another method uses the piercing tool to push the pipe through the borehole. A third method uses a pipe pulling adapter attached to the piercing tool to advance the pipe at the same time as the piercing tool advances the bore.

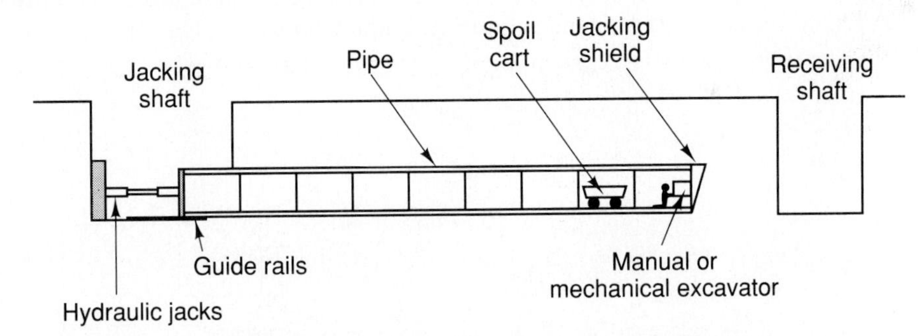

Figure 3-4 Installing a utility line by pipe jacking.

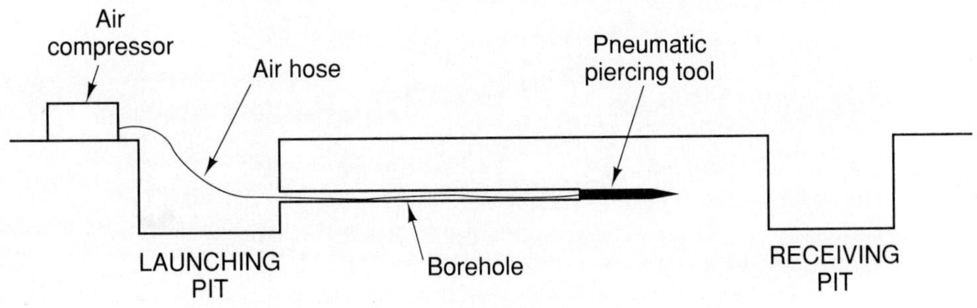

Figure 3-5 Installing a utility line by horizontal earth boring.

Utility tunneling is similar to the conventional tunneling described in Section 6-1 except for the tunnel size and use. Since the tunnels are used as conduit for utility systems rather than for vehicle passage, they are normally smaller than road or rail tunnels. They differ from other trenchless excavation methods in their use of a conventional tunnel liner instead of using the pipe itself as a liner.

3-2 SHOVELS

Operation and Employment

The *hydraulic shovel* illustrated in Figure 3-6 is also called a *front shovel* or *hydraulic excavator-front shovel.* Its major components are identified in Figure 3-7. The hydraulic shovel digs with a combination of crowding force and breakout (or prying) force as illustrated in Figure 3-8. Crowding force is generated by the stick cylinder and acts at the bucket edge on a tangent to the arc of the radius from point *A*. Breakout force is generated by the bucket cylinder and acts at the bucket edge on a tangent to the arc of the radius through point *B*. After the bucket has penetrated and filled with material, it is rolled up to reduce spillage during the swing cycle.

Both front-dump and bottom-dump buckets are available for hydraulic shovels. Bottom-dump buckets are more versatile, provide greater reach and dump clearance, and produce less spillage. However, they are heavier than front-dump buckets of equal capacity, resulting in a lower bucket capacity for equal bucket weight. Hence front-dump buckets usually have a slight production advantage. In addition, front-dump buckets cost less and require less maintenance.

Figure 3-6 Hydraulic shovel. (Courtesy of Kobelco America, Inc.)

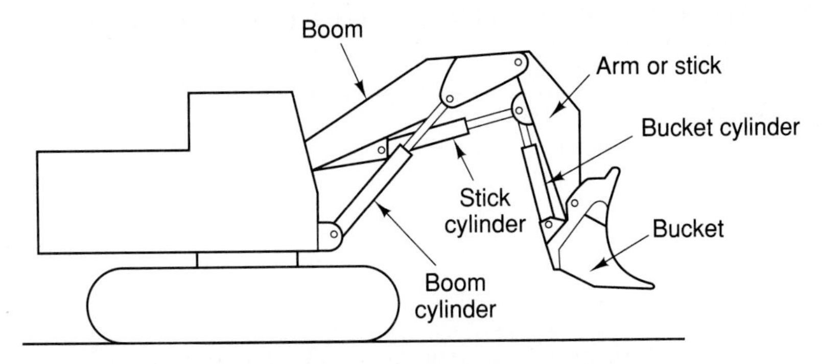

Figure 3-7 Components of a hydraulic shovel.

Although the shovel has a limited ability to dig below track level, it is most efficient when digging above track level. The shovel should have a vertical face to dig against for most effective digging. This surface, known as the *digging face,* is easily formed when excavating a bank or hillside. Thus embankment digging with the material dumped to one side (sidecast) or loaded into haul units provides the best application of the shovel. The ability of the shovel to form its own roadway as it advances is a major advantage. Other possible applications of the shovel include dressing slopes, loading hoppers, and digging shallow trenches.

Production Estimating

Production for hydraulic shovels may be estimated using Equation 3-1 together with Table 3-3, which has been prepared from manufacturers' data.

$$\text{Production (LCY/h or Lm}^3\text{/h)} = C \times S \times V \times B \times E \qquad (3\text{-}1)$$

where C = cycles/h (Table 3-3)
S = swing factor (Table 3-3)
V = heaped bucket volume (LCY or Lm^3)
B = bucket fill factor (Table 3-2)
E = job efficiency

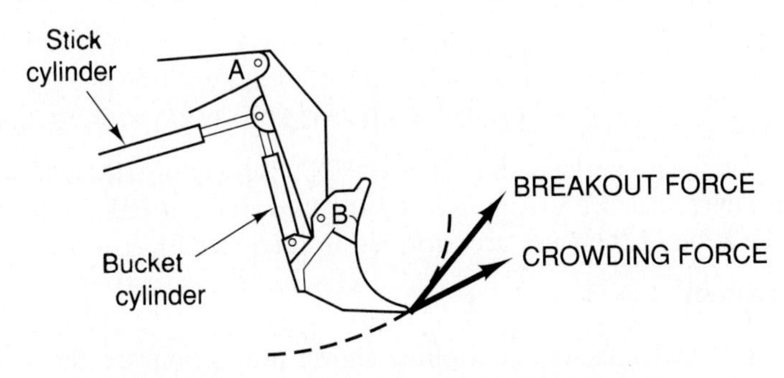

Figure 3-8 Digging action of a hydraulic shovel.

Table 3-3 Standard cycles per hour for hydraulic shovels

	Machine Size					
	Small Under 5 yd ($3.8\ m^3$)		*Medium 5–10 yd ($3.8–7.6\ m^3$)*		*Large Over 10 yd ($7.6\ m^3$)*	
Material	*Bottom Dump*	*Front Dump*	*Bottom Dump*	*Front Dump*	*Bottom Dump*	*Front Dump*
Soft (sand, gravel, coal)	190	170	180	160	150	135
Average (common earth, soft clay, well-blasted rock)	170	150	160	145	145	130
Hard (tough clay, poorly blasted rock)	150	135	140	130	135	125

Adjustment for Swing Angle

	Angle of Swing (deg)					
	45	*60*	*75*	*90*	*120*	*180*
Adjustment factor	1.16	1.10	1.05	1.00	0.94	0.83

Example 3-2

PROBLEM Find the expected production in loose cubic yards (Lm^3) per hour of a 3-yd ($2.3\text{-}m^3$) hydraulic shovel equipped with a front-dump bucket. The material is common earth with a bucket fill factor of 1.0. The average angle of swing is 75° and job efficiency is 0.80.

SOLUTION

$$\begin{aligned}
\text{Standard cycles} &= 150/60 \text{ min (Table 3-3)} \\
\text{Swing factor} &= 1.05 \text{ (Table 3-3)} \\
\text{Bucket volume} &= 3.0 \text{ LCY } (2.3\ Lm^3) \\
\text{Bucket fill factor} &= 1.0 \\
\text{Job efficiency} &= 0.80 \\
\text{Production} &= 150 \times 1.05 \times 3.0 \times 1.0 \times 0.80 = 378 \text{ LCY/h} \\
&[= 150 \times 1.05 \times 2.3 \times 1.0 \times 0.80 = 290\ Lm^3/h]
\end{aligned}$$

For cable-operated shovels, the PCSA Bureau of CIMA has developed production tables that are widely used by the construction industry. These tables and an explanation of their use are provided in reference 3.

Job Management

The two major factors controlling shovel production are the swing angle and lost time during the production cycle. Therefore, the angle of swing between digging and dumping positions should always be kept to a minimum. Haul units must be positioned to minimize the time lost as units enter and leave the loading position. When only a single loading position is available, the shovel operator should utilize the time between the departure of one haul unit and the arrival of the next to move up to the digging face and to smooth the excavation area. The floor of the cut should be kept smooth to provide an even footing for the shovel and to facilitate movement in the cut area. The shovel should be moved up frequently to keep it at an optimum distance from the working face. Keeping dipper teeth sharp will also increase production.

3-3 DRAGLINES

Operation and Employment

The *dragline* is a very versatile machine that has the longest reach for digging and dumping of any member of the crane-shovel family. It can dig from above machine level to significant depths in soft to medium-hard material. The components of a dragline are shown in Figure 3-9.

Bucket teeth and weight produce digging action as the drag cable pulls the bucket across the ground surface. Digging is also controlled by the position at which the drag chain is attached to the bucket (Figure 3-10). The higher the point of attachment, the greater the angle at which the bucket enters the soil. During hoisting and swinging, material is retained in the bucket by tension on the dump cable. When tension on the drag cable is released, tension is removed from the dump cable, allowing the bucket to dump. Buckets are available in a wide range of sizes and weights, solid and perforated. Also available are archless buckets which eliminate the front cross-member connecting the bucket sides to provide easier flow of material into and out of the bucket.

While the dragline is very versatile excavator, it does not have the positive digging action or lateral control of the shovel. Hence the bucket may bounce or move sideways during hard digging. Also, more spillage must be expected in loading operations than would occur with a shovel. While a skilled dragline operator can overcome many of these limitations, the size of haul units used for dragline loading

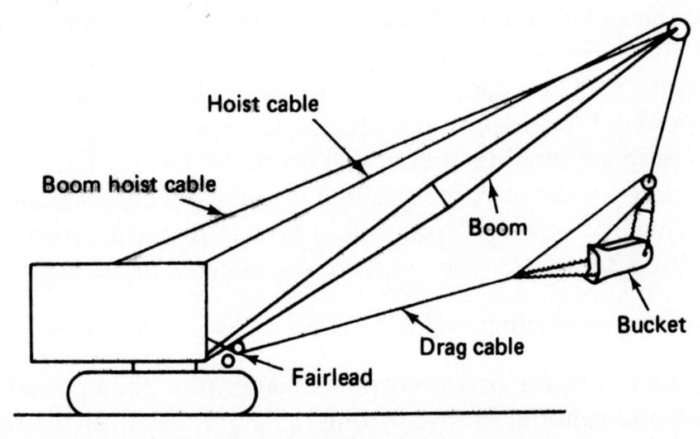

Figure 3-9 Components of a dragline.

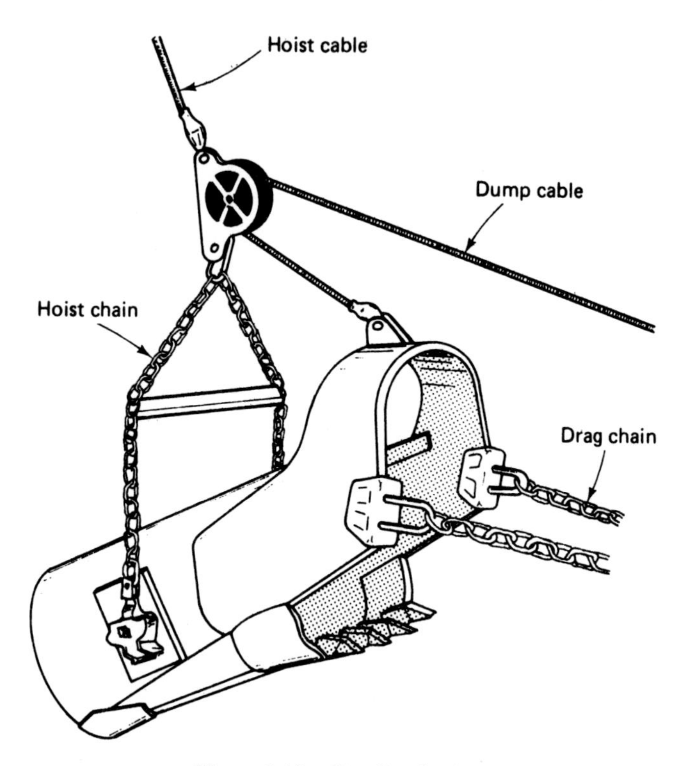

Figure 3-10 Dragline bucket.

should be greater than that of those used with a similar-size shovel. The maximum bucket size to be used on a dragline depends on machine power, boom length, and material weight. Therefore, use the dragline capacity chart provided by the manufacturer instead of the machine's lifting capacity chart to determine maximum allowable bucket size.

Production Estimating

The Construction Industry Manufacturers Association (CIMA), through its PCSA Bureau, has made studies of cable-operated dragline operations and has developed production tables that are widely used by the construction industry. Tables 3-4 to 3-6 are based on PCSA data. Note, however, that these tables are applicable only to diesel-powered, cable-operated draglines.

To estimate dragline production using the tables, determine the ideal output of the dragline for the machine size and material (Table 3-4), then adjust this figure by multiplying it by a swing-depth factor (Table 3-6) and a job efficiency factor, as shown in Equation 3-2. Notice the conditions applicable to Table 3-4 given in the footnote.

$$\text{Expected production} = \text{Ideal output} \times \text{Swing-depth factor} \times \text{Efficiency} \quad (3\text{-}2)$$

To use Table 3-6 it is first necessary to determine the optimum depth of cut for the machine and material involved from Table 3-5. Next, divide the actual depth of cut

Table 3-4 Ideal dragline output—short boom [BCY/h (Bm^3/h)]*

	Bucket Size [cu yd (m^3)]										
Type of Material	¾ (0.57)	1 (0.75)	1¼ (0.94)	1½ (1.13)	1¾ (1.32)	2 (1.53)	2½ (1.87)	3 (2.29)	3½ (2.62)	4 (3.06)	5 (3.82)
Light moist clay	130	160	195	220	245	265	305	350	390	465	540
or loam	(99)	(122)	(149)	(168)	(187)	(203)	(233)	(268)	(298)	(356)	(413)
Sand and gravel	125	155	185	210	235	255	295	340	380	455	530
	(96)	(119)	(141)	(161)	(180)	(195)	(226)	(260)	(291)	(348)	(405)
Common earth	105	135	165	190	210	230	265	305	340	375	445
	(80)	(103)	(126)	(145)	(161)	(176)	(203)	(233)	(260)	(287)	(340)
Tough clay	90	110	135	160	180	195	230	270	305	340	410
	(69)	(84)	(103)	(122)	(138)	(149)	(176)	(206)	(233)	(260)	(313)
Wet, sticky clay	55	75	95	110	130	145	175	210	240	270	330
	(42)	(57)	(73)	(84)	(99)	(111)	(134)	(161)	(183)	(206)	(252)

*Based on 100% efficiency, 90° swing, optimum depth of cut, material loaded into haul units at grade level. Based on PCSA data.

Table 3-5 Optimum depth of cut for draglines [ft (m)]*

	Bucket Size [cu yd (m³)]										
Type of Material	¾ (0.57)	1 (0.75)	1¼ (0.94)	1½ (1.13)	1¾ (1.32)	2 (1.53)	2½ (1.87)	3 (2.29)	3½ (2.62)	4 (3.06)	5 (3.82)
Light moist clay, loam, sand, and gravel	6.0 (1.8)	6.6 (2.0)	7.0 (2.1)	7.4 (2.2)	7.7 (2.3)	8.0 (2.4)	8.5 (2.6)	9.0 (2.7)	9.5 (2.9)	10.0 (3.0)	11.0 (3.3)
Common earth	7.4 (2.3)	8.0 (2.4)	8.5 (2.6)	9.0 (2.7)	9.5 (2.9)	9.9 (3.0)	10.5 (3.2)	11.0 (3.3)	11.5 (3.5)	12.0 (3.7)	13.0 (4.0)
Wet, sticky clay	8.7 (2.7)	9.3 (2.8)	10.0 (3.0)	10.7 (3.2)	11.3 (3.4)	11.8 (3.6)	12.3 (3.7)	12.8 (3.9)	13.3 (4.1)	13.8 (4.2)	14.3 (4.4)

*Based on PCSA data.

Table 3-6 Swing-depth factor for draglines*

Depth of Cut (% of Optimum)	Angle of Swing (deg)							
	30	*45*	*60*	*75*	*90*	*120*	*150*	*180*
20	1.06	0.99	0.94	0.90	0.87	0.81	0.75	0.70
40	1.17	1.08	1.02	0.97	0.93	0.85	0.78	0.72
60	1.25	1.13	1.06	1.01	0.97	0.88	0.80	0.74
80	1.29	1.17	1.09	1.04	0.99	0.90	0.82	0.76
100	1.32	1.19	1.11	1.05	1.00	0.91	0.83	0.77
120	1.29	1.17	1.09	1.03	0.98	0.90	0.82	0.76
140	1.25	1.14	1.06	1.00	0.96	0.88	0.81	0.75
160	1.20	1.10	1.02	0.97	0.93	0.85	0.79	0.73
180	1.15	1.05	0.98	0.94	0.90	0.82	0.76	0.71
200	1.10	1.00	0.94	0.90	0.87	0.79	0.73	0.69

*Based on PCSA data.

by the optimum depth and express the result as a percentage. The appropriate swing-depth factor is then obtained from Table 3-6, interpolating as necessary. The method of calculating expected hourly production is illustrated in Example 3-3.

Example 3-3

PROBLEM Determine the expected dragline production in loose cubic yards (Lm^3) per hour based on the following information.

Dragline size = 2 cu yd (1.53 m^3)
Swing angle = 120°
Average depth of cut = 7.9 ft (2.4 m)
Material = common earth
Job efficiency = 50 min/h
Soil swell = 25%

SOLUTION

Ideal output = 230 BCY/h (176 Bm^3/h) (Table 3-4)
Optimum depth of cut = 9.9 ft (3.0 m) (Table 3-5)
Actual depth/optimum depth = $7.9/9.9 \times 100 = 80\%$
[$= 2.4/3.0 \times 100 = 80\%$]
Swing-depth factor = 0.90 (Table 3-6)
Efficiency factor = $50/60 = 0.833$
Volume change factor = $1 + 0.25 = 1.25$
Estimated production = $230 \times 0.90 \times 0.833 \times 1.25 = 216$ LCY/h
[$= 176 \times 0.90 \times 0.833 \times 1.25 = 165$ Lm^3/h]

Job Management

Trial operations may be necessary to select the boom length, boom angle, bucket size and weight, and the attachment position of the drag chain that yield maximum production. As in shovel operation, maximum production is obtained with a minimum swing angle. In general, the lightest bucket capable of satisfactory digging should be used, since this increases the allowable bucket size and reduces cycle time. It has been found that the most efficient digging area is located within 15° forward and back of a vertical line through the boom point, as illustrated in Figure 3-11. Special bucket hitches are available which shorten the drag distance necessary to obtain a full bucket load. Deep cuts should be excavated in layers whose thickness is as close to the optimum depth of cut as possible.

3-4 BACKHOES

Operation and Employment

A *backhoe* (or simply *hoe*) is an excavator designed primarily for excavation below grade. The components of a hydraulic backhoe or hydraulic excavator-backhoe are identified in Figure 3-12. It digs by pulling the dipper back toward the machine; hence the name backhoe. The backhoe shares the characteristics of positive digging action and precise lateral control with the shovel. Cable-operated backhoes exist but are largely being replaced by hydraulic models because of their superior speed and ease of control. Backhoe attachments are also available for loaders and tractors.

The backhoe is widely utilized for trenching work. In addition to excavating the trench, it can perform many other trenching functions, such as laying pipe bedding, placing pipe, pulling trench shields, and backfilling the trench. In trench excavation the best measure of production is the length of trench excavated per unit of time. Therefore, a dipper width should be chosen which matches the required trench width as closely as possible. For this reason, dippers are available in a wide range of sizes and widths. Side cutters are also available to increase the cutting width of dippers. Others suitable backhoe applications include excavating basements, cleaning roadside ditches, and grading embankments.

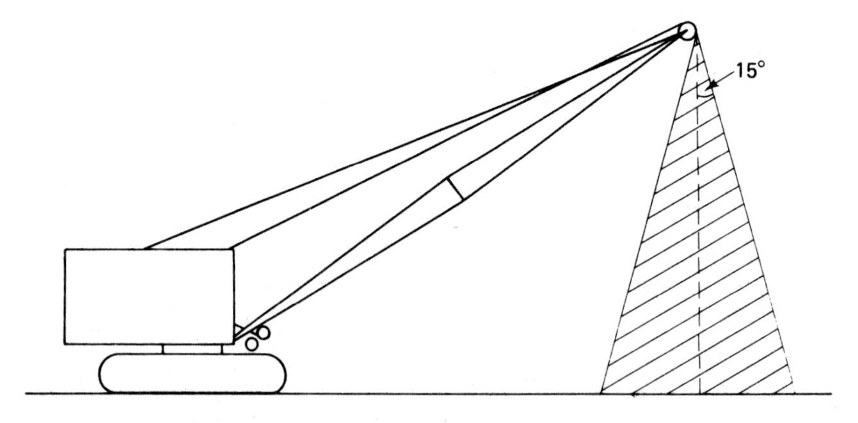

Figure 3-11 Most efficient digging area for a dragline.

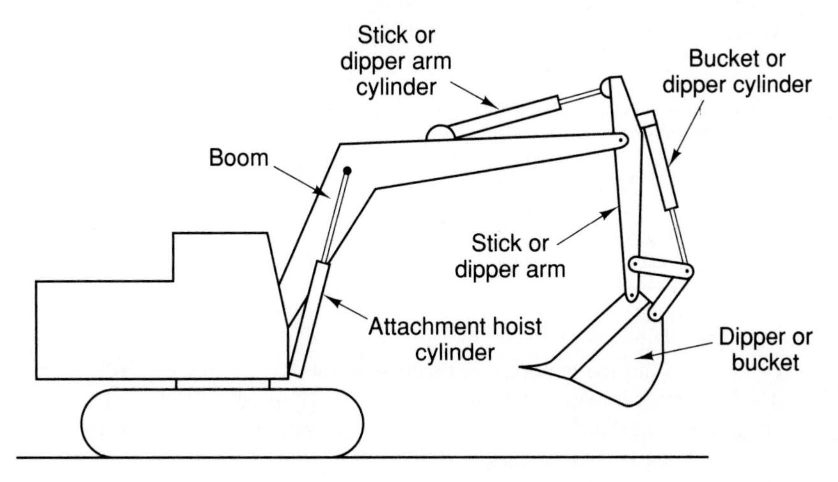

Figure 3-12 Components of a hydraulic excavator-backhoe.

Production Estimating

No PCSA production tables have been prepared for the backhoe. However, production may be estimated by using Equation 3-3 together with Tables 3-7 and 3-8 which have been prepared from manufacturers' data.

$$\text{Production (LCY/h)} = C \times S \times V \times B \times E \qquad (3\text{-}3)$$

where C = cycles/h (Table 3-7)
S = swing-depth factor (Table 3-8)
V = heaped bucket volume (LCY or Lm^3)
B = bucket fill factor (Table 3-2)
E = job efficiency

In trenching work a fall-in factor should be applied to excavator production to account for the work required to clean out material that falls back into the trench from the trench walls. Normal excavator production should be multiplied by the appropriate value from Table 3-9 to obtain the effective trench production.

Table 3-7 Standard cycles per hour for hydraulic backhoes

	Machine Size			
Type of Material	*Wheel Tractor*	*Small Excavator: 1 yd (0.76 m³) or Less*	*Medium Excavator: 1¼–2¼ yd (0.94–1.72 m³)*	*Large Excavator: Over 2½ yd (1.72 m³)*
Soft (sand, gravel, loam)	170	250	200	150
Average (common earth, soft clay)	135	200	160	120
Hard (tough clay, rock)	110	160	130	100

Table 3-8 Swing-depth factor for backhoes

Depth of Cut (% of Maximum)	*Angle of Swing (deg)*					
	45	*60*	*75*	*90*	*120*	*180*
30	1.33	1.26	1.21	1.15	1.08	0.95
50	1.28	1.21	1.16	1.10	1.03	0.91
70	1.16	1.10	1.05	1.00	0.94	0.83
90	1.04	1.00	0.95	0.90	0.85	0.75

Example 3-4

PROBLEM Find the expected production in loose cubic yards (Lm^3) per hour of a small hydraulic excavator. Heaped bucket capacity is ¾ cu yd (0.57 m^3). The material is sand and gravel with a bucket fill factor of 0.95. Job efficiency is 50 min/h. Average depth of cut is 14 ft (4.3 m). Maximum depth of cut is 20 ft (6.1 m) and average swing is 90°.

SOLUTION

$$\text{Cycle output} = 250 \text{ cycles/60 min} \quad \text{(Table 3-7)}$$
$$\text{Swing-depth factor} = 1.00 \quad \text{(Table 3-8)}$$
$$\text{Bucket volume} = 0.75 \text{ LCY } (0.57 \text{ m}^3)$$
$$\text{Bucket fill factor} = 0.95$$
$$\text{Job efficiency} = 50/60 = 0.833$$
$$\text{Production} = 250 \times 1.00 \times 0.75 \times 0.95 \times 0.833 = 148 \text{ LCY/h}$$
$$[= 250 \times 1.00 \times 0.57 \times 0.95 \times 0.833 = 113 \text{ Lm}^3\text{/h}]$$

Job Management

In selecting the proper backhoe for a project, consideration must be given to the maximum depth, working radius, and dumping height required. Check also for adequate clearance for the carrier, superstructure, and boom during operation.

Although the backhoe will excavate fairly hard material, do not use the bucket as a sledge in attempting to fracture rock. Light blasting, ripping, or use of a power hammer may be necessary to loosen rock sufficiently for excavation. When lifting pipe into place do not exceed load given in the manufacturer's safe capacity chart for the situation.

Table 3-9 Adjustment factor for trench production

Type of Material	*Adjustment Factor*
Loose (sand, gravel, loam)	0.60–0.70
Average (common earth)	0.90–0.95
Firm (firm plastic soils)	0.95–1.00

3-5 CLAMSHELLS

When the crane-shovel is equipped with a crane boom and clamshell bucket, it becomes an excavator known as a *clamshell.* The clamshell is capable of excavating to great depths but lacks the positive digging action and precise lateral control of the shovel and backhoe. Clamshells are commonly used for excavating vertical shafts and footings, unloading bulk materials from rail cars and ships, and moving bulk material from stockpiles to bins, hoppers, or haul units. The components of a cable-operated clamshell are identified in Figure 3-13. Clamshell attachments are also available for the hydraulic excavator.

A clamshell bucket is illustrated in Figure 3-14. Notice that the bucket halves are forced together by the action of the closing line against the sheaves. When the closing line is released, the counterweights cause the bucket halves to open as the bucket is held by the holding line. Bucket penetration depends on bucket weight assisted by the bucket teeth. Therefore, buckets are available in light, medium, and heavy weights, with and without teeth. Heavy buckets are suitable for digging medium soils. Medium buckets are used for general-purpose work, including the excavation of loose soils. Light buckets are used for handling bulk materials such as sand and gravel.

The orange peel bucket illustrated in Figure 3-15 is principally utilized for underwater excavation and for rock placement. Because of its circular shape, it is also well suited to excavating piers and shafts. It operates on the same principle as does the clamshell.

Production Estimating

No standard production tables are available for the clamshell. Thus production estimation should be based on the use of Equation 2-1. The procedure is illustrated in Example 3-5.

Example 3-5

PROBLEM Estimate the production in loose cubic yards per hour for a medium-weight clamshell excavating loose earth. Heaped bucket capacity is 1 cu yd (0.75 m^3). The soil is

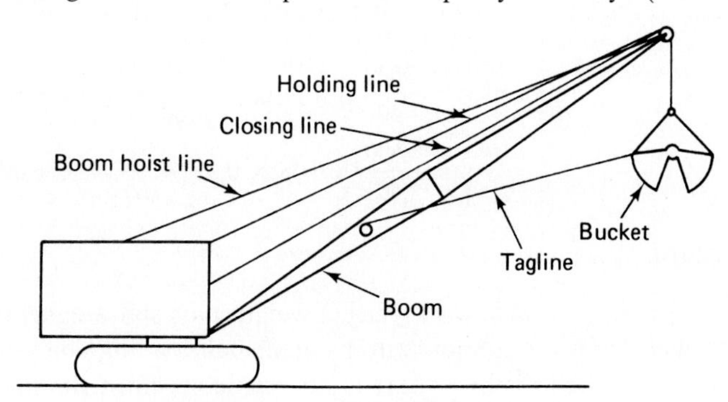

Figure 3-13 Components of a clamshell.

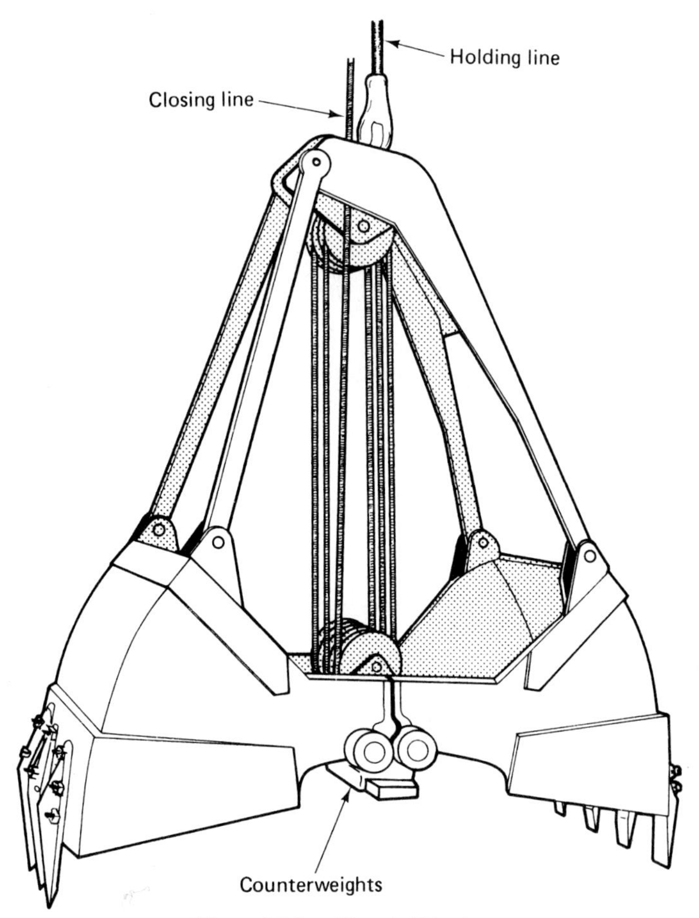

Figure 3-14 Clamshell bucket.

common earth with a bucket fill factor of 0.95. Estimated cycle time is 40 s. Job efficiency is estimated at 50 min/h.

SOLUTION

$$\text{Production} = \frac{3600}{40} \times 1 \times 0.95 \times \frac{50}{60} = 71 \text{ LCY/h}$$

$$\left[= \frac{3600}{40} \times 0.75 \times 0.95 \times \frac{50}{60} = 53 \text{ Lm}^3\text{/h} \right]$$

Job Management

The maximum allowable load (bucket weight plus soil weight) on a clamshell should be obtained from the manufacturer's clamshell loading chart for continuous operation. If a clamshell loading chart is not available, limit the load to 80% of the safe lifting capacity given by the crane capacity chart for rubber-tired equipment or 90% for crawler-mounted equipment. Since the machine load includes the weight of the

Figure 3-15 Orange peel bucket. (Courtesy of ESCO Corporation)

bucket as well as its load, use of the lightest bucket capable of digging the material will enable a larger bucket to be used and will usually increase production. Tests may be necessary to determine the size of bucket that yields maximum production in a particular situation. Cycle time is reduced by organizing the job so that the dumping radius is the same as the digging radius. Keep the machine level to avoid swinging uphill or downhill. Nonlevel swinging is hard on the machine and usually increases cycle time.

3-6 CRANES

Operation and Employment

Cranes are used primarily for lifting, lowering, and transporting loads. They move loads horizontally by swinging or traveling. Most mobile cranes consist of a crane-shovel carrier and superstructure equipped with a boom and hook as illustrated in Figure 3-16. The current trend toward the use of hydraulically operated equipment includes hydraulically powered telescoping boom cranes. The mobile telescoping-boom crane shown in Figure 3-17 is capable of lifting loads to the top of a 24-story building. Some mobile cranes are intended to be used only as cranes and do not have the capability of using the crane-shovel front-end attachments described earlier. Another special type of crane is the *tower crane,* illustrated in Figure 3-18. The tower crane is widely used on building construction projects because of its wide operating radius and almost unlimited height capability. The majority of tower cranes are of the saddle-jib or horizontal boom type shown in Figure 3-18. However, luffing jib (inclined boom) models (see Figure 13-8) are available which have the ability to operate in areas of restricted horizontal clearance not suitable for conventional tower cranes with their fixed jibs and counterweights. Types of tower crane by method of mounting include static (fixed mount) tower cranes, rail-mounted tower cranes, mobile tower cranes, and climbing cranes. Climbing cranes are supported by completed building floors and are capable of raising themselves from floor to floor as the building is erected. Most tower cranes incorporate self-raising masts. That is, they can raise themselves section by section until the mast or tower reaches the desired height. A typical procedure is as follows (refer to Figure 3-19). The crane lifts an additional tower section together with a monorail beam and trolley (A). The monorail beam is fastened to the crane's turntable base and the new section is trolleyed close to the tower. The turntable base is unbolted from the tower. The climbing frame's hydraulic cylinders lift the climbing frame and the new section is inserted into the climbing frame using the monorail beam trolley (B). The climbing frame is then lowered and the new section is bolted to the tower and the turntable base (C).

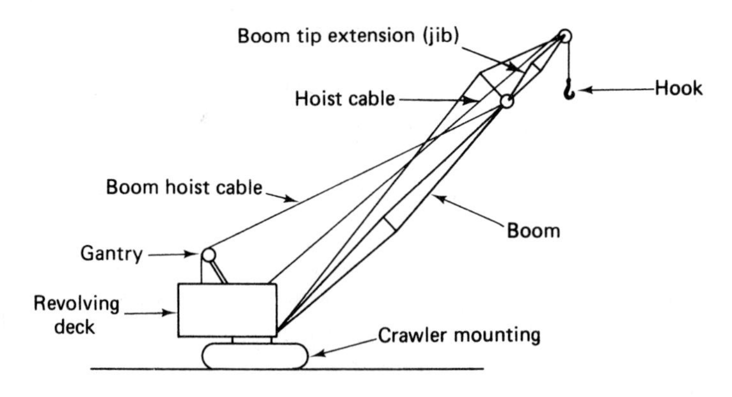

Figure 3-16 Components of a crane.

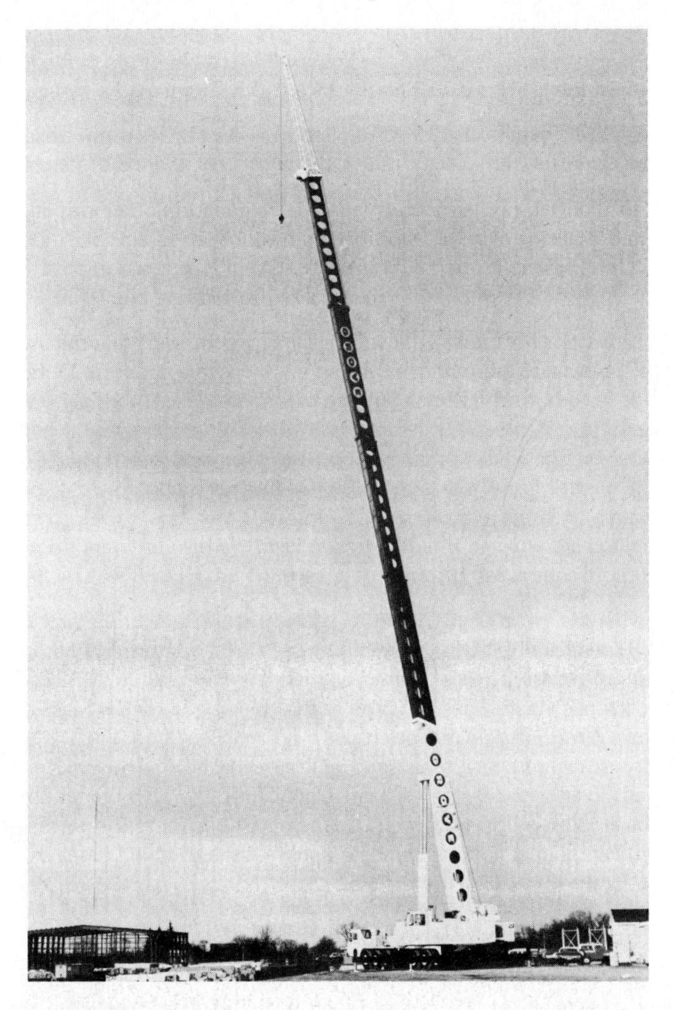

Figure 3-17 Large mobile hydraulic crane with telescoping boom. (Courtesy of Grove Manufacturing Company)

The major factor controlling the load that may safely be lifted by a crane is its *operating radius* (horizontal distance from the center of rotation to the hook). For other than horizontal boom tower cranes, this is a function of boom length and boom angle above the horizontal. Some of the other factors influencing a crane's safe lifting capacity include the position of the boom in relation to the carrier, whether or not *outriggers* (beams that widen the effective base of a crane) are used, the amount of counterweight, and the condition of the supporting surface. Safety regulations limit maximum crane load to a percentage of the *tipping load* (load that will cause the crane to actually begin to tip). Crane manufacturers provide charts such as that shown in Figure 3-20 giving the safe load capacity of the machine under various conditions. Notice that hook blocks and other load-handling devices are considered part of the load and their weight must be included in the maximum safe load

Figure 3-18 Tower crane on a building site. (Courtesy of Potain Tower Cranes, Inc.)

capacity calculation. Electronic load indicators are available that measure the actual load on the crane and provide a warning if the safe load is being exceeded.

A standard method of rating the capacity of mobile cranes has been adopted by the PCSA Bureau of the Construction Industry Manufacturers Association. Under this system, a nominal capacity rating is assigned which indicates the safe load capacity (with outriggers set) for a specified operating radius [usually 12 ft (3.6 m) in the direction of least stability]. The PCSA class number following the nominal rating consists of two number symbols. The first number indicates the operating radius for the nominal capacity. The second number gives the rated load in hundreds of pounds at a 40-ft (12.2-m) operating radius using a 50-ft (15.2-m) boom. Thus the crane whose capacity chart is shown in Figure 3-20 has a nominal capacity of 22 tons (19.9 t) at a 10-ft (3-m) operating radius. Therefore, this crane should be able to safely lift a load of 22 tons (19.9 t) at a radius of 10 ft (3 m) and a load of 8000 lb (3629 kg) at an operating radius of 40 ft (12.2 m) with a 50-ft (15.2-m) boom. Both capacities require outriggers to be set and apply regardless of the position of the boom relative to the carrier.

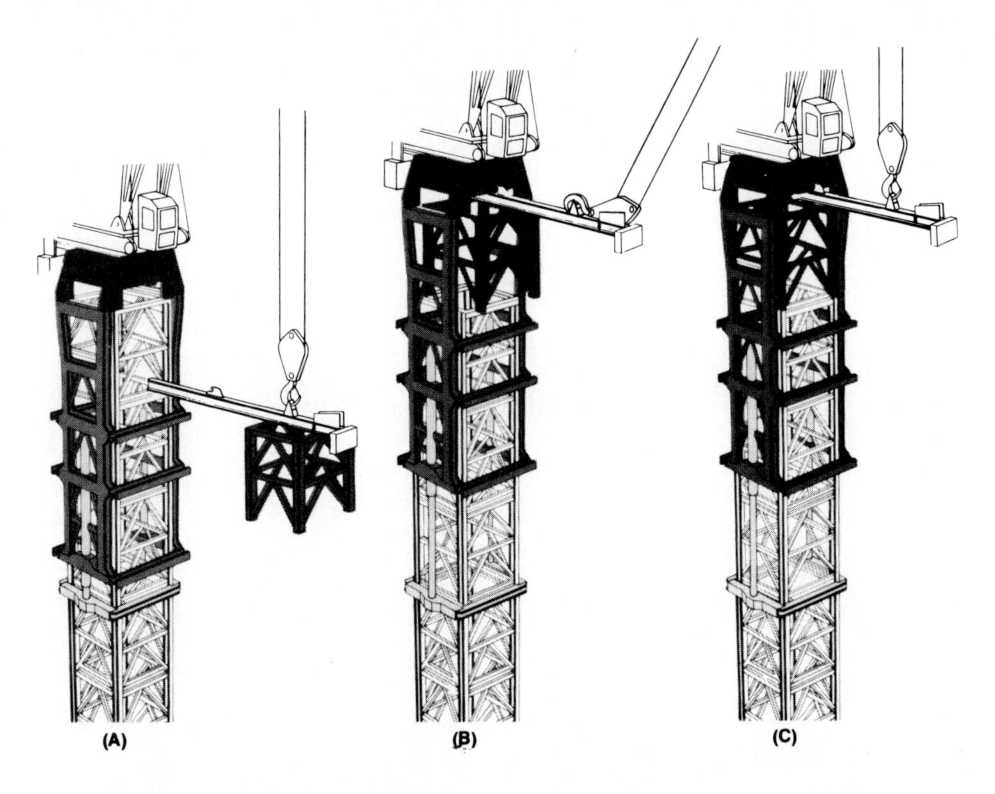

Figure 3-19 Self-raising tower crane mast. (Courtesy of FMC Construction Equipment Group)

Job Management

There are a number of attachments besides the basic hook available to assist the crane in lifting and transporting various types of loads. A number of these attachments are illustrated in Figure 3-21. Among these attachments, concrete buckets, slings, and special hooks are most often used in construction applications.

High-voltage lines present a major safety hazard to crane operations. U.S. Occupational Safety and Health Act (OSHA) regulations prohibit a crane or its load from approaching closer than 10 ft (3 m) to a high-voltage line carrying 50 kV or less. An additional 0.4 in. (1 cm) must be added for each kilovolt over 50 kV. These safety clearances must be maintained unless the line is deenergized and visibly grounded at the work site or unless insulating barriers not attached to the crane are erected which physically prevent contact with the power line.

Crane accidents occur all too frequently in construction work, particularly when lifting near-capacity loads and when operating with long booms. Some suggestions for safe crane operations include the following:

- Carefully set outriggers on firm supports.
- The crane base must be level. Safe crane capacity is reduced as much 50% when the crane is out of level by only 3° and operating with a long boom at minimum radius.

RATED LIFTING CAPACITIES IN POUNDS

28 ft. - 70 ft. BOOM

ON OUTRIGGERS FULLY EXTENDED - OVER FRONT

Radius in Feet	Boom Length in Feet 28	34	40	46	52	58	64	70
10	44,000 (64)	36,000 (69)	36,000 (73)					
12	40,000 (59.5)	36,000 (65.5)	36,000 (70)	35,000 (73)				
15	31,000 (51.5)	31,000 (59.5)	30,700 (65)	29,850 (69)	29,150 (72)	28,600 (74.5)		
20	23,200 (36.5)	23,200 (49)	23,200 (57)	23,200 (62)	23,003 (66)	22,600 (69.5)	22,150 (72)	20,500 (74)
25	17,950 (6)	17,950 (36)	17,950 (47.5)	17,950 (54.5)	17,950 (60)	17,950 (64)	17,950 (67)	17,650 (69.5)
30		15,350 (15.5)	15,350 (36.5)	15,350 (46.5)	15,350 (53)	15,150 (58)	14,950 (62)	14,750 (65)
35			11,900 (20)	11,900 (36.5)	11,900 (45.5)	11,900 (51.5)	11,900 (56.5)	11,900 (60)
40	See Warning Note 16			9,410 (23)	9,410 (36.5)	9,410 (45)	9,410 (50.5)	9,410 (55)
45					7,720 (25)	7,720 (37)	7,720 (44.5)	7,720 (49.5)
50						6,410 (26.5)	6,410 (37)	6,410 (43.5)
55						5,410 (3.5)	5,410 (28)	5,410 (37)
60							4,530 (13)	4,530 (28.5)
65								3,780 (15.5)
Min. boom angle (deg.) for indicated length [No Load]								0
Max. boom length (ft.) at 0 degree boom angle [No Load]								70.0

NOTE: Boom Angles are in degrees. A6-829-003704 & -003716A

ON OUTRIGGERS FULLY EXTENDED - 360°

Radius in Feet	Boom Length in Feet 28	34	40	46	52	58	64	70
10	44,000 (64)	36,000 (69)	36,000 (73)					
12	40,000 (59.5)	36,000 (65.5)	36,000 (70)	35,000 (73)				
15	31,000 (51.5)	31,000 (59.5)	30,700 (65)	29,850 (69)	29,150 (72)	28,600 (74.5)		
20	23,200 (36.5)	23,200 (49)	23,200 (57)	23,200 (62)	23,000 (66)	22,600 (69.5)	22,150 (72)	20,500 (74)
25	17,950 (6)	17,950 (36)	17,950 (47.5)	17,950 (54.5)	17,950 (60)	17,950 (64)	17,950 (67)	17,650 (69.5)
30		13,470 (15.5)	13,470 (36.5)	13,470 (46.5)	13,470 (53)	13,470 (58)	13,470 (62)	13,470 (65)
35			10,220 (20)	10,220 (36.5)	10,220 (45.5)	10,220 (51.5)	10,220 (56.5)	10,220 (60)
40	See Warning Note 16			8,010 (23)	8,010 (36.5)	8,010 (45)	8,010 (50.5)	8,010 (55)
45					6,530 (25)	6,530 (37)	6,530 (44.5)	6,530 (49.5)
50						5,430 (26.5)	5,430 (37)	5,430 (43.5)
55						4,440 (3.5)	4,440 (28)	4,440 (37)
60							3,620 (13)	3,620 (28.5)
65								2,980 (15.5)
Min. boom angle (deg.) for indicated length [No Load]								0
Max. boom length (ft.) at 0 degree boom angle [No Load]								70.0

NOTE: Boom Angles are in degrees. A6-829-003710 & -003716A

ON RUBBER CAPACITIES 16.00x25 TIRES

Radius in Feet	Stationary Capacity: Defined Arc Over Front (3)	Stationary Capacity: 360° Arc	Pick & Carry Capacity Up to 2.5 MPH: Boom Centered Over Front (7)
10	29,150 (a)	22,260 (a)	30,460 (a)
12	24,030 (a)	16,550 (a)	26,320 (a)
15	20,150 (a)	11,780 (b)	21,670 (a)
20	14,650 (b)	6,780 (c)	15,130 (a)
25	9,760 (b)	3,970 (d)	8,290(b)
30	7,130 (c)	2,450 (e)	6,550 (c)
35	5,270 (d)	1,630 (f)	5,140 (d)
40	4,110 (e)	920 (f)	4,110 (d)
45	3,180 (f)		3,180 (e)
50	2,410 (g)		2,410 (f)
55	1,850 (g)		1,850 (f)
60	1,410 (h)		
65	1,090 (h)		

A6-829-003756B

Maximum Permissible Boom Length:
(a) 28.0 ft. (c) 40.0 ft. (e) 52.0 ft. (g) 64.0 ft.
(b) 34.0 ft. (d) 46.0 ft. (f) 58.0 ft. (h) 70.0 ft.

NOTES FOR RUBBER CAPACITIES

		Main Boom 70 ft.	Main Boom w/23 ft. Jib
Front (No Load)	Minimum boom angle for indicated boom length	0°	0°
	Maximum boom length at 0° boom angle	70 ft.	93 ft.
360° (No Load)	Minimum boom angle for indicated boom length	42°	51°
	Maximum boom length at 0° boom angle	52 ft.	57 ft.

1. Capacities are in pounds and do not exceed 85% of tipping loads as determined by test in accordance with SAE J-765.
2. Capacities are applicable to machines equipped with 16.00x25 (20 ply) bias ply tires, at 95 PSI cold inflation pressure (80 PSI for 2.5 MPH pick & carry capacities.)
3. (Defined Arc) - Over front includes ±6° on either side of longitudinal centerline of machine.
4. Capacities appearing above bold line are based on structural strength and tipping should not be relied upon as a capacity limitation.
5. Capacities are applicable only with machine on a firm level surface.
6. On rubber lifting with jib not permitted.
7. For pick and carry operation, boom must be centered over front of machine and mechanical swing lock engaged. When handling loads in the structural range with capacities close to maximum rating, travel should be reduced to creep speed.
8. Axle lockouts must be functioning before lifting on rubber. (Check automatic lockout system for proper functioning. Refer to "Operation and Maintenance Manual" for description of a proper functioning axle lockout system.)
9. All lifting depends on proper tire inflation, capacity and condition. Capacities must be reduced for lower tire inflation pressures. See lifting capacity chart for tire used. Damaged tires are hazardous to safe operation of crane.

LINE PULLS AND REEVING INFORMATION

HOISTS	CABLE SPECS.	PERMISSIBLE LINE PULLS
MAIN Model 15	5/8 in. 6x41, EIPS, IWRC 5/8 in. 19x7, IPS, Strand	7,926 lbs. 7,926 lbs.
MAIN & AUX. Model 15 & 25	1/2 in. 6x41, EIPS, IWRC 1/2 in. 19x7, EIPS, Strand	7,600 lbs. 6,150 lbs.

For multiple reeving with main hoist use one line for each 7,333 lbs. of load or portion thereof.

19x7 is a non-spin rope intended for single line operation and is not recommended for multiple part reeving.

IDENTIFICATION

RT522

PCSA CLASS 10 - 80

SERIAL NUMBER

NOTES FOR LIFTING CAPACITIES

GENERAL:

1. Rates loads as shown on lift chart pertain to this machine as originally manufactured and equipped. Modifications to the machine or use of optional equipment other than that specified can result in a reduction of capacity.
2. Construction equipment can be hazardous if improperly operated or maintained. Operation and maintenance of this machine shall be in compliance with the information in the operator's, parts, and safety manuals supplied with this machine. If these manuals are missing, order replacements from the manufacturer through the distributor.
3. The operator and other personnel associated with this machine shall fully acquaint themselves with the latest applicable American National Standards Institute (ANSI) Safety Standards for cranes.

SETUP:

1. The machine shall be leveled on a firm supporting surface. Depending on the nature of the supporting surface, it may be necessary to have structural supports under the outrigger floats or tires to spread the load to a larger bearing surface.
2. For outrigger operation, outriggers shall be fully extended with tires raised free of crane weight before operating the boom or lifting loads.
3. If machine is equipped with front jack cylinder, the front jack cylinder shall be set in accordance with written procedure.
4. If machine is equipped with extendable counterweight, the counterweight shall be fully extended before operation.
5. Tires shall be inflated to the recommended pressure before lifting on rubber.
6. With certain boom and hoist tackle combinations, maximum capacities may not be obtainable with standard cable lengths.

OPERATION:

1. Rated loads at rated radius shall not be exceeded. Do not tip the machine to determine allowable loads. For clamshell or concrete bucket operation, weight of bucket and load must not exceed 80% of rated lifting capacities.
2. Rated loads do not exceed 85% of the tipping load as determined by SAE Crane Stability Test Code J-765a.
3. Rated loads include the weight of hook block, slings and auxiliary lifting devices and their weights shall be subtracted from the listed ratings to obtain the net load to be lifted.
4. Load ratings are based on freely suspended loads. No attempt shall be made to move a load horizontally on the ground in any direction.
5. Rated loads do not account for wind on lifted load or boom. It is recommended when wind velocity is above 20 mph (32 km/h), rated loads and boom lengths shall be appropriately reduced.
6. Rated loads are for lift crane service only.
7. Do not operate at a radius or boom length where capacities are not listed. At these positions, the machine may overturn without any load on the hook.
8. The maximum load which can be telescoped is not definable because of variations in loadings and crane maintenance, but it is safe to attempt retraction and extension within the limits of the capacity chart.
9. When either boom length or radius or both are between values listed, the smallest load shown at either the next larger radius or boom length shall be used.
10. For safe operation, the user shall make due allowances for his particular job conditions, such as: soft or uneven ground, out of level conditions, high winds, side loads, pendulum action, jerking or sudden stopping of loads, hazardous conditions, experience of personnel, two machine lifts, traveling with loads, electric wires, etc. Side pull on boom or jib is extremely dangerous.
11. Power telescoping boom sections must be extended equally at all times.
12. Handling of personnel from the boom is not authorized except with equipment furnished and installed by Grove Manufacturing Company.
13. Keep load handling devices a minimum of 12 inches (30 cm) below boom head when lowering or extending boom.
14. Loaded boom angles give an approximation of the operating radius at specified boom lengths. The boom angle before loading should be greater to account for deflection.
15. Capacities appearing above the bold line are based on structural strength and tipping should not be relied upon as a capacity limitation.
16. Capacities for the 28 ft. (8.6m) boom length shall be lifted with the boom fully retracted. If boom is not fully retracted, capacities shall not exceed those shown for the 34 ft. (10.4m) boom length.

DEFINITIONS:

1. Operating Radius: Horizontal distance from a projection of the axis of rotation to the supporting surface before loading to the center of the vertical hoist line or tackle with load applied.
2. Loaded Boom Angle (Shown in Parenthesis on Main Boom Capacity Chart): is the angle between the boom base section and the horizontal, after lifting the rated load at the rated radius.
3. Working Area: Areas measured in a circular arc about the center line of rotation as shown on the working area diagram.
4. Freely Suspended Load: Load hanging free with no direct external force applied except by the lift cable.
5. Side Load: Horizontal force applied to the lifted load either on the ground or in the air.

Figure 3-20 Crane load capacity chart. (Courtesy of Grove Manufacturing Company)

JIB CAPACITIES IN POUNDS
23 ft. - 38 ft. TELE JIB

Boom Angle	23 ft. Jib Length (Fully Retracted) 0° Offset Radius (Ref.) ft.	0° Offset Cap. lbs.	15° Offset Radius (Ref.) ft.	15° Offset Cap. lbs.	30° Offset Radius (Ref.) ft.	30° Offset Cap. lbs.	33 ft. Jib Length 0° Offset Radius (Ref.) ft.	0° Offset Cap. lbs.	15° Offset Radius (Ref.) ft.	15° Offset Cap. lbs.	30° Offset Radius (Ref.) ft.	30° Offset Cap. lbs.	38 ft. Jib Length (Fully Extended) 0° Offset Radius (Ref.) ft.	0° Offset Cap. lbs.	15° Offset Radius (Ref.) ft.	15° Offset Cap. lbs.	30° Offset Radius (Ref.) ft.	30° Offset Cap. lbs.
75°	27.5	12,500	31.4	7,300	35.0	4,500	29.0	7,600	35.3	4,900	41.5	2,900	31.0	5,000	39.0	3,750	45.4	2,230
70	33.3	9,390	37.8	6,390	40.6	4,150	35.9	6,500	42.5	4,270	48.8	2,650	37.9	4,650	45.6	3,300	51.8	1,990
65	40.2	6,670	44.7	5,750	47.2	3,900	43.9	5,300	50.2	3,820	56.1	2,440	46.3	4,470	53.7	2,950	59.3	1,870
60	47.0	5,020	51.3	4,630	53.6	3,680	51.6	4,300	57.5	3,450	62.8	2,330	54.3	3,550	61.2	2,640	66.4	1,770
55	53.2	3,860	57.3	3,420	59.5	3,120	58.8	3,320	64.3	2,770	69.2	2,230	62.0	2,910	68.4	2,450	72.9	1,680
50	59.2	3,080	62.9	2,790	65.1	2,650	65.7	2,590	70.7	2,190	74.9	1,910	69.2	2,430	75.0	2,030	78.9	1,620
45	64.7	2,450	68.0	2,280	69.9	2,180	71.9	2,060	76.5	1,730	80.2	1,600	75.8	1,920	81.1	1,660	84.3	1,500
40	69.6	1,980	72.6	1,870	74.2	1,750	77.7	1,640	81.7	1,400	84.7	1,360	81.8	1,480	86.4	1,360	89.0	1,240
35	74.0	1,580	76.6	1,530	77.9	1,440	82.8	1,300	86.2	1,150	88.6	1,130	87.2	1,080	91.2	1,020	93.0	980
30	77.8	1,290	80.1	1,270	81.0	1,230	87.3	1,020	90.2	940	91.8	920	92.0	860	95.2	840	96.3	830

A6-829-003907C

No load stability on outriggers 360° with 23' - 38' tele-jib installed:

	Tele-jib fully Retracted 93'	33' Tele-jib Length 103'	Tele-jib fully Extended 108'
Minimum boom angle for indicated boom length	0°	0°	0°
Maximum boom length including jib for 0° boom angle	93'	103'	108'

23-38 ft. (7.1-11.6m) TELE JIB CAPACITY NOTES:

1. 23 ft. (7.1m) tele jib length may be used for double line lifting service. 33 ft. (10.1m) and 38 ft. (11.6m) jib lengths may be used for single line lifting service only. Capacities are based on structural strength of 23-28 ft. (7.1-11.6m) tele jib at a given main boom angle regardless of main boom length.
2. WARNING: Operation of machine with heavier loads than the capacities listed strictly prohibited. Machine tipping with jib occurs rapidly and without advance warning.
3. Capacities listed are with fully extended outriggers only.
4. WARNING: Lifting on rubber with jib is prohibited.
5. Reference radii listed are for fully extended boom only 70 ft. (21.2m).

JIB CAPACITIES IN POUNDS
23 ft. A-FRAME JIB

MAIN BOOM ANGLE	0° OFFSET Radius (Ref.)	0° OFFSET Cap. lbs.	15° OFFSET Radius (Ref.)	15° OFFSET Cap. lbs.	30° OFFSET Radius (Ref.)	30° OFFSET Cap. lbs.
75°	27.0	12,000	32.5	7,700	35.7	5,070
70	33.3	10,400	38.1	7,000	41.2	4,800
65	40.2	8,300	44.9	6,300	47.8	4,500
60	47.0	5,870	51.3	5,450	54.0	4,300
55	53.2	4,450	57.3	4,080	59.8	3,690
50	59.2	3,560	62.9	3,170	65.1	3,030
45	64.7	2,910	68.0	2,610	69.9	2,590
40	69.6	2,400	72.6	2,230	74.2	2,160
35	74.0	2,020	76.6	1,920	77.9	1,880
30	77.8	1,730	80.1	1,680	81.0	1,670

A6-829-003755D

23 ft. (7.1m) JIB CAPACITY NOTES:

1. 23 ft. (7.1m) jib may be used for double line lifting service. Capacities are based on structural strength of 23 ft. (7.1m) jib at a given main boom angle regardless of main boom length.
2. WARNING: Operation of machine with heavier loads than the capacities listed strictly prohibited. Machine tipping with jib occurs rapidly and without advance warning.
3. Capacities listed are with fully extended outriggers only.
4. WARNING: Lifting on rubber with jib is prohibited.
5. Reference radii listed are for fully extended main boom only.
6. No load stability on outriggers with 23 ft. (7.1m) jib installed.
   a. Minimum boom angle for fully extended main boom = 0°.
   b. Maximum boom length at 0° main boom angle = 93ft. (28.3m).

RANGE DIAGRAM

WEIGHT REDUCTIONS FOR LOAD HANDLING DEVICES

23-38 ft. TELE. JIB with 28-70 ft. BOOM	
*Stowed	604 lbs.
*Erected (Retracted)	3,659 lbs.
*Erected (Extended)	4,583 lbs.

23 ft. JIB with 28-70 ft. BOOM	
*Stowed	381 lbs.
*Erected	1,950 lbs.

*Reduction of main boom capacities.

HOOK BLOCKS	
22 Ton, 3 Sheave (12 1/8" OD)	320 lbs.
22 Ton, 3 Sheave (15 7/8" OD)	455 lbs.
15 Ton, 2 Sheave	298 lbs.
12 Ton, 1 Sheave (15 7/8" OD)	400 lbs.
12 Ton, 1 Sheave (12 1/8" OD)	285 lbs.
Auxiliary Boom Head	100 lbs.
5 Ton Headache Ball	150 lbs.

NOTE: All Load Handling Devices and Boom Attachments are Considered Part of the Load and Suitable Allowances MUST BE MADE for Their Combined Weights. Weights are for Grove furnished equipment.

Figure 3-20 (Continued)

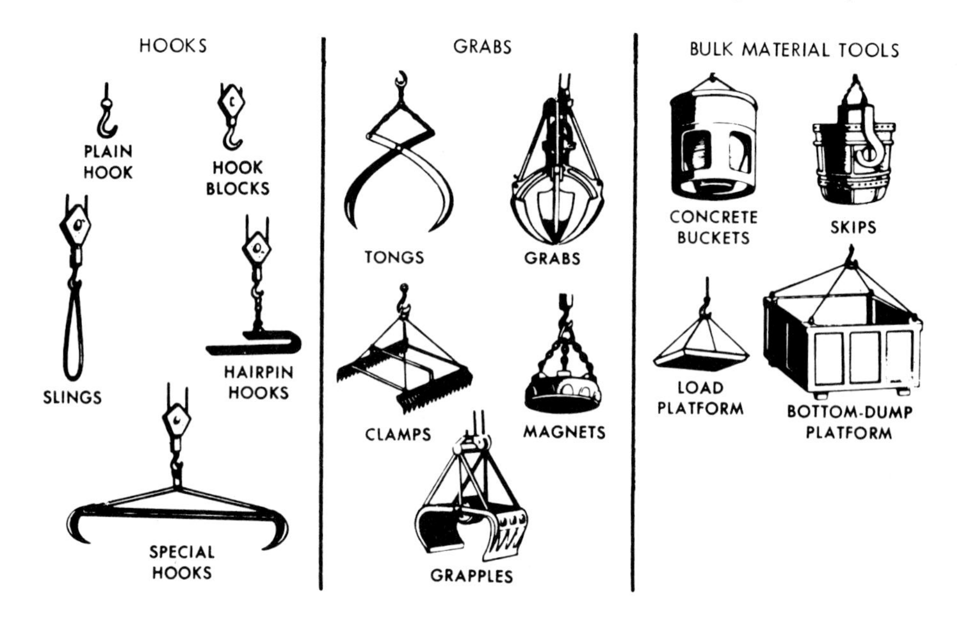

Figure 3-21 Lifting attachments for the crane. [Permission to reproduce this material has been granted by the Power Crane & Shovel Assn. (PCSA), a bureau of the Construction Industry Manufacturers Assn. (CIMA). Neither PCSA nor CIMA can assume responsibility for the accuracy of the reproduction.]

- Use a communications system or hand signals when the crane operator cannot see the load. Make sure that all workers involved in the operation know the signals to be used.
- Provide *tag lines* (restraining lines) when there is any danger due to swinging loads.
- Ensure that crane operators are well trained and know the capability of their machines.
- Check safe-lifting-capacity charts for the entire range of planned swing before starting a lift. Use a load indicator if possible.

PROBLEMS

1. Estimate the actual bucket load in bank measure for a hydraulic excavator-backhoe whose heaped bucket capacity is 2.10 cu yd (1.6 m^3). The machine is excavating sand and gravel.
2. Estimate the volume of common earth in bank measure carried by a hydraulic shovel bucket whose heaped capacity is 5.0 cu yd (3.82 m^3).
3. Estimate the hourly production in bank measure of a hydraulic shovel loading sand and gravel into trucks. The machine is equipped with a front dump bucket having a heaped bucket capacity of 3.0 cu yd (2.3 m^3). Average swing angle is 75°, job conditions are good, and management conditions are excellent.

4. A 4-yd (3.06-m^3) (heaped) hydraulic shovel with a bottom dump bucket is excavating tough clay. The swing angle is 150° and job efficiency is 50 min/h. Estimate the shovel's hourly production in bank measure.

5. A 3-yd (2.29-m^3) dragline is excavating and stockpiling sand and gravel. Average depth of cut is 5.4 ft (1.65 m), average swing angle is 75°, and job efficiency is 45 min/h. Estimate the dragline's hourly production in bank measure.

6. Estimate the time required to load 500 cu yd (382 m^3) of gravel into trucks using a clamshell having a heaped bucket capacity of 0.75 cu yd (0.57 m^3). Estimated cycle time is 20 seconds. Job efficiency is estimated to be 50 min/h.

7. A small hydraulic excavator will be used to dig a trench in soft clay (bucket fill factor = 0.90). The minimum trench size is 24 in. (61 cm) wide by 6 ft (1.83 m) deep. The excavator bucket available is 30 in. (76 cm) wide and has a heaped capacity of ¾ cu yd (0.57 m^3). The maximum digging depth of the excavator is 17.5 ft (5.3 m). The average swing angle is expected to be 90°. Estimate the hourly trench production in linear feet (meters) if job efficiency is 50 min/h.

8. What is the maximum net load that can be safely lifted over a 360° swing by the crane of Figure 3-20 under the following conditions? The crane is equipped with a 23- to 38-ft telescoping jib (stowed) with a 15-ton, two-sheave block hook, the boom length is 52 ft (15.9 m), and the operating radius is 25 ft (7.6 m). What restrictions must be observed to lift this load safely?

9. Write a computer program to estimate the trenching production of a hydraulic backhoe based on Equation 3-3 and Tables 3-7, 3-8, and 3-9. Input should include heaped bucket capacity, type of material, bucket width, trench width and depth, average angle of swing, maximum digging depth of the backhoe, soil load factor, bucket fill factor, and job efficiency. Output should provide hourly production in both bank measure and trench linear measure.

10. Write a computer program to estimate the production of a hydraulic shovel based on Equation 3-1 and Table 3-3. Input should include rated shovel size, type of material, angle of swing, heaped bucket capacity, bucket fill factor, soil load factor, and job efficiency. Output should be in bank measure if a soil load factor is input; otherwise, in loose measure.

REFERENCES

1. *Caterpillar Performance Handbook,* 21st ed. Caterpillar Inc., Peoria, Ill., 1990.
2. *Hydraulic Excavators and Telescoping-Boom Cranes.* PCSA Bureau of Construction Industry Manufacturers Association, Milwaukee, Wis., 1974.
3. NUNNALLY, S.W. *Managing Construction Equipment.* Englewood Cliffs, N. J.: Prentice Hall, 1977.
4. *Operating Safety: Link-Belt Cranes and Excavators.* FMC Corporation, Cedar Rapids, Iowa, 1975.
5. *Production and Cost Estimating of Material Movement with Earthmoving Equipment.* Terex Corporation, Hudson, Ohio, 1981.
6. *Technical Bulletin No. 1: Man the Builder.* PCSA Bureau of Construction Industry Manufacturers Association, Milwaukee, Wis., 1971.

4
Loading and Hauling

4-1
ESTIMATING EQUIPMENT TRAVEL TIME

In calculating the time required for a haul unit to make one complete cycle, it is customary to break the cycle down into fixed and variable components.

$$\text{Cycle time} = \text{Fixed time} + \text{Variable time} \tag{4-1}$$

Fixed time represents those components of cycle time other than travel time. It includes spot time (moving the unit into position to begin loading), load time, maneuver time, and dump time. Fixed time can usually be closely estimated for a particular type of operation.

Variable time represents the travel time required for a unit to haul material to the unloading site and return. As you would expect, travel time will depend on the vehicle's weight and power, the condition of the haul road, the grades encountered, and the altitude above sea level. This section represents methods for calculating a vehicle's resistance to movement, its maximum speed, and its travel time. Methods for estimating fixed times are presented in Sections 4-2 to 4-5, which describe specific types of hauling equipment.

Rolling Resistance

To determine the maximum speed of a vehicle in a specific situation, it is necessary to determine the total resistance to movement of the vehicle. The resistance that a vehicle encounters in traveling over a surface is made up of two components, rolling resistance and grade resistance.

$$\text{Total resistance} = \text{Grade resistance} + \text{Rolling resistance} \quad (4\text{-}2)$$

Resistance may be expressed in either pounds per ton of vehicle weight (kilograms per metric ton) or in pounds (kilograms). To avoid confusion, the term *resistance factor* will be used in this chapter to denote resistance in lb/ton (kg/t). *Rolling resistance* is primarily due to tire flexing and penetration of the travel surface. The rolling resistance factor for a rubber-tired vehicle equipped with conventional tires moving over a hard, smooth, level surface has been found to be about 40 lb/ton of vehicle weight (20 kg/t). For vehicles equipped with radial tires, the rolling resistance factor may be as low as 30 lb/ton (15 kg/t). It has been found that the rolling resistance factor increases about 30 lb/ton (15 kg/t) for each inch (2.5 cm) of tire penetration. This leads to the following equation for estimating rolling resistance factors:

$$\text{Rolling resistance factor (lb/ton)} = 40 + (30 \times \text{in. penetration}) \quad (4\text{-}3A)$$

$$\text{Rolling resistance factor (kg/t)} = 20 + (5.9 \times \text{cm penetration}) \quad (4\text{-}3B)$$

The rolling resistance in pounds (kilograms) may be found by multiplying the rolling resistance factor by the vehicle's weight in tons (metric tons). Table 4-1 provides typical values for the rolling resistance factor in construction situations.

Crawler tractors may be thought of as traveling over a road created by their own tracks. As a result, crawler tractors are usually considered to have no rolling resistance when calculating vehicle resistance and performance. Actually, of course, the rolling resistance of crawler tractors does vary somewhat between different surfaces. However, the standard method for rating crawler tractor power (drawbar horsepower) measures the power actually produced at the hitch when operating on a standard surface. Thus the rolling resistance of the tractor over the standard surface has already been subtracted from the tractor's performance. Although a crawler tractor is considered to have no rolling resistance, when it tows a wheeled vehicle (such as a

Table 4-1 Typical values of rolling resistance factor

	Rolling Resistance Factor	
Type of Surface	*lb/ton*	*kg/t*
Concrete or asphalt	40 (30)*	20 (15)
Firm, smooth, flexing slightly under load	64 (52)	32 (26)
Rutted dirt roadway, 1–2 in. penetration	100	50
Soft, rutted dirt, 3–4 in. penetration	150	75
Loose sand or gravel	200	100
Soft, muddy, deeply rutted	300–400	150–200

*Values in parentheses are for radial tires.

scraper or compactor) the rolling resistance of the towed vehicle must be considered in calculating the total resistance of the combination.

Grade Resistance

Grade resistance represents that component of vehicle weight which acts parallel to an inclined surface. When the vehicle is traveling up a grade, grade resistance is positive. When traveling downhill, grade resistance is negative. The exact value of grade resistance may be found by multiplying the vehicle's weight by the sine of the angle that the road surface makes with the horizontal. However, for the grades usually encountered in construction, it is sufficiently accurate to use the approximation of Equation 4-4. That is, a 1% grade (representing a rise of 1 unit in 100 units of horizontal distance) is considered to have a grade resistance equal to 1% of the vehicle's weight. This corresponds to a grade resistance factor of 20 lb/ton (10 kg/t) for each 1% of grade.

$$\text{Grade resistance factor (lb/ton)} = 20 \times \text{grade (\%)} \quad (4\text{-}4A)$$

$$\text{Grade resistance factor (kg/t)} = 10 \times \text{grade (\%)} \quad (4\text{-}4B)$$

Grade resistance (lb or kg) may be calculated using Equation 4-5 or 4-6.

$$\text{Grade resistance (lb)} = \text{Vehicle weight (tons)} \times \text{Grade resistance factor (lb/ton)} \quad (4\text{-}5A)$$

$$\text{Grade resistance (kg)} = \text{Vehicle weight (t)} \times \text{Grade resistance factor (kg/t)} \quad (4\text{-}5B)$$

$$\text{Grade resistance (lb)} = \text{Vehicle weight (lb)} \times \text{Grade} \quad (4\text{-}6A)$$

$$\text{Grade resistance (kg)} = \text{Vehicle weight (kg)} \times \text{Grade} \quad (4\text{-}6B)$$

Effective Grade

The total resistance to movement of a vehicle (the sum of its rolling resistance and grade resistance) may be expressed in pounds or kilograms. However, a somewhat simpler method for expressing total resistance is to state it as a grade (%), which would have a grade resistance equivalent to the total resistance actually encountered. This method of expressing total resistance is referred to as *effective grade, equivalent grade,* or *percent total resistance* and is often used in manufacturers' performance charts. Effective grade may be easily calculated by use of Equation 4-7.

$$\text{Effective grade (\%)} = \text{Grade (\%)} + \frac{\text{Rolling resistance factor (lb/ton)}}{20} \quad (4\text{-}7A)$$

$$\text{Effective grade (\%)} = \text{Grade (\%)} + \frac{\text{Rolling resistance factor (kg/t)}}{10} \quad (4\text{-}7B)$$

Example 4-1

PROBLEM A wheel tractor-scraper weighing 100 tons (91 t) is being operated on a haul road with a tire penetration of 2 in. (5 cm). What is the total resistance (lb and kg) and effective grade when (a) the scraper is ascending a slope of 5%; (b) the scraper is descending a slope of 5%?

SOLUTION

$$\text{Rolling resistance factor} = 40 + (30 \times 2) = 100 \text{ lb/ton} \quad (4\text{-}3A)$$
$$[= 20 + (6 \times 5) = 50 \text{ kg/t}] \quad (4\text{-}3B)$$
$$\text{Rolling resistance} = 100 \text{ (lb/ton)} \times 100 \text{ (tons)} = 10{,}000 \text{ lb}$$
$$[= 50 \text{ (kg/t)} \times 91 \text{ (t)} = 4550 \text{ kg}]$$

(a)
$$\text{Grade resistance} = 100 \text{ (tons)} \times 2000 \text{ (lb/ton)} \times 0.05 = 10{,}000 \text{ lb} \quad (4\text{-}6A)$$
$$[= 91 \text{ (t)} \times 1000 \text{ (kg/t)} \times 0.05 = 4550 \text{ kg}] \quad (4\text{-}6B)$$
$$\text{Total resistance} = 10{,}000 \text{ lb} + 10{,}000 \text{ lb} = 20{,}000 \text{ lb}$$
$$[= 4550 \text{ kg} + 4550 \text{ kg} = 9100 \text{ kg}] \quad (4\text{-}2)$$
$$\text{Effective grade} = 5 + \frac{100}{20} = 10\% \quad (4\text{-}7A)$$

(b)
$$\text{Grade resistance} = 100 \text{ (tons)} \times 2000 \text{ (lb/ton)} \times (-0.05) = -10{,}000 \text{ lb} \quad (4\text{-}6A)$$
$$[= 91 \text{ (t)} \times 1000 \text{ (kg/t)} \times (-0.05) = -4550 \text{ kg}] \quad (4\text{-}6B)$$
$$\text{Total resistance} = -10{,}000 \text{ lb} + 10{,}000 \text{ lb} = 0 \text{ lb}$$
$$[= -4550 \text{ kg} + 4550 \text{ kg} = 0 \text{ kg}] \quad (4\text{-}2)$$
$$\text{Effective grade} = -5 + \frac{100}{20} = 0\% \quad (4\text{-}7A)$$
$$\left[= -5 + \frac{50}{10} = 0\%\right] \quad (4\text{-}7B)$$

Example 4-2

PROBLEM A crawler tractor weighing 80,000 lb (36 t) is towing a rubber-tired scraper weighing 100,000 lb (45.5 t) up a grade of 4%. What is the total resistance (lb and kg) of the combination if the rolling resistance factor is 100 lb/ton (50 kg/t)?

SOLUTION

$$\text{Rolling resistance (neglect crawler)} = \frac{100{,}000 \text{ (lb)}}{2000 \text{ (lb/ton)}} \times 100 \text{ (lb/ton)} = 5000 \text{ lb}$$
$$[= 45.5 \text{ (t)} \times 50 \text{ (kg/t)} = 2275 \text{ kg}]$$
$$\text{Grade resistance} = 180{,}000 \times 0.04 = 7200 \text{ lb} \quad (4\text{-}6A)$$
$$[= 81.5 \times 1000 \text{ kg/t} \times 0.04 = 3260 \text{ kg}] \quad (4\text{-}6B)$$
$$\text{Total resistance} = 5000 + 7200 = 12{,}200 \text{ lb}$$
$$[= 2275 + 3260 = 5535 \text{ kg}] \quad (4\text{-}2)$$

Effect of Altitude

All internal combustion engines lose power as their elevation above sea level increases because of the decreased density of air at higher elevations. There is some variation in the performance of two-cycle and four-cycle naturally aspirated and turbocharged diesel engines. However, engine power decreases approximately 3% for each 1000 ft (305 m) increase in altitude above the maximum altitude at which full rated power is delivered. Turbocharged engines are more efficient at higher altitude than are naturally aspirated engines and may deliver full rated power up to an altitude of 10,000 ft (3050 m) or more.

Manufacturers use a *derating factor* to express percentage of reduction in rated vehicle power at various altitudes. Whenever possible, use the manufacturer's derating table for estimating vehicle performance. However, when derating tables are not available, the derating factor obtained by the use of Equation 4-8 is sufficiently accurate for estimating the performance of naturally aspirated engines.

$$\text{Derating factor (\%)} = 3 \times \left[\frac{\text{Altitude (ft)} - 3000^*}{1000}\right] \tag{4-8A}$$

$$\text{Derating factor (\%)} = \frac{\text{Altitude (m)} - 915^*}{102} \tag{4-8B}$$

The percentage of rated power available is, of course, 100 minus the derating factor. The use of derating factors in determining maximum vehicle power is illustrated in Example 4-3.

Effect of Traction

The power available to move a vehicle and its load is expressed as *rimpull* for wheel vehicles and *drawbar pull* for crawler tractors. Rimpull is the pull available at the rim of the driving wheels under rated conditions. Since it is assumed that no slippage of the tires on the rims will occur, this is also the power available at the surface of the tires. Drawbar pull is the power available at the hitch of a crawler tractor operating under standard conditions. Operation at increased altitude may reduce the maximum pull of a vehicle, as explained in the previous paragraph. Another factor limiting the usable power of a vehicle is the maximum traction that can be developed between the driving wheels or tracks and the road surface. Traction depends on the coefficient of traction and the weight on the drivers as expressed by Equation 4-9. This represents the maximum pull that a vehicle can develop, regardless of vehicle horsepower.

$$\text{Maximum usable pull} = \text{Coefficient of traction} \times \text{Weight on drivers} \tag{4-9}$$

For crawler tractors and all-wheel-drive rubber-tired equipment, the weight on the drivers is the total vehicle weight. For other types of vehicles, consult the manufacturer's specifications to determine the weight on the drivers. Typical values of coefficient of traction for common surfaces are given in Table 4-2.

Example 4-3

PROBLEM A four-wheel drive tractor weighs 44,000 lb (20 000 kg) and produces a maximum rimpull of 40,000 lb (18 160 kg) at sea level. The tractor is being operated at an altitude of 10,000 ft (3050 m) on wet earth. A pull of 22,000 lb (10 000 kg) is required to move the tractor and its load. Can the tractor perform under these conditions? Use Equation 4-8 to estimate altitude deration.

*Substitute maximum altitude for rated performance, if known.

Table 4-2 Typical values of coefficient of traction

Type of Surface	*Rubber Tires*	*Tracks*
Concrete, dry	0.90	0.45
Concrete, wet	0.80	0.45
Earth or clay loam, dry	0.60	0.90
Earth or clay loam, wet	0.45	0.70
Gravel, loose	0.35	0.50
Quarry pit	0.65	0.55
Sand, dry, loose	0.25	0.30
Sand, wet	0.40	0.50
Snow, packed	0.20	0.25
Ice	0.10	0.15

SOLUTION

$$\text{Derating factor} = 3 \times \left[\frac{10{,}000 - 3000}{1000}\right] = 21\% \qquad (4\text{-}8A)$$

$$\left[= \frac{3050 - 915}{102} = 21\%\right] \qquad (4\text{-}8B)$$

$$\text{Percent rated power available} = 100 - 21 = 79\%$$

$$\text{Maximum available power} = 40{,}000 \times 0.79 = 31{,}600 \text{ lb}$$

$$[= 18\,160 \times 0.79 = 14\,346 \text{ kg}]$$

$$\text{Coefficient of traction} = 0.45 \quad \text{(Table 4-2)}$$

$$\text{Maximum usable pull} = 0.45 \times 44{,}000 = 19{,}800 \text{ lb} \qquad (4\text{-}9)$$

$$[= 0.45 \times 20\,000 = 9000 \text{ kg}]$$

Because the maximum pull as limited by traction is less than the required pull, the tractor *cannot perform under these conditions.* For the tractor to operate, it would be necessary to reduce the required pull (total resistance), increase the coefficient of traction, or increase the tractor's weight on the drivers.

Use of Performance and Retarder Curves

Crawler tractors may be equipped with direct-drive (manual gearshift) transmissions. The drawbar pull and travel speed of this type of transmission are determined by the gear selected. For other types of transmissions, manufacturers usually present the speed versus pull characteristics of their equipment in the form of performance and retarder charts. A *performance chart* indicates the maximum speed that a vehicle can maintain under rated conditions while overcoming a specified total resistance. A *retarder chart* indicates the maximum speed at which a vehicle may descend a slope when the total resistance is negative without using brakes. Retarder charts derive their name from the vehicle retarder, which is a hydraulic device used for controlling vehicle speed on a downgrade.

Figure 4-1 illustrates a relatively simple performance curve of the type often used for crawler tractors. Rimpull or drawbar pull is shown on the vertical scale and

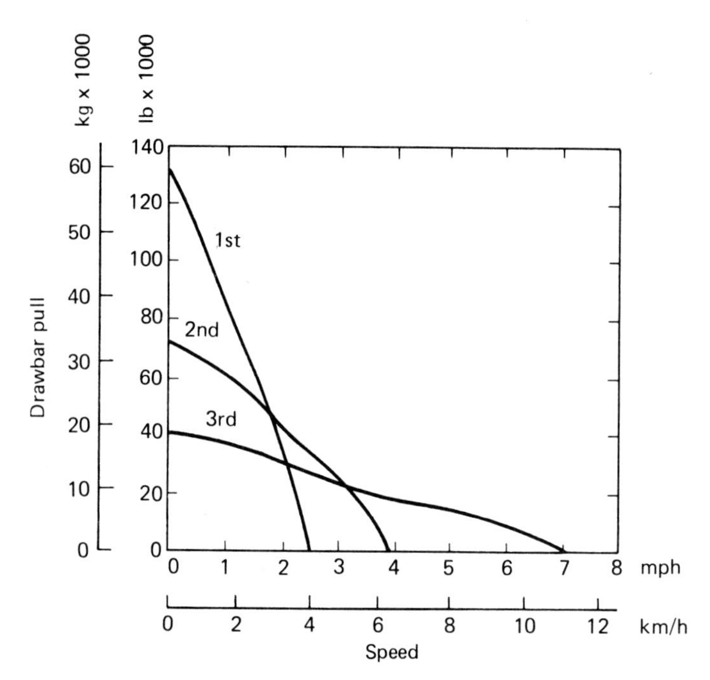

Figure 4-1 Typical crawler tractor performance curve.

maximum vehicle speed on the horizontal scale. The procedure for using this type of curve is to first calculate the required pull or total resistance of the vehicle and its load (lb or kg). Then enter the chart on the vertical scale with the required pull and move horizontally until you intersect one or more gear performance curves. Drop vertically from the point of intersection to the horizontal scale. The value found represents the maximum speed that the vehicle can maintain while developing the specified pull. When the horizontal line of required pull intersects two or more curves for different gears, use the point of intersection farthest to the right, because this represents the maximum speed of the vehicle under the given conditions.

Example 4-4

PROBLEM Use the performance curve of Figure 4-1 to determine the maximum speed of the tractor when the required pull (total resistance) is 60,000 lb (27 240 kg).

SOLUTION Enter Figure 4-1 at a drawbar pull of 60,000 lb (27 240 kg) and move horizontally until you intersect the curves for first and second gears. Read the corresponding speeds of 1.0 mi/h (1.6 km/h) for second gear and 1.5 mi/h (2.4 km/h) for first gear. The maximum possible speed is therefore 1.5 mi/h (2.4 km/h) in first gear.

Figure 4-2 represents a more complex performance curve of the type frequently used by manufacturers of tractor-scrapers, trucks, and wagons. In addition to curves of speed versus pull, this type of chart provides a graphical method for calculating the required pull (total resistance). To use this type of curve, enter the top scale at the actual weight of the vehicle (empty or loaded as applicable). Drop vertically until you intersect the diagonal line corresponding to the percent total resistance (or ef-

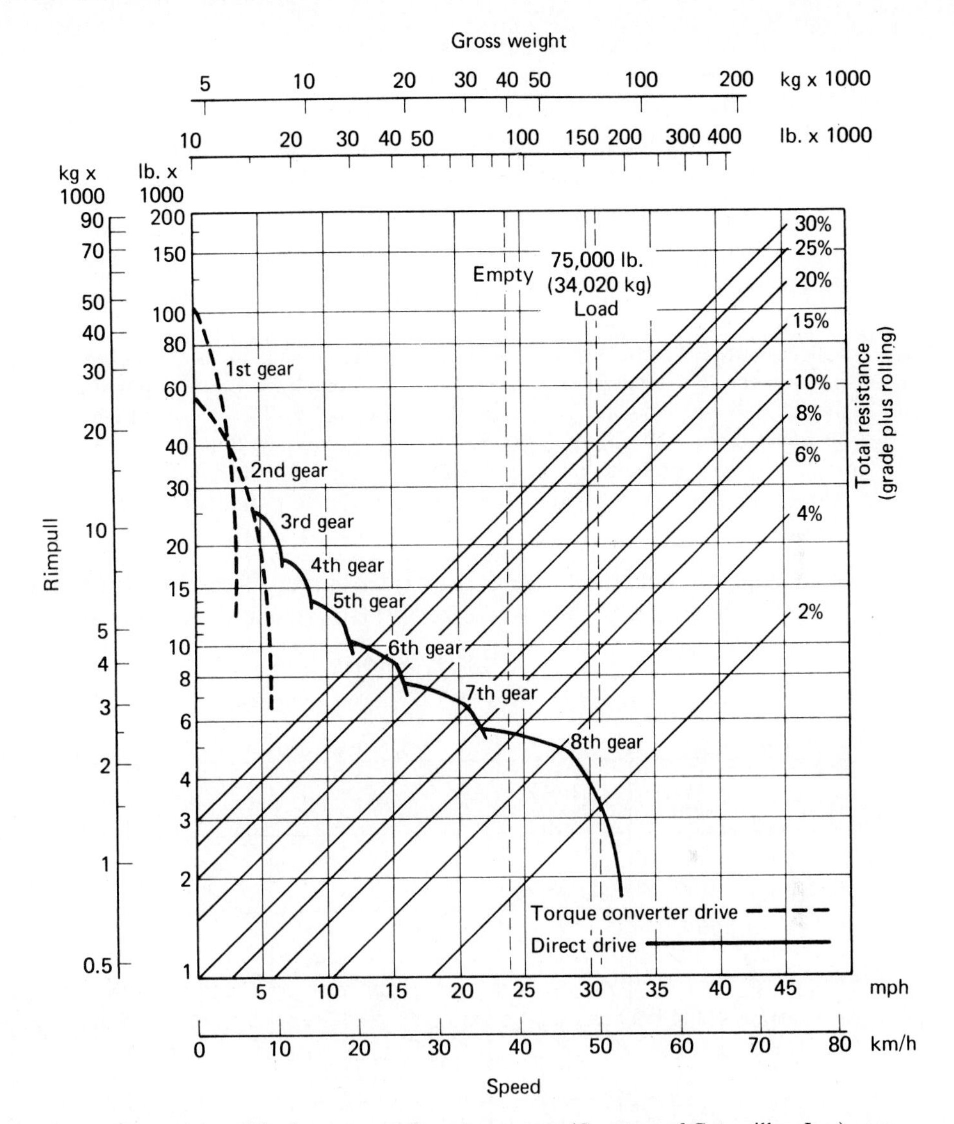

Figure 4-2 Wheel scraper performance curve. (Courtesy of Caterpillar, Inc.)

fective grade), interpolating as necessary. From this point move horizontally until you intersect one or more performance curves. From the point of intersection, drop vertically to find the maximum vehicle speed.

When altitude adjustment is required, the procedure is modified slightly. In this case, start with the gross weight on the top scale and drop vertically until you intersect the total resistance curve. Now, however, move horizontally all the way to the left scale to read the required pull corresponding to vehicle weight and effective grade. Next, divide the required pull by the quantity "1 − derating factor (expressed as a decimal)" to obtain an adjusted required pull. Now, from the adjusted value of required pull on the left scale move horizontally to intersect one or more gear curves and drop vertically to find the maximum vehicle speed. This procedure is equivalent to saying that when a vehicle produces only one-half of its rated power due to altitude effects, its maximum speed can be found from its standard performance curve by doubling the actual required pull. The procedure is illustrated in Example 4-5.

Example 4-5

PROBLEM Using the performance curve of Figure 4-2, determine the maximum speed of the vehicle if its gross weight is 150,000 lb (68 000 kg), the total resistance is 10%, and the altitude derating factor is 25%.

SOLUTION Start on the top scale with a weight of 150,000 lb (68 000 kg), drop vertically to intersect the 10% total grade line, and move horizontally to find a required pull of 15,000 lb (6800 kg) on the left scale. Divide 15,000 lb (6800 kg) by 0.75 (1 − derating factor) to obtain an adjusted required pull of 20,000 lb (9080 kg). Enter the left scale at 20,000 lb (9080 kg) and move horizontally to intersect the first, second, and third gear curves. Drop vertically from the point of intersection with the third gear curve to find a maximum speed of 6 mi/h (10 km/h).

Figure 4-3 illustrates a typical retarder curve. In this case, it is the retarder curve for the tractor-scraper whose performance curve is shown in Figure 4-2. The

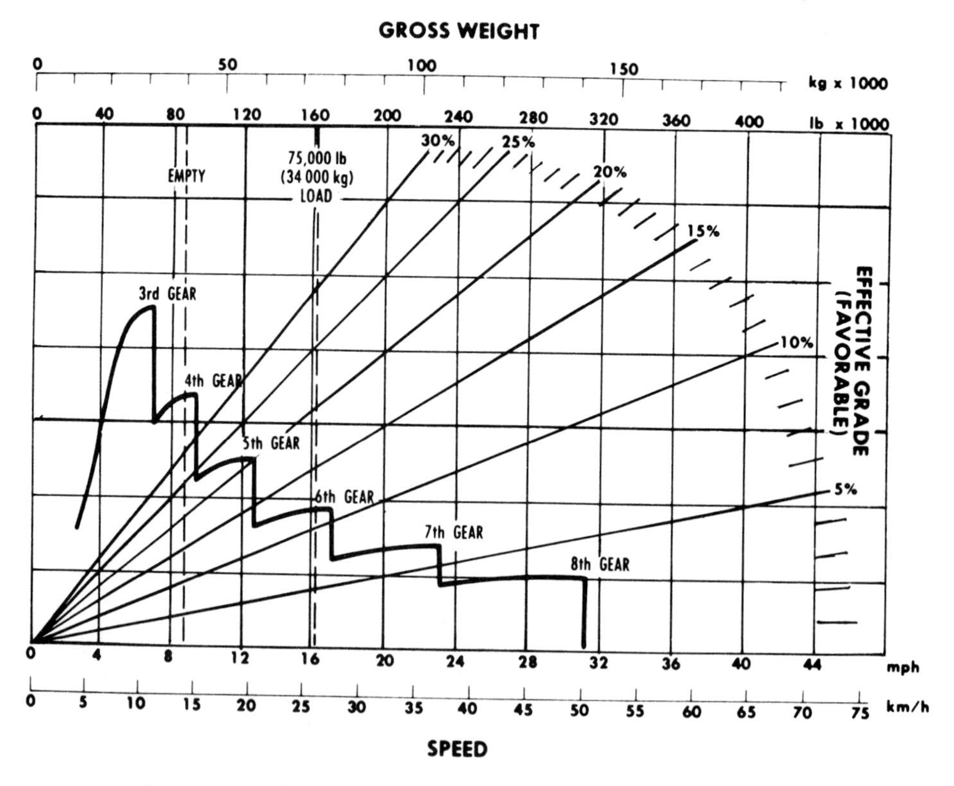

Figure 4-3 Wheel scraper retarder curve. (Courtesy of Caterpillar, Inc.)

retarder curve is read in a manner similar to the performance curve. Remember, however, that in this case the vertical scale represents *negative* total resistance. After finding the intersection of the vehicle weight with effective grade, move horizontally until you intersect the retarder curve. Drop vertically from this point to find the maximum speed at which the vehicle should be operated.

Estimating Travel Time

The maximum speed that a vehicle can maintain over a section of the haul route cannot be used for calculating travel time over the section, because it does not include vehicle acceleration and deceleration. One method for accounting for acceleration and deceleration is to multiply the maximum vehicle speed by an average speed factor from Table 4-3 to obtain an average vehicle speed for the section. Travel time for the section is then found by dividing the section length by the average vehicle speed. When a section of the haul route involves both starting from rest and coming to a stop, the averge speed factor from the first column of Table 4-3 should be applied twice (i.e., use the square of the table value) for that section.

A second method for estimating travel time over a section of haul route is to use the travel-time curves provided by some manufacturers. Separate travel-time curves are prepared for loaded (rated payload) and empty conditions, as shown in Figures 4-4 and 4-5. As you see, travel time for a section of the haul route may be read directly from the graph given section length and effective grade. However, travel-time curves cannot be used when the effective grade is negative. In this case, the average speed method must be used along with the vehicle retarder curve. To adjust for altitude deration when using travel-time curves, multiply the time obtained from the curve by the quantity "1 + derating factor" to obtain the adjusted travel time.

The use of both the average speed and the travel-time curve method is illustrated in the example problems of this chapter.

Table 4-3 Average speed factors

Length of Haul Section (ft)	Length of Haul Section (m)	*Starting from 0 or Coming to a Stop*	*Increasing Maximum Speed from Previous Section*	*Decreasing Maximum Speed from Previous Section*
150	46	0.42	0.72	1.60
200	61	0.51	0.76	1.51
300	92	0.57	0.80	1.39
400	122	0.63	0.82	1.33
500	153	0.65	0.84	1.29
700	214	0.70	0.86	1.24
1000	305	0.74	0.89	1.19
2000	610	0.86	0.93	1.12
3000	915	0.90	0.95	1.08
4000	1220	0.93	0.96	1.05
5000	1525	0.95	0.97	1.04

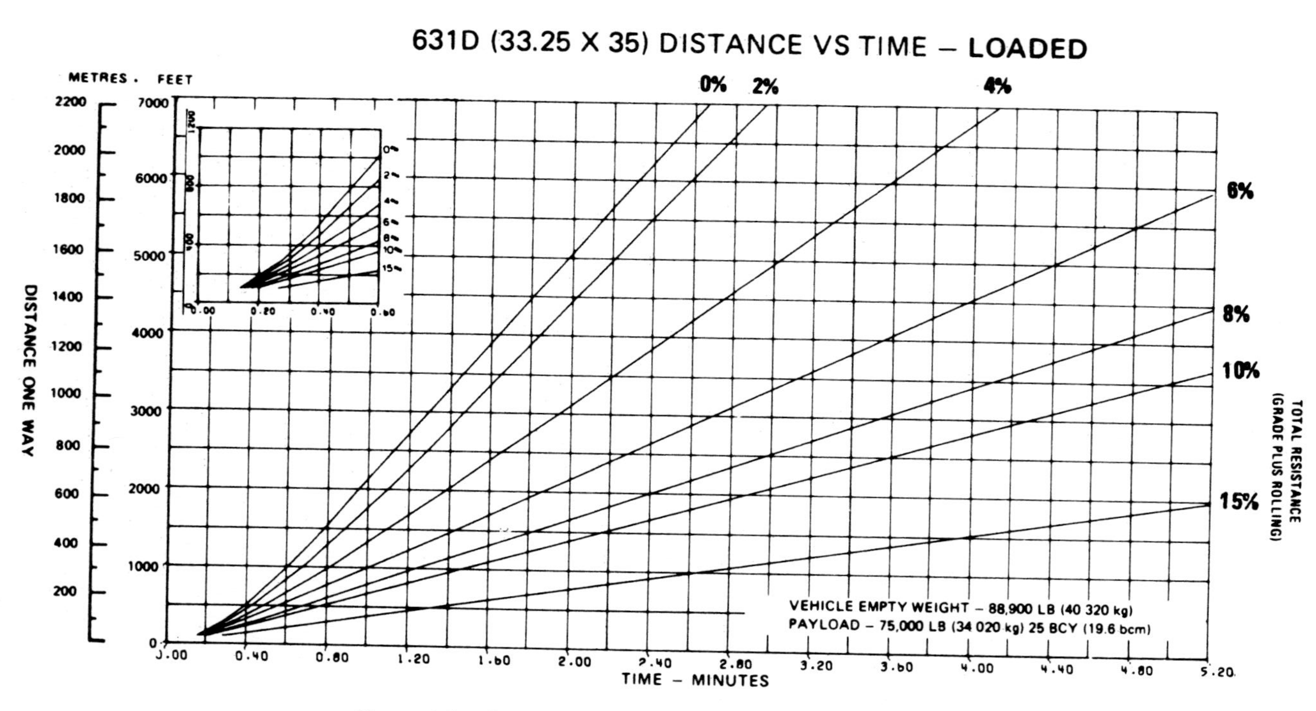

Figure 4-4 Scraper travel time—loaded. (Courtesy of Caterpillar, Inc.)

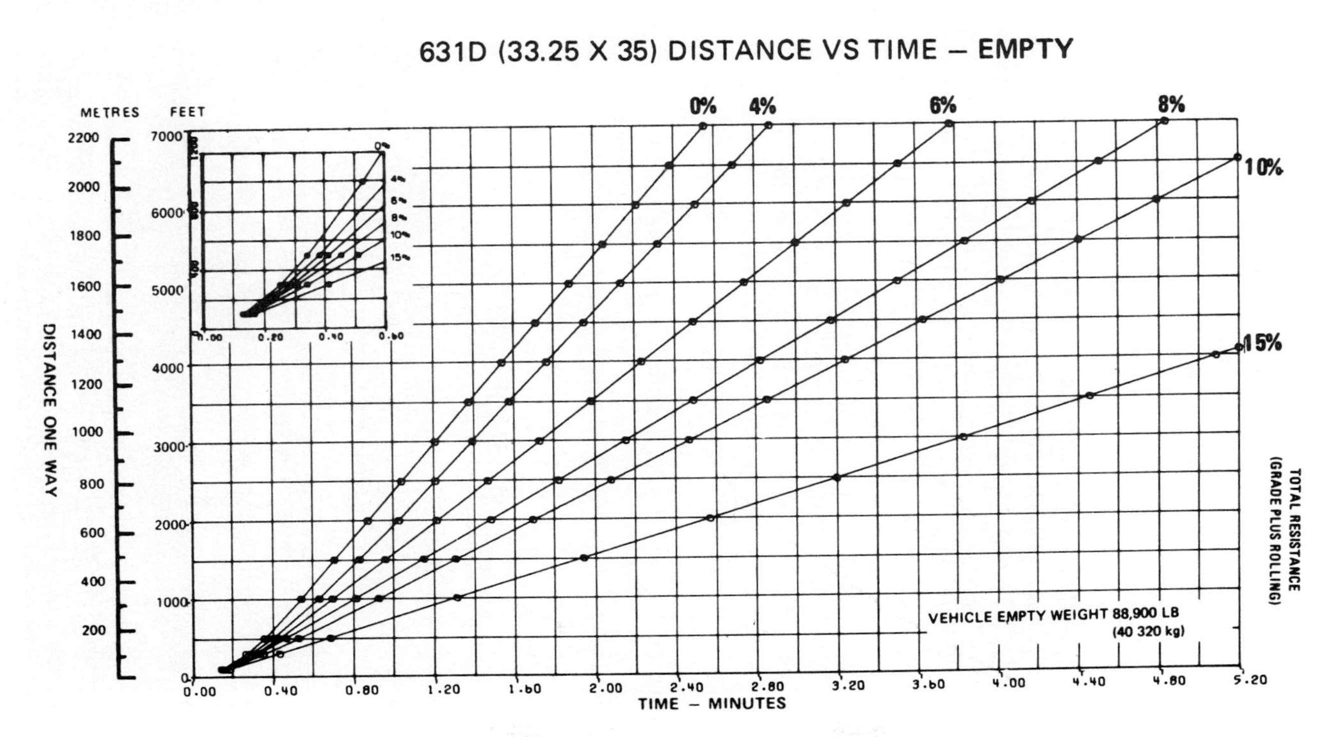

Figure 4-5 Scraper travel time—empty. (Courtesy of Caterpillar, Inc.)

4-2 DOZERS

Tractors and Dozers

A tractor equipped with a front-mounted earthmoving blade is known as a *dozer* or *bulldozer.* A dozer moves earth by lowering the blade and cutting until a full blade load of material is obtained. It then pushes the material across the ground surface to the required location. The material is unloaded by pushing it over a cliff or into a hopper or by raising the blade to form a spoil pile.

Both rubber-tired (or wheel) dozers and crawler (or track) dozers are available. Because of its excellent traction and low ground pressure (typically 6 to 9 lb/sq in.; 0.4 to 0.6 bar), crawler dozers (Figure 4-6) are well suited for use in rough terrain or areas of low trafficability. Low-ground-pressure models with extra-wide tracks are available having ground pressures as low as 3 lb/sq in. (0.2 bar). Crawler dozers can operate on steeper side slopes and climb greater grades than can wheel dozers. Wheel dozers, on the other hand, operate at higher speed than do crawlers dozers. Wheel dozers are also capable of operating on paved roads without damaging the surface. While the wheel tractor's dozing ability is limited somewhat by its lower traction and high ground pressure (25 to 35 lb/sq in.; 1.7 to 2.4 bars), its high ground pressure makes it an effective soil compactor.

Either rubber-tired or crawler tractors may be equipped with attachments other than dozer blades. These include rakes used for gathering up brush and small fallen trees, and plows, rippers, and scarifiers, which are used to break up hard surfaces.

Figure 4-6 Crawler tractor dozer. (Courtesy of Fiatallis North America, Inc.)

Tractors are also used to tow many items of construction equipment, such as compactors, scrapers, and wagons. Towing applications are discussed in succeeding chapters.

Dozers may be equipped with direct-drive, power-shift, or hydrostatic transmissions. *Hydrostatic transmissions* utilize individual hydraulic motors to drive each track. Therefore, the speed of each track may be infinitely varied, forward or reverse. As a result, it is possible for a dozer equipped with a hydrostatic drive to turn in its own length by moving one track forward while the other track moves in reverse.

Dozer Blades

There are a number of types of dozer blades available, and the four most common types are illustrated in Figure 4-7. The three types of adjustments that may be made to dozer blades are illustrated in Figure 4-8. Tilting the blade is useful for ditching and breaking up frozen or crusty soils. Pitching the blade forward reduces blade penetration and causes the loosened material to roll in front of the blade, whereas pitching the blade backward increases penetration. Angling the blade is helpful when side-hill cutting, ditching, and moving material laterally. All the blades shown in Figure 4-7 may be tilted except the cushion blade. However, only the angle blade may be angled.

The two indicators of potential dozer performance are based on the ratio of tractor power to blade size. These indicators are horsepower per foot of cutting edge and horsepower per loose cubic yard. A blade's *horsepower per foot of cutting edge* provides a measure of the blade's ability to penetrate hard soils. The *horsepower per loose cubic yard* rating provides an indication of the blade's ability to push material once the blade is loaded.

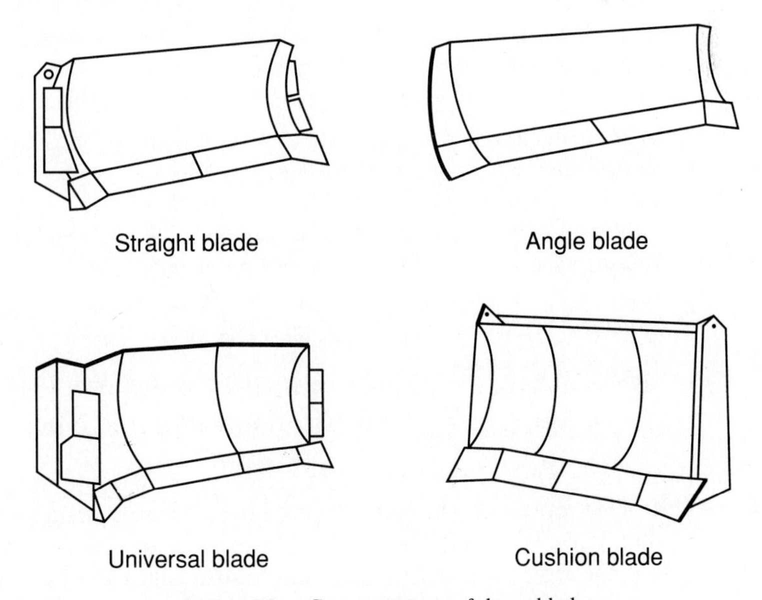

Figure 4-7 Common types of dozer blades.

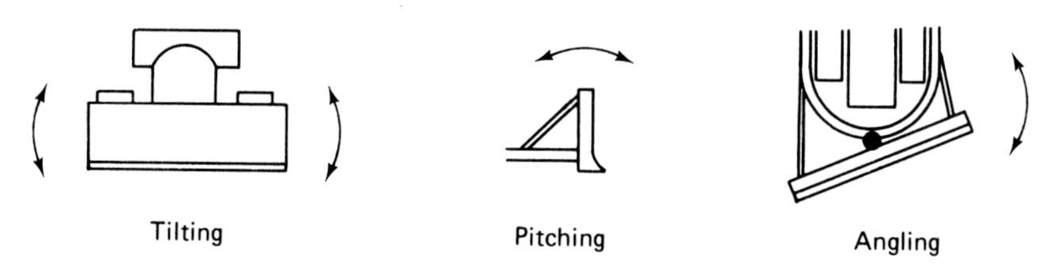

Figure 4-8 Dozer blade adjustments.

The wings on the universal blade (Figure 4-7) enable it to push a large volume of material over long distances. However, its low horsepower per foot of cutting edge and per cubic yard limit its ability to penetrate hard soils or to move heavy materials. The straight blade is considered the most versatile dozer blade. Its smaller size gives it good penetrating and load pushing ability. The ability of angle blades to angle approximately 25° to either side make them very effective in sidehill cutting, ditching, and backfilling. They may also be used for rough grading and for moving material laterally. The cushion blade is reinforced and equipped with shock absorbers to enable it to push-load scrapers. It may also be used for cleanup of the loading or dumping areas and for general dozing when not push-loading scrapers. Other available types of dozer blades include light-material U-blades, special clearing blades, and ripdozer blades (blades equipped with ripper shanks on each end).

Estimating Dozer Production

The basic earthmoving production equation (Equation 2-1) may be applied in estimating dozer production. This method requires an estimate of the average blade load and the dozer cycle time. There are several methods available for estimating average blade load, including the blade manufacturer's capacity rating, previous experience under similar conditions, and actual measurement of several typical loads. A suggested method for calculating blade volume by measuring blade load is as follows:

- Doze a full blade load, then lift the blade while moving forward on a level surface until an even pile is formed.
- Measure the width of the pile (W) perpendicular to the blade and in line with the inside of each track or wheel. Average the two measurements.
- Measure the height (H) of the pile in a similar manner.
- Measure the length of the pile parallel to the blade.
- Calculate blade volume using Equation 4-10.

$$\text{Blade load (LCY)} = 0.0139 \times H\ (\text{ft}) \times W\ (\text{ft}) \times L\ (\text{ft}) \qquad (4\text{-}10A)$$

$$\text{Blade load (Lm}^3) = 0.375 \times H\ (\text{m}) \times W\ (\text{m}) \times L\ (\text{m}) \qquad (4\text{-}10B)$$

Total dozer cycle time is the sum of its fixed cycle time and variable cycle time. *Fixed cycle time* represents the time required to maneuver, change gears, start loading, and dump. Table 4-4 may be used to estimate dozer fixed cycle time. *Variable cycle time* is the time required to doze and return. Since the haul distance is rela-

Table 4-4 Typical dozer fixed cycle times

Operating Conditions	*Time (min)*
Power-shift transmission	0.05
Direct-drive transmission	0.10
Hard digging	0.15

tively short, a dozer usually returns in reverse gear. Table 4-5 provides typical operating speeds for dozing and return. Some manufacturers provide dozer production estimating charts for their equipment.

Example 4-6

PROBLEM A power-shift crawler tractor has a rated blade capacity of 10 LCY (7.65 Lm^3). The dozer is excavating loose common earth and pushing it a distance of 200 ft (61 m). Maximum reverse speed in third range is 5 mi/h (8 km/h). Estimate the production of the dozer if job efficiency is 50 min/h.

SOLUTION

$$\text{Fixed time} = 0.05 \text{ min} \quad \text{(Table 4-4)}$$

$$\text{Dozing speed} = 2.5 \text{ mi/h (4.0 km/h)} \quad \text{(Table 4-5)}$$

$$\text{Dozing time} = \frac{200}{2.5 \times 88} = 0.91 \text{ min}$$

$$\left[= \frac{61}{4 \times 16.7} = 0.91 \text{ min} \right]$$

Note: 1 mi/h = 88 ft/min; 1 km/h = 16.7 m/min.

Table 4-5 Typical dozer operating speeds

Operating Conditions	*Speeds*
Dozing	
Hard materials, haul 100 ft (30 m) or less	1.5 mi/h (2.4 km/h)
Hard materials, haul over 100 ft (30 m)	2.0 mi/h (3.2 km/h)
Loose materials, haul 100 ft (30 m) or less	2.0 mi/h (3.2 km/h)
Loose materials, haul over 100 ft (30 m)	2.5 mi/h (4.0 km/h)
Return	
100 ft (30 m) or less	Maximum reverse speed in second range (power shift) or reverse speed in gear used for dozing (direct drive)
Over 100 ft (30 m)	Maximum reverse speed in third range (power shift) or highest reverse speed (direct drive)

$$\text{Return time} = \frac{200}{5 \times 88} = 0.45 \text{ min}$$

$$\left[= \frac{61}{8 \times 16.7} = 0.45 \text{ min} \right]$$

$$\text{Cycle time} = 0.05 + 0.91 + 0.45 = 1.41 \text{ min}$$

$$\text{Production} = 10 \times \frac{50}{1.41} = 355 \text{ LCY/h}$$

$$\left[= 7.65 \times \frac{50}{1.41} = 271 \text{ Lm}^3\text{/h} \right]$$

Job Management

Some techniques used to increase dozer production include downhill dozing, slot dozing, and blade-to-blade dozing. By taking advantage of the force of gravity, downhill dozing enables blade load to be increased or cycle time to be reduced compared to dozing on the level. Slot dozing utilizes a shallow trench (or slot) cut between the loading and dumping areas to increase the blade capacity that can be carried on each cycle. Under favorable conditions, slot dozing may increase dozer production as much as 50%. The slot dozing technique may be applied to the excavation of large cut areas by leaving uncut sections between slots. These uncut sections are removed after all other material has been excavated. Blade-to-blade dozing involves two dozers operating together with their blades almost touching. This technique results in a combined blade capacity considerably greater than that of two single blades. However, the technique is not efficient for use over short dozing distances because of the extra maneuvering time required. Mechanically coupled side-by-side (S × S) dozers equipped with a single large blade are available and are more productive than are blade-to-blade dozers.

4-3 LOADERS

A tractor equipped with a front-end bucket is called a *loader, front-end loader,* or *bucket loader.* Both wheel loaders (Figure 4-9) and track loaders (Figure 4-10) are available. Loaders are used for excavating soft to medium-hard material, loading hoppers and haul units, stockpiling material, backfilling ditches, and moving concrete and other construction materials.

Wheel loaders possess excellent job mobility and are capable of over-the-road movement between jobs at speeds of 25 mi/h or higher. While their ground pressure is relatively low and may be varied by the use of different-size tires and by changing inflation pressures, they do not have the all-terrain capability of track loaders. Most modern wheel loaders are *articulated.* That is, they are hinged between the front and rear axles to provide greater maneuverability.

Track loaders are capable of overcoming steeper grades and side slopes than are wheel loaders. Their low ground pressure and high tractive effort enable them to op-

Figure 4-9 Large wheel loader. (Courtesy of Terex Corporation)

erate in all but the lowest trafficability soils. However, because of their lower speed, their production is less than that of a wheel loader over longer haul distances.

Attachments available for the loader include augers, backhoes, crane booms, dozer and snow blades, and forklifts in addition to the conventional loader bucket.

Figure 4-10 Track loader. (Courtesy of Fiatallis North America, Inc.)

Estimating Loader Production

Loader production may be estimated as the product of average bucket load multiplied by cycles per hour (Equation 2-1). Basic cycle time for a loader includes the time required for loading, dumping, making four reversals of direction, and traveling a minimum distance (15 ft or less for track loaders). Table 4-6 presents typical values of basic cycle time for wheel and track loaders. While manufacturers' performance curves should be used whenever possible, typical travel-time curves for wheel loaders are presented in Figure 4-11.

Federal Highway Administration (FHWA) studies have shown little variation in basic cycle time for wheel loaders up to a distance of 80 ft (25 m) between loading and dumping position. Therefore, travel time should not be added until one-way distance exceeds this distance.

Loader bucket capacity is rated in heaped (loose) volume, as shown in Table 3-1. Bucket capacity should be adjusted by a bucket fill factor (Table 3-2) to obtain the best estimate of actual bucket volume.

Example 4-7

PROBLEM Estimate the hourly production in loose volume (LCY and Lm^3) of a 3½-yd ($2.68\text{-}m^3$) wheel loader excavating sand and gravel (average material) from a pit and moving it to a stockpile. The average haul distance is 200 ft (61 m), the effective grade is 6%, the bucket fill factor is 1.00, and job efficiency is 50 min/h.

SOLUTION

$$\text{Bucket volume} = 3.5 \times 1 = 3.5 \text{ LCY } (2.68 \text{ Lm}^3)$$

$$\text{Basic cycle time} = 0.50 \text{ min} \quad \text{(Table 4-6)}$$

$$\text{Travel time} = 0.30 \text{ min} \quad \text{(Figure 4-11)}$$

$$\text{Cycle time} = 0.50 + 0.30 = 0.80 \text{ min}$$

$$\text{Production} = 3.5 \times \frac{50}{0.80} = 219 \text{ LCY/h}$$

$$\left[= 2.68 \times \frac{50}{0.80} = 168 \text{ Lm}^3/\text{h} \right]$$

Job Management

Some considerations involved in choosing a loader for a project have already been presented. Cutting of tires is a major problem when loading shot rock with a wheel loader. Type L-5 tires (rock, extra deep tread) should be used to increase tire life

Table 4-6 Basic loader cycle time

	Basic Cycle Time (min)	
Loading Conditions	*Articulated Wheel Loader*	*Track Loader*
Loose materials	0.35	0.30
Average material	0.50	0.35
Hard materials	0.65	0.45

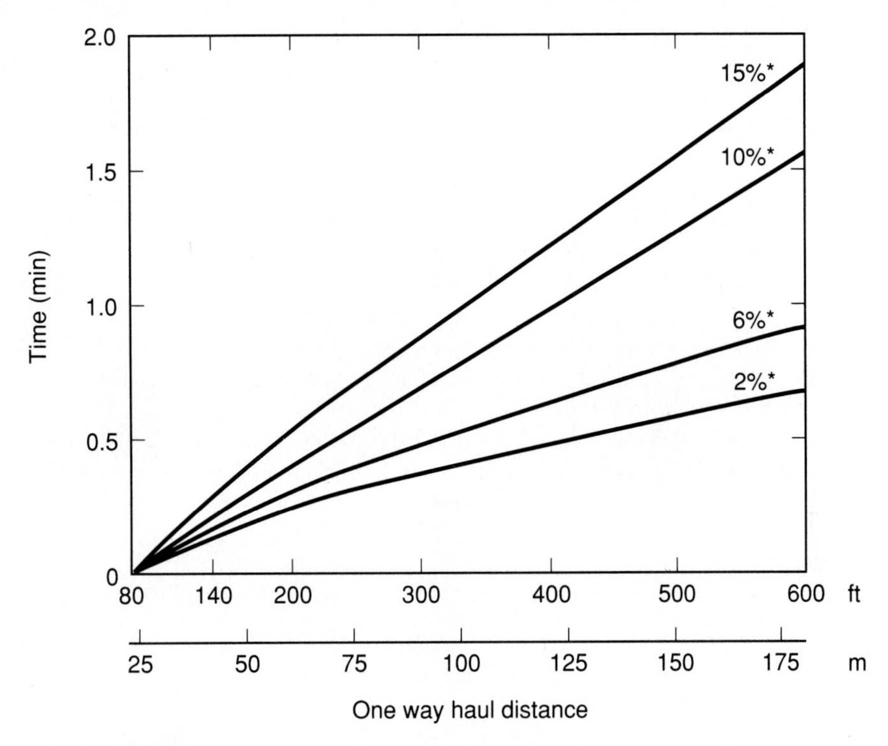

Figure 4-11 Travel time, wheel loader (haul + return).

when loading rock. The pit must be kept well drained, because water acts as a lubricant to increase the cutting action of rock on rubber tires.

Because of tipping load limitations, the weight of the material being handled may limit the size of the bucket that may be used on a loader. In selection of a loader, consideration must also be given to the clearances required during loading and dumping. Like excavators, optimum positioning of the loader and haul units will minimize loading, maneuver, and dump times. Multisegment buckets, also called 4-in-1 buckets and multipurpose buckets (Figure 4-12), are capable of performing as a clamshell, dozer, or scraper, as well as a conventional loader. Such buckets are often more effective than are conventional buckets in handling wet, sticky materials. Blasting or ripping hard materials before attempting to load them will often increase loader production in such materials.

4-4 SCRAPERS

Operation and Employment

Scrapers are capable of excavating, loading, hauling, and dumping material over medium to long haul distances. However, only the elevating scraper is capable of achieving high efficiency in loading without the assistance of a pusher tractor or

Figure 4-12 Multisegment loader bucket.

another scraper. Loading procedures are discussed later in this section. The scraper excavates (or cuts) by lowering the front edge of its bowl into the soil. The bowl front edge is equipped with replaceable cutting blades, which may be straight, curved, or extended at the center (stinger arrangement). Both the stinger arrangement and curved blades provide better penetration than does a straight blade. However, straight blades are preferred for finish work.

Although there are a number of different types of scrapers, principal types include single-engine overhung (two-axle) scrapers, three-axle scrapers, twin-engine all-wheel-drive scrapers (Figure 4-13), elevating scrapers (Figure 4-14), auger scrapers, and push-pull or twin-hitch scrapers (Figure 4-15). *Two-axle* or *overhung scrapers* utilize a tractor having only one axle (Figure 4-13). Such an arrangement has a lower rolling resistance and greater maneuverability than does a *three-axle scraper* that is pulled by a conventional four-wheel tractor. However, the additional stability of the three-axle scraper permits higher operating speeds on long, relatively flat haul roads. *All-wheel-drive scrapers,* as the name implies, utilize drive wheels on both the tractor and scraper. Normally, such units are equipped with twin engines. The additional power and drive wheels give these units greater tractive effort than that of conventional scrapers. *Elevating scrapers* utilize a ladder-type elevator to assist in cutting and lifting material into the scraper bowl. Elevating scrapers are not designed to be push-loaded and may be damaged by pushing. *Auger scrapers* are self-loading scrapers that use a rotating auger (similar to a posthole auger) located in the center of the scraper bowl to help lift material into the bowl. *Push-pull* or *twin-hitch scrapers* are all-wheel-drive scrapers equipped with coupling devices that enable two scrapers to assist each other in loading. Their operation is described later in this section.

Estimating Scraper Production

Scraper cycle time is estimated as the sum of fixed cycle time and variable cycle time. Fixed cycle time in this case includes spot time, load time, and maneuver and dump time. Spot time represents the time required for a unit to position itself in the

Figure 4-13 Twin-engine all-wheel drive scraper. (Courtesy of Caterpillar Inc.)

Figure 4-14 Elevating scraper. (Courtesy of Caterpillar Inc.)

cut and begin loading, including any waiting for a pusher. Table 4-7 provides typical values of fixed cycle time for scrapers.

Variable cycle time, or travel time, includes haul time and return time. As usual, haul and return times are estimated by the use of travel-time curves or by using the average-speed method with performance and retarder curves. It is usually necessary

Figure 4-15 Twin-hitch scraper loading. (Courtesy of Terex Corporation)

Table 4-7 Scraper fixed time (min)

	Spot Time	
	Single Pusher	*Tandem Pusher*
Favorable	0.2	0.1
Average	0.3	0.2
Unfavorable	0.5	0.5

	Load Time				
	Single Pusher	*Tandem Pusher*	*Elevating Scraper*	*Auger*	*Push-Pull**
Favorable	0.5	0.4	0.8	0.7	0.7
Average	0.6	0.5	1.0	0.9	1.0
Unfavorable	1.0	0.9	1.5	1.3	1.4

	Maneuver and Dump Time	
	Single Engine	*Twin Engine*
Favorable	0.3	0.3
Average	0.7	0.6
Unfavorable	1.0	0.9

*Per pair of scrapers.

to break a haul route up into sections having similar total resistance values. The total travel time required to traverse all sections is found as the sum of the section travel times.

In determining the payload per scraper cycle, it is necessary to check both the rated weight payload and the heaped volume capacity. The volume corresponding to the lesser of these two values will, of course, govern. The method of estimating production is illustrated in Examples 4-8 and 4-9.

Example 4-8

PROBLEM Estimate the production of a single engine two-axle tractor scraper whose travel-time curves are shown in Figures 4-4 and 4-5 based on the following information.

Maximum heaped volume = 31 LCY (24 Lm^3)

Maximum payload = 75,000 lb (34 020 kg)

Material: Sandy clay, 3200 lb/BCY (1898 kg/Bm^3), 2650 lb/LCY (1571 kg/Lm^3), rolling resistance 100 lb/ton (50 kg/t)

Job efficiency = 50 min/h

Operating conditions = average

Single pusher

Haul route:

Section 1. Level loading area

Section 2. Down a 4% grade, 2000 ft (610 m)

Section 3. Level dumping area
Section 4. Up a 4% grade, 2000 ft (610 m)
Section 5. Level turnaround, 600 ft (183 m)

SOLUTION Load per cycle:

$$\text{Weight of heaped capacity} = 31 \times 2650 = 82{,}150 \text{ lb}$$
$$[= 24 \times 1571 = 37\,794 \text{ kg}]$$

Weight exceeds rated payload of 75,000 lb (34 020 kg), therefore, maximum capacity is

$$\text{Load} = \frac{75{,}000}{3200} = 23.4 \text{ BCY/load}$$
$$\left[= \frac{34\,020}{1898} = 17.9 \text{ Bm}^3\text{/load}\right]$$

Effective grade:

$$\text{Haul} = -4.0 + \frac{100}{20} = +1\%$$
$$\left[= -4.0 + \frac{50}{10} = +1\%\right]$$
$$\text{Return} = 4.0 + \frac{100}{20} = +9\%$$
$$\left[= 4.0 + \frac{50}{10} = +9\%\right]$$
$$\text{Turnaround} = 0 + \frac{100}{20} = 5\%$$
$$\left[= 0 + \frac{50}{10} = +5\%\right]$$

Travel time:

$$\text{Section 2} = 1.02 \text{ min} \quad \text{(Figure 4-4)}$$
$$\text{Section 4} = 1.60 \text{ min} \quad \text{(Figure 4-5)}$$
$$\text{Section 5} = \underline{0.45 \text{ min}} \quad \text{(Figure 4-5)}$$
$$\text{Total} = 3.07 \text{ min}$$

Fixed cycle (Table 4-7):

$$\text{Load spot} = 0.3 \text{ min}$$
$$\text{Load} = 0.6 \text{ min}$$
$$\text{Maneuver and dump} = \underline{0.7 \text{ min}}$$
$$\text{Total} = 1.6 \text{ min}$$
$$\text{Total cycle time} = 3.07 + 1.6 \text{ min} = 4.67 \text{ min}$$
$$\text{Estimated production} = 23.4 \times \frac{50}{4.67} = 251 \text{ BCY/h}$$
$$\left[= 17.9 \times \frac{50}{4.67} = 192 \text{ Bm}^3\text{/h}\right]$$

Example 4-9

PROBLEM Solve the problem of Example 4-8 using the average-speed method and the performance curves of Figure 4-2.

SOLUTION

$$\text{Payload} = 23.4 \text{ BCY } (17.9 \text{ Bm}^3) \text{ from Example 4-8}$$

Effective grades from Example 4-8:

$$\begin{aligned} \text{Haul} &= +1.0\% \\ \text{Return} &= +9.0\% \\ \text{Turnaround} &= +5.0\% \end{aligned}$$

Maximum speed (Figure 4-2):

$$\begin{aligned} \text{Haul} &= 32 \text{ mi/h } (52 \text{ km/h}) \\ \text{Return} &= 16 \text{ mi/h } (26 \text{ km/h}) \\ \text{Turnaround} &= 28 \text{ mi/h } (45 \text{ km/h}) \end{aligned}$$

Average speed factor (Table 4-3):

$$\begin{aligned} \text{Haul} &= 0.86 \times 0.86 = 0.74 \\ \text{Return} &= 0.86 \\ \text{Turnaround} &= 0.68 \end{aligned}$$

Average speed:

$$\begin{aligned} \text{Haul} &= 32 \times 0.74 = 24 \text{ mi/h } (38 \text{ km/h}) \\ \text{Return} &= 16 \times 0.86 = 13 \text{ mi/h } (22 \text{ km/h}) \\ \text{Turnaround} &= 28 \times 0.68 = 19 \text{ mi/h } (31 \text{ km/h}) \end{aligned}$$

Travel time:

$$\begin{aligned} \text{Haul} &= \frac{2000}{24 \times 88} = 0.95 \text{ min} \\ &\left[= \frac{610}{38 \times 16.7} = 0.95 \text{ min} \right] \\ \text{Return} &= \frac{2000}{13 \times 88} = 1.75 \text{ min} \\ &\left[= \frac{610}{21 \times 16.7} = 1.75 \text{ min} \right] \\ \text{Turnaround} &= \frac{600}{19 \times 88} = 0.36 \text{ min} \\ &\left[= \frac{183}{31 \times 16.7} = 0.36 \text{ min} \right] \end{aligned}$$

Total = 3.06 min

Fixed cycle = 1.6 min (Example 4-8)

Total cycle time = 4.66 min

$$\text{Estimated production} = 23.4 \times \frac{50}{4.66} = 251 \text{ BCY/h}$$

$$\left[= 17.9 \times \frac{50}{4.66} = 192 \text{ Bm}^3\text{/h} \right]$$

Note: The travel-time curves of Figures 4-4 and 4-5 assume acceleration from an initial velocity of 2.5 mi/h (4 km/h) upon leaving the cut and fill and deceleration to 2.5 mi/h (4 km/h) upon entering the cut and fill. The result of adding together the travel times for several sections will, because of an excessive allowance for acceleration and deceleration, yield a travel time greater than that obtained by the use of the average-speed method. The time estimate obtained by the use of the average-speed method should be more realistic.

Push-Loading

Except for elevating and push-pull scrapers, wheel scrapers require the assistance of pusher tractors to obtain maximum production. The three basic methods of push-loading scrapers are illustrated in Figure 4-16. The back-track method is most com-

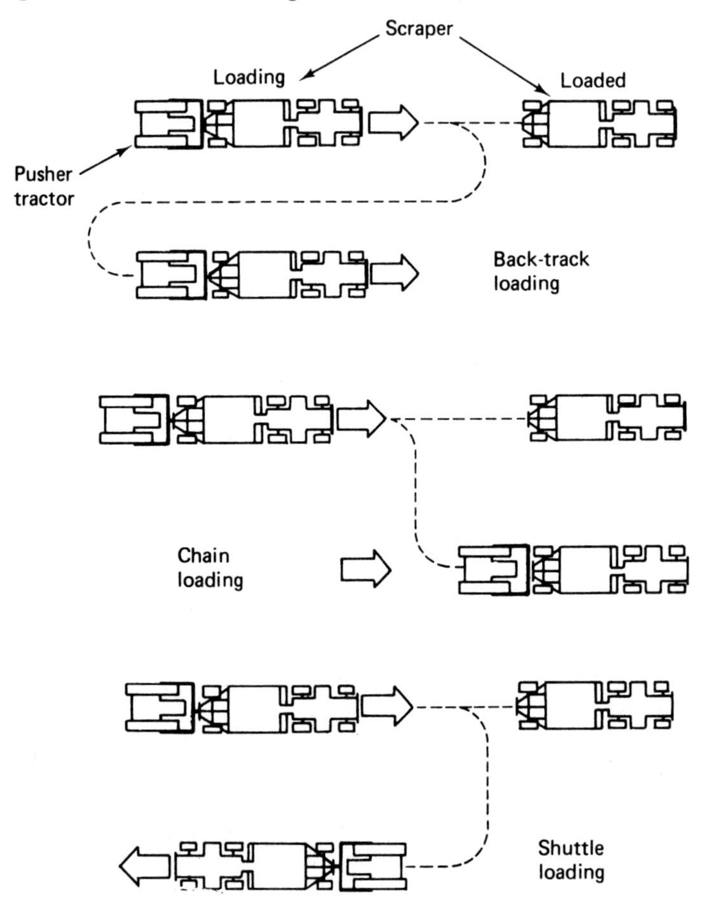

Figure 4-16 Methods of push-loading scrapers.

monly used since it permits all scrapers to load in the same general area. However, it is also the slowest of the three methods because of the additional pusher travel distance. Chain loading is suitable for a long, narrow cut area. Shuttle loading requires two separate fill areas for efficient operations.

A complete pusher cycle consists of maneuver time (while the pusher moves into position and engages the scraper), load time, boost time (during which the pusher assists in accelerating the scraper out of the cut), and return time. Tandem pushing involves the use of two pusher tractors operating one behind the other during loading and boosting. The use of tandem pushers reduces scraper load time and frequently results in obtaining larger scraper loads. The dual tractor described in Section 4-2 is a more efficient pusher than tandem tractors because the dual tractor is controlled by a single operator and no time is lost in coordination between two operators.

Calculating the Number of Pushers Required

The number of scrapers that can theoretically be handled by one pusher without a scraper having to wait for a pusher can be calculated by the use of Equation 4-11. The number of pushers required to fully service a given scraper feet may then be determined from Equation 4-12. It is suggested that the result obtained from Equation 4-11 be rounded down to one decimal place for use in Equation 4-12. The result obtained from Equation 4-12 must be rounded up to the next whole number to ensure that scrapers do not have to wait for a pusher. Methods for estimating scraper cycle time have already been presented. Table 4-8 may be used for estimating pusher cycle time.

$$\text{Number of scrapers served} = \frac{\text{Scraper cycle time}}{\text{Pusher cycle time}} \tag{4-11}$$

$$\text{Number of pushers required} = \frac{\text{Number of scrapers}}{\text{Number served by one pusher}} \tag{4-12}$$

When the number of pushers actually used is less than the number required to fully serve the scraper fleet, expected production is reduced to that obtained using Equation 4-13. In performing this calculation, use the precise number of pushers required, not the integer value.

$$\text{Production} = \frac{\text{No. of pushers}}{\text{Required number}} \times \text{No. of scrapers} \times \text{Production per scraper} \tag{4-13}$$

Table 4-8 Typical pusher cycle time (min)

Loading Method	*Single Pusher*	*Tandem Pusher*
Back-track	1.5	1.4
Chain or shuttle	1.0	0.9

Example 4-10

PROBLEM The estimated cycle time for a wheel scraper is 6.5 min. Calculate the number of pushers required to serve a fleet of nine scrapers using single pushers. Determine the result for both back-track and chain loading methods.

SOLUTION

Number of scrapers per pusher (Equation 4-11):

$$\text{Back-track} = \frac{6.5}{1.5} = 4.3$$

$$\text{Chain} = \frac{6.5}{1.0} = 6.5$$

Number of pushers required (Equation 4-12):

$$\text{Back-track} = \frac{9}{4.3} = 2.1 = 3$$

$$\text{Chain} = \frac{9}{6.5} = 1.4 = 2$$

Example 4-11

PROBLEM Find the expected production of the scraper fleet of Example 4-10 if only one pusher is available and the chain loading method is used. Expected production of a single scraper assuming adequate pusher support is 226 BCY/h (173 Bm^3/h).

SOLUTION

Number of pushers required to fully serve fleet = 1.4

$$\text{Production} = \frac{1}{1.4} \times 9 \times 226 = 1453 \text{ BCY/h} \qquad (4\text{-}13)$$

$$\left[= \frac{1}{1.4} \times 9 \times 173 = 1112 \text{ Bm}^3/\text{h} \right] \qquad (4\text{-}13)$$

Push-Pull Loading

In *push-pull* or *twin-hitch* scraper loading, two all-wheel-drive scrapers assist each other to load without the use of pusher tractors. The scrapers are equipped with special push blocks and coupling devices, as shown in Figure 4-16. The sequence of loading operations is as follows:

1. The first scraper to arrive in the cut starts to self-load.
2. The second scraper arrives, makes contact, couples, and pushes the front scraper to assist it in loading.
3. When the front scraper is loaded, the operator raises its bowl. The second scraper then begins to load with the front scraper pulling to assist in loading.
4. The two scrapers uncouple and separate for the haul to the fill.

Although there are a number of advantages claimed for this method of loading, basically it offers the loading advantages of self-loading scrapers while retaining the hauling advantages of standard scrapers. No pusher tractor or its operator is required. There is no problem of pusher-scraper mismatch and no lost time due to pusher downtime. However, scrapers must operate in pairs so that if one scraper breaks down, its partner must be diverted to a different operation. Conditions favoring push-pull operations include long, straight hauls with relatively easy to load materials. An adequate number of spreading and compacting units must be available at the fill, since two scrapers dump almost simultaneously.

Job Management

The type of scraper that may be expected to yield the lowest cost per unit of production is a function of the total resistance and the haul distance, as shown in Figure 4-17. Elevating scrapers can use their self-loading ability effectively for short hauls. However, their additional weight puts them at a disadvantage on long hauls. Of the conventional scrapers, single-engine overhung units are best suited to medium distances on relatively flat haul roads where maneuverability is important and adequate pusher power is available. Three-axle units are faster on long hauls and uneven surfaces. All-wheel-drive tandem-powered units are favored for conditions of high total resistance at all but the shortest haul distances. Notice that push-pull or twin-hitch scrapers overlap the entire all-wheel-drive zone of Figure 4-17 and extend into the elevating and conventional zones.

Some techniques for maximizing scraper production include:

- Use downhill loading whenever possible to reduce the required pusher power and load time.

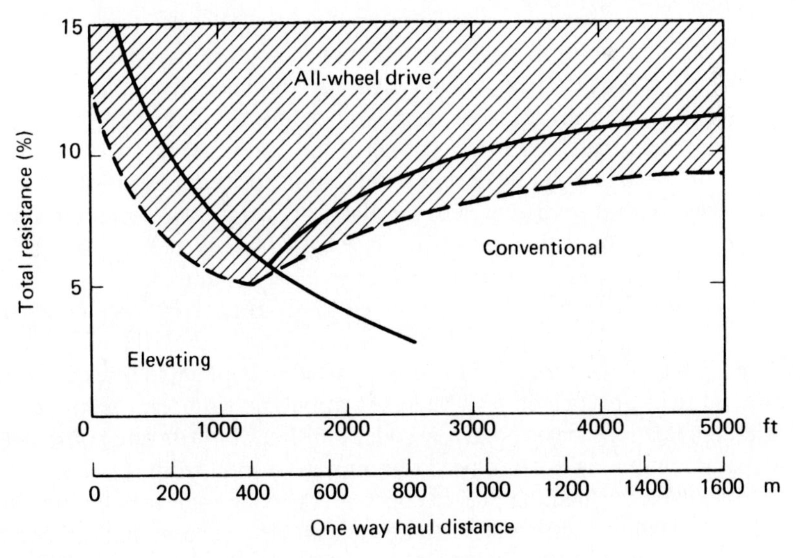

Figure 4-17 Scraper application zones.

- Use chain or shuttle loading methods if possible.
- Use rippers or scarifiers to loosen hard soils before attempting to load.
- Have pushers give scrapers an adequate boost to accelerate units out of the cut.
- Keep the cut in good condition by using pushers during their idle time or by employing other equipment. Provide adequate drainage in the cut to improve trafficability.
- Maintain the haul road in the best possible condition. Full-time use of a motor grader on the haul road will usually pay off in increased scraper production.
- Make the haul road wide enough to permit high-speed hauling without danger. One-way haul roads should be utilized whenever possible.
- Keep the fill surface smooth and compacted to minimize scraper time in the fill.
- Boost scrapers on the fill if spreading time is excessive.

Supervisors must carefully control operations in the cut, on the haul road, and in the fill to maximize production. Scrapers must be kept evenly spaced throughout their cycle to avoid interference between units. Scrapers that break down or cannot maintain their place in the cycle must be repaired promptly or replaced by standby units.

4-5 TRUCKS AND WAGONS

Operation and Employment

Because hauling (or the transportation of excavation) is a major earthmoving activity, there are many different types of hauling equipment available to the constructor. In addition to the dozer, loader, and scraper already described, hauling equipment includes trucks, wagons, conveyor belts, and trains. Most of the belt-type conveyors used in construction are portable units used for the movement of bulk construction materials within a small area or for placing concrete. However, conveyors are capable of moving earth and stone relatively long distances at high speed. Their ability to move earth for highway construction has been demonstrated in Great Britain. In the United States, they have been utilized on a number of large construction projects, such as dams. Their application is primarily limited by their large capital cost.

Conventional freight trains may be used to haul earth or rock over long distances when tracks are located near the excavation and fill areas. However, most construction applications involve narrow-gauge rail lines built in the construction area. This type of equipment is often used to remove the spoil from tunneling. Special rail cars are available for hauling plastic concrete. Although not usually thought of as a piece of earthmoving equipment, a dredge is capable of excavating soil and fractured rock and transporting it through pipelines in the form of a slurry.

Trucks and wagons are still the most common forms of construction hauling equipment. The heavy-duty rear-dump truck is most widely used because of its flexibility of use and the ability of highway models to move rapidly between job sites. There are a wide variety of types and sizes of dump truck available. Trucks may be powered by diesel or gasoline engines, have rear axle or all-wheel drive, have two or three axles, be equipped with standard or rock bodies, and so on. Trucks used for hauling on public highways are limited by transportation regulations in their maximum width, gross weight, and axle load. There is a growing trend toward the use of off-highway models that can be larger and heavier and carry payloads up to several hundred tons. Figure 4-18 shows a 40-ton rear-dump truck being loaded by a shovel. The all-wheel-drive articulated dump truck illustrated in Figure 4-19 (also called an *articulated hauler*) is finding increasing usage because of its ability to carry large loads over low-trafficability soils.

Wagons are tractors equipped with earthmoving semitrailers. Wagons are available in end-dump and side-dump models as well as the more common bottom-dump model. Bottom-dump models are preferred for moving earth and crushed rock because of their ability to dump and spread while moving at a relatively high speed.

Determining the Number of Haul Units Needed

The components of the truck or wagon cycle are similar to those of the scraper described in Section 4-4. Thus total cycle time is the sum of the fixed time (spot, load, maneuver, and dump) and the variable time (haul and return). The fixed time ele-

Figure 4-18 40-ton rear-dump truck. (Courtesy of Terex Corporation)

Figure 4-19 All-wheel drive articulated dump truck. (Courtesy of Terex Corporation)

ments of spot, maneuver, and dump may be estimated by the use of Table 4-9. Loading time, however, should be calculated by the use of Equation 4-14 or 4-15.

$$\text{Load time} = \frac{\text{Haul unit capacity}}{\text{Loader production at 100\% efficiency}} \tag{4-14}$$

$$\text{Load time} = \text{Number of bucket loads} \times \text{Excavator cycle time} \tag{4-15}$$

The reason for using an excavator loading rate based on 100% excavator efficiency in Equation 4-14 is that excavators have been found to operate at or near 100% efficiency when actually loading. Thus the use of the 100% efficiency loading rate is intended to ensure that an adequate number of trucks is provided so that the excavator will not have to wait for a truck. Either bank or loose measure may be used in Equation 4-14, but the same unit must be used in both numerator and denominator.

The number of trucks theoretically required to keep a loader fully occupied and thus obtain the full production of the loader may be calculated by the use of Equa-

Table 4-9 Spot, maneuver, and dump time for trucks and wagons (min)

	Bottom Dump	*Rear Dump*
Favorable	1.1	0.5
Average	1.6	1.1
Unfavorable	2.0	2.5

tion 4-16. Although this method gives reasonable values for field use, it should be recognized that some instances of the loader waiting for haul units will occur in the field when this method is used. This is due to the fact that some variance in loader and hauler cycle time will occur in the real-world situation. More realistic results may be obtained by the use of computer simulation techniques or the mathematical technique known as queueing theory (see reference 5).

$$\text{Number of haulers required } (N) = \frac{\text{Haul unit cycle time}}{\text{Load time}} \tag{4-16}$$

The result obtained from Equation 4-16 must be rounded up to the next integer. Using this method, the expected production of the loader/hauler system is the same as though the excavator were simply excavating and stockpiling. Reviewing the procedure, system output is assumed to equal normal loader output, including the usual job efficiency factor. However, the number of haul units required is calculated using 100% loader efficiency.

If more than the theoretically required number of trucks is supplied, no increase in system production will occur, because system output is limited to excavator output. However, if less than the required number of trucks is supplied, system output will be reduced, because the excavator will at times have to wait for a haul unit. The expected production in this situation may be calculated by the use of Equation 4-17. In performing this calculation, use the precise value of N, not its integer value.

$$\begin{matrix}\text{Expected production} \\ (\text{no. units less than } N)\end{matrix} = \frac{\text{Actual number of units}}{N} \times \begin{matrix}\text{Excavator} \\ \text{production}\end{matrix} \tag{4-17}$$

Example 4-12

PROBLEM Given the following information on a shovel/truck operation, (a) calculate the number of trucks theoretically required and the production of this combination; (b) calculate the expected production if two trucks are removed from the fleet.

Shovel production at 100% efficiency = 371 BCY/h (283 Bm^3/h)
Job efficiency = 0.75
Truck capacity = 20 BCY (15.3 Bm^3)
Truck cycle time, excluding loading = 0.5 h

SOLUTION

(a)

$$\text{Load time} = \frac{20}{371} = 0.054 \text{ h} \tag{4-14}$$

$$\left[= \frac{15.3}{283} = 0.054 \text{ h} \right]$$

$$\text{Truck cycle time} = 0.5 + 0.054 = 0.554 \text{ h}$$

$$\text{Number of trucks required} = \frac{0.554}{0.054} = 10.3 = 11 \tag{4-16}$$

$$\text{Expected production} = 371 \times 0.75 = 278 \text{ BCY/h}$$

$$[= 283 \times 0.75 = 212 \text{ Bm}^3\text{/h}]$$

(b) With nine trucks available,

$$\text{Expected production} = \frac{9}{10.3} \times 278 = 243 \text{ BCY/h} \tag{4-17}$$

$$\left[= \frac{9}{10.3} \times 212 = 186 \text{ Bm}^3/\text{h} \right]$$

Job Management

An important consideration in the selection of excavator/haul unit combinations is the effect of the size of the target that the haul unit presents to the excavator operator. If the target is too small, excessive spillage will result and excavator cycle time will be increased. Studies have found that the resulting loss of production may range from 10 to 20%. As a rule, haul units loaded by shovels, backhoes, and loaders should have a capacity of 3 to 5 times excavator bucket capacity. Because of their less precise control, clamshells and draglines require larger targets. A haul unit capacity of 5 to 10 times excavator bucket capacity is recommended for these excavators. Haul units that hold an integer number of bucket loads are also desirable. Using a partially filled bucket to top off a load is an inefficient operation.

Time lost in spotting haul units for loading is another major cause of inefficiency. As discussed under excavator operations, reducing the excavator swing angle between digging and loading will increase production. The use of two loading positions, one on each side of the excavator, will reduce the loss of excavator production during spotting. When haul units are required to back into loading position, bumpers or spotting logs will assist the haul unit operator in positioning his vehicle in the minimum amount of time.

Some other techniques for maximizing haul unit production include:

- If possible, stagger starting and quitting times so that haul units do not bunch up at the beginning and end of the shift.
- Do not overload haul units. Overload results in excessive repair and maintenance.
- Maintain haul roads in good condition to reduce travel time and minimize equipment wear.
- Develop an efficient traffic pattern for loading, hauling, and dumping.
- Roads must be wide enough to permit safe travel at maximum speeds.
- Provide standby units (about 20% of fleet size) to replace units that break down or fail to perform adequately.
- Do not permit speeding. It is a dangerous practice; it also results in excessive equipment wear and upsets the uniform spacing of units in the haul cycle.

In unit price earthmoving contracts, payment for movement of soil or rock from cut to fill that exceeds a specified distance is termed *overhaul*. Overhaul can be minimized by selection of an optimum design surface elevation (grade) and by use of borrow and waste areas at appropriate locations.

PROBLEMS

1. An off-highway truck weighs 40,000 lb (18 144 kg) empty and carries a payload of 64,000 lb (29 030 kg). The loaded truck travels up a 5% grade over a road having a rolling resistance factor of 100 lb/ton (50 kg/t). What is the total resistance (lb or kg) and effective grade?

2. The tractor-scraper whose travel-time curves are shown in Figures 4-4 and 4-5 hauls its rated payload 2000 ft (610 m) up a 4% grade from the cut to the fill and returns empty over the same route. The rolling resistance factor for the haul road is 160 lb/ton (80 kg/t). Estimate the scraper travel time.

3. A wheel tractor-scraper whose weight on the driving wheels is 36,000 lb (16 330 kg) has a gross weight of 64,000 lb (29 030 kg). If the road surface is wet earth having a rolling resistance factor of 120 lb lb/ton (60 kg/t), what is the maximum grade the scraper could ascend?

4. The performance and retarder curves for a wheel tractor-scraper are shown in Figures 4-2 and 4-3. Using the average speed method, estimate the travel time for the situation in Problem 2.

5. A power-shift crawler tractor is excavating tough clay and pushing it a distance of 95 ft (29 m). Maximum reverse speeds are: first range, 3 mi/h (4.8 km/h); second range, 5 mi/h (8.1 km/h); and third range, 8 mi/h (12.9 km/h). Rated blade capacity is 10 LCY (7.65 Lm^3). Estimate dozer production if the job efficiency factor is 0.83.

6. How many hours should it take an articulated wheel loader equipped with a 5-yd (3.82-m^3) bucket to load 5000 cu yd (3823 m^3) of gravel from a stockpile into rail cars if the average haul distance is 400 ft (122 m) one way? The area is level with a rolling resistance factor of 200 lb/ton (100 kg/t). Job efficiency is estimated at 45 min/h.

7. A hydraulic shovel will be used to excavate sandy clay and load it into 14-BCY (10.7-Bm^3) dump trucks. The shovel's production at 100% efficiency is estimated at 360 BCY/h (275 Bm^3/h), and job efficiency is 0.83. Truck travel time is estimated at 10 min, and truck fixed cycle time (excluding loading) is estimated at 2.6 min. Equipment costs for the shovel and trucks are $30/h and $16/h, respectively.

   (a) How many trucks are theoretically required to obtain maximum production?
   (b) What is the expected production of the system in bank measure using this number of trucks?
   (c) What is the expected unit loading and hauling cost ($/BCY or $/$Bm^3$)?

8. Using the data of Problem 7, calculate the expected production and unit cost of loading and hauling if the truck fleet consists of 4 trucks.

9. Write a computer program to calculate the number of pushers required to serve a specified scraper fleet. Input should include scraper cycle time, method of push loading, and whether single or tandem pushers will be used. Using your program, verify the solution given for Example 4-10.

10. Write a computer program to calculate the number of haul units required, the system production, and unit production cost for an excavator/haul unit system. Input should include excavator loading rate, haul unit capacity, haul unit cycle time less loading time, hourly equipment costs, and job efficiency. Solve Problem 7 using your program.

REFERENCES

1. BERNARD, D. A., T. R. FERRAGUT, AND D. L. NEUMANN. *Production Efficiency Study on Large-Capacity, Rubber-Tired Front-End Loaders* (Report No. FHWA-RDDP-PC-520). U.S. Department of Transportation, Federal Highway Administration, Arlington, Va., 1973.
2. *Caterpillar Performance Handbook,* 21st ed. Caterpillar Inc., Peoria, Ill., 1990.
3. CLEMMENS, J. P., AND R. J. DILLMAN. *Production Efficiency Study on Rubber-Tired Scrapers* (Report No. FHWA-DP-PC-920). U.S. Department of Transportation, Arlington, Va., 1977.
4. *Euclid Hauler Handbook,* 15th ed. VME Americas, Inc., Cleveland, Ohio, 1982.
5. NUNNALLY, S.W. *Managing Construction Equipment.* Englewood Cliffs, N.J.: Prentice Hall, 1977.
6. *Production and Cost Estimating of Material Movement with Earthmoving Equipment.* Terex Corporation, Hudson, Ohio. 1981.

5
Compacting and Finishing

5-1
PRINCIPLES OF COMPACTION

The Compaction Process

Compaction is the process of increasing the density of a soil by mechanically forcing the soil particles closer together, thereby expelling air from the void spaces in the soil. Compaction should not be confused with *consolidation,* which is an increase in soil density of a cohesive soil resulting from the expulsion of water from the soil's void spaces. Consolidation may require months or years to complete, whereas compaction is accomplished in a matter of hours.

Compaction has been employed for centuries to improve the engineering properties of soil. Improvements include increased bearing strength, reduced compressibility, improved volume change characteristics, and reduced permeability. Although the compaction principles are the same, the equipment and methods employed for compaction in building construction are usually somewhat different from these employed in heavy and highway construction. Some of the building construction characteristics producing these differences include the limited differential settlement that can be tolerated by a building foundation, the necessity for working in confined areas close to structures, and the smaller quantity of earthwork involved.

The degree of compaction that may be achieved in a particular soil depends on the soil's physical and chemical properties (see Chapter 2), the soil's moisture content, the compaction method employed, the amount of compactive effort, and the thickness of the soil layer being compacted (lift thickness). The four basic compaction forces are static weight, manipulation (or kneading), impact, and vibration. Although all compactors utilize *static weight* to achieve compaction, most compactors combine this with one or more of the other compaction forces. For example, a plate vibrator combines static weight with vibration. *Manipulation* of soil under pressure to produce compaction is most effective in plastic soils. The forces involved in impact and vibration are similar except for their frequency. *Impact or tamping* involves blows delivered at low frequencies, usually about 10 cycles per second (Hz), and is most effective in plastic soils. *Vibration* involves higher frequencies, which may extend to 80 cycles per second (Hz) or more. Vibration is particularly effective in compaction of cohesionless soils such as sand and gravel. The selection and employment of compaction equipment is discussed in Section 5-2.

Optimum Moisture Content

Although soil moisture content is only one of the five factors influencing compaction results, it is a very important one. As a result, a standard laboratory test called a *Proctor test* has been developed to evaluate a soil's moisture-density relationship under a specified compaction effort. Actually, there are two Proctor tests which have been standardized by the American Society for Testing and Materials (ASTM) and the American Association of State Highway and Transportation Officials (AASHTO). These are the Standard Proctor Test (ASTM D 698, AASHTO T-99) and the Modified Proctor Test (ASTM D 1557, AASHTO T-180). Characteristics of these two tests are given in Table 5-1. Since the modified test was developed for use where high design loads were involved (such as airport runways), the compactive effort for the modified test is more than four times as great as for the standard test.

To determine the maximum density of a soil using Proctor test procedures, compaction tests are performed over a range of soil moisture contents. The results are then plotted as dry density versus moisture content as illustrated in Figure 5-1. The peak of each curve represents the maximum density obtained under the compactive effort supplied by the test. As you might expect, Figure 5-1 shows that the maximum density achieved under the greater compactive effort of the modified test is higher than the density achieved in the standard test. Note the line labeled "zero air voids" on Figure 5-1. This curve represents the density at which all air voids have been eliminated; that is, all void spaces are completely filled with water. Thus it represents the maximum possible soil density for any specified water content. Actual density will always be somewhat less than the zero air voids density, because it is virtually impossible to remove all air from the soil's void spaces.

The moisture content at which maximum dry density is achieved under a specific compaction effort is referred to as the *optimum moisture content of the soil.* Referring to Figure 5-1, we see that for the Standard Proctor Test the optimum moisture content for this soil is about 20% of the soil's dry weight. Notice, however, that the optimum moisture content for the modified test is only about 15%. This relationship is typical for most soils. That is, a soil's optimum moisture content decreases as the compactive effort is increased. If tests are run at several different levels of compactive

Table 5-1 Characteristics of Proctor compaction tests

Test Details	*Standard*	*Modified*
Diameter of mold		
in.	6	6
cm	15.2	15.2
Height of sample		
in.	5 cut to 4.59	5 cut to 4.59
cm	12.7 cut to 11.7	12.7 cut to 11.7
Number of layers	3	5
Blows per layer	25	25
Weight of hammer		
lb	5.5	10
kg	2.5	4.5
Diameter of hammer		
in.	2	2
cm	5.1	5.1
Height of hammer drop		
in.	12	18
cm	30.5	45.7
Volume of sample		
cu ft	$\frac{1}{30}$	$\frac{1}{30}$
l	0.94	0.94
Compactive effort		
ft-lb	12,400	56,200
kg-m	1715	7772

effort, a line of optimum moisture contents may be drawn as shown in Figure 5-1 to illustrate the variation of optimum moisture with compactive effort. The effect of soil type on compaction test results is illustrated in Figure 5-2. While most soils display a similar characteristic shape, notice the rather flat curve obtained when compacting uniform fine sands (curve 5). The compaction curve for heavy clays (curve 7) is intermediate between that of uniform fine sands and those of the more typical soils.

The importance of soil moisture content to field compaction practice can be demonstrated using Figure 5-1. Suppose that specifications require a density of 100 lb/cu ft (1.6 g/cm^3) for this soil and that the compactive effort being used is equal to that of the Standard Proctor Test. From Figure 5-1 it can be seen that the required density may be achieved at any moisture content between 13 and 24%. However, a density of 105 lb/cu ft (1.68 g/cm^3) can only be achieved at a moisture content of 20%. Relationships between field and laboratory results for various soils, different types of equipment, and varying levels of compactive effort are discussed in Section 5-2.

Compaction Specifications

Compaction specifications are intended to ensure that the compacted material provides the required engineering properties and a satisfactory level of uniformity. To ensure that the required engineering properties are provided, it is customary to pre-

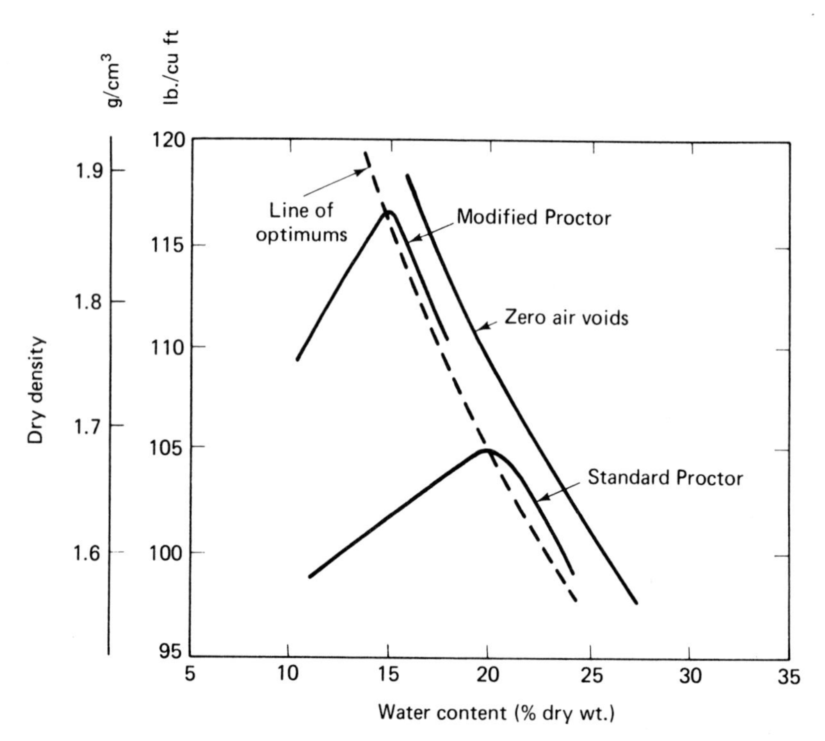

Figure 5-1 Typical compaction test results.

scribe the characteristics of the material to be used and a minimum dry density to be achieved. If the natural site material is to be compacted, only a minimum density requirement is needed. The Proctor test is widely used for expressing the minimum density requirement. That is, the specification will state that a certain percentage of Standard Proctor or Modified Proctor density must be obtained. For the soil of Figure 5-1, 100% of Standard Proctor density corresponds to a dry density of 105 lb/cu ft (1.68 g/cm^3). Thus a specification requirement for 95% of Standard Proctor density corresponds to a minimum dry density of 99.8 lb/cu ft (1.60 g/cm^3).

Typical density requirements range from 90% of Standard Proctor to 100% of Modified Proctor. For example, 95% of Standard Proctor is often specified for embankments, dams, and backfills. A requirement of 90% of Modified Proctor might be used for the support of floor slabs. For the support of structures and for pavement base courses where high wheel loads are expected, requirement of 95 to 100% of Modified Proctor are commonly used.

A lack of uniformity in compaction may result in differential settlement of structures or may produce a bump or depression in pavements. Therefore, it is important that uniform compaction be obtained. Uniformity is commonly controlled by specifying a maximum variation of density between adjacent areas. Compaction specifications may range from performance specifications in which only a minimum dry density is prescribed to method specifications that prescribe the exact equipment and procedures to be used. (See Section 17-4 for a discussion of specifications.)

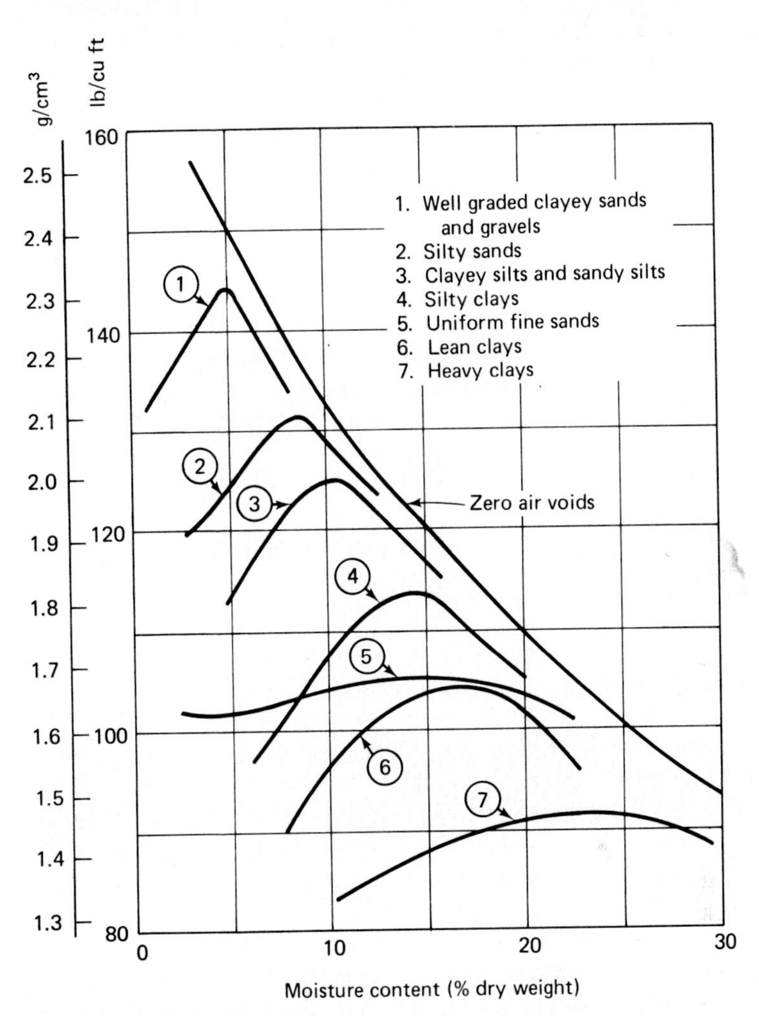

Figure 5-2 Modified Proctor Test results for various soils. (Courtesy of Dr. Harvey E. Wahls)

Measuring Field Density

To verify the adequacy of compaction the soil density actually obtained in the field must be measured and compared with the specified soil density. The methods available for performing in-place density tests include a number of traditional methods (liquid tests, sand tests, etc.) and nuclear density devices. All the traditional test methods involve removing a soil sample, measuring the volume of the hole produced, and determining the dry weight of the material removed. Density is then found as the dry weight of soil removed divided by the volume of the hole.

Liquid tests measure the volume of soil removed by measuring the volume of liquid required to fill the hole. A viscous fluid such as engine oil is poured from a calibrated container directly into the hole when testing relatively impermeable soils such as clays and silts. A method used for more permeable soils involves forcing water from a calibrated container into a rubber balloon inserted into the hole. *Sand tests* involve filling both the hole and an inverted funnel placed over the hole with a uniform fine sand. The volume of the hole and funnel is found by dividing the weight of sand used by its density. The volume of the funnel is then subtracted to yield the hole volume.

Nuclear density devices measure the amount of radioactivity from a calibrated source that is reflected back from the soil to determine both soil density and moisture content. When properly calibrated and operated, these devices produce accurate results in a fraction of the time required to perform traditional density tests. The increasing size and productivity of earthmoving equipment has greatly increased the need for rapid determination of the soil density being achieved. As a result, the use of nuclear density devices is becoming widespread.

5-2 COMPACTION EQUIPMENT AND PROCEDURES

Types of Compaction Equipment

Principal types of compaction equipment include tamping foot rollers, grid or mesh rollers, vibratory compactors, smooth steel drum rollers, pneumatic rollers, segmented pad rollers, and tampers or rammers (see Figure 5-3).

Tamping foot rollers utilize a compaction drum equipped with a number of protruding feet. Tamping foot rollers are available in a variety of foot sizes and shapes, including the sheepsfoot roller. During initial compaction, roller feet penetrate the loose material and sink to the lower portion of the lifts. As compaction proceeds, the roller rises to the surface or "walks out" of the soil. All tamping foot rollers utilize static weight and manipulation to achieve compaction. Therefore, they are most effective on cohesive soils. While the sheepsfoot roller produces some impact force, it tends to displace and tear the soil as the feet enter and leave the soil. Newer types of tamping foot rollers utilize a foot designed to minimize displacement of soil during entry and withdrawal. These types of rollers more effectively utilize impact forces. High-speed tamping foot rollers may operate at speeds of 10 mi/h (16 km/h) or more. At these speeds they deliver impacts at a frequency approaching vibration.

Grid or *mesh rollers* utilize a compactor drum made up of a heavy steel mesh. Because of their design, they can operate at high speed without scattering the material being compacted. Compaction is due to static weight and impact plus limited manipulation. Grid rollers are most effective in compacting clean gravels and sands. They can also be used to break up lumps of cohesive soil. They are capable of both crushing and compacting soft rock (rock losing 20% or more in the Los Angeles Abrasion Test).

Vibratory compactors are available in a wide range of sizes and types. In size they range from small hand-operated compactors (Figure 5-4) through towed rollers to large self-propelled rollers (Figure 5-5). By type they include plate compactors,

SMOOTH, STEEL WHEEL ROLLER

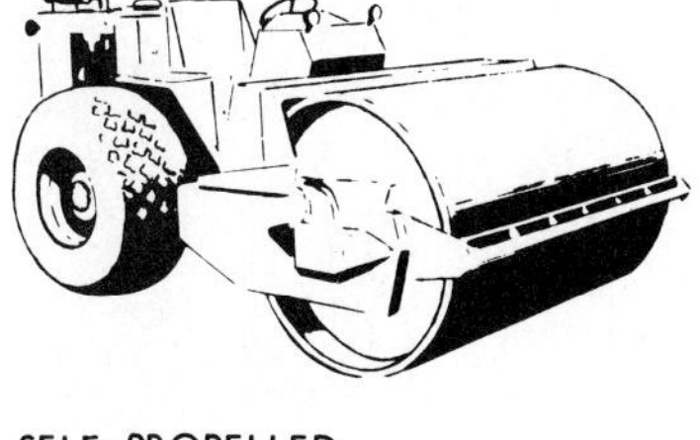

SELF-PROPELLED VIBRATING ROLLER

SMALL, MULTITIRED PNEUMATIC ROLLER

HEAVY PNEUMATIC ROLLER

SELF-PROPELLED TAMPING FOOT ROLLER

SELF-PROPELLED SEGMENTED STEEL WHEEL ROLLER

TOWED SHEEPSFOOT ROLLER

GRID ROLLER

Figure 5-3 Major types of compaction equipment. (Reprinted by permission of Caterpillar Inc., © 1971)

Figure 5-4 Walk-behind vibratory plate compactor. (Courtesy of Wacker Corp.)

smooth drum rollers, and tamping foot rollers. Small walk-behind vibratory plate compactors and vibratory rollers are used primarily for compacting around structures and in other confined areas. Vibratory plate compactors are also available as attachments for hydraulic excavators. The towed and self-propelled units are utilized in general earthwork. Large self-propelled smooth drum vibratory rollers are often used for compacting bituminous bases and pavements. While vibratory compactors are most effective in compacting noncohesive soils, they may also be effective in compacting cohesive soils when operated at low frequency and high amplitude. Many vibratory compactors can be adjusted to vary both the frequency and amplitude of vibration.

Steel wheel or *smooth drum rollers* are used for compacting granular bases, asphaltic bases, and asphalt pavements. Types available include towed rollers and self-propelled rollers. Self-propelled rollers include three-wheel (two-axle) and two- and three-axle tandem rollers. The compactive force involved is primarily static weight.

Rubber-tired or *pneumatic rollers* are available as light- to medium-weight multi-tired rollers and heavy pneumatic rollers. Wobble-wheel rollers are multitired rollers with wheels mounted at an angle so that they appear to wobble as they travel. This

Figure 5-5 Vibratory tamping foot compactor. [Courtesy of BOMAG (USA)]

imparts a kneading action to the soil. Heavy pneumatic rollers weighing up to 200 tons are used for dam construction, compaction of thick lifts, and proof rolling. Pneumatic rollers are effective on almost all types of soils but are least effective on clean sands and gravels.

Segmented pad rollers are somewhat similar to tamping foot rollers except that they utilize pads shaped as segments of a circle instead of feet on the roller drum. As a result, they produce less surface disturbance than do tamping foot rollers. Segmented pad rollers are effective on a wide range of soil types.

Rammers or *tampers* are small impact-type compactors which are primarily used for compaction in confined areas. Some rammers, like the one shown in Figure 5-6, are classified as vibratory rammers because of their operating frequency.

Figure 5-6 Small vibratory rammer. (Courtesy of Wacker Corp.)

Rammers are also available that mount on the end of a backhoe boom. These have excellent maneuverability and are especially useful in compacting material in deep excavations such as trenches.

Selection of Compaction Equipment

The proper selection of compaction equipment is an important factor in obtaining the required soil density with a minimum expenditure of time and effort. Table 5-2 provides a guide to the selection of suitable compaction equipment based on the Unified System of soil classification (see Chapter 2). It also indicates typical maximum dry densities for each soil based on the Modified Proctor Test.

Compaction Operations

After selecting appropriate compaction equipment, a compaction plan must be developed. The major variables to be considered include soil moisture content, lift thickness, number of passes used, ground contact pressure, compactor weight, and compactor speed. For vibratory compactors, it is also necessary to consider the frequency and amplitude of vibration to be employed.

Table 5-2 Soil compaction guide (Unified System)

Soil Type	Compaction Equipment* Recommended	Suitable	Maximum Dry Density Modified Proctor lb/cu ft	g/cm^3
GW	VR, VP	PH, SW, SP, GR, CT	125–140	2.00–2.24
GP	VR, VP	PH, SW, SP, GR, CT	110–140	1.76–2.24
GM	VR, PH, SP	VP, SW, GR, CT	115–145	1.84–2.32
GC	PH, SP	SW, VR, VP, TF, GR, CT	130–145	2.08–2.32
SW	VR, VP	PH, SW, SP, GR, CT	110–130	1.76–2.08
SP	VR, VP	PH, SW, SP, GR, CT	105–135	1.68–2.16
SM	VR, PH, SP	VP, SW, GR, CT	100–135	1.60–2.16
SC	PH, SP	SW, VR, VP, TF, GR, CT	100–135	1.60–2.16
ML	PH, SP	TF, SW, VR, VP, GR, CT	90–130	1.44–2.08
CL	PH, SP	TF, SW, VR, GR, CT	90–130	1.44–2.08
OL	PH, SP	TF, SW, VR, GR, CT	90–105	1.44–1.68
MH	PH, SP	TF, SW, VR, GR, CT	80–105	1.28–1.68
CH	TF, PH, SP	VR, GR, SW	90–115	1.44–1.84
OH	TF, PH, SP	VR, GR, SW	80–110	1.28–1.76
Pt	Compaction not practical			

*Symbols
CT = Crawler Tractor 0-30T
GR = Grid Roller 5-15T
PH = Pneumatic Roller 10-50T
SP = Segmented Pad 5-30T
SW = Smooth Wheel 3-15T
TF = Tamping Foot 5-30T
VP = Vibrating Plate < 1T
VR = Vibrating Roller 3-25T

The general concepts of optimum moisture content as related to compaction effort and soil density have been discussed in Section 5-1. However, it must be recognized that the compactive effort delivered by a piece of compaction equipment will seldom be exactly the same as that of either the standard or modified compaction test. As a result, the field optimum moisture content for a particular soil/compactor combination will seldom be the same as the laboratory optimum. This is illustrated by Figure 5-7, where only one of the four compactors has a field optimum moisture content close to the laboratory optimum. For plastic soils it has been observed that the field optimum moisture content is close to the laboratory Standard Proctor optimum for pneumatic rollers. However, the field optimum is appreciably lower than laboratory optimum for tamping foot rollers. For nonplastic soils, the field optimum for all nonvibratory equipment appears to run about 80% of the laboratory Standard Proctor optimum. The vibratory compactor appears to be most effective in all types of soil when the field moisture is appreciably lower than laboratory optimum.

Lifts should be kept thin for most effective compaction. For all rollers except vibratory rollers and heavy pneumatic rollers, a maximum lift thickness of 5 to 8 in. (15 to 20 cm) is suggested. Lift thicknesses of 12 in. (30 cm) or more may be satisfactory using heavy pneumatic rollers. However, precompaction with a light roller may be required to prevent rutting when heavy pneumatic rollers are used on thick lifts. The maximum lift thickness for effective vibratory compaction depends on the static weight of the compactor. Appropriate lift thicknesses for clean granular soils

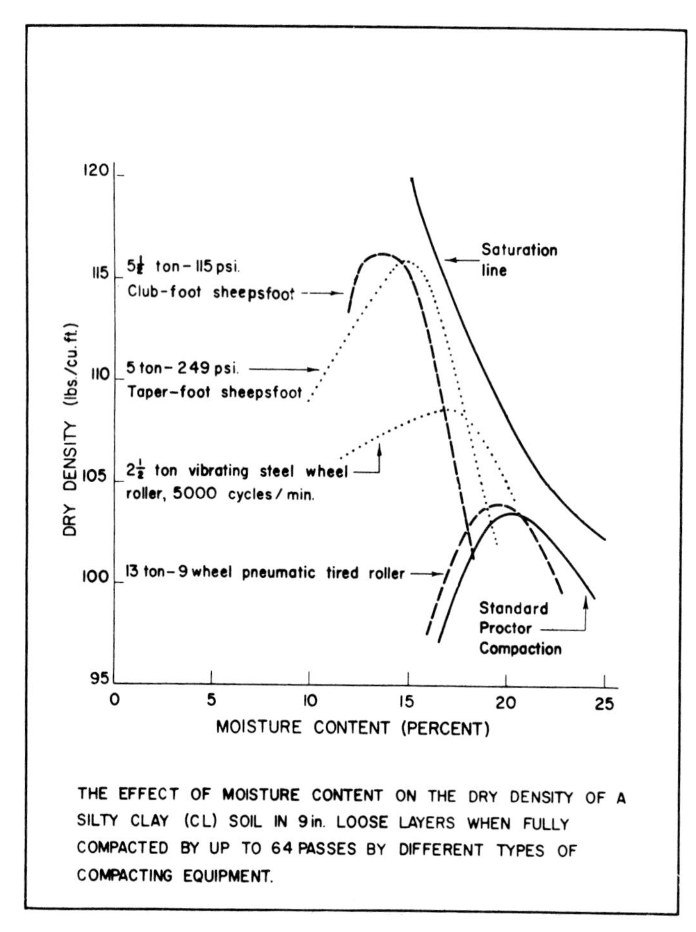

Figure 5-7 Variation of optimum moisture content with roller type. (From reference 6)

may range from 8 in. (20 cm) for a 1-ton compactor to 48 in. (122 cm) for a 15-ton (13.6 t) compactor. Heavy vibratory rollers have successful compacted rock using lift thicknesses of 7 ft (2.1 m).

The compaction achieved by repeated passes of a compactor depends on the soil/compactor combination utilized. For some combinations (such as a tamping foot roller compacting a clayey gravel), significant increases in density may continue to occur beyond 50 passes. However, as shown in Figure 5-8, the increase in density is relatively small after about 10 passes for most soil/compactor combinations.

Ground contact pressure may vary from 30 lb/sq in. (207 kPa) for a pneumatic roller to 300 lb/sq in. (2070 kPa) or more for a tamping foot roller. Within these ranges it has been found that total roller weight has a much more pronounced effect on the compaction achieved than does contact pressure. Thus increasing the foot size on a tamping foot roller while maintaining a constant contact pressure will in-

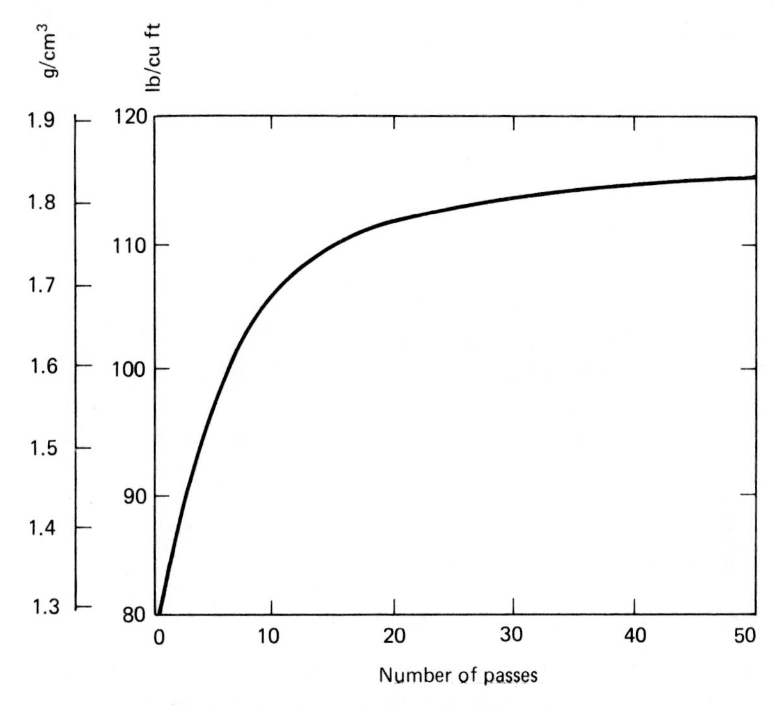

Figure 5-8 Typical effect of number of passes.

crease both the soil density and the surface area covered in one pass. Likewise, increasing the weight of a pneumatic roller at constant tire pressure will increase the effective depth of compaction. The use of excessive ground contact pressure will result in shearing and displacement of the soil being compacted.

If a tamping foot roller fails to "walk out" to within 1 in. (2.5 cm) of the surface after about five passes, it usually indicates that either the contact pressure or the soil moisture content is too high.

Tests have shown little relationship between compactor travel speed and the compaction achieved, except for vibratory compactors. In the case of vibratory equipment, travel speed (at a fixed operating frequency) determines the number of vibrations that each point on the ground surface will receive. Therefore, when using vibratory equipment, tests should be performed to determine the compactor speed that results in the highest compactor productivity. For conventional equipment the highest possible speed should be utilized that does not result in excessive surface displacement.

Estimating Compactor Production

Equation 5-1 may be used to calculate compactor production based on compactor speed, lift thickness, and effective width of compaction. The accuracy of the result obtained will depend on the accuracy in estimating speed and lift thickness. Trial operations will usually be necessary to obtain accurate estimates of these factors. Typical compactor operating speeds are given in Table 5-3.

$$\text{Production (CCY/h)} = \frac{16.3 \times W \times S \times L \times E}{P} \tag{5-1A}$$

$$\text{Production (Cm}^3\text{/h)} = \frac{10 \times W \times S \times L \times E}{P} \tag{5-1B}$$

where P = number of passes required
W = width compacted per pass (ft or m)
S = compactor speed (mi/h or km/h)
L = compacted lift thickness (in. or cm)
E = job efficiency

The power required to tow rollers depends on the roller's total resistance (grade plus rolling resistance). The rolling resistance of tamping foot rollers has been found to be approximately 450 to 500 lb/ton (225 to 250 kg/t). The rolling resistance of pneumatic rollers and the maximum vehicle speed may be calculated by the methods of Chapter 4.

Job Management

After applying the principles explained above, trial operations are usually required to determine the exact values of soil moisture content, lift thickness, and roller weight that yield maximum productivity while achieving the specified soil density. The use of a nuclear density device to measure the soil density actually being obtained during compaction is strongly recommended.

Table 5-3 Typical operating speed of compaction equipment

	Speed	
Compactor	*mi/h*	*km/h*
Tamping foot, crawler-towed	3–5	5–8
Tamping foot, wheel-tractor-towed	5–10	8–16
High-speed tamping foot		
First two or three passes	3–5	5–8
Walking out	8–12	13–19
Final passes	10–14	16–23
Heavy pneumatic	3–5	5–8
Multitired pneumatic	5–15	8–24
Grid roller		
Crawler-towed	3–5	5–8
Wheel-tractor-towed	10–12	16–19
Segmented pad	5–15	8–24
Smooth wheel	2–4	3–6
Vibratory		
Plate	0.6–1.2	1–2
Roller	1–2	2–3

Traffic planning and control is an important factor in compaction operations. Hauling equipment must be given the right-of-way without unduly interfering with compaction operations. The use of high-speed compaction equipment may be necessary to avoid conflicts between hauling and compacting equipment.

5-3 GRADING AND FINISHING

Grading is the process of bringing earthwork to the desired shape and elevation (or grade). *Finish grading,* or simply *finishing,* involves smoothing slopes, shaping ditches, and bringing the earthwork to the elevation required by the plans and specification. Finishing usually follows closely behind excavation, compaction, and grading. Finishing, in turn, is usually followed closely by seeding or sodding to control soil erosion. The piece of equipment most widely used for grading and finishing is the motor grader (Figure 5-9). Grade trimmers and excavators are frequently used on large highway and airfield projects because their operating speed is greater than that of the motor grader.

In highway construction, the process of cutting down high spots and filling in low spots of each roadway layer is called *balancing. Trimming* is the process of bringing each roadway layer to its final grade. Typical tolerances allowed for final roadway grades are ½ in. per 10 ft (1.25 cm/3m) for subgrades and subbases and ⅛ in. per 10 ft (0.3 cm/3m) for bases. Typical roadway components are illustrated in Figure 5-10.

Finishing is seldom a pay item in a construction contract because the quantity of earthwork involved is difficult to measure. As a result, the planning of finishing operations is usually rudimentary. However, studies have shown that the careful planning and execution of finishing operations can pay large dividends.

Figure 5-9 Modern motor grader. (Courtesy of Fiatallis North America, Inc.)

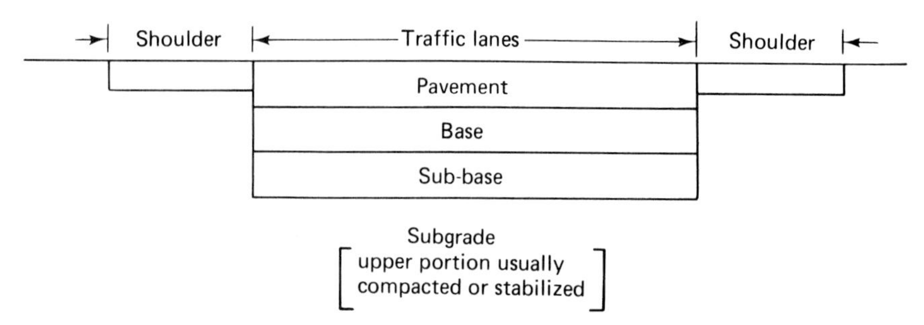

Figure 5-10 Typical roadway components.

Motor Grader

The *motor grader* is one of the most versatile items of earthmoving equipment. It can be used for light stripping, grading, finishing, trimming, bank sloping, ditching, backfiling, and scarifying. It is also capable of mixing and spreading soil and asphaltic mixtures. It is used on building construction projects as well as in heavy and highway construction. It is frequently used for the maintenance of highways and haul roads.

The blade of a motor grader is referred to as a *moldboard* and is equipped with replaceable cutting edges and end pieces (end bits). The wide range of possible blade positions is illustrated in Figure 5-11. The pitch of the blade may be changed in a manner similar to dozer blades. Pitching the blade forward results in a rolling action of the excavated material and is used for finishing work and for blending materials. Pitching the blade backward increases cutting action but may allow material to spill over the top of the blade. Blade cutting edges are available in flat, curved, or serrated styles. Flat edges produce the least edge wear, but curved edges are recommended for cutting hard materials and for fine grading. Serrated edges are used for breaking up packed gravel, frozen soil, and ice.

Motor graders are available with articulated frames that increase grader maneuverability. The three possible modes of operation for an articulated grader are illustrated in Figure 5-12. The machine may operate in the conventional manner when in the straight mode (Figure 5-12A). The articulated mode (Figure 5-12B) allows the machine to turn in a short radius. Use of the crab mode (Figure 5-12C) permits the rear driving wheels to be offset so that they remain on firm ground while the machine cuts banks, side slopes, or ditches. The front wheels of both conventional and articulated graders may be leaned from side to side. Wheels are leaned away from the cut to offset the side thrust produced by soil pressure against the angled blade. Wheel lean may also be used to assist in turning the grader.

Graders are available with automatic blade control systems that permit precise grade control. Such graders utilize a sensing system that follows an existing surface, string line, or laser beam to automatically raise or lower the blade as required to achieve the desired grade. A scarifier is shown directly behind the moldboard of the grader in Figure 5-9. Scarifiers are used to loosen hard soils before grading and to break up asphalt pavements and frozen soil. Their operation is similar to that of the ripper described in Chapter 6. However, scarifiers are not intended for heavy-duty

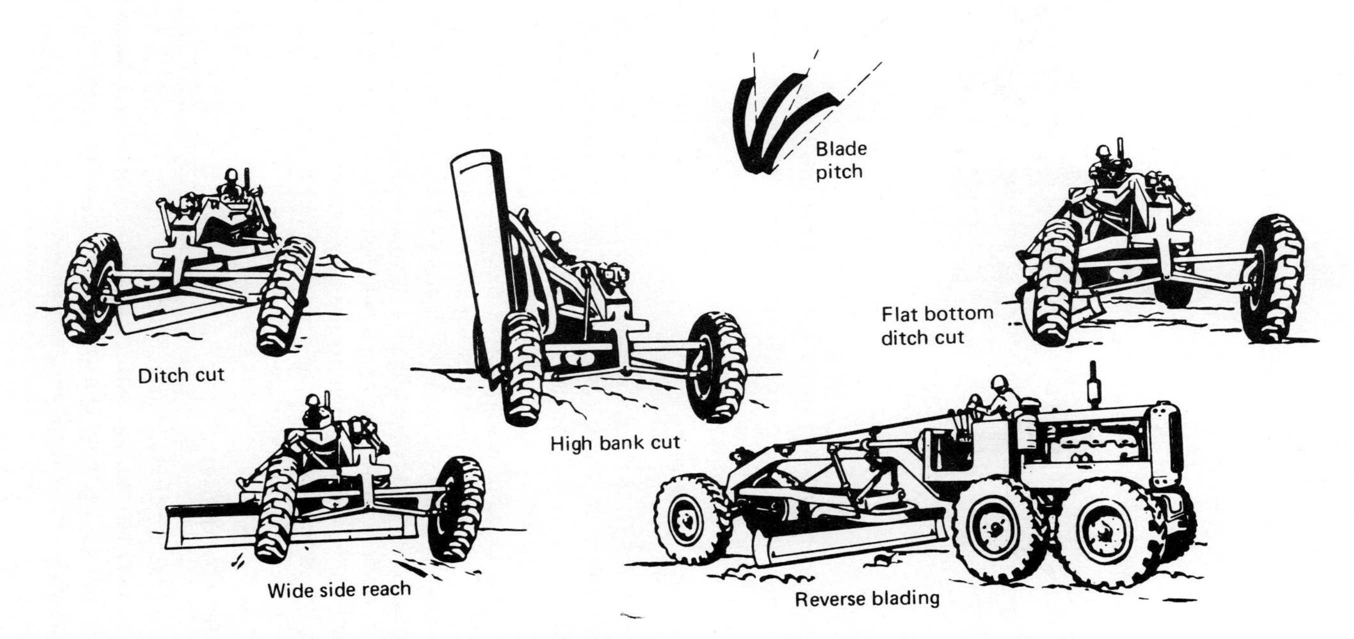

Figure 5-11 Blade positions for the motor grader. (U.S. Department of the Army)

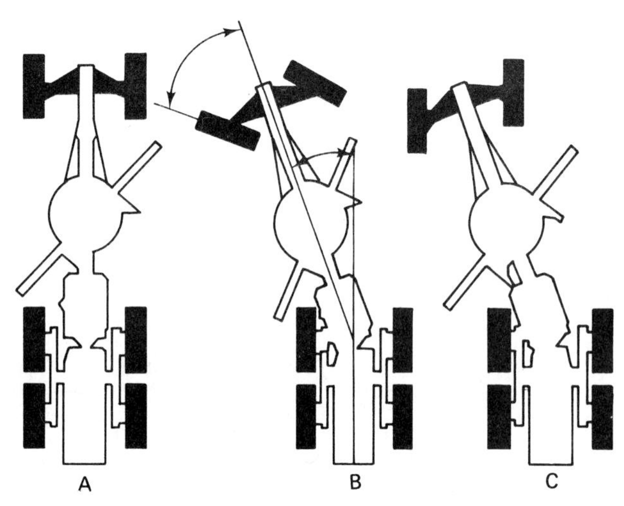

Figure 5-12 Articulated grader positions.

use as are rippers. While rippers are available for graders, their ripping ability is limited by the weight and power of the grader. Grader rippers are usually mounted on the rear of the machine.

Grade Excavators and Trimmers

Grade excavators or *trimmers* are machines that are capable of finishing roadway and airfield subgrades and bases faster and more accurately than can motor graders. Many of these machines also act as reclaimers. That is, they are capable of scarifying and removing soil and old asphalt pavement. Trimmers and reclaimers are usually equipped with integral belt conveyors that are used for loading excavated material into haul units or depositing it in windrows outside the excavated area. The large grade trimmer/reclaimer shown in Figure 5-13 is also capable of compacting base material, laying asphalt, or acting as a slipform power.

While grade trimmers lack the versatility of motor graders, their accuracy and high speed make them very useful on large roadway and airfield projects. Their large size often requires that they be partially disassembled and transported between job sites on heavy equipment trailers.

Estimating Grader Production

Grader production is usually calculated on a linear basis (miles or kilometers completed per hour) for roadway projects and on an area basis (square yards or square meters per hour) for general construction projects. The time required to complete a roadway project may be estimated as follows:

$$\text{Time (h)} = \left[\sum \frac{\text{Number of passes} \times \text{Section length (mi or km)}}{\text{Average speed for section (mi/h or km/h)}} \right] \times \frac{1}{\text{Efficiency}} \quad (5\text{-}2)$$

Figure 5-13 Large grade trimmer/reclaimer/paver. (Courtesy of CMI Corp.)

Average speed will depend on operator skill, machine characteristics, and job conditions. Typical grader speeds for various types of operations are given in Table 5-4.

Example 5-1

PROBLEM Fifteen miles (24.1 km) of gravel road require reshaping and leveling. You estimate that six passes of a motor grader will be required. Based on operator skill, machine characteristics, and job conditions, you estimate two passes at 4 mi/h (6.4 km/h), two passes at 5 mi/h (8.0 km/h), and two passes at 6 mi/h (9.7 km/h). If job efficiency is 0.80, how many grader hours will be required for this job?

Table 5-4 Typical grader operating speed

	Speed	
Operation	*mi/h*	*km/h*
Bank sloping	2.5	4.0
Ditching	2.5–4.0	4.0–6.4
Finishing	4.0–9.0	6.5–14.5
Grading and road maintenance	4.2–6.0	6.4–9.7
Mixing	9.0–20.0	14.5–32.2
Snow removal	12.0–20.0	19.3–32.3
Spreading	6.0–9.0	9.7–14.5

SOLUTION

$$\text{Time (h)} = \left(\frac{2 \times 15}{4.0} + \frac{2 \times 15}{5.0} + \frac{2 \times 15}{6.0}\right) \times \frac{1}{0.80} = 23.1 \text{ h} \qquad (5\text{-}2)$$

$$\left[= \left(\frac{2 \times 24.1}{6.4} + \frac{2 \times 24.1}{8.9} + \frac{2 \times 24.1}{9.7}\right) \times \frac{1}{0.80} = 23.1 \text{ h}\right]$$

Job Management

Careful job planning, the use of skilled operators, and competent supervision is required to maximize grader production efficiency. Use the minimum possible number of grader passes to accomplish the work. Eliminate as many turns as possible. For working distances less than 1000 ft (305 m), have the grader back up rather than turn around. Grading in reverse may be used for longer distances when turning is difficult or impossible. Several graders may work side by side if sufficient working room is available. This technique is especially useful for grading large areas.

PROBLEMS

1. The data in the accompanying table resulted from performing Modified Proctor Tests on a soil. Plot the data and determine the soil's laboratory optimum moisture content. What is the 100% Modified AASHTO density of this soil?

Dry Density		*Moisture Content*
lb/cu ft	*g/cm³*	*(%)*
109	1.746	10
112	1.794	12
116	1.858	14
115	1.842	16
110	1.762	18
106	1.698	20

2. The data in the accompanying table resulted from performing Modified Proctor Tests on a soil. What minimum field density must be achieved to meet job specifications that require compaction to 95% of Modified Proctor Density?

Dry Density		*Moisture Content*
lb/cu ft	*g/cm³*	*(%)*
120	1.922	3
125	2.003	5
129	2.067	7
130	2.083	9
127	2.035	11

3. If you were going to employ a compactor that delivered approximately one-half of the compactive effort of the Modified Proctor Test, how would you expect the field optimum moisture content to compare with the Modified Proctor optimum moisture content?

4. Briefly explain why a soil's field optimum moisture content may differ from its laboratory optimum moisture content.

5. Estimate the production in compacted cubic yards (meters) per hour of a self-propelled tamping foot roller under the following conditions: average speed = 6 mi/h (9.6 km/h), compacted lift thickness = 8 in. (20.32 cm), effective roller width = 12 ft (3.7 m), job efficiency = 0.83, and number of passes = 6.

6. A highway contractor has opened a cut in fine silty sand (SM). Because of the cut location and lack of drainage, surface water has drained into the excavation, leaving the material very wet. Rubber-tired scrapers are hauling material from the cut to an adjacent fill. The material is being placed in the fill in 8- to 10-in. (20- to 25-cm) lifts and compacted by a heavy sheepsfoot roller towed by a crawler tractor. It is apparent that the specified compaction (95% Standard AASHTO density) is not being attained. The sheepsfoot roller is not "walking out" and the scraper tires are causing deep rutting and displacement of the fill material as they travel over it. What is causing the compaction problem? What would you suggest to the contractor to improve the compaction operation?

7. A highway embankment is being constructed of low-plasticity clay (CL) whose natural moisture content is slightly below the Standard AASHTO optimum. Specifications require the use of a tamping foot roller for compaction. However, the roller being used does not always "walk out" as compaction proceeds, and even when it does, it often fails to produce the required soil density. The contractor and the project engineer disagree over appropriate corrective measures. The contractor wants to increase the weight of the roller by adding more ballast, thus increasing the contact pressure of the roller feet. The project engineer disagrees and suggests that the weight of the roller is adequate and that the foot size should be increased. Which recommendation is more nearly correct? Why?

8. Ten miles (16 km) of gravel road require reshaping and leveling. You estimate that a motor grader will require two passes at 4 mi/h (6.4 km/h), two passes at 5 mi/h (8.0 km/h), and one pass at 6 mi/h (9.7 km/h) to accomplish the work. How many grader hours will be required for this work if the job efficiency factor is 0.80?

9. Write a computer program to calculate the expected hourly production (compacted measure) of a compactor. Input should include average compactor speed, effective compactor width, compacted lift thickness, number of passes required to achieve the desired compaction, and the job efficiency. Solve Problem 5 using your program.

10. Write a computer program to calculate the time required (hours) for a motor grader to perform a finishing operation. Input should include section length, number of passes at each expected speed, and job efficiency. Solve Problem 8 using your computer program.

REFERENCES

1. *Compaction Handbook.* Hyster Company, Kewanee, Ill., 1978.
2. *Compaction Industry Vibratory Roller Handbook.* Construction Industry Manufacturers Association, Milwaukee, Wis., 1978.

3. *Fundamentals of Compaction,* 2nd ed. Caterpillar Inc., Peoria, Ill., 1979.
4. *Handbook of Soil Compactionology.* Bros, Inc., Brooklyn Park, Minn., 1977.
5. Nunnally, S.W. *Managing Construction Equipment.* Englewood Cliffs, N.J.: Prentice Hall, 1977.
6. Townsend, D. L. *The Performance and Efficiency of Standard Compacting Equipment* (Engineering Report No. 6). Queens University, Kingston, Ontario, 1959.
7. Wahls, H. E., C. P. Fisher, and L. J. Langfelder. *The Compaction of Soil and Rock Materials for Highway Purposes* (Report No. FHWA-RD-73-8). U.S. Department of Transportation, Washington, D.C., 1966.

6
Rock Excavation

6-1
INTRODUCTION

Rock Characteristics

Rock may be classified as igneous, sedimentary, or metamorphic, according to its origin. *Igneous rock* formed when the earth's molten material cooled. Because of its origin, it is quite homogeneous and is therefore the most difficult type of rock to excavate. Examples of igneous rock are granite and basalt. *Sedimentary rock* was formed by the precipitation of material from water or air. As a result, it is highly stratified and has many planes of weakness. Thus it is the most easily excavated type of rock. Examples include sandstone, shale, and limestone. *Metamorphic rock* originated as igneous or sedimentary rock but has been changed by heat, pressure, or chemical action into a different type of rock. Metamorphic rock is intermediate between igneous rock and sedimentary rock in its difficulty of excavation. Examples of metamorphic rock include slate, marble, and schist.

The difficulty involved in excavating rock depends on a number of factors in addition to the rock type. Some of these factors include the extent of fractures and other planes of weakness, the amount of weathering that has occurred, the predominant grain size, whether the rock has a crystalline structure, rock brittleness, and rock hardness.

Rock Investigation

Relative hardness is measured on *Moh's scale* from 1 (talc) to 10 (diamond). As a rule, any rock that can be scratched by a knife blade (hardness about 5) can be easily excavated by ripping or other mechanical methods. For harder rock, additional investigation is required to evaluate the rock characteristics described above. The principal methods for investigating subsurface conditions include drilling, excavating test pits, and making seismic measurements. Drilling may be used to remove core samples from the rock or to permit visual observation of rock conditions. Core samples may be visually inspected as well as tested in the laboratory. Observation in a test pit or inspection by TV cameras placed into drilled holes will reveal layer thickness, the extent of fracturing and weathering, and the presence of water. Use of the refraction seismograph permits a rapid determination of rock soundness by measuring the velocity at which sound travels through the rock.

In performing a seismic refraction test, a sound source and a number of receivers (geophones) are set up, as illustrated in Figure 6-1. The time required for a sound wave to travel from the sound source to each receiver is measured and plotted against the distance from the sound source, as illustrated in Figure 6-2. In this plot the slope of each segment of the curve represents the sound velocity in the corresponding subsurface layer. This velocity has been found to range from about 1000 ft/s (305 m/s) in loose soil to about 20,000 ft/s (6100 m/s) in sound rock. Since the relationship between the angle of incidence and the angle of refraction of a sound wave crossing the interface of two rock layers is a function of their respective sound velocities, the seismic refraction test method may also be used to determine the thickness of rock layers. Equation 6-1 may be used to determine the thickness of the upper layer when the sound velocity increases with layer depth. That is when the velocity in the top layer is less than the velocity in the second layer, which is the usual case in the field.

$$H_1 = \frac{D_1}{2}\left[\frac{V_2 - V_1}{V_2 + V_1}\right]^{1/2} \tag{6-1}$$

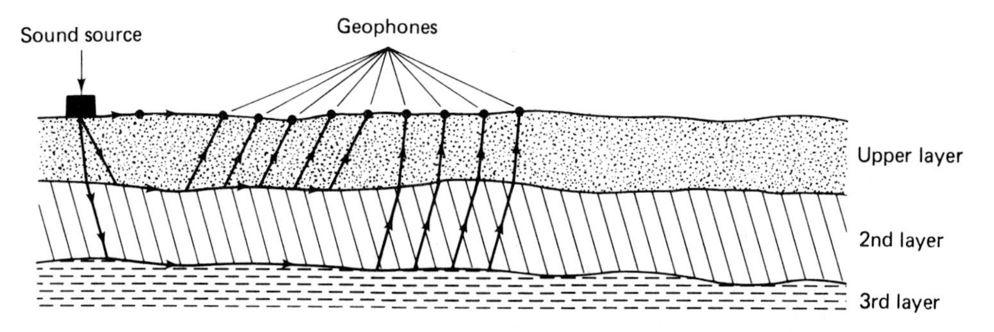

Figure 6-1 Schematic representation of seismic refraction test.

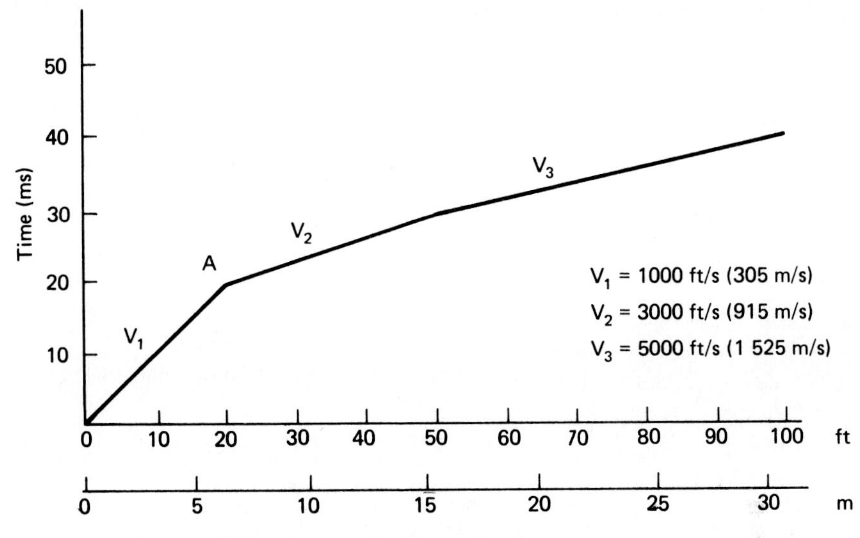

Figure 6-2 Graph of refraction test data.

where H_1 = thickness of upper layer (ft or m)
D_1 = distance from sound source to first intersection of lines on time-distance graph (ft or m) (point A, Figure 6-2)
V_1 = velocity in upper layer (ft/s or m/s)
V_2 = velocity in second layer (ft/s or m/s)

Example 6-1

PROBLEM Find the seismic wave velocity and depth of the upper soil layer based on the following refraction seismograph data:

Distance from Sound Source to Geophone		
ft	*m*	*Time (ms)*
10	3.05	5
20	6.10	10
30	9.15	15
40	12.20	20
50	15.25	22
60	18.30	24
70	21.35	26
80	24.40	28

SOLUTION Plot time of travel against distance from sound source to geophone as shown in Figure 6-3.

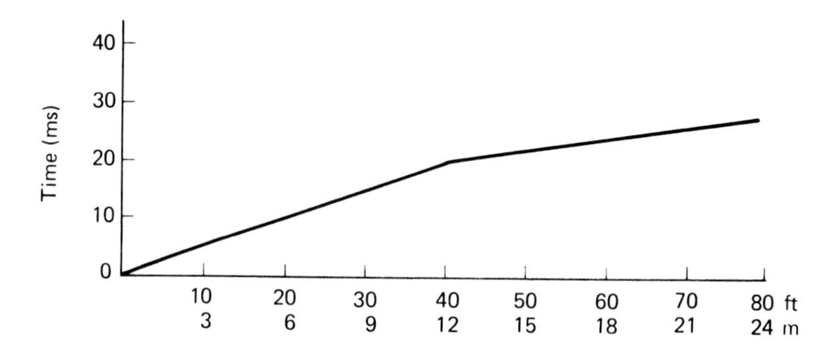

Figure 6-3 Graph of refraction test data, Example 6-1.

$$V_1 = \frac{40 - 0}{0.020 - 0} = 2000 \text{ ft/s}$$

$$\left[= \frac{12.2 - 0}{0.020 - 0} = 610 \text{ m/s} \right]$$

$$V_2 = \frac{80 - 40}{0.028 - 0.020} = 5000 \text{ ft/s}$$

$$\left[= \frac{24.4 - 12.2}{0.028 - 0.020} = 1525 \text{ m/s} \right]$$

$$H_1 = \frac{40}{2}\left(\frac{5000 - 2000}{5000 + 2000}\right)^{1/2} = 13.1 \text{ ft}$$

$$\left[= \frac{12.2}{2}\left(\frac{1525 - 610}{1525 + 610}\right)^{1/2} = 4.0 \text{ m} \right]$$

Rock-Handling Systems

The process of rock moving may be considered in four phases: loosening, loading, hauling, and compacting. The methods employed for rock compacting are discussed in Chapter 5. Therefore, this discussion of rock-handling systems will be limited to the phases of loosening, loading, and hauling. The principal methods and equipment available for accomplishing each of these phases are listed in Table 6-1.

The traditional method for excavating rock involves drilling blastholes in the rock, loading the holes with explosives, detonating the explosives, loading the fractured rock into haul units with power shovels, and hauling the rock away in trucks or wagons. Newer alternatives include the use of tractor-mounted rippers to loosen rock, the use of wheel loaders to load fractured rock into haul units, and the use of reinforced (or "special application") scrapers to both load and haul fractured rock. The equipment and procedures utilized are explained in Sections 6-2 to 6-4. The selection of the rock-handling system to be employed in a particular situation should be based on maximizing the contractor's profit from the operation.

Table 6-1 Principal rock-handling systems

Operation	*Equipment and Process*
Loosen	Drill and blast
	Rip
Load	Shovel
	Wheel loader
Haul	Truck
	Wagon
Load and haul	Reinforced scraper

Tunneling

Tunneling in rock is a specialized form of rock excavation that has traditionally been accomplished by drilling and blasting. Recently, however, *tunneling machines* or mechanical *moles* equipped with multiple cutter heads and capable of excavating to full tunnel diameter have come into increasing use. The tunneling machine shown in Figure 6-4 weighs 285 tons (258 t), produces a thrust of 1,850,000 lb (8229 kN), and drills a 19-ft (5.8-m)-diameter hole.

Some of the specialized terms used in tunneling include jumbos, hydraulic jumbos, and mucking machines. A *jumbo* is a large mobile frame equipped with platforms at several elevations to enable drills and workers to work on the full tunnel

Figure 6-4 Large tunneling machine. (Courtesy of The Robbins Company)

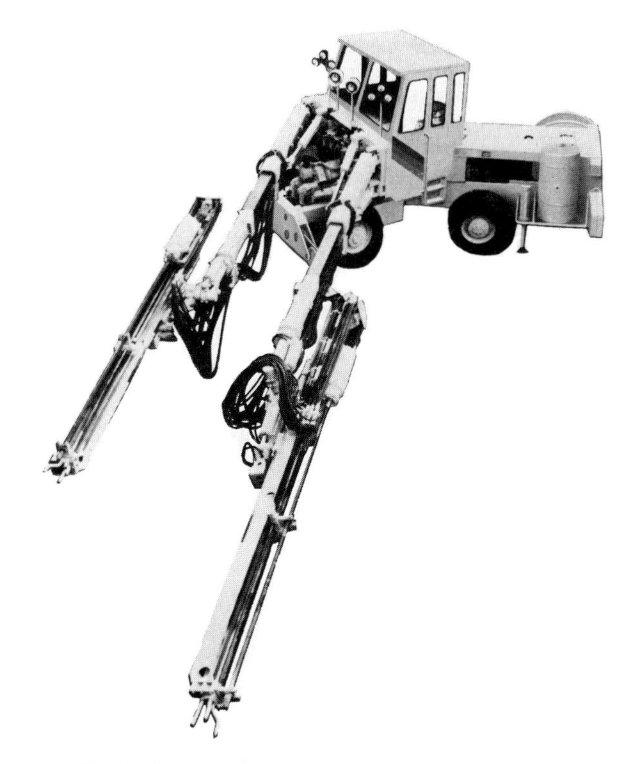

Figure 6-5 Hydraulic jumbo. (Courtesy of Ingersoll-Rand Company)

face at one time. The newer *hydraulic jumbo* illustrated in Figure 6-5 is a self-propelled machine equipped with a number of hydraulic drills, each mounted on its own hydraulic boom. Such a machine is capable of drilling blastholes across the full tunnel face at one time. A *mucking machine* is a form of shovel especially designed for loading fractured rock into haul units during tunnel excavation.

6-2 DRILLING

Drilling Equipment

Common types of drilling equipment include percussion drills, rotary drills, and rotary-percussion drills, as listed in Table 6-2. *Percussion drills* penetrate rock by impact action alone. While the bits of these drills rotate to assist in cleaning the hole and to provide a fresh surface for each impact, rotation takes place on the upstroke. Thus no cutting is accomplished during rotation. Common types of percussion drills include the hand-held rock drill (or jackhammer); the drifter, which is mounted on a frame or column; the wheel-mounted wagon drill; and the crawler-mounted track drill. Wagon drills are basically large drifter drills mounted on wheels for mobility.

Table 6-2 Typical characteristics of rock drilling equipment

	Maximum Drill Diameter		Maximum Depth	
Type of Drill	*in.*	*cm*	*ft*	*m*
Percussion drill				
Jack hammer	2.5	6.3	20	6.1
Drifter	4.5	11.4	15	4.6
Wagon drill	6.0	15.2	50	15.3
Track drill	6.0	15.2	50	15.3
Rotary drill	72	183	1000+	305+
Rotary-percussion	6.0	15.2	150	46

They have been used extensively for construction and quarry drilling but are largely being displaced by track drills. Track drills (Figure 6-6) are large, self-propelled, crawler-mounted drills usually equipped with a hydraulically powered boom for positioning the drill feed.

Rotary drills (Figure 6-7) cut by turning a bit under pressure against the rock face. *Rotary-percussion drills* combine rotary and percussion cutting action to pene-

Figure 6-6 Hydraulic track drill. (Courtesy of LeRoi Division, Dresser Industries)

Figure 6-7 Rotary blast hole drill. (Courtesy of Ingersoll-Rand Company)

trate rock several times as fast as a comparable percussion drill. Downhole drills (Figure 6-8) utilize a percussion drilling device mounted directly above the drill bit at the bottom of the hole. Downhole drills have several advantages over conventional percussion drills: drill rod life is longer, because drill rods are not subjected to impact; less air is required for cleaning the hole, because drill exhaust may be used for

Figure 6-8 Downhole drill mounted on a rotary drill. (Courtesy of Ingersoll-Rand Company)

this purpose; noise level is lower, for the same reason; and there is little loss of energy between the drill and the bit. Although downhole drills are available for track drills, they are usually mounted on rotary drilling machines, since relatively large diameter holes are necessary to provide sufficient space for the downhole drill body. The combination of a rotary drill mechanism and a downhole drill permits such a machine to function as a rotary drill, a percussion drill, or a rotary-percussion drill.

Drilling rate (rate of penetration) depends on rock hardness, drill type and energy, and the type of drill bit used. Table 6-3 provides representative drilling rates for common drilling equipment.

Increased air pressure at the drill will result in increased drill production (penetration/h) as shown in Figure 6-9. However, for safety, pressure at the drill must not exceed the maximum safe operating pressure specified by the drill manufacturer. In addition, the use of increased air pressure will also shorten the life of drills, drill steel, and bits, as well as increase drill maintenance and repair costs. Thus, field tests should be run to determine the drilling pressure that results in minimum cost per unit of rock excavation.

Increasingly, large-diameter vertical and inclined holes are being drilled for rescue of miners, to create blastholes for spherical charges in mining operations, and for constructing penstocks and other shafts. In addition to the use of large-diameter downhole and rotary drills, raise boring machines have been used to drill vertical shafts over 20 ft (6 m) in diameter. *Raise boring* is a drilling technique in which the large-diameter hole is drilled upward from the bottom. The procedure is as follows. First, a pilot hole is drilled from the surface into a mine or other underground cavity. Next, a large reaming head is attached to the lower end of a raise boring machine drill string which has been lowered into the mine. The large-diameter hole is then created as the reaming head is rotated and raised by the raise boring machine sitting on the surface. Raise boring machines are capable of creating an upward thrust of 1 million pounds (44.5 MN) or more and a torque exceeding 400,000 ft-lb (542 kN m).

Rock drills have traditionally been powered by compressed air. However, the current trend is toward the use of hydraulically powered drills (Figure 6-6). Hydraulic drills penetrate faster and consume less energy than do comparable compressed air drills. Hydraulic drills also produce less noise and dust than do compressed air drills.

Table 6-3 Representative drilling rates (carbide bit)

Bit Size		*Hand Drill*		*Wagon Drill*		*Crawler Drill*		*Hydraulic Drill*	
in.	*cm*	*ft/h*	*m/h*	*ft/h*	*m/h*	*ft/h*	*m/h*	*ft/h*	*m/h*
1⅝	4.1	14–24	4.3–7.3						
1¾	4.4	11–21	3.4–6.4						
2	5.1	10–19	3.1–5.8	25–50	7.6–15.3	125–290	38.1–88.5	185–425	56.4–129.6
2½	6.4			19–48	5.8–14.9	80–180	24.4–54.9	120–280	36.6–85.4
3	7.6			18–46	5.5–14.0	75–160	22.9–48.8	100–240	30.5–73.2
4	10.2			10–35	3.1–10.7	50–125	15.3–31.1	80–180	24.4–54.9
6	15.2					20–50	6.1–15.3	30–75	9.2–22.9

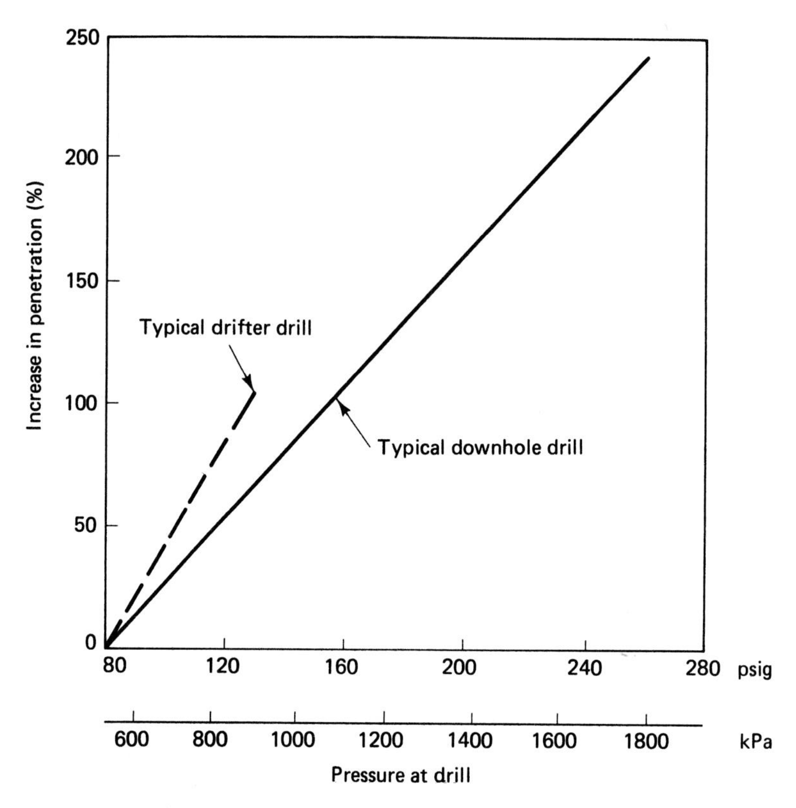

Figure 6-9 Drill penetration versus air pressure.

Drill Bits and Steel

Major types of rock drill bits are illustrated in Figure 6-10. Principal percussion drill bits include cross-type bits (Figure 6-10a), X-type bits, and button bits (Figure 6-10b). *X-type bits* differ from *cross-type bits* only in that the points of the X-type bits are not spaced at 90° angles but rather form the shape of an X. X-type bits tend to drill straighter holes than do cross-type bits. Both types of bits are available with either solid steel or tungsten carbide cutting edges. Since tungsten carbide bits cut faster and last longer than do solid steel bits, carbide bits are more widely used. *Button-type bits* have a higher penetration rate than X or cross-type bits and are less likely to jam in the hole. Button bits do not normally require grinding or sharpening. Bits used with rotary-percussion drills are similar to X-type percussion drill bits but utilize special designs for the cutting edges.

Common types of bits for rotary drills include coring bits (Figure 6-10c) and rolling cutter or cone bits (Figure 6-10d). *Coring bits* are available as diamond drill bits and shot drill bits. Diamond drill bits utilize small diamonds set in a matrix on the bit body as the cutting agent to achieve rapid rock penetration. Diamond bits are also available in other shapes, such as concave bits, to drill conventional holes. The shape of a shot drill coring bit is similar to that of the diamond coring bit. However,

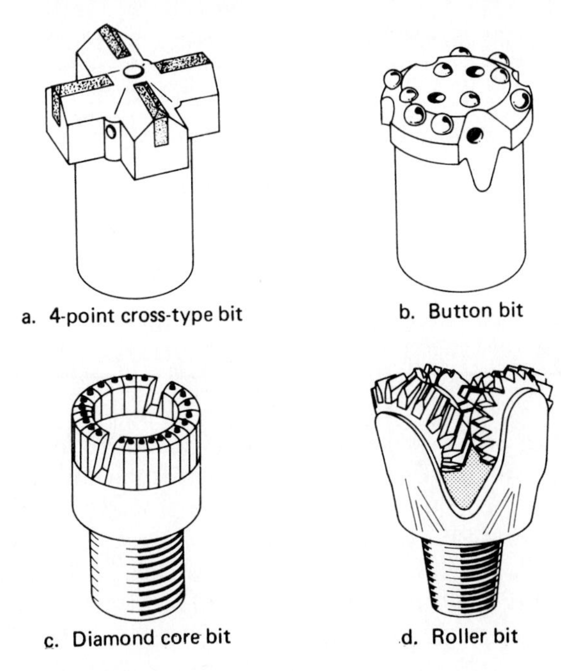

Figure 6-10 Major types of rock drill bits.

chilled steel shot fed into the hole around the bit replaces diamond as the cutting agent. The lower end of the shot drill bit is slotted to assist in retaining the shot between the bit and rock as the bit rotates. *Rolling cutter bits* use several cutters (usually three) shaped like gears to penetrate the rock as the drill bit rotates.

The steel rod connecting a percussion drill and its bit is referred to as *drill steel* or simply *steel.* Drill steel is available in sections ⅞ in. (2.2 cm) to 2 in. (5.1 cm) in diameter and 2 ft (0.61 m) to 20 ft (6.1 m) in length. Drill steel sections are fitted with threaded ends so that sections may be added as the bit penetrates. Sections are hollow to allow air flow to the bit for hole cleaning. The drill rod used for rotary drilling is called a *drill pipe.* It is available in length increments of 5 ft (1.5 m) starting with 10 ft (3.1 m) and threaded on each end. Drill pipe is also hollow to permit the flow of compressed air or drilling fluid to the bottom of the hole.

Drilling Patterns and Rock Yield

The choice of hole size, depth, and spacing, as well as the amount of explosive used for each hole, depends on the degree of rock break desired, rock type and soundness, and the type of explosive utilized. In general, small holes closely spaced will yield small rock particles while large holes widely spaced will yield large rock particles. Although equations have been developed for estimating hole spacing based on rock strength and explosive pressure, test blasts are usually necessary to determine optimum hole spacing and quantity of explosive per hole. Table 6-4 indicates typical drill hole spacing.

Table 6-4 Typical drill hole spacing (rectangular pattern) [ft (m)]

	Hole Diameter [in. (cm)]							
Rock Type	*2¼ (5.7)*	*2½ (6.4)*	*3 (7.6)*	*3½ (8.9)*	*4 (10.2)*	*4½ (11.4)*	*5 (12.7)*	*5½ (14.0)*
Strong rock (granite, basalt)	4.5 (1.4)	5.0 (1.5)	6.0 (1.8)	7.0 (2.1)	8.5 (2.6)	9.0 (2.7)	10.0 (3.0)	11.0 (3.4)
Medium rock (limestone)	5.0 (1.5)	6.0 (1.8)	7.0 (2.1)	8.0 (2.4)	9.0 (2.7)	10.0 (3.0)	11.5 (3.5)	12.5 (3.8)
Weak rock (sandstone, shale)	6.0 (1.8)	6.5 (2.0)	8.0 (2.4)	10.0 (3.0)	11.0 (3.4)	12.0 (3.7)	13.5 (4.1)	15.5 (4.7)

The two principal drilling patterns used for rock excavation are illustrated in Figure 6-11. While the rectangular pattern is most widely used, the staggered pattern reduces the amount of oversized rock produced. For a rectangular pattern the volume of blasted rock produced per hole may be computed by the use of Equation 6-2. The effective depth of a blast hole is the average depth of the excavation area after the blast, not the original hole depth.

$$\text{Volume/hole (cu yd)} = \frac{S^2 \times H}{27} \tag{6-2A}$$

$$\text{Volume/hole (m}^3\text{)} = S^2 \times H \tag{6-2B}$$

where S = pattern spacing (ft or m)
H = effective hole depth (ft or m)

Effective hole depth should be determined by trial blasting. However, effective depth has been found to average about 90% of original hole depth.

The rock produced per hole may be divided by the original hole depth to yield rock volume per unit of hole drilled:

$$\text{Volume/ft of hole} = \frac{\text{Volume per hole (cu yd)}}{\text{Drilled hole depth (ft)}} \tag{6-3A}$$

$$\text{Volume/m of hole} = \frac{\text{Volume per hole (m}^3\text{)}}{\text{Drilled hole depth (m)}} \tag{6-3B}$$

The amount of drilling required to produce a unit volume of blasted rock may then be calculated as the reciprocal of the volume per unit of drilled depth.

Example 6-2

PROBLEM Trial blasting indicates that a rectangular pattern of drilling using 3-in. (7.6-cm) holes spaced on 9-ft (2.75-m) centers, 20 ft (6.1 m) deep will produce a satisfactory rock break with a particular explosive loading. The effective hole depth resulting from the blast is 18 ft (5.5 m). Determine the rock volume produced per foot (meter) of drilling.

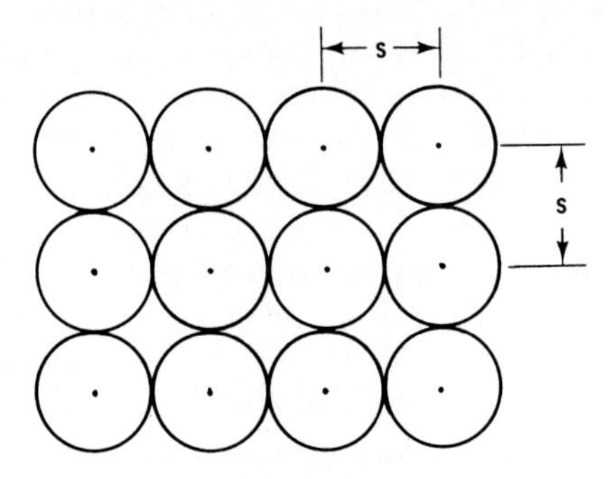

a. Rectangular pattern

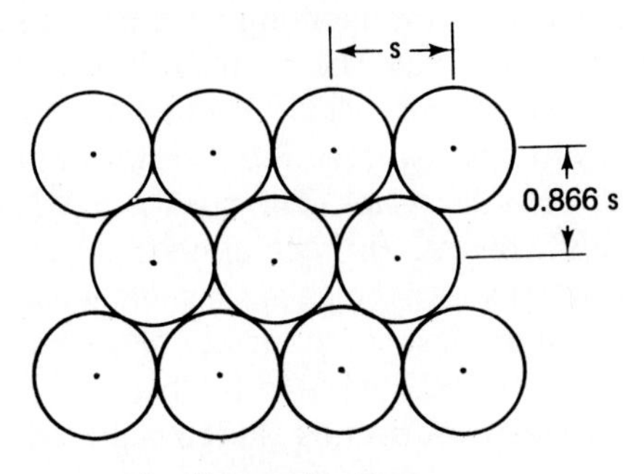

b. Staggered pattern

Figure 6-11 Principal drilling patterns.

SOLUTION

$$\text{Volume/hole} = \frac{9^2 \times 18}{27} = 54 \text{ cu yd} \qquad (6\text{-}2A)$$

$$[= 2.75^2 \times 5.5 = 41.6 \text{ m}^3] \qquad (6\text{-}2B)$$

$$\text{Volume/unit depth} = \frac{54}{20} = 2.7 \text{ cu yd/ft} \qquad (6\text{-}3A)$$

$$\left[= \frac{41.6}{6.1} = 6.8 \text{ m}^3/\text{m}\right] \qquad (6\text{-}3B)$$

6-3 BLASTING

Explosives

The principal explosives used for rock excavation include dynamite, ammonium nitrate, ammonium nitrate in fuel oil (ANFO), and slurries. For construction use dynamite has largely been replaced by ammonium nitrate, ANFO, and slurries because these explosives are lower in cost and easier to handle than dynamite. Ammonium nitrate and ANFO are the least expensive of the explosives listed. ANFO is particularly easy to handle, because it is a liquid that may simply be poured into the blasthole. However, ammonium nitrate explosives are not water resistant and they require an auxiliary explosive (primer) for detonation. *Slurries* are mixtures of gels, explosives, and water. They may also contain powered metals (metalized slurries) to increase blast energy. Slurries are cheaper than dynamite but are more expensive than the ammonium nitrate explosives. Water resistance and greater power are their principal advantages over ammonium nitrate explosives. Slurries are available as liquids or packaged in plastic bags. Slurries also require a primer for detonation.

Detonators used to initiate an explosion include both electric and nonelectric caps. Electric blasting (EB) caps are most widely used and are available as instantaneous caps or with delay times from a few milliseconds up to several seconds. For less-sensitive explosives such as ammonium nitrates and slurries, caps are used to initiate *primers,* which in turn initiate the main explosive. Primers may be small charges of high explosives or *primacord,* which is a high explosive in cord form.

The amount of explosive required to produce the desired rock fracture is usually expressed as a *powder factor.* The powder factor represents the weight of explosive used per unit volume of rock produced (lb/BCY or kg/Bm^3). Except in specialized applications, blastholes are usually loaded with a continuous column of explosive to within a few feet of the surface. *Stemming* (an inert material used to confine and increase the effectiveness of the blast) is placed in the top portion of the hole above the explosive. A primed charge is placed near the bottom of the hole for blast initiation.

Electric Blasting Circuits

Electric blasting caps may be connected in series, parallel, or parallel-series circuits, as illustrated in Figure 6-12. The type of insulated wires used to make up these circuits include legwires, buswires, connecting wires, and firing lines. *Legwires* are the two wires that form an integral part of each electric blasting cap. *Buswires* are wires used to connect the legwires of caps into parallel or parallel series circuits. *Connecting wires,* when used, are wires that connect legwires or buswires to the firing line. The *firing line* consists of two parallel conductors that connect the power source to the remainder of the blasting circuit. Note the use of a "reverse hookup" for the buswires shown in Figure 6-12. In the reverse hookup the firing lines are connected to opposite ends of the two buswires. Such a hookup has been found to provide a more equal distribution of current to all caps in the circuit than would be the case if the firing lines were connected to the same ends of the two buswires. Power sources include ac and dc power lines and blasting machines. *Blasting machines* are small dc generators or capacitive discharge (CD) units designed especially for firing electric cap circuits.

When designing or analyzing an electric blasting circuit, it is necessary to determine the total circuit resistance and the current required to safely fire all caps. Recall that in a series circuit the same current flows through all elements of the circuit,

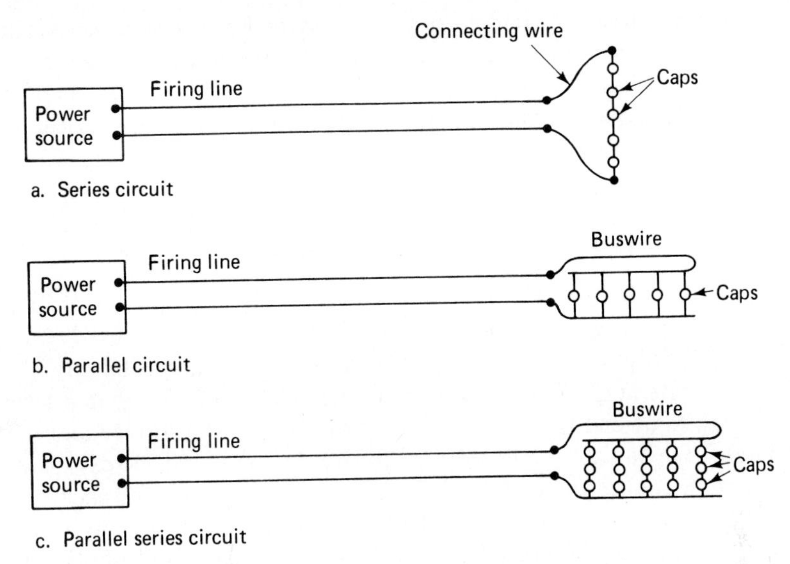

Figure 6-12 Types of electric blasting circuits.

whereas in a parallel circuit the current in the main (firing) line is the sum of the currents in all the parallel branches. Ohm's law (Equation 6-4) may then be used to find the power source voltage required to safely detonate all caps.

$$\text{Voltage (volts)} = \text{Current (amperes)} \times \text{Resistance (ohms)} \tag{6-4}$$

or

$$E = IR$$

Conversely, if the power source voltage and circuit resistance are known, Equation 6-4 may be used to determine whether sufficient current will be produced to fire the blast. Representative current requirements for firing electric blasting caps are given in Table 6-5. Notice that a maximum current per cap is specified to prevent arcing

Table 6-5 Representative current requirements for firing electric blasting caps

	Minimum Current (A)				
	Normal		*Leakage*		
Type of Circuit	*dc*	*ac*	*dc*	*ac*	*Maximum Current*
Single cap	0.5	0.5	—	—	—
Series	1.5	2.0	3.0	4.0	—
Parallel	1.0/cap	1.0/cap	—	—	10/cap
Parallel–series	1.5/series	2.0/series	3.0/series	4.0/series	—

of parallel circuits. Arcing results when excessive heat is produced in a cap by high current. Arcing may result in misfires or erratic timing of delay caps and, therefore, must be avoided.

If the same number and type of caps are used in each series, the resistance of the cap circuit may be calculated as follows:

$$R_c = \frac{\text{Number of caps/series} \times \text{Resistance/cap}}{\text{Number of parallel branches}} \tag{6-5}$$

Representative values of individual cap resistance are presented in Table 6-6. Note that cap resistance depends on the length of cap legwire used. Caps are normally available with legwires ranging from 4 to 100 ft (1.2 to 30.5 m). Caps with longer legwires are available on special order. Each legwire of a cap in a series must be long enough to extend from the bottom of the hole to the surface and then half the distance to the adjacent hole. Minimum legwire length may be calculated on this basis. Buswire, connecting wire, and firing-line resistance may be calculated using the resistance factors of Table 6-7. The effective length of buswire to be used in resistance calculations may be taken as the distance along the wire from the firing-line end of one buswire to the center of the cap pattern and from there to the firing-line end of the other buswire.

The steps in performing a blasting circuit analysis are listed below. This procedure is illustrated by Example 6-3.

1. Determine the resistance of the cap circuit.
2. Determine the resistance of buswires and connecting wires.
3. Determine the resistance of the firing line.

Table 6-6 Representative resistance of electric blasting caps

Legwire Length		
ft	*m*	*Nominal Resistance (Ω)*
4	1.2	1.5
6	1.8	1.6
8	2.4	1.7
10	3.1	1.8
12	3.7	1.8
16	4.9	1.9
20	6.1	2.1
24	7.3	2.3
28	8.5	2.4
30	9.2	2.2
40	12.2	2.3
50	15.3	2.6
60	18.3	2.8
80	24.4	3.3
100	30.5	3.8

Table 6-7 Resistance of solid copper wire

American Wire Gauge Number	Ohms/1000 ft (305 m)
6	0.395
8	0.628
10	0.999
12	1.59
14	2.53
16	4.02
18	6.38
20	10.15

4. Compute the total circuit resistance as the sum of cap, buswire, connecting wire, and firing-line resistance.
5. Determine the minimum total current requirement (and maximum, if applicable).
6. Solve Equation 6-4 for required power voltage. If the power source voltage is known, solve Equation 6-4 for the circuit current and compare with the values in step 5.

Example 6-3

PROBLEM Analyze the cap circuit shown in Figure 6-13 to determine circuit resistance and the minimum power source current and voltage necessary to fire the blasts. Caps are equipped with 30-ft (9.2-m) legwires. Assume normal firing conditions.

SOLUTION

$$R_{\text{cap}} = \frac{3 \times 2.22}{3} = 2.20\ \Omega \qquad (6\text{-}5)$$

$$\text{Effective length of buswire } (A - B) + (C - D) = 80 \text{ ft (24.4 m)}$$

$$R_{\text{bus}} = \frac{80 \times 6.38}{1000} = 0.51\ \Omega$$

$$R_{\text{firing line}} = \frac{2 \times 1000 \times 1.59}{1000} = 3.18\ \Omega$$

$$R_{\text{total}} = 2.20 + 0.51 + 3.18 = 5.89\ \Omega$$

$$\text{Minimum current per cap} = 1.5 \text{ A} \quad \text{(Table 6-5)}$$

$$\text{Minimum current per series} = 1.5 \text{ A}$$

$$\text{Required current in firing line} = 3 \times 1.5 = 4.5 \text{ A}$$

$$\text{Minimum voltage for power source} = 4.5 \times 5.89 = 26.5 \text{ V} \qquad (6\text{-}4)$$

The current leakage conditions indicated in Table 6-5 may exist when legwire insulation is damaged or bare connections are allowed to touch wet ground. To test for current leakage conditions, use a blasting ohmmeter to measure the resistance from one end of the circuit to a good ground (metal rod driven into wet earth). If resistance is less than 10,000 Ω, use the current-leakage conditions values of Table 6-5.

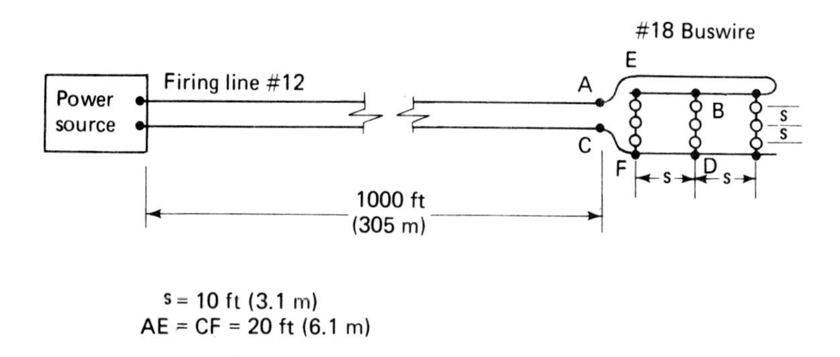

Figure 6-13 Circuit for Example 6-3.

Nonelectric Blasting Circuits

There has long been an interest in developing a blasting system which combines the safety of nonelectric explosives with the precise timing and flexibility of electric blasting systems. Several such nonelectric systems have been put on the market in recent years. Although a number of proprietary systems exist, most of these fall into one of the three categories described below. With all of these systems, two firing paths are normally used to ensure firing of all charges in the event of a break in the firing circuit.

The first and most conventional type of system utilizes detonating cord, either conventional or low energy, for the trunkline (main firing line), branch lines, and downlines (lines extending into individual blast holes). When using conventional detonating cord downlines, special primer assemblies that accept variable-delay detonator inserts are placed on the downline at one or several depths. Thus several charges having differing delays may be placed at different depths into a single blast hole. However, in this case, cap-insensitive explosives must be used to avoid premature detonation of charges by the detonating cord itself. When low-energy detonating cord is used for downlines, a separate downline with an in-hole delay detonator is used for each charge. Due to the low sensitivity of low-energy detonating cord, special starter detonators must be used to fire the cord and propagate the explosion between trunklines and downlines. Although detonating cord firing circuits are easy to check visually, no positive circuit check can be made prior to firing. In addition, such systems may not be compatible with all types of explosives.

A second type of system uses a special shock/signal tube in lieu of detonating cord to initiate in-hole delay detonators. The hollow plastic shock/signal tube used may be uncharged or may contain a small thread of explosive. The tubes permit sharp bends and kinks without misfiring. The trunkline may consist of conventional detonating cord or shock/signal tubes. Shock/signal tubes that incorporate delay detonators are used for downlines. Special connectors are required to connect shock/signal tubes to each other or to conventional detonating cord. The use of shock/signal tubes for both trunklines and downlines results in a noiseless initiation system but leaves a residue (plastic tubing) that must be removed after the blast. Again, no positive circuit check is possible prior to firing.

The third type of system utilizes a firing circuit (trunklines and downlines) made up of inert plastic tubing, manifolds, tees, and connectors. Immediately prior to use, the system is charged with an explosive gas mixture. The gas mixture is then ignited to initiate in-hole delay detonators or conventional detonating cord. The firing circuit may be checked prior to firing using a pressure meter in a manner similar to the blasting galvanometer check of an electric blasting circuit. This is the only nonelectric blasting system that permits such a positive circuit check prior to firing. However, the requirement for specialized tubing, fittings, and test equipment, together with the use of hazardous gases makes the system more complex than other nonelectric blasting systems. Plastic tubing residue must be removed after the blast.

Controlled and Secondary Blasting

Secondary blasting is blasting used to break up boulders and oversized fragments resulting from primary blasting. The two principal methods of secondary blasting are block holing and mud capping. *Block holing* utilizes conventional drilling and blasting techniques to further fragment the rock. A hole is drilled to the center of the rock, an explosive charge is inserted, the hole is stemmed, and the charge is detonated. *Mud capping* utilizes an explosive charge placed on the surface of the rock and tamped with an inert material such as mud (wet clay). While mud capping requires 10 to 15 times as much explosive as block holing, the time and expense of drilling are eliminated.

Nonexplosive techniques for breaking up boulders and oversized rock fragments include rock splitting and the use of percussion hammers. Rock splitters are hydraulically powered devices which are expanded inside a drilled hole to shatter the rock. Percussion hammers available for fragmenting rock include pneumatic handheld paving breakers and larger hydraulically powered units that may be attached to backhoes or other machines. Mechanical methods of rock breaking eliminate the safety hazards and liability problems associated with blasting.

Controlled blasting is utilized to reduce disturbance to nearby structures, to reduce rock throw, to obtain a desired fracture line, or to reduce the cost of overbreak (fracture of rock beyond the line required for excavation). The principal controlled blasting technique is called *presplitting*. Presplitting involves drilling a line of closely spaced holes, loading the holes with light charges evenly distributed along the hole depth, and exploding these charges before loading and shooting the main blast. Typical presplitting procedures involve 2½- to 3-in. (5.4- to 7.6-cm)-diameter holes spaced 18 to 30 in. (45.7 to 76.2 cm) apart. Detonation of the presplitting charges results in a narrow fracture line that serves to reflect shock waves from the main blast and leaves a relatively smooth surface for the final excavation.

A related technique for producing a smooth fracture line is called *line drilling*. In this technique a single row of closely spaced unloaded holes [4 to 12 in. (10 to 30 cm) on center] is drilled along the desired fracture line. In preparing the main blast, blastholes adjacent to the line drilling holes are moved close together and loaded lighter than usual. The line drilling technique normally produces less disturbance to adjacent structures than does presplitting. However, drilling costs for line drilling are high and the results can be unpredictable except in very homogeneous rock formations.

Blasting Safety

Blasting is a dangerous procedure that is controlled by a number of governmental regulations. Following are a few of the major safety precautions that should be observed.

- Storage magazines for explosives should be located in isolated areas, properly marked, sturdily constructed, and resistant to theft. Detonators (caps) must be stored in separate containers and magazines from other explosives.
- Electrical blasting circuits should not be connected to the power source and should be kept short-circuited (except for testing) until ready to fire.
- Permit no radio transmission in the vicinity of electric blasting circuits and discontinue work if there is evidence of an approaching electrical storm.
- Protect detonators and primed charges in the work area from all physical harm.
- Check blastholes before loading, because hot rock or a piece of hot drill steel in the hole can cause an explosion.
- Do not drop or tamp primed charges or drop other charges directly on them.
- Use only nonmetallic tools for tamping.
- Employ simple, clear signals for blasting and ensure that all persons in the work area are familiar with the signals.
- Make sure that the blasting area is clear before a blast is fired.
- Misfires are particularly dangerous. Wait at least 1 hour before investigating a misfire. Allow only well-trained personnel to investigate and dispose of misfires.

6-4 ROCK RIPPING

Employment of Rippers

Rippers have been utilized since ancient times to break up hard soils. However, only since the advent of the heavy-duty tractor-mounted ripper has it become feasible to rip rock. The availability of powerful heavy tractors such as the one shown in Figure 6-14 now make it possible to rip all but the toughest rock. Where ripping can be satisfactorily employed, it is usually cheaper than drilling and blasting. Additional advantages of ripping over drilling and blasting include increased production, fewer safety hazards, and reduced insurance cost.

Ripping Equipment

Although other types of rippers are available, most modern rippers are of the adjustable parallelogram type, shown in Figure 6-15. This type of ripper maintains a constant angle with the ground as it is raised and lowered. However, the upper hydraulic cylinder permits the tip angle to be varied as desired to obtain optimum results. The tip angle that produces the best surface penetration is usually different

Figure 6-14 Heavy duty crawler-mounted ripper. (Courtesy of Caterpillar Inc.)

from the tip angle that produces optimum rock breakage after penetration. Automatic ripper control systems are available that control ripping depth and automatically vary tip angle as the ripper is raised or lowered.

Impact rippers utilize a hydraulic mechanism to impart a hammering action to a single shank ripper. As a result, impact rippers are able to effectively rip tougher rock than can conventional rippers and usually produce a significant increase in ripper production. Some typical values for the increased performance provided by impact rippers include a 5 to 15% increase in the maximum seismic velocity for rippability and a 10 to 45% increase in hourly ripper production.

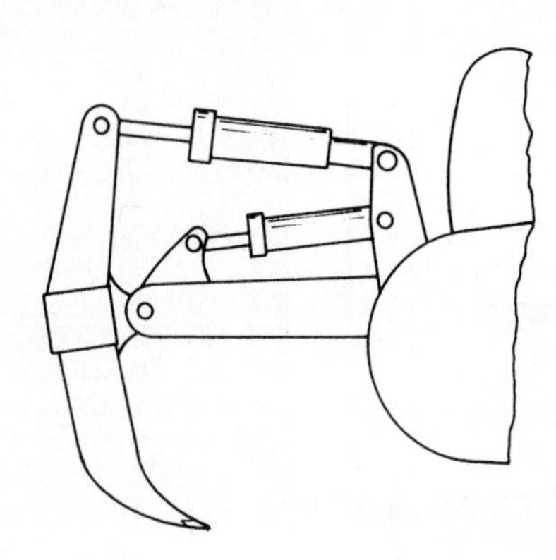

Figure 6-15 Adjustable parallelogram ripper.

Ripper shanks and tips are available in several different styles and in a variety of lengths, for use in different types of material. Shank protectors, which fit on the front of the ripper shank immediately above the tip, are used to reduce wear on the shank itself. Both tips and shank protectors are designed to be easily replaced when necessary.

Ripper Production

The seismic velocity of a rock formation (Section 6-1) provides a good indication of the rock's ripability. Charts such as the one shown in Figure 6-16 have been prepared to provide a guide to the ripping ability of a particular tractor/ripper combination in various types of rock over a range of seismic velocities. When ripping conditions are marginal, the use of two tractors to power the ripper (tandem ripping) will often produce a substantial increase in production and reduce unit excavation cost.

Equation 6-6 may be used to predict ripper production when effective ripping width, depth, and speed can be established. Trial operations are usually required to

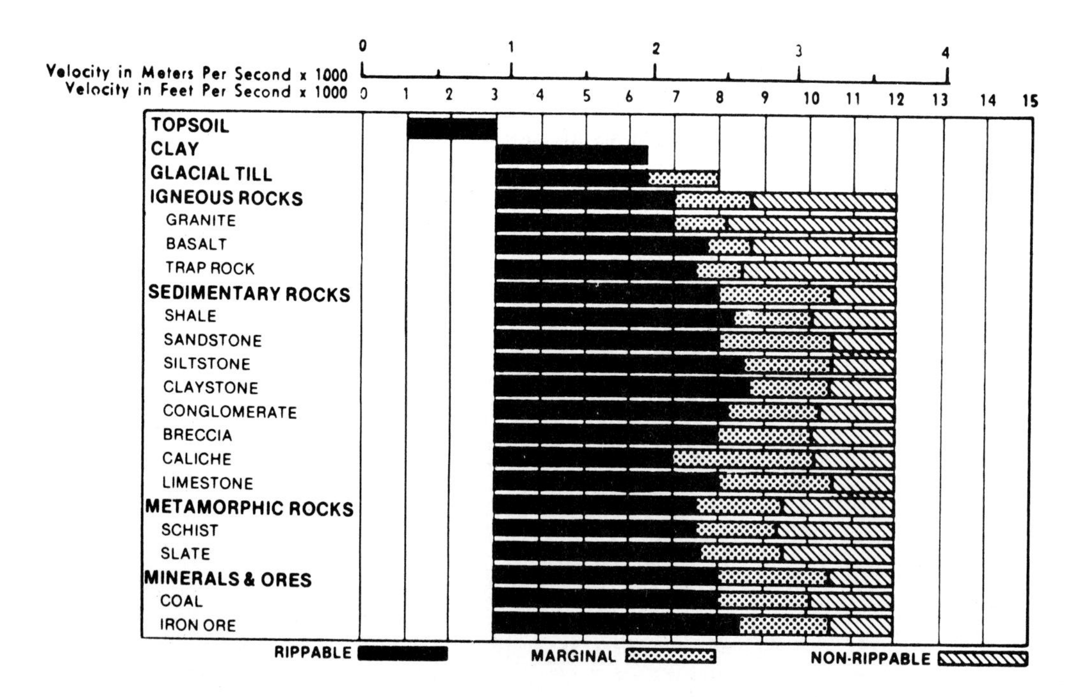

Figure 6-16 Ripper performance versus seismic velocity. (Courtesy of Caterpillar Inc.)

accurately estimate these values unless such data are available from previous operations under similar conditions.

$$\text{Production (BCY/h)} = \frac{2.22 \times D \times W \times L \times E}{T} \tag{6-6A}$$

$$\text{Production (m}^3\text{/h)} = \frac{60 \times D \times W \times L \times E}{T} \tag{6-6B}$$

where D = average penetration (ft or m)
W = average width loosened (ft or m)
L = length of pass (ft or m)
E = job efficiency factor
T = time for one ripper pass, including turn (min)

Considerations in Ripping

Ripping speed and depth, spacing of ripper passes, and the number of shanks to be used for maximum ripper production depend on rock type and soundness, and tractor power. The presence and inclination of laminations will also affect ripping procedures. In general, rip downhill to take advantage of gravity. However, when ripping laminated material, it may be necessary to rip uphill to enable the ripper teeth to penetrate under the layers. The number of shanks to be used in a ripper should be the number that yields the desired penetration without straining the tractor. Depth of ripping will depend on the number of shanks used and the tractor power. In stratified material try to match ripping depth to layer thickness. Ripping speed should be kept low to reduce wear on ripper teeth and shanks. First gear is usually used when ripping tough materials. The spacing of ripper passes will depend on rock hardness and the degree of fracture desired.

When loading ripped material, always leave a thin layer of material on the surface to provide a good working surface for the tractor. When ripping extremely hard rock, production may be increased and unit cost lowered if the rock is loosened by light blasting before ripping.

6-5 ESTIMATING PRODUCTION AND COST

The procedure for estimating rock excavation production and cost for a typical project is illustrated by Examples 6-4 and 6-5. Example 6-4 employs conventional drilling and blasting procedures for rock loosening, while Example 6-5 employs a tractor-mounted ripper.

Example 6-4

PROBLEM Estimate the hourly production and the unit cost of the rock excavation involved in preparing an industrial building site by drilling and blasting. The site is 300 ft (91.4 m) by 400 ft (121.9 m) and must be excavated an average depth of 12 ft (3.658 m). The material to be excavated is a thinly laminated shale with a sonic velocity of 4000 ft/s (1220 m/s). The drilling equipment to be used will consist of an air-powered track drill and air compressor.

The average drilling rate, including steel changes, moves, and delays, is estimated at 100 ft/h (30.5 m/h).

Trial blasting indicates that 3-in. (7.6-cm) holes drilled in a 12-ft (3.658-m) rectangular pattern will provide adequate fracturing. A hole depth of 13.5 ft (4.115 m) must be drilled to yield a 12-ft (3.658-m) effective depth. The blasting agent is ANFO. One-half pound (0.23 kg) of primer with an electric blasting cap will be used in each hole. The powder factor is 0.5 lb/BCY (0.297 kg/Bm^3).

A labor force of one drill operator and one compressor operator will be used for drilling. One blaster and one helper will be employed in blasting.

Cost information:

Labor:	Blaster	= $18.00/h
	Helper	= $15.00/h
	Drill operator	= $17.50/h
	Compressor operator	= $18.00/h
Equipment:	Track drill and compressor	= $63.00/h
Material:	ANFO	= $0.32/lb ($0.705/kg)
	Primer, cap, and stemming	= $3.00/hole

SOLUTION

(a) Production

$$\text{Total volume} = \frac{300 \times 400 \times 12}{27} = 53{,}333 \text{ BCY}$$

$$[= 91.4 \times 121.9 \times 3.66 = 40\,778 \text{ Bm}^3]$$

$$\text{Yield} = \frac{12^2 \times 12}{27} = 64.0 \text{ BCY/hole}$$

$$[= (3.658)^2 \times 3.658 = 48.9 \text{ Bm}^3\text{/hole}]$$

$$\text{Drilling yield} = \frac{64.0}{13.5} = 4.74 \text{ BCY/ft}$$

$$\left[= \frac{48.9}{4.115} = 11.88 \text{ Bm}^3\text{/m}\right]$$

$$\text{Production} = 4.74 \text{ BCY/ft} \times 100 \text{ ft/h} = 474 \text{ BCY/h}$$

$$[= 11.88 \times 30.5 = 362.3 \text{ Bm}^3\text{/h}]$$

$$\text{Time required} = \frac{53{,}333 \text{ BCY}}{474 \text{ BCY/h}} = 112.5 \text{ h}$$

$$\left[= \frac{40\,778}{362.3} = 112.5 \text{ h}\right]$$

(b) Drilling cost

$$\text{Labor} = \$35.50\text{/h}$$

$$\text{Equipment} = \$63.00\text{/h}$$

$$\text{Drilling cost/volume} = \frac{\$98.50\text{/h}}{474 \text{ BCY/h}} = \$0.208\text{/BCY}$$

$$\left[= \frac{\$98.50}{362.3} = \$0.272/\text{Bm}^3\right]$$

(c) Blasting cost

Material:

$$\text{ANFO} = 0.5\ \text{lb/BCY} \times 64\ \text{BCY} \times \$0.32/\text{lb} = \$10.24/\text{hole}$$
$$[= 0.297 \times 48.9 \times \$0.705 = \$10.24/\text{hole}]$$
$$\text{Primer, cap, and stemming} = \$3.00/\text{hole}$$
$$\frac{\$13.24}{64\ \text{BCY}} = \$0.207/\text{BCY}$$
$$\left[\frac{\$13.24}{48.9} = \$0.271/\text{Bm}^3\right]$$

Labor:

$$\text{Unit cost} = \frac{\$33.00/\text{h}}{474\ \text{BCY/h}} = \$0.070/\text{BCY}$$
$$\left[= \frac{\$33.00}{362.3} = \$0.091/\text{Bm}^3\right]$$
$$\text{Total blasting} = \$0.207 + \$0.070 = \$0.277/\text{BCY}$$
$$[= \$0.271 + \$0.091 = \$0.362/\text{Bm}^3]$$

(d) $\text{Total cost} = \$0.208 + \$0.277 = \$0.485/\text{BCY}$
$[= \$0.272 + \$0.362 = \$0.634/\text{Bm}^3]$

Example 6-5

PROBLEM Estimate the hourly production and the unit cost of rock excavation by ripping for the problem of Example 6-4. Field tests indicate that a D7G dozer with ripper can obtain satisfactory rock fracturing to a depth of 27 in. (0.686 m) with two passes of a single ripper shank at 3-ft (0.914-m) intervals. Average speed, including turns, is estimated at 82 ft/min (25 m/min).

Cost information:

Labor (operator) = \$20.00/h
Equipment (D7G with ripper, including ripper tips, shanks, and shank protectors) = \$75.00/h

SOLUTION

(a) Production

$$\text{Volume} = 53{,}333\ \text{BCY}\ (40\,778\ \text{Bm}^3)$$
$$\text{Production} = \frac{27\ \text{in.} \times 3\ \text{ft} \times 82\ \text{ft/min} \times 50\ \text{min/h}}{2\ \text{passes} \times 12\ \text{in./ft} \times 27\ \text{cu ft/cu yd}} = 512\ \text{BCY/h}$$
$$\left[= \frac{0.686 \times 0.914 \times 25 \times 50}{2} = 392\ \text{Bm}^3/\text{h}\right]$$
$$\text{Time required} = \frac{53{,}333\ \text{BCY}}{512\ \text{BCY/h}} = 104\ \text{h}$$
$$\left[= \frac{40778}{392} = 104\ \text{h}\right]$$

(b) Cost

$$\text{Labor} = \$20.00/\text{h}$$
$$\text{Equipment} = \$75.00/\text{h}$$
$$\text{Unit cost} = \frac{\$95.00/\text{h}}{512 \text{ BCY/h}} = \$0.186/\text{BCY}$$
$$\left[= \frac{\$95.00}{392} = \$0.242/\text{Bm}^3 \right]$$

PROBLEMS

1. Given the refraction seismograph test data in the accompanying table, calculate the seismic wave velocity in each soil layer. Find the depth of the upper layer.

Distance		*Time*
ft	*m*	*(ms)*
10	3.05	2.5
20	6.10	5.0
30	9.15	7.5
40	12.20	8.8
50	15.25	10.0
60	18.30	11.3
70	21.35	12.5

2. Using the seismograph test data in the accompanying table, find the seismic wave velocity and the depth of the upper layer.

Distance		*Time*
ft	*m*	*(ms)*
10	3.05	10.0
20	6.10	20.0
30	9.15	23.3
40	12.20	26.7
50	15.25	30.0
60	18.30	33.3
70	21.35	35.0
80	24.40	36.7

3. Trial operations indicate that a rectangular pattern with holes 21 ft (6.4 m) deep spaced on 10-ft (3.05-m) centers will yield a satisfactory rock break with an effective depth of 19 ft (5.8 m). Determine the rock volume produced per foot (meter) of drilling.

4. A parallel series electric blasting circuit consists of nine branches of five caps each. Holes are spaced 15 ft (4.6 m) apart in a rectangular pattern, resulting in a total pattern length of 120 ft (36.6 m). Cap resistance is 2.4 ohms per cap. A No. 20 gauge reverse hookup bus having a total length of 360 ft (110 m) is used. The lead wires are No. 14 gauge with a one-way length of 750 ft (229 m). Find the minimum dc current and voltage required to safely fire this blast under current leakage conditions.

5. A parallel–series electric blasting circuit consists of five branches of 10 caps each. Holes are spaced 10 ft (3.1 m) apart in a rectangular pattern. Cap legwire length is 24 ft (7.3 m). Each side of the cap circuit is connected by a No. 16 gauge bus wire 40 ft (12.2 m) long. The lead wires are No. 12 gauge 1200 ft (336 m) long one-way. Find the minimum dc current and voltage required to fire this blast safely under normal conditions.

6. A tractor-mounted ripper will be used for excavating a limestone having a seismic velocity of 6000 ft/s (1830 m/s). Field tests indicate that the ripper can obtain satisfactory rock fracturing to a depth of 2 ft (0.61 m) with one pass of a single ripper shank at 3 ft (0.91 m) intervals. Average ripping speed for each 300 ft (91.4 m) pass is 1 mi/h (1.6 km/h). Maneuver and turn time at the end of each pass averages 0.25 min. Job efficiency is estimated at 0.75. Machine cost including the operator is $124/h. Estimate the hourly production and unit cost of excavation.

7. You measure a seismic velocity of 9000 ft/s (2743 m/s) in shale. Would you expect this rock to be rippable by a D9H tractor equipped with a ripper (Figure 6-16)? If so, would you recommend using a single, tandem, or impact ripper in this situation? Explain your recommendation.

8. Estimate the hourly production and unit cost of rock excavation by drilling and blasting. The rock is a limestone having a seismic velocity of 6000 ft/s (1830 m/s). Trial blasting indicates that 3½-in. (8.9-cm) holes drilled in an 8-ft (2.44-m) rectangular pattern will provide the desired fracturing. A hole depth of 20 ft (6.1 m) yields an effective depth of 18 ft (5.5 m). The average drilling rate is estimated at 70 ft/h (21.4 m/h). A powder factor of 1 lb/BCY (0.59 kg/Bm^3) of ANFO will be used.

Cost information:

Labor:	Drilling crew	$35.00/h
	Blasting crew	$33.00/h
	Drilling equipment	$60.00/h
Material:	ANFO	$0.32/lb ($0.705/kg)
	Primer, caps, and stemming	$3.50/hole

9. Write a computer program to estimate hourly production and unit cost of rock excavation by ripping. Input should include length of each pass, average ripper speed, effective depth of ripping, spacing of passes, turning time at the end of each pass, job efficiency, hourly operator cost, and hourly machine cost, including ripper tips, shanks, and shank protectors. Solve Problem 6 using your computer program.

10. Write a computer program to estimate hourly production and unit cost of rock excavation by drilling and blasting. Input should include hole spacing, drilled hole depth, effective depth, average drilling rate including job efficiency, hourly labor and equipment cost, powder factor, unit cost of primary explosive, and cost per hole for primers, caps, and stemming. Solve Problem 8 using your computer program.

REFERENCES

1. *Blasters Handbook.* E. I. du Pont de Nemours and Co. (Inc.), Wilmington, Del., 1966.
2. CHURCH, H. K. *Excavation Handbook.* New York: McGraw-Hill, 1980.
3. *Crawair Drill Production Calculator.* Ingersoll-Rand Company, Phillipsburg, N. J., 1977.
4. *Handbook of Electric Blasting.* Atlas Powder Company, Dallas, Tex., 1985.
5. *Handbook of Ripping,* 7th ed. Caterpillar Inc., Peoria, Ill., 1983.
6. PUGLIESE, J. M. *Designing Blast Patterns Using Empirical Formulas* (Bureau of Mines Information Circular No. 8550). Washington, D. C.: U.S. Government Printing Office, 1972.
7. ROLLINS, JOHN P., ed. *Compressed Air and Gas,* 5th ed. Englewood Cliffs, N. J.: Prentice Hall, 1988.

7
Compressed Air and Water Systems

7-1 INTRODUCTION

Construction Applications

Compressed air is widely used as a power source for construction tools and equipment. While hydraulic power is gradually replacing compressed air as the power source for rock drills (see Chapter 6), compressed air is still required for cleaning out the drill hole produced by a hydraulic drill. Some of the other uses for compressed air in construction include paint spraying, the pneumatic application of concrete (shotcrete), conveying cement, pumping water, and operating pneumatic tools. Common pneumatic construction tools include spaders (or trench diggers), concrete vibrators, drills (steel and wood), grinders, hammers, paving breakers, sandblasting guns, saws (circular, chain, and reciprocating), staple guns, tampers, and wrenches.

Pumps and water supply systems are utilized in construction to dewater excavations and to supply water for cleaning equipment and aggregates, for mixing and curing concrete, for aiding soil compaction, and for jetting piles into place.

Construction Manager's Responsibilities

The construction manager must be able to select the appropriate type and size of air compressor or pump for a construction operation and to design the associated air or water supply system. Sections 7-2 and 7-3 provide guidance in performing these tasks.

7-2
COMPRESSED AIR SYSTEMS

Types of Compressors

Air compressors may be classified as positive displacement compressors or dynamic compressors according to the method by which they compress air. *Positive displacement compressors* achieve compression by reducing the air volume within a confined space. Positive displacement compressor types include reciprocating compressors, rotary vane compressors, and rotary screw compressors. *Dynamic compressors* achieve compression by using fans or impellers to increase air velocity and pressure. The principal type of dynamic compressor used in construction is the centrifugal compressor. Rotary compressors, both positive displacement and dynamic, are smaller, lighter, and quieter than are reciprocating compressors of similar capacity. As a result, most compressors used in construction are rotary compressors.

A schematic diagram of a rotary vane air compressor system is shown in Figure 7-1. This compressor is classified as a two-stage, oil-flooded, sliding-vane rotary compressor. Oil is injected into each compressor stage for lubrication and cooling. An *oil separator* removes the oil from the output air. The oil is then cooled and returned for reuse. The output of the compressor's first stage is cooled by an *intercooler* to increase the efficiency of the second-stage compressor. The *receiver* serves as a compressed air reservoir, provides additional cooling of the air leaving the compressor, reduces pressure fluctuations in the output, and permits water to settle out of the compressed air. Compressors can sometimes operate satisfactorily without a receiver in the system.

The principle of operation of a sliding-vane rotary compressor is as follows (refer to Figure 7-1). As the compressor rotor turns, centrifugal force causes the vanes to maintain contact with the cylinder. Air intake occurs while the volume of air trapped between two adjacent vanes is increasing, creating a partial vacuum. As rotation continues, the volume of air trapped between adjacent vanes decreases, compressing the trapped air. The compressed air is exhausted on the opposite side of the cylinder as the volume trapped between vanes approaches a minimum.

Compressors are available as portable units (skid- or wheel-mounted) or stationary units. Although portable units are most often used in construction work, stationary compressors may be employed in quarries and similar permanent installations. Portable units are available in capacities from 75 to over 2000 cu ft/min (2.1 to 56.6 m^3/min). Figure 7-2 shows a small portable air compressor being used to power two pneumatic paving breakers. Since pneumatic tools normally require air at 90 psig (lb/sq in. gauge) (621 kPa) to deliver rated performance, compressors usually operate in the pressure range 90 to 125 psig (621 to 862 kPa).

A rotary screw or helical screw air compressor utilizes two mating rotating helical rotors to achieve compression (see Figure 7-3). The main or male rotor is driven by the power source. The mating gate or female rotor is usually driven by timing gears attached to the main rotor but may be driven directly by the main rotor in an oil-flooded unit. The principle of operation is as follows. Air enters at the inlet end,

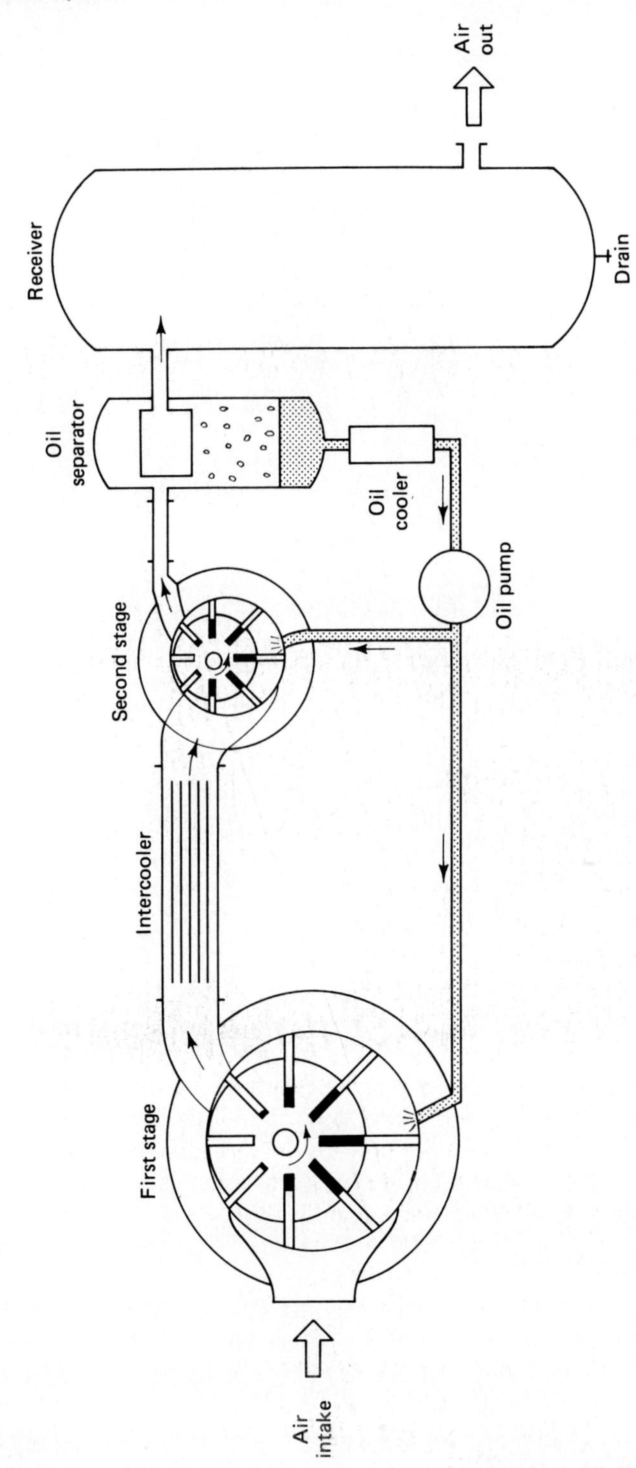

Figure 7-1 Schematic of sliding-vane rotary air compressor.

Figure 7-2 Small portable air compressor powering pneumatic tools. (Courtesy of Ingersoll-Rand Company)

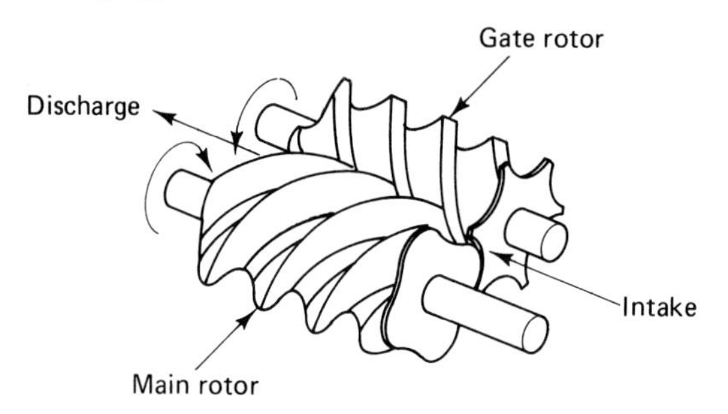

Figure 7-3 Rotary screw air compressor.

where the volume between mating lobes is large. As the rotors turn, the trapped volume becomes smaller and moves toward the discharge end. The trapped volume reaches a minimum as it lines up with the discharge port and the compressed air is exhausted. The advantages of rotary screw compressors include high efficiency, few moving parts, low maintenance, and long life.

Required Compressor Capacity

Air compressor ratings indicate capacity as the volume of "standard" air or "free" air at standard conditions which the compressor will deliver at a specified discharge pressure. Standard air is defined as air at a temperature of 68°F (20°C), a pressure of 14.7 psia (lb/sq in. absolute) (100 kPa absolute), and 36% relative humidity. Since atmospheric conditions at a construction site rarely correspond exactly to standard air conditions, it may be necessary to adjust the rated compressor capacity to correspond to actual conditions. Since air is less dense at altitudes above sea level, altitude has the most pronounced effect on compressor capacity. A method for adjusting actual air demand at an altitude above sea level to standard conditions is explained later.

The quantity of compressed air required to supply a construction site is found by summing up the air demand of all individual tools and equipment. Representative air consumption values for common pneumatic construction equipment are given in Table 7-1. However, since all tools of a particular type will seldom operate simultaneously, a tool load factor or diversity factor should be applied to the total theoretical air requirement for each type of tool. Table 7-2 provides suggested values of the tool load factor to be used for various numbers of tools of the same type. Where a system is being designed to operate several different types of tools or equipment simultaneously, it may be appropriate to apply a second or job load factor to the sum

Table 7-1 Representative air consumption of pneumatic construction equipment

Type of Equipment	*Size*	*Air Consumption*	
		cu ft/min	*m³/min*
Drills, rock			
Hand-held	35-lb class	55– 75	1.6– 2.1
	45-lb class	50– 100	1.4– 2.8
	55-lb class	90– 110	2.5– 3.1
Drifter and wagon	Light (3½-in. bore)	190– 250	5.1– 7.1
	Medium (4-4½-in. bore)	225– 300	6.4– 8.5
	Heavy (5-in. bore)	350– 500	9.9–14.2
Track	Medium-weight	600–1000	17.0–28.3
	Heavyweight	900–1300	25.5–36.8
Paint sprayers		8– 15	0.2– 0.4
Paving breakers		40– 60	1.1– 1.7
Pumps, submersible	Low head	75– 100	2.1– 2.8
	High head	150– 300	4.2– 8.5
Shotcrete guns		200– 300	5.7– 8.5
Spaders/trench diggers		30– 50	0.8– 1.4
Tampers, backfill		35– 50	1.0– 1.4

Table 7-2 Suggested values of tool load factor for pneumatic construction tools

Number of Tools of Same Type	*Diversity Factor*
1– 6	1.00
7– 9	0.94
10–14	0.89
15–19	0.84
20–29	0.80
30 or more	0.77

of the adjusted air demand for the different types of tools. If used, the job load factor should be based on the probability that the various types of tools will operate simultaneously. Finally, an allowance should be made for leakage in the air supply system. It is customary to add 5 to 10% to the estimated air demand as a leakage loss.

After the total air consumption including leakage has been determined, total air demand must be adjusted for altitude before selecting the nominal or rated size of air compressor required to supply the system. The altitude adjustment factor may be calculated as the ratio of the compression ratio at the specified altitude to the compression ratio at sea level. These adjustment factors are listed in Table 7-3.

Table 7-3 Air compressor altitude-adjustment factors

Altitude		
ft	*m*	*Adjustment Factor*
1,000	305	1.03
2,000	610	1.06
3,000	915	1.10
4,000	1220	1.14
5,000	1525	1.17
6,000	1830	1.21
7,000	2135	1.25
8,000	2440	1.30
9,000	2745	1.34
10,000	3050	1.39
11,000	3355	1.44
12,000	3660	1.49
13,000	3965	1.55
14,000	4270	1.61
15,000	4575	1.67

The procedure for determining the rated capacity of the air compressor required for a project is illustrated in Example 7-1.

Example 7-1

PROBLEM Determine the rated size of air compressor required to operate the following tools at an altitude of 6000 ft (1830 m). Assume a 10% leakage loss and a job load factor of 0.80.

	Rated Consumption		
Item	*cu ft/min*	*m^3/min*	*Number*
Drifter drills	215	6.1	2
Hand-held drills	90	2.5	8
Trench diggers	30	0.8	4
Tampers	40	1.1	6
Submersible pumps	80	2.3	2

SOLUTION Use the tool diversity factors from Table 7-2.

Item	Tool Consumption cu ft/min	Tool Consumption m^3/min
Drifter = 2 × 215 × 1.00 =	430	12.2
Hand-held drills = 8 × 90 × 0.94 =	677	19.2
Trench diggers = 4 × 30 × 1.00 =	120	3.4
Tampers = 6 × 40 × 1.00 =	240	6.8
Pumps = 2 × 80 × 1.00 =	160	4.5
	1627	46.1

Applying the job diversity factor:

$$\begin{aligned}\text{Expected consumption} &= 1627 \times 0.80 = 1302 \text{ cu ft/min } (36.9 \text{ m}^3/\text{min})\\ \text{Leakage} &= 1302 \times 0.10 = 130 \text{ cu ft/min } (3.7 \text{ m}^3/\text{min})\\ \text{Total consumption} &= 1302 + 130 = 1432 \text{ cu ft/min } (40.6 \text{ m}^3/\text{min})\\ \text{Altitude adjustment factor} &= 1.21 \quad \text{(Table 7-3)}\\ \text{Minimum rated capacity} &= 1432 \times 1.21 = 1733 \text{ cu ft/min } (49.1 \text{ m}^3/\text{min})\end{aligned}$$

Friction Losses in Supply Systems

As compressed air travels through a supply system, the pressure of the air gradually drops as a result of friction between the air and the pipe, hose, and fittings. The pressure drop in a pipe is a function of air flow, pipe size, initial pressure, and pipe length. Table 7-4 indicates the pressure drop per 1000 ft (305 m) of clean, smooth pipe for various flows at an initial pressure of 100 psig (690 kPa). If the initial pressure is greater or less than 100 psig (690 kPa), multiply the value from Table 7-4 by the appropriate correction factor from Table 7-5. Notice that the pressure loss due to friction decreases as initial pressure increases.

The pressure drop due to pipe fittings is most conveniently calculated by converting each fitting to an equivalent length of straight pipe of the same nominal diameter. Figure 7-4 provides a nomograph for finding the equivalent length of common pipe fittings. The equivalent length of all fittings is added to the actual length of straight pipe to obtain the total effective length of pipe. This pipe length is then used to calculate pressure drop in the pipe and fittings.

Pressure drop in hoses is calculated in the same manner as pressure drop in pipe except that values from Table 7-6 are used. Table 7-6 indicates the pressure drop per 50-ft (15.3-m) length of hose at an initial pressure of 100 psig (690 kPa). The pressure drop in manifolds or other special fittings should be based on manufacturers' data or actual measurements.

The procedure for calculating total pressure drop from receiver to individual tool is illustrated in Example 7-2.

Table 7-4 Pressure drop due to pipe friction*

Flow		Nominal Diameter [in. (cm)]					
cu ft/min	m^3/min	½ (1.3)	¾ (1.9)	1 (2.5)	1¼ (3.2)	1½ (3.8)	2 (5.1)
10	0.3	5.5 (38)	1.0 (6.9)	0.3 (2.1)	0.1 (0.7)		
20	0.6	25.9 (178)	3.9 (27)	1.1 (7.6)	0.3 (2.1)	0.1 (0.7)	
30	0.8	58.5 (403)	9.0 (62)	2.5 (17)	0.6 (4.1)	0.3 (2.1)	
40	1.1		16.1 (111)	4.5 (31)	1.0 (6.9)	0.5 (3.4)	
50	1.4		25.1 (173)	7.0 (48)	1.6 (11)	0.7 (4.8)	0.2 (1.4)
60	1.7		36.2 (250)	10.0 (69)	2.3 (16)	1.0 (6.9)	0.3 (2.1)
70	2.0		49.4 (341)	13.7 (94)	3.2 (22)	1.4 (9.7)	0.4 (2.8)
80	2.3		64.5 (445)	17.8 (123)	4.1 (28)	1.8 (12)	0.5 (3.4)
90	2.5			22.6 (159)	5.2 (36)	2.3 (16)	0.6 (4.1)
100	2.8			27.9 (192)	6.4 (45)	2.9 (20)	0.8 (5.5)
125	3.5			48.6 (335)	10.2 (70)	4.5 (31)	1.2 (8.3)
150	4.2			62.8 (433)	14.6 (101)	6.4 (44)	1.7 (12)
175	5.0				19.8 (137)	8.7 (60)	2.4 (17)
200	5.7				25.9 (179)	11.5 (79)	3.1 (21)
250	7.1				40.4 (279)	17.9 (123)	4.8 (33)
300	8.5				58.2 (401)	25.9 (178)	6.9 (48)
350	9.9					35.1 (242)	9.4 (64)
400	11.3					45.8 (316)	12.1 (83)
450	12.7					58.0 (400)	15.4 (106)
500	14.2						19.2 (132)
600	17.0						27.6 (397)
700	19.8						37.7 (260)
800	22.6						49.0 (338)
900	25.5						60.0 (414)
1000	28.3						
1500	42.5						
2000	56.6						
2500	70.8						
3000	84.9						
4000	113						
5000	141						
10,000	283						
15,000	425						
20,000	566						
30,000	849						

*Psi (kPa) for 1000 ft (305 m) of pipe at 100 psig (690 kPa) initial pressure.

Table 7-4 *(continued)*

Nominal Diameter [in. (cm)]							
2½ (6.4)	3 (7.6)	4 (10.2)	5 (12.7)	6 (15.2)	8 (20.3)	10 (25.4)	12 (30.5)
0.1 (0.7)							
0.2 (1.4)							
0.2 (1.4)							
0.3 (2.1)							
0.5 (3.4)							
0.7 (4.8)	0.2 (1.4)						
0.9 (6.2)	0.3 (2.1)						
1.2 (8.3)	0.4 (2.8)						
1.8 (12)	0.6 (4.1)						
2.7 (18)	0.8 (5.5)						
3.6 (24)	1.1 (7.6)	0.3 (2.1)					
4.8 (33)	1.5 (10.3)	0.4 (2.8)					
6.0 (41)	1.9 (13)	0.5 (3.4)					
7.4 (51)	2.3 (16)	0.6 (4.1)					
10.7 (74)	3.4 (23)	0.8 (5.5)					
14.6 (101)	4.6 (32)	1.1 (7.6)	0.3 (2.1)				
19.0 (131)	6.0 (41)	1.4 (9.6)	0.4 (2.8)				
24.1 (166)	7.6 (52)	1.8 (12)	0.5 (3.4)				
29.7 (205)	9.4 (65)	2.2 (15)	0.7 (4.8)	0.2 (1.4)			
67.0 (462)	21.0 (145)	5.0 (34)	1.5 (10)	0.6 (4.1)			
	37.4 (258)	8.9 (61)	2.7 (19)	1.0 (6.9)	0.2 (1.4)		
	58.4 (403)	13.9 (96)	4.2 (29)	1.6 (11)	0.4 (2.8)		
		20.0 (138)	6.0 (41)	2.3 (16)	0.5 (3.4)		
		35.5 (245)	10.7 (74)	4.0 (28)	0.9 (6.2)	0.3 (2.1)	
		55.5 (383)	16.8 (116)	6.3 (43)	1.5 (10)	0.4 (2.8)	
			67.1 (463)	25.1 (173)	5.9 (41)	1.8 (12)	0.7 (4.8)
				56.7 (391)	13.2 (91)	4.0 (28)	1.5 (10)
					23.6 (163)	7.1 (49)	2.7 (19)
					52.1 (359)	15.9 (110)	6.2 (43)

Table 7-5 Correction factors for friction losses

Inlet Pressure		
psig	kPa (gauge)	Correction Factor
80	552	1.211
90	621	1.096
100	690	1.000
110	758	0.920
120	827	0.852
130	896	0.793

Table 7-6 Pressure drop due to hose friction* [psi (kPa) per 50 ft (15.3 m) of hose]

Flow		Hose Size [in. (cm)]				
cu ft/min	m³/min	½ (1.3)	¾ (1.9)	1 (2.5)	1¼ (3.2)	1½ (3.8)
20	0.6	1.8 (12)				
30	0.8	4.0 (28)				
40	1.1	6.8 (47)	1.0 (6.9)			
50	1.4	10.4 (72)	1.4 (9.7)			
60	1.7		2.0 (14)			
80	2.3		3.5 (24)			
100	2.8		5.2 (36)	1.2 (8.3)		
120	3.4		7.4 (51)	1.8 (12)		
140	4.0		9.9 (68)	2.4 (17)		
160	4.5		12.7 (88)	3.1 (21)		
180	5.1			3.8 (26)		
200	5.7			4.6 (32)	1.6 (11)	
220	6.2			5.5 (38)	1.9 (13)	
240	6.8			6.5 (45)	2.2 (15)	
250	7.1			7.0 (48)	2.4 (17)	
300	8.5			9.9 (68)	3.4 (23)	1.4 (9.7)
350	9.9			13.3 (92)	4.5 (31)	1.9 (13)
400	11.3				5.8 (40)	2.4 (17)
450	12.7				7.3 (50)	3.0 (21)
500	14.2				8.9 (61)	3.7 (26)
550	15.6				10.7 (74)	4.4 (30)
600	17.0				12.6 (87)	5.2 (36)

*Clean, dry air (no line lubricator), hose inlet pressure of 100 psig (690 kPa). Based on Ingersoll-Rand Company data.

Example 7-2

PROBLEM The compressed air system illustrated in Figure 7-5 is being operated with a receiver pressure of 110 psig (758 kPa). Pressure drop in the manifold is determined to be 2 psig (13.8 kPa). If all three drills are operated simultaneously, what is the pressure at the tools? Assume no line leakage.

A simple way to account for the resistance offered to flow by valves and fittings is to add to the length of pipe in the line a length which will give a pressure drop equal to that which occurs in the valves and fittings in the line. The chart on this page can be used to find the additional length which must be added for each resistance.

Example: The dotted line shows that the resistance of a 6-inch standard elbow is equivalent to approximately 16 feet of 6-inch standard pipe.

Note: For sudden enlargements or sudden contractions, use the smaller diameter, d, on the pipe size scale.

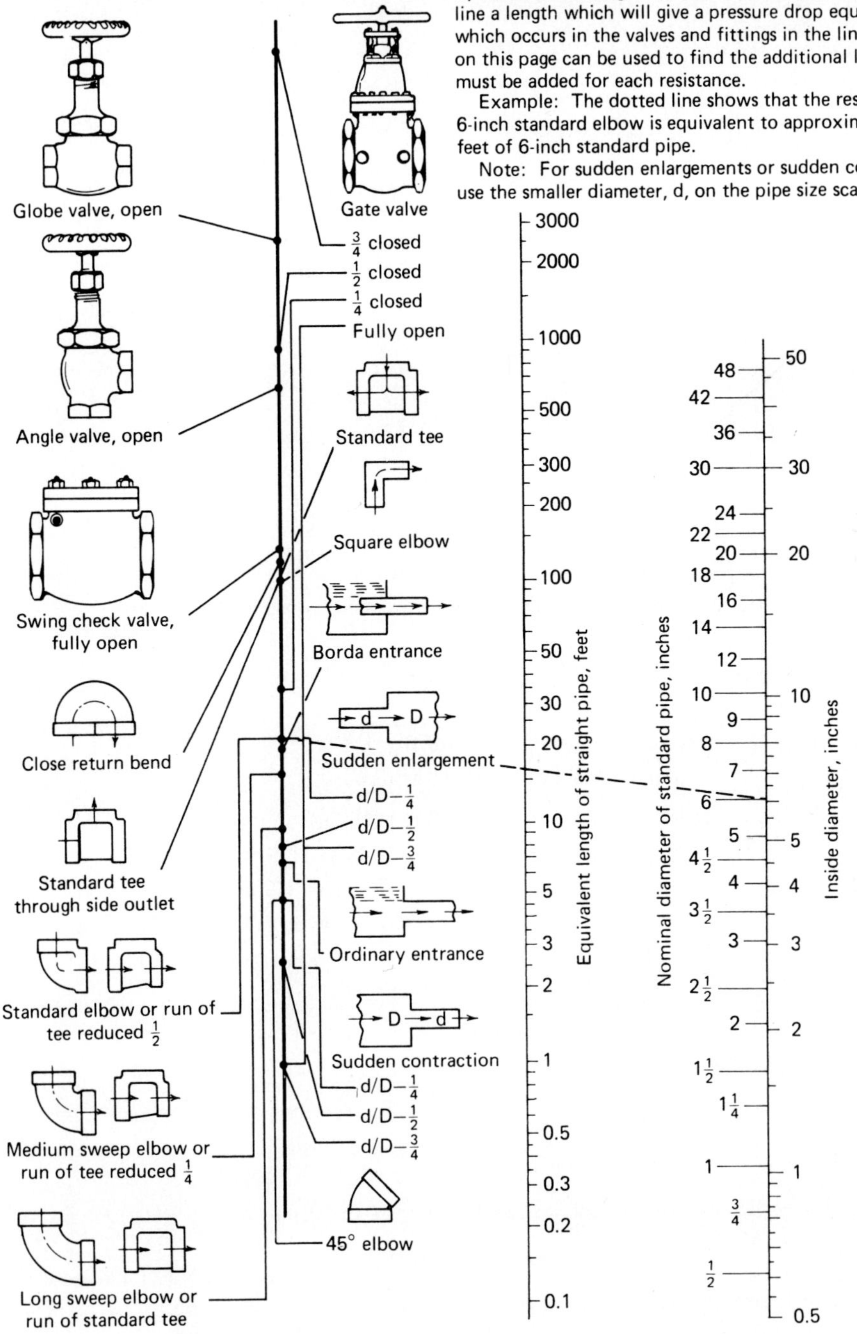

Figure 7-4 Resistance of valves and fittings to fluid flow. (Adapted from a monograph appearing in *Contractor's Pump Manual,* 1976. Copyright © Crane Company, by permission)

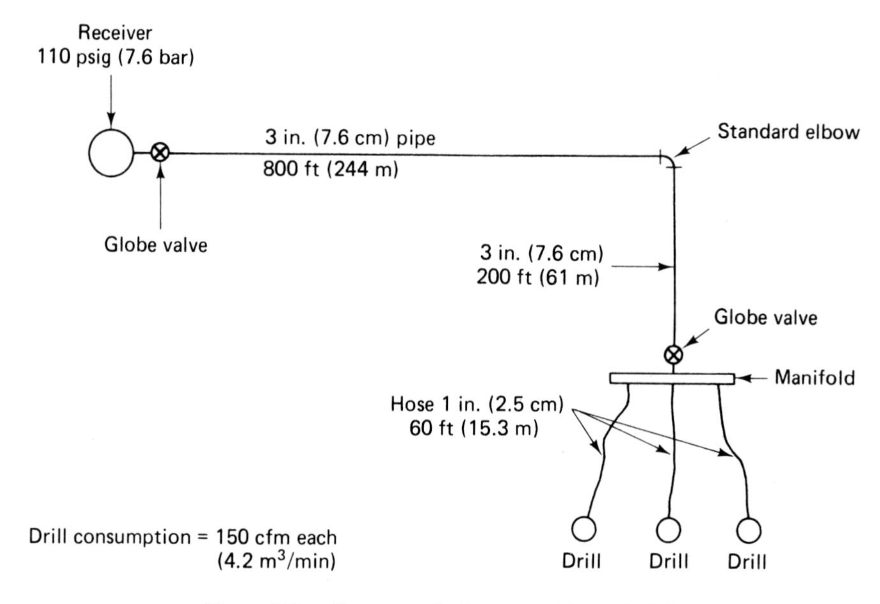

Figure 7-5 Compressed air system, Example 7-2.

SOLUTION Equivalent length of fittings (Figure 7-4):

$$\begin{aligned}
\text{2 globe valves at 85 ft} &= \text{170 ft (51.8 m)}\\
\text{1 standard ell at 8 ft} &= \underline{\text{8 ft (2.4 m)}}\\
&\ \text{178 ft (54.2 m)}\\
\text{Length of pipe plus fittings} &= 1000 + 178 = 1178\ \text{ft}\\
[&= 304.8 + 54.2 = 359\ \text{m}]\\
\text{Total flow} &= 3 \times 150 = 450\ \text{cu ft/min (12.7 m}^3\text{/min)}\\
\text{Pressure loss per 1000 ft (305 m)} &= \text{1.9 psig (13.1 kPa)}\quad \text{(Table 7-4)}\\
\text{Correction factor} &= 0.92\quad \text{(Table 7-5)}
\end{aligned}$$

Pressure loss in pipe and fittings (p_f):

$$P_f = \frac{1178}{1000} \times 1.9 \times 0.92 = 2.1\ \text{psig}$$

$$\left[= \frac{359}{305} \times 13.1 \times 0.92 = 14.2\ \text{kPa} \right]$$

Pressure at hose inlet (P_h):

$$P_h = 110 - (2.1 + 2) = 105.9\ \text{psig}$$

$$[= 758 - (14.2 + 13.8) = 730\ \text{kPa}]$$

Hose pressure correction factor = 0.95 (interpolate in Table 7-5)

Pressure loss in 60-ft (18.3-m) hose (p_f'):

$$p_f' = \frac{60}{50} \times 2.7 \times 0.95 = 3.1 \text{ psig}$$

$$\left[= \frac{18.3}{15.3} \times 18.6 \times 0.95 = 21.1 \text{ kPa} \right]$$

Pressure at tool (P_t):

$$P_t = 105.9 - 3.1 = 102.8 \text{ psig}$$

$$[= 730 - 21.1 = 708.9 \text{ kPa (gauge)}]$$

Compressed Air Costs

The cost of providing compressed air power may be calculated using the methods of Chapter 16. Air cost is usually expressed in dollars per 1000 cu ft (or per cubic meter). Typical production and distribution costs range from $0.10 to $0.25/1000 cu ft ($0.0036 to $0.0088/m^3).

Leaks in the air system can be costly. For example, a ⅛-in. (0.3-cm) hole in a 100-psig (690-kPa) supply line would waste about 740,000 cu ft (20 942 m^3) of air per month. At a cost of $0.20/1000 cu ft ($0.0071/m^3), this amounts to almost $150 per month.

7-3 WATER SUPPLY SYSTEMS

Principal Types of Pumps

The principal types of pumps by method of operation include displacement pumps and centrifugal (or dynamic) pumps. *Displacement pumps* include reciprocating pumps and diaphragm pumps. Although *reciprocating pumps* are not often used in construction operations, it is well to understand the terminology used for such pumps. Double-acting reciprocating pumps have a chamber on each end of the piston so that water is pumped as the piston moves in either direction; single-acting pumps move water only when the piston travels in one direction. A simplex pump has one cylinder, a duplex pump has two cylinders, and a triplex pump has three cylinders. Thus a single-acting duplex pump is a two-cylinder reciprocating pump in which pumping occurs during only one-half of the piston travel.

Diaphragm pumps utilize flexible circular disks or diaphragms and appropriate valves to pump water. As the diaphragm is pushed back and forth, the size of the pump chamber increases and decreases to produce the pumping action. Diaphragm pumps are self-priming, capable of pumping water containing a high percentage of sand or trash, and can handle large volumes of air along with water. Hence diaphragm pumps are widely used for dewatering excavations that contain large quantities of mud or trash or have an unsteady influx of water. Standard sizes of diaphragm pump include 2 in. (5.1 cm), 3 in. (7.6 cm), 4 in. (10.2 cm), and double 4 in. (10.2 cm). Pump size designates the nominal diameter of the intake and discharge openings. A gasoline-engine-powered diaphragm pump is shown in Figure 7-6.

Figure 7-6 Gas-powered diaphragm pump. (Courtesy of Peabody Barnes, Inc.)

Centrifugal pumps are available in a number of models and types. Conventional centrifugal pumps must have the impeller surrounded by water before they will operate. Self-priming centrifugal pumps utilize a water reservoir built into the pump housing to create sufficient pumping action to remove air from the suction line and fully prime the pump. Most centrifugal pumps utilized in construction are of the self-priming variety.

In the United States the Contractors Pump Bureau (CPB) has developed standards for self-priming centrifugal pumps, submersible pumps, and diaphragm pumps designed for construction service. Self-priming centrifugal pumps certified by the Contractors Pump Bureau include M-, MT-, and MTC-rated pumps. M-rated pumps are available in sizes of 1½ in. (3.8 cm) to 10 in. (25.4 cm), with capacities of 5000 to 200,000 gal (18 925 to 757 000 ℓ) per hour. A contractors 3-in. self-priming centrifugal pump is shown in Figure 7-7. M-rated pumps are required to pass spherical solids having a diameter equal to 25% of the nominal pump size and to handle up to 10% solids by volume. MT-rated or trash pumps are designed to handle up to 40% solids by volume. The maximum diameter of spherical solids that they can handle ranges from 1 in. (2.5 cm) for a 1½-in. (3.8-cm) pump to 2½ in. (6.4 cm) for a 6-in. (15.2-cm) pump. MTC-rated pumps are compact lightweight trash pumps designed for easy portability. Sizes range from 1½ in. (3.8 cm) to 4 in. (10.2 cm) with capacities of 5000 to 23,000 gal/h (18 925 to 83 220 ℓ/h). The maximum diameter of spherical particles which they must pass is 50% of the nominal pump size. A compact 2-in. centrifugal trash pump is shown in Figure 7-8.

Submersible pumps are centrifugal pumps designed to operate within the body of water which they are pumping. Submersible pumps may be powered by compressed air or electricity and are available in both low-head and high-head models. Trash-type submersible pumps are also available. Since submersible pumps operate submerged, suction lines and priming problems are eliminated and pump noise is reduced. An electrically powered submersible pump is illustrated in Figure 9-22.

Figure 7-7 Self-priming centrifugal pump at dam construction site. (Courtesy of Peabody Barnes, Inc.)

Figure 7-8 Compact 2-in. centrifugal trash pump. (Courtesy of Stow Manufacturing Co.)

Determining Required Head

In water supply systems, pressure is expressed as the equivalent height of a column of water (feet or meters). This unit of measure is called *head.* The total head that a pump must overcome is the sum of the static head (difference in elevation between two points) and the friction head (loss of pressure due to friction). For the usual pumping system, *static head* is equal to the difference in elevation between the surface of the source (*not* the elevation of the inlet) and the point of free discharge. If the outlet pipe is below the discharge surface, static head is the difference in elevation between the surface of the source and the surface of the discharge water.

Friction head is calculated in a manner similar to pressure loss in a compressed air system. That is, all fittings are converted to an equivalent pipe length using Figure 7-4. The length of pipe plus the equivalent length of fittings is then multiplied by the appropriate friction factor from Table 7-7. The friction loss for any hose in the system must be calculated separately by multiplying their length by the appropriate factor from Table 7-8. Total friction head is then found as the sum of pipe friction and hose friction. Static head is then added to obtain total head. The procedure is illustrated in Example 7-3.

Example 7-3

PROBLEM Water must be pumped from a pond to an open discharge 40 ft (12.2 m) above the pump. The line from the pump to discharge consists of 340 ft (103.7 m) of 4-in. (10.2-cm) pipe equipped with a check valve and three standard elbows. The pump is located 10 ft (3.1 m) above the pond water level. The intake line consists of a 20-ft (6.1-m) hose 4 in. (10.2 cm) in diameter equipped with a foot valve. Find the total head that the pump must produce for a flow of 280 gal/min (1060 ℓ min). The equivalent length of the foot valve is 70 ft (21.4 m).

SOLUTION Equivalent length of fittings (Figure 7-4):

$$
\begin{aligned}
\text{3 elbows at 11 ft (3.4 m)} &= \text{33 ft (10.1 m)}\\
\text{1 foot valve at 70 ft (21.4 m)} &= \text{70 ft (21.4 m)}\\
\text{1 check valve at 25 ft (7.6 m)} &= \underline{\text{25 ft (7.6 m)}}\\
\text{Total} &= \text{128 ft (39.1 m)}\\
\text{Length of straight pipe} &= \underline{\text{340 ft (103.7 m)}}\\
\text{Total equivalent pipe length} &= \text{468 ft (142.8 m)}
\end{aligned}
$$

$$h_f(\text{pipe}) = \frac{468}{100} \times 5.4 = 25.3 \text{ ft}$$

$$\left[= \frac{142.8}{100} \times 5.4 = 7.7 \text{ m} \right]$$

$$h_f(\text{hose}) = \frac{20}{100} \times 4.3 = 0.9 \text{ ft}$$

$$\left[= \frac{6.1}{100} \times 4.3 = 0.3 \text{ m} \right]$$

$$\text{Static head} = 40 + 10 = 50 \text{ ft (15.3 m)}$$

$$\text{Total head} = 50 + 25.3 + 0.9 = 76.2 \text{ ft}$$

$$[= 15.3 + 7.7 + 0.3 = 23.3 \text{ m}]$$

Pump Selection

The capacity of a centrifugal pump depends on the pump size and horsepower, the resistance of the system (total head), and the elevation of the pump above the source water level. After the total head and the height of the pump above water have been established, the minimum size of rated self-priming centrifugal pump that will provide the required capacity may be selected from the minimum capacity tables published by the Contractors Pump Bureau. The capacity table for M-rated pumps is reproduced in Table 7-9. Linear interpolation may be used to estimate capacity for values of total head and height of pump not shown in the table. The procedures for pump selection is illustrated in Example 7-4.

Example 7-4

PROBLEM Determine the minimum size of M-rated centrifugal pump required to pump 280 gal/min (1060 ℓ/min) through the system of Example 7-3.

SOLUTION

$$\text{Height of pump above source water} = 10 \text{ ft } (3.1 \text{ m})$$

$$\text{Required total head} = 76.2 \text{ ft } (23.3 \text{ m})$$

$$\text{Capacity of 40-M pump} = 535 - \left(\frac{76.2 - 70.0}{80 - 70}\right)(535 - 465)$$

$$= 535 - 43 = 492 \text{ gal/min}$$

$$\left[= 2025 - \left(\frac{23.3 - 21.3}{24.4 - 21.3}\right)(2025 - 1760) \right.$$

$$\left. = 2025 - 171 = 1854 \text{ ℓ/min} \right]$$

The 40-M pump is satisfactory.

Effect of Altitude and Temperature

The maximum lift (height of pump above the source water level) at which a centrifugal pump will theoretically operate is equal to the atmospheric pressure minus the vapor pressure of water at the prevailing temperature. At a temperature of 68 °F (20 °C), for example, the maximum theoretical lift at sea level is 33.1 ft (10.1 m). The maximum practical lift is somewhat less and is about 23 ft (7.0 m) at sea level at a temperature at 68 °F (20 °C). Figure 7-9 illustrates the effect of temperature and altitude on the maximum practical suction lift.

Special Types of Pumps

It is sometimes necessary to pump water in situations where the vertical distance between the source water level and the ground surface exceeds the maximum practical lift for conventional pumps. In such a situation submersible pumps are a logical choice. When construction-type submersible pumps cannot be used, deep-well submersible pumps capable of operating in wells up to 500 ft (152.5 m) deep are available.

Jet pumps and air-lift pumps are two other types of pumps capable of lifting water more than 33 ft (10 m). *Jet pumps* recirculate a portion of the pump output to

Table 7-7 Water friction (head) in pipe* (Adapted from Contractors Pump Bureau)

Flow		Pipe Size [in. (cm)]													
gal/min	ℓ/min	½ (1.3)	¾ (1.9)	1 (2.5)	1¼ (3.2)	1½ (3.8)	2 (5.1)	2½ (6.4)	3 (7.6)	4 (10.2)	5 (12.7)	6 (15.2)	8 (20.3)	10 (25.4)	12 (30.5)
2	8	4.8	1.2												
3	11	10.2	2.7	0.82											
4	15	17.4	4.5	1.39	0.37										
5	19	26.5	6.8	2.11	0.55										
10	38	95.0	24.7	7.61	1.98	0.93	0.31	0.11							
15	57		52.0	16.3	4.22	1.95	0.70	0.23							
20	76		88.0	27.3	7.21	3.38	1.18	0.40							
25	95			41.6	10.8	5.07	1.75	0.60	0.25						
30	114			57.8	15.3	7.15	2.45	0.84	0.35						
35	132			77.4	20.3	9.55	3.31	1.1	0.46						
40	151				26.0	12.2	4.29	1.4	0.59	0.14					
45	170				32.5	15.1	5.33	1.8	0.75	0.18					
50	189				39.0	18.5	6.43	2.2	0.90	0.22					
60	227				56.8	26.6	9.05	3.0	1.3	0.32					
70	265				73.5	35.1	11.9	4.0	1.7	0.41	0.14				
75	284					39.0	13.6	4.6	2.0	0.48	0.16				
80	303					44.8	15.4	5.0	2.3	0.58	0.18				
90	341					55.5	18.9	6.3	2.7	0.68	0.22				
100	379					66.3	23.3	7.8	3.2	0.79	0.27	0.09			
125	473						35.1	11.8	4.9	1.2	0.42	0.18			
150	578						49.4	16.6	6.8	1.7	0.57	0.21			
175	662						66.3	22.0	9.1	2.2	0.77	0.31			
200	757							28.0	11.6	2.9	0.96	0.40			
225	852							35.5	14.5	3.5	1.2	0.48			
250	946							43.0	17.7	4.4	1.5	0.60	0.15		

*Values are ft/100 ft or m/100 m.

Table 7-7 *(continued)*

Flow		Pipe Size [in. (cm)]													
gal/min	ℓ/min	½ (1.3)	¾ (1.9)	1 (2.5)	1¼ (3.2)	1½ (3.8)	2 (5.1)	2½ (6.4)	3 (7.6)	4 (10.2)	5 (12.7)	6 (15.2)	8 (20.3)	10 (25.4)	12 (30.5)
275	1 041								21.2	5.2	1.8	0.75	0.18		
300	1 136								24.7	6.1	2.0	0.84	0.21		
325	1 230								29.1	7.0	2.3	0.92	0.24		
350	1 325								33.8	8.0	2.7	1.0	0.27		
375	1 419									9.2	3.1	1.2	0.31		
400	1 514									10.4	3.5	1.4	0.35		
450	1 703									12.9	4.4	1.7	0.45	0.14	
500	1 893									15.6	5.3	2.2	0.53	0.18	0.08
600	2 271									22.4	6.2	3.1	0.74	0.25	0.10
700	2 650									30.4	9.9	4.1	1.0	0.34	0.14
800	3 028											5.2	1.3	0.44	0.18
900	3 407											6.6	1.6	0.54	0.22
1000	3 785											7.8	2.0	0.65	0.27
1100	4 164											9.3	2.3	0.78	0.32
1200	4 542											10.8	2.7	0.95	0.37
1300	4 921											12.7	3.1	1.1	0.42
1400	5 299											14.7	3.6	1.2	0.48
1500	5 678											16.8	4.1	1.4	0.55
1600	6 056												4.7	1.6	0.65
1800	6 813												5.6	2.0	0.78
2000	7 570												7.0	2.4	0.93
2500	9 463													3.5	1.5
3000	11 355													5.1	2.1
3500	13 248													6.5	2.7
4000	15 140														3.5
4500	17 033														4.5
5000	18 925														5.5

Table 7-8 Water friction (head) in hose* (Adapted from Contractors Pump Bureau)

Flow		Hose Size [in. (cm)]											
gal/min	ℓ/min	3/8 (1.0)	3/4 (1.9)	1 (2.5)	1¼ (3.2)	1½ (3.8)	2 (5.1)	2½ (6.4)	3 (7.6)	4 (10.2)	5 (12.7)	6 (15.2)	8 (20.3)
1.5	5.7	2.3	0.97										
2.5	9.5	6.0	2.5										
5	19	21.4	8.9	2.2	0.74	0.3							
10	38	76.8	31.8	7.8	2.64	1.0	0.2						
15	57		68.5	16.8	5.7	2.3	0.5						
20	76			28.7	9.6	3.9	0.9	0.32					
25	95			43.2	14.7	6.0	1.4	0.51					
30	114			61.2	20.7	8.5	2.0	0.70	0.3				
35	132			80.5	27.6	11.2	2.7	0.93	0.4				
40	151				35.0	14.3	3.5	1.2	0.5				
45	170				43.0	17.7	4.3	1.5	0.6				
50	189				52.7	21.8	5.2	1.8	0.7				
60	227				73.5	30.2	7.3	2.5	1.0				
70	265					40.4	9.8	3.3	1.3				
80	303					52.0	12.6	4.3	1.7				
90	341					64.2	15.7	5.3	2.1	0.5			
100	379					77.4	18.9	6.5	2.6	0.6			
125	473						28.6	9.8	4.0	0.9			
150	568						40.7	13.8	5.6	1.3			
175	662						53.4	18.1	7.4	1.8			
200	757						68.5	23.4	9.6	2.3	0.8	0.32	
225	852							29.0	11.9	2.9	1.0	0.40	
250	946							35.0	14.8	3.5	1.2	0.49	
275	1 041							42.0	17.2	4.2	1.4	0.58	
300	1 136							49.0	20.3	4.9	1.7	0.69	

*Values are ft/100 ft or m/100 m.

Table 7-8 *(continued)*

Flow		Hose Size [in. (cm)]											
gal/min	ℓ/min	3/8 (1.0)	3/4 (1.9)	1 (2.5)	1¼ (3.2)	1½ (3.8)	2 (5.1)	2½ (6.4)	3 (7.6)	4 (10.2)	5 (12.7)	6 (15.2)	8 (20.3)
325	1 230								23.5	5.7	2.0	0.80	
350	1 325								27.0	6.6	2.3	0.90	
375	1 419								30.7	7.4	2.6	1.0	
400	1 514									8.4	2.9	1.1	0.28
450	1 703									10.5	3.6	1.4	0.35
500	1 893									12.7	4.3	1.7	0.43
600	2 271									17.8	6.1	2.4	0.60
700	2 650									23.7	8.1	3.3	0.80
800	3 028										10.3	4.2	1.1
900	3 407										12.8	5.2	1.3
1000	3 785										15.6	6.4	1.6
1100	4 164										18.5	7.6	1.9
1200	4 542											9.2	2.3
1300	4 921											10.0	2.6
1400	5 299											11.9	3.0
1500	5 678											13.6	3.3
1600	6 056												3.7
1800	6 813												4.7
2000	7 570												5.7
2500	9 463												8.6
3000	11 355												12.2

Table 7-9 Minimum capacity [gal (ℓ)/min] of M-rated centrifugal pumps (Courtesy of Contractors Pump Bureau)

Model 5-M (1½-in.)

Total Head Including Friction	*Height of Pump Above Water [ft (m)]*			
	10 (3.0)	*15 (4.6)*	*20 (6.1)*	*25 (7.6)*
15 (4.6)	85 (321.7)	—	—	—
20 (6.1)	84 (317.9)	68 (257.4)	—	—
25 (7.6)	82 (310.4)	67 (253.6)	—	—
30 (9.1)	79 (299.0)	66 (249.8)	49 (185.5)	35 (132.5)
40 (12.2)	71 (268.7)	60 (227.1)	46 (174.1)	33 (124.9)
50 (15.2)	59 (223.3)	52 (196.8)	41 (155.2)	28 (106.0)
60 (18.3)	42 (159.0)	40 (151.4)	32 (121.1)	22 (83.3)
70 (21.3)	22 (83.3)	22 (83.3)	20 (75.0)	12 (45.4)

Model 8-M (2-in.)

Total Head Including Friction	*Height of Pump Above Water [ft (m)]*			
	10 (3.0)	*15 (4.6)*	*20 (6.1)*	*25 (7.6)*
20 (6.1)	135 (511.0)	—	—	—
25 (7.6)	134 (507.2)	117 (442.8)	—	—
30 (9.1)	132 (499.6)	115 (435.3)	93 (352.0)	65 (246.0)
40 (12.2)	123 (465.6)	109 (412.6)	88 (333.1)	63 (238.5)
50 (15.2)	109 (412.6)	99 (374.7)	81 (306.6)	59 (223.3)
60 (18.3)	90 (340.7)	84 (317.9)	70 (265.0)	51 (193.0)
70 (21.3)	66 (249.8)	65 (246.0)	57 (215.7)	41 (155.2)
80 (24.4)	40 (151.4)	40 (151.4)	40 (151.4)	28 (106.0)

Model 7-M (2-in.)

Total Head Including Friction	*Height of Pump Above Water [ft (m)]*			
	10 (3.0)	*15 (4.6)*	*20 (6.1)*	*25 (7.6)*
20 (6.1)	117 (442.8)	—	—	—
30 (9.1)	116 (439.1)	102 (386.1)	82 (310.4)	—
40 (12.2)	105 (397.4)	100 (378.5)	80 (302.8)	58 (219.5)
50 (15.2)	92 (348.2)	90 (340.7)	76 (287.7)	55 (208.2)
60 (18.3)	70 (265.0)	70 (265.0)	70 (265.0)	55 (208.2)
70 (21.3)	40 (151.4)	40 (151.4)	40 (151.4)	40 (151.4)

Model 10-M (2-in.)

Total Head Including Friction	*Height of Pump Above Water [ft (m)]*			
	10 (3.0)	*15 (4.6)*	*20 (6.1)*	*25 (7.6)*
25 (7.6)	166 (628.3)	—	—	—
30 (9.1)	165 (624.5)	140 (529.9)	110 (416.4)	—
40 (12.2)	158 (598.0)	140 (529.9)	110 (416.4)	75 (283.9)
50 (15.2)	145 (548.8)	130 (492.1)	106 (401.2)	70 (265.0)
60 (18.3)	126 (476.9)	117 (442.8)	97 (367.1)	68 (257.4)
70 (21.3)	102 (386.1)	100 (378.5)	85 (321.7)	60 (227.1)
80 (24.4)	74 (280.1)	74 (280.1)	68 (257.4)	48 (181.7)
90 (27.4)	40 (151.4)	40 (151.4)	40 (151.4)	32 (121.1)

Table 7-9 *(continued)*

Model 15-M (3-in.)

Total Head Including Friction	Height of Pump Above Water [ft (m)]			
	10 (3.0)	15 (4.6)	20 (6.1)	25 (7.6)
20 (6.1)	259 (980.3)	—	—	—
30 (9.1)	250 (946.3)	210 (794.9)	200 (757.0)	—
40 (12.2)	241 (912.2)	207 (783.5)	177 (669.9)	160 (605.6)
50 (15.2)	225 (851.6)	202 (764.6)	172 (651.0)	140 (529.9)
60 (18.3)	197 (745.6)	197 (745.6)	169 (639.7)	140 (529.9)
70 (21.3)	160 (605.6)	160 (605.6)	160 (605.6)	138 (522.3)
80 (24.4)	125 (473.1)	125 (473.1)	125 (473.1)	125 (473.1)
90 (27.4)	96 (363.4)	96 (363.4)	96 (363.4)	96 (363.4)

Model 20-M (3-in.)

Total Head Including Friction	Height of Pump Above Water [ft (m)]			
	10 (3.0)	15 (4.6)	20 (6.1)	25 (7.6)
30 (9.1)	333 (1,260.0)	280 (1,059.8)	235 (889.5)	165 (624.5)
40 (12.2)	315 (1,192.3)	270 (1,022.0)	230 (870.6)	162 (613.2)
50 (15.2)	290 (1,097.7)	255 (965.2)	220 (832.7)	154 (582.9)
60 (18.3)	255 (965.2)	235 (889.5)	205 (775.9)	143 (541.3)
70 (21.3)	212 (802.4)	209 (791.1)	184 (696.4)	130 (492.1)
80 (24.4)	165 (624.5)	165 (624.5)	157 (594.2)	114 (431.5)
90 (27.4)	116 (439.1)	116 (439.1)	116 (439.1)	94 (355.8)
100 (30.5)	60 (227.1)	60 (227.1)	60 (227.1)	60 (227.1)

Model 18-M (3-in.)

Total Head Including Friction	Height of Pump Above Water [ft (m)]			
	10 (3.0)	15 (4.6)	20 (6.1)	25 (7.6)
25 (7.6)	301 (1,139.3)	—	—	—
30 (9.1)	295 (1,116.6)	255 (965.2)	200 (757.0)	—
40 (12.2)	276 (1,044.7)	250 (946.3)	200 (757.0)	162 (613.2)
50 (15.2)	250 (946.3)	237 (897.0)	198 (749.4)	159 (601.8)
60 (18.3)	216 (817.6)	212 (802.4)	182 (688.9)	146 (552.6)
70 (21.3)	174 (658.6)	174 (658.6)	158 (598.0)	127 (480.7)
80 (24.4)	129 (488.3)	129 (488.3)	125 (473.1)	104 (393.6)
90 (27.4)	82 (310.4)	82 (310.4)	82 (310.4)	74 (280.1)
95 (29.0)	57 (215.7)	57 (215.7)	57 (215.7)	57 (215.7)

Model 40-M (4-in.)

Total Head Including Friction	Height of Pump Above Water [ft (m)]			
	10 (3.0)	15 (4.6)	20 (6.1)	25 (7.6)
25 (7.6)	665 (2,517.0)	—	—	—
30 (9.1)	660 (2,498.1)	575 (2,176.4)	475 (1,797.9)	355 (1,343.7)
40 (12.2)	645 (2,441.3)	565 (2,138.5)	465 (1,760.0)	350 (1,324.8)
50 (15.2)	620 (2,346.7)	545 (2,062.8)	455 (1,722.2)	345 (1,305.8)
60 (18.3)	585 (2,214.2)	510 (1,930.3)	435 (1,646.5)	335 (1,268.0)
70 (21.3)	535 (2,025.0)	475 (1,797.9)	410 (1,551.9)	315 (1,192.3)
80 (24.4)	465 (1,760.0)	410 (1,551.9)	365 (1,381.5)	280 (975.8)
90 (27.4)	375 (1,419.4)	325 (1,230.1)	300 (1,135.5)	220 (832.7)
100 (30.5)	250 (946.3)	215 (813.8)	195 (738.1)	145 (548.8)
110 (33.5)	65 (246.0)	60 (227.1)	50 (189.2)	40 (151.4)

Table 7-9 *(continued)*

Model 90-M (6-in.)

Total Head Including Friction	Height of Pump Above Water [ft (m)]			
	10 (3.0)	15 (4.6)	20 (6.1)	25 (7.6)
25 (7.6)	1,500 (5,677.5)	—	—	—
30 (9.1)	1,480 (5,601.8)	1,280 (4,844.8)	1,050 (3,974.3)	790 (2,990.1)
40 (12.2)	1,430 (5,412.6)	1,230 (4,655.6)	1,020 (3,860.7)	780 (2,952.3)
50 (15.2)	1,350 (5,109.8)	1,160 (4,390.6)	970 (3,671.5)	735 (2,782.0)
60 (18.3)	1,225 (4,636.6)	1,050 (3,974.2)	900 (3,406.5)	690 (2,611.7)
70 (21.3)	1,050 (3,974.2)	900 (3,406.5)	775 (2,933.4)	610 (2,308.9)
80 (24.4)	800 (3,028.0)	680 (2,573.8)	600 (2,271.0)	490 (1,854.7)
90 (27.4)	450 (1,703.3)	400 (1,514.0)	365 (1,381.5)	300 (1,135.5)
100 (30.5)	100 (378.5)	100 (378.5)	100 (378.5)	100 (378.5)

Model 200-M (10-in.)

Total Head Including Friction	Height of Pump Above Water [ft (m)]			
	10 (3.0)	15 (4.6)	20 (6.1)	25 (7.6)
20 (6.1)	3,350 (12,679.8)	3,000 (11,355.0)	—	—
30 (9.1)	3,000 (11,355.0)	2,800 (10,598.0)	2,500 (9,462.5)	1,550 (5,866.8)
40 (12.2)	2,500 (9,462.5)	2,500 (9,462.5)	2,250 (8,516.3)	1,500 (5,677.5)
50 (15.2)	2,000 (7,570.0)	2,000 (7,570.0)	2,000 (7,570.0)	1,350 (5,109.8)
60 (18.3)	1,300 (4,920.5)	1,300 (4,920.5)	1,300 (4,920.5)	1,150 (4,352.8)
70 (21.3)	500 (1,892.5)	500 (1,892.5)	500 (1,892.5)	500 (1,892.5)

Model 125-M (8-in.)

Total Head Including Friction	Height of Pump Above Water [ft (m)]			
	10 (3.0)	15 (4.6)	20 (6.1)	25 (7.6)
25 (7.6)	2,100 (7,948.5)	1,850 (7,002.3)	1,570 (5,942.5)	—
30 (9.1)	2,060 (7,797.1)	1,820 (6,888.7)	1,560 (5,904.6)	1,200 (4,542.0)
40 (12.2)	1,960 (7,418.6)	1,740 (6,585.9)	1,520 (5,753.2)	1,170 (4,428.5)
50 (15.2)	1,800 (6,813.0)	1,620 (6,131.7)	1,450 (5,488.3)	1,140 (4,314.9)
60 (18.3)	1,640 (6,207.4)	1,500 (5,677.5)	1,360 (5,147.6)	1,090 (4,125.7)
70 (21.3)	1,460 (5,526.1)	1,340 (5,071.9)	1,250 (4,731.3)	1,015 (3,840.8)
80 (24.4)	1,250 (4,731.3)	1,170 (4,428.5)	1,110 (4,201.4)	950 (3,595.8)
90 (27.4)	1,020 (3,860.7)	980 (3,709.3)	940 (3,557.9)	840 (3,179.4)
100 (30.5)	800 (3,028.0)	760 (2,876.6)	710 (2,687.4)	680 (2,573.8)
110 (33.5)	570 (2,157.5)	540 (2,043.9)	500 (1,892.5)	470 (1,779.0)
120 (36.6)	275 (1,040.9)	245 (927.3)	240 (908.4)	240 (908.4)

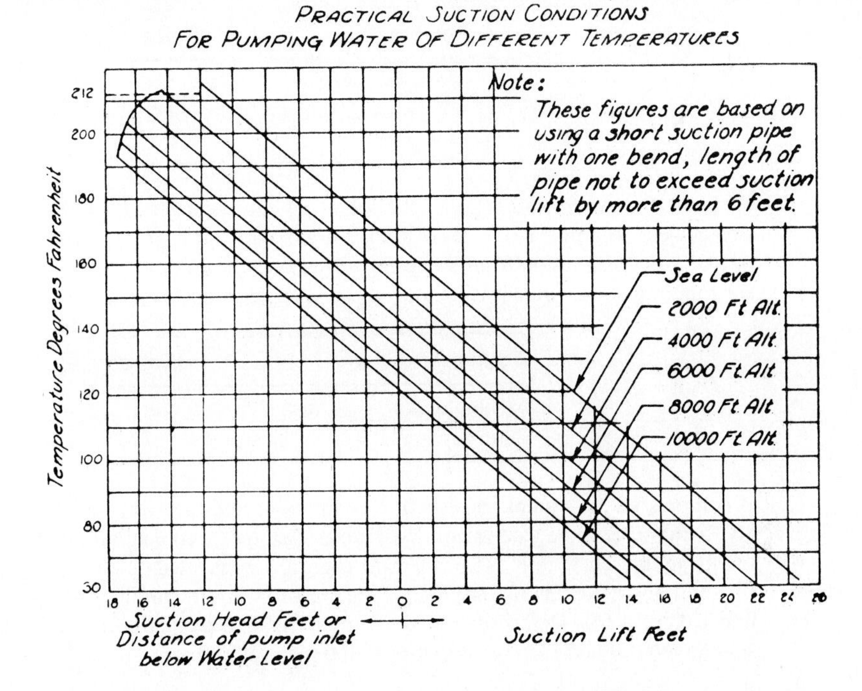

Figure 7-9 Maximum practical suction lift for centrifugal pumps. (Courtesy of Contractors Pump Bureau)

a venturi tube located below the source water level and then back to the pump inlet. Low pressure in the venturi tube draws water from the source into the recirculating line, where it flows to the pump inlet. While jet pumps are relatively inefficient, they are capable of lifting water 100 ft (30.5 m) or more. Other advantages of jet pumps include simplicity, ease of maintenance, the ability to operate in wells having diameters as small as 2 in. (5.1 cm), and the ability to locate the pump mechanism on the surface or even at some distance from the well. *Airlift pumps* discharge compressed air below the source water level inside a discharge line. The air bubbles formed within the discharge line lower the specific gravity of the water and air mixture enough to cause the mixture to flow up through the discharge line. The pump discharge must be fed into an open tank for deaeration if the water is to be repumped by a conventional pump. Air-lift pumps have low efficiency and a limited lift capability. They are principally used for testing new wells and for cleaning and dewatering drilled pier excavations.

PROBLEMS

1. Estimate the volume of compressed air required to operate the following tools. Assume a 5% leakage loss, a job load factor of 0.90, and maximum values of tool air consumption from Table 7-1. Determine the rated size of air compressor required to supply this equipment at an altitude of 7000 ft (2135 m).

Equipment	*Number*
Light wagon drill	4
45-lb hand-held drill	4
Medium wagon drill	1

2. Estimate the air consumption of the following equipment to be used on a rock excavation project. Assume a 10% leakage loss, a job load factor of 0.85, and minimum values of tool air consumption from Table 7-1. What is the minimum rated size of air compressor required if the project is located at an altitude of 4000 ft (1220 m)?

Equipment	*Number*
Medium-weight track drill	3
45-lb hand-held drill	7

3. A compressed air system consists of a compressor and receiver, 1000 ft (305 m) of 4-in. (10.2-cm) pipe, three gate valves, five standard elbows, and a manifold. Three rock drills requiring 400 cu ft/min (11.3 m^3/min) each are connected to the manifold by 1¼-in. (3.2-cm) hoses 100 ft (30.5 m) long. Pressure drop in the manifold is 3 psi (20 kPa) and line leakage is 5%. Determine the pressure at the drill when all three drills are operating simultaneously and receiver pressure is 100 psig (690 kPa).

4. What minimum air pressure is usually required for pneumatic tools to deliver their rated performance? What is the effect on rock drill performance and operating cost when air pressure is increased above this minimum (see Chapter 6)?

5. You are designing an air delivery system to supply air from a receiver operated at 100 psig (690 kPa) through 1600 ft (488 m) of pipe and a manifold to three hand-held rock drills. Each drill requires 100 cu ft/min (2.8 m^3/min) of compressed air and is connected to the manifold by 100 ft (30.5 m) of 1-in (2.5 cm) hose. Manifold pressure loss is rated at 2 psig (14 kPa). Determine the minimum size of pipe required to maintain a pressure of at least 90 psig (621 kPa) at the drills when all drills are operating simultaneously. Assume a 5% line leakage.

6. Why are diaphrahm pumps often used for dewatering construction excavations?

7. An M-rated self-priming centrifugal pump will be used to dewater a trench during pipeline construction. The required pumping volume is estimated at 200 gal/min (757 /min). The pump will be located 10 ft (3.1 m) above the trench bottom. The suction line will consist of 25 ft (7.6 m) of 3-in. (7.6-cm) hose and the discharge line will consist of 75 ft (22.8 m) of 3-in. (7.6-cm) hose. Water will be discharged 42 ft (12.8 m) above pump level. What is the total head developed? What is the minimum-size pump required?

8. What is the maximum practical suction lift for a centrifugal pump located at an altitude of 8000 ft (2440 m) when the temperature is 100°F (38°C)?

9. Water must be pumped from a stream to a water tank 500 ft (153 m) away. The discharge point is 30 ft (9.2 m) above the stream. The pipeline from the pump to the tank will consist of 550 ft (168 m) of 4-in. (10.2-cm) straight pipe, four standard elbows, two

45° elbows, and a gate valve. The pump will be located 10 ft (3.0 m) above the stream. The suction line will consist of 25 ft (7.5 m) of 4-in. (10.2-cm) hose equipped with a strainer [equivalent length = 5 ft (1.5 m)]. If the required flow is 200 gal/min (757 ℓ/min), find the total head that the pump must overcome. What is the minimum-size M-rated pump required for this system?

10. Write a computer program that will determine the minimum rated size of air compressor required to service a compressed air system. Input should include (for each tool type) the type of tool, the number of tools, and the expected air consumption per tool. Additional input should include the job load factor, the leakage allowance, and the altitude adjustment factor. Solve Problem 1 using your computer program.

REFERENCES

1. *Contractors Pump Manual.* Contractors Pump Bureau, Rockville, Md., 1976.
2. Crocker, Sabin, and R. C. King. *Piping Handbook,* 5th ed. New York: McGraw-Hill, 1967.
3. Loomis, A.W., ed. *Compressed Air and Gas Data,* 3rd ed. Ingersoll-Rand Company, Woodcliff, N.J., 1982.
4. Rollins, John P., ed. *Compressed Air and Gas,* 5th ed. Englewood Cliffs, N.J.: Prentice Hall, 1988.

8
Paving

8-1
PRODUCTION OF AGGREGATE

The production of high-quality concrete and asphalt mixes, which is discussed in succeeding sections, requires a supply of aggregate (gravel, sand, and mineral filler) meeting the specified gradation and other requirements. Sometimes a blend of natural materials from several different sources will satisfy specification requirements. In this case, the minimum aggregate production equipment required will include excavating, loading, screening, and transporting equipment. Washing equipment may also be required to remove associated plastic fine-grained soils from the natural materials. More often, rock produced at a quarry is processed by a sophisticated aggregate processing plant which includes crushing, screening, conveying, storage, weighing, and loading facilities.

8-2
PRODUCTION OF CONCRETE

Concrete is produced by mixing portland cement, aggregates (sand and gravel), and water. In addition, a fourth component, an additive, may be added to improve the workability or other properties of the concrete mix. The construction operations involved in the production of concrete include batching, mixing, transporting, placing, consolidating, finishing, and curing. The equipment and methods used in producing concrete and for constructing pavements are described in this chapter. Concrete materials and their properties, as well as construction practices involved in general concrete construction, are explained in Chapter 10.

Batching and Mixing

The process of proportioning cement, water, aggregates, and additives prior to mixing concrete is called *batching.* Since concrete specifications commonly require a batching accuracy of 1 to 3%, depending on the mix component, materials should be carefully proportioned by weight. Central batching plants that consist of separate aggregate and cement batching units are often used for servicing truck mixers and for feeding central mixing plants. In such batching plants cement is usually handled in bulk. The addition of water to the mix may be controlled by the batching plant or it may be controlled by the mixer operator. Batching for small construction mixers is accomplished by loading the required quantity of cement and aggregate directly into the skip (hopper) of the mixer. Water is added by the mixer operator. Cement is usually measured by the sack (94-lb or 42.6-kg) when batching small mixers.

A standard classification system consisting of a number followed by a letter is used in the United States to identify mixer type and capacity. In this system, the number indicates the rated capacity of the mixer in cubic feet (0.028 m^3) of plastic concrete. Satisfactory mixing should be obtained as long as the volume of material in the mixer does not exceed its rated capacity by more than 10%. The letter in the rating symbol indicates the mixer type: S is a construction mixer, E is a paving mixer, and M is a mortar mixer. Thus the symbol "34E" indicates a 34-cu ft (0.96-m^3) paving mixer, "16S" indicates a 16-cu ft (0.45-m^3) construction mixer, and so on.

Construction mixers are available as wheel-mounted units, trailer-mounted units, and stationary units. Mixer drums may be single or double, tilting or nontilting. Mixer capacity ranges from 3½ cu ft (0.1 m^3) to over 12 cu yd (9.2 m^3). The wheel-mounted 16-cu ft (type 16S) construction mixer is often used on small construction projects where ready-mixed concrete is not available. Large central mix plants are used to supply concrete for projects such as dams, which require large quantities of concrete.

Truck mixers are truck-mounted concrete mixers capable of mixing and transporting concrete. The product they deliver is referred to as *ready-mixed concrete.* The usual procedure is to charge the truck mixer with cement and aggregate at a central batch plant, then add water to the mix when ready to begin mixing. Truck mixers are also capable of operating as agitator trucks for transporting plastic concrete from a central mix plant. A truck mixer used as an agitator truck can haul a larger quantity of concrete than it is capable of mixing. While a unit's capacity when used as an agitator truck is established by the equipment manufacturer, agitating capacity is about one-third greater than mixing capacity. Standard truck mixer capacity ranges from 6 cu yd (4.6 m^3) to over 15 cu yd (11.5 m^3).

Paving mixers are self-propelled concrete mixers especially designed for concrete paving operations. They are equipped with a boom and a bucket which enable them to place concrete at any desired point within the roadway. With the increasing use of slipform pavers, paving mixers are now often used to supply slipform pavers or to operate as stationary mixers. Dual-drum paving mixer production is almost double that of a single-drum mixer. When operated as a stationary plant, a type 34E dual drum paving mixer is capable of producing about 100 cu yd (76.5 m^3) of concrete per hour.

A minimum mixing time of 1 min plus ¼ min for each cubic yard (0.76 m^3) over 1 cu yd (0.76 m^3) is often specified. However, the time required for a complete mixer cycle has been found to average 2 to 3 min. A mixing procedure that has been found to help clean the mixer drum and provide uniform mixing is to add 10% of the mix water before charging the drum, 80% during charging, and the remaining 10% when charging is completed. Timing of the mixing cycle should not begin until all solid materials are placed into the drum. All water should be added before one-fourth of the mixing time has passed. Standards of the Truck Mixer Manufacturers Bureau require truck mixers to mix concrete for 70 to 100 revolutions at mixing speed after all ingredients, including water, have been added. Any additional rotation must be at agitating speed. Concrete in truck mixers should be discharged within 1½ h after the start of mixing and before the drum has revolved 300 times.

Estimating Mixer Production

After a concrete mix design has been established (see Chapter 10), the volume of plastic concrete produced by the mix may be calculated by the *absolute-volume method.* In this method the volume of one batch is calculated by summing up the absolute volume of all mix components. The absolute volume of each component may be found as follows:

$$\text{Volume (cu ft)} = \frac{\text{Weight (lb)}}{62.4 \times \text{specific gravity}} \tag{8-1A}$$

$$\text{Volume (m}^3) = \frac{\text{Weight (kg)}}{1000 \times \text{specific gravity}} \tag{8-1B}$$

When calculating the absolute volume of aggregate using Equation 8-1, aggregate weight must be based on the saturated, surface-dry (SSD) condition. Such aggregate will neither add nor subtract water from the mix. If aggregate contains free water, a correction must be made in the quantity of water to be added to the mix. Example 8-1 illustrates these procedures.

Example 8-1

PROBLEM (a) Calculate the volume of plastic concrete that will be produced by the mix design given in the table.

Component	Specific Gravity	Quantity	
		lb	kg
Cement	3.15	340	154
Sand (SSD)	2.65	940	426
Gravel (SSD)	2.66	1210	549
Water	1.00	210	95

(b) Determine the actual weight of each component to be added if the sand contains 5% excess moisture and the gravel contains 2% excess moisture.

(c) Determine the weight of each component required to make a three-bag mix and the mix volume.

SOLUTION

(a) $$\text{Cement volume} = \frac{340}{3.15 \times 62.4} = 1.7 \text{ cu ft}$$

$$\left[= \frac{154}{3.15 \times 1000} = 0.05 \text{ m}^3 \right]$$

$$\text{Sand volume} = \frac{940}{2.65 \times 62.4} = 5.7 \text{ cu ft}$$

$$\left[= \frac{426}{2.65 \times 1000} = 0.16 \text{ m}^3 \right]$$

$$\text{Gravel volume} = \frac{1210}{2.66 \times 62.4} = 7.3 \text{ cu ft}$$

$$\left[= \frac{549}{2.66 \times 1000} = 0.21 \text{ m}^3 \right]$$

$$\text{Water volume} = \frac{210}{1.00 \times 62.4} = 3.4 \text{ cu ft}$$

$$\left[= \frac{95}{1.00 \times 1000} = 0.09 \text{ m}^3 \right]$$

$$\text{Mix volume} = 1.7 + 5.7 + 7.3 + 3.4 = 18.1 \text{ cu ft}$$

$$[= 0.05 + 0.16 + 0.21 + 0.09 = 0.51 \text{ m}^3]$$

(b) $$\text{Excess water in sand} = 940 \times 0.05 = 47 \text{ lb}$$

$$[= 426 \times 0.05 = 21 \text{ kg}]$$

$$\text{Excess water in gravel} = 1210 \times 0.02 = 24 \text{ lb}$$

$$[= 549 \times 0.02 = 11 \text{ kg}]$$

$$\text{Total excess water} = 47 + 24 = 71 \text{ lb}$$

$$[= 21 + 11 = 32 \text{ kg}]$$

Field mix quantities:

$$\text{Water} = 210 - 71 = 139 \text{ lb}$$

$$[= 95 - 32 = 63 \text{ kg}]$$

$$\text{Sand} = 940 + 47 = 987 \text{ lb}$$
$$[= 426 + 21 = 447 \text{ kg}]$$
$$\text{Gravel} = 1210 + 24 = 1234 \text{ lb}$$
$$[= 549 + 11 = 560 \text{ kg}]$$

(c) Adjusting to a three-bag mix:

$$\text{Cement} = 3 \times 94 = 282 \text{ lb}$$
$$[= 3 \times 42.6 = 127.8 \text{ kg}]$$
$$\text{Sand} = \frac{282}{340} \times 987 = 819 \text{ lb}$$
$$\left[= \frac{127.8}{154} \times 447 = 370 \text{ kg}\right]$$
$$\text{Gravel} = \frac{282}{340} \times 1234 = 1023 \text{ lb}$$
$$\left[= \frac{127.8}{154} \times 560 = 464 \text{ kg}\right]$$
$$\text{Water} = \frac{282}{340} \times 139 = 115 \text{ lb}$$
$$\left[= \frac{127.8}{154} \times 63 = 52 \text{ kg}\right]$$
$$\text{Mix volume} = \frac{282}{340} \times 18.1 = 15.0 \text{ cu ft}$$
$$\left[= \frac{127.8}{154} \times 0.51 = 0.42 \text{ m}^3\right]$$

After the batch volume has been calculated, mixer production may be estimated as follows:

$$\text{Mixer production (cu yd/h)} = \frac{2.22 \times V \times E}{T} \tag{8-2A}$$

$$\text{Mixer production (m}^3\text{/h)} = \frac{60 \times V \times E}{T} \tag{8-2B}$$

where V = batch volume (cu ft or m^3)
T = cycle time (min)
E = job efficiency

Transporting and Handling

A number of different items of equipment are available for moving concrete from the mixer to its final position. Equipment commonly used includes wheelbarrows, buggies, chutes, conveyors, pumps, buckets, and trucks. Regardless of the equipment used, care must be taken to avoid segregation when handling plastic concrete. The height of free fall should be limited to about 5 ft (1.5 m) unless downpipes or ladders are used to prevent segregation. Downpipes having a length of at least 2 ft (0.6 m) should be used at the end of concrete conveyors.

Wheelbarrows have a very limited capacity (about 1½ cu ft or 0.04 m^3) but are often used for transporting and placing small amounts of concrete. Push buggies that carry 6 to 11 cu ft (0.17 to 0.31 m^3) and powered buggies carrying up to ½ cu yd (0.38 m^3) are often employed on building construction projects. However, these items of equipment are gradually being replaced by concrete pumps capable of moving concrete from a truck directly into final position up to heights of 500 ft (152 m) or more. Truck-mounted concrete pumps equipped with placement booms such as that shown in Figure 8-1 are widely used in building construction.

Concrete conveyors are available to move concrete either horizontally or vertically. Chutes are widely used for moving concrete from the mixer to haul units and for placing concrete into forms. Truck mixers are equipped with integral retracting chutes that may be used for discharging concrete directly into forms within the radius of the chute. When chuting concrete, the slope of the chute must be high enough to keep the chute clean but not high enough to produce segregation of the concrete. Concrete buckets attached to cranes are capable of lifting concrete to the top of highrise buildings and of moving concrete over a wide area. Concrete buckets are equipped with a bottom gate and a release mechanism for unloading concrete at the desired location. The unloading mechanism may be powered or may be operated

Figure 8-1 Concrete pump and truck mixer. (Courtesy of Challenge-Cook Bros., Inc.)

manually. The use of remotely controlled power-operated bucket gates reduces the safety hazard involved in placing concrete above ground level.

Although truck mixers are most often employed for hauling plastic concrete to the job site, dump trucks equipped with special concrete bodies are also available for hauling concrete. The bodies of such trucks are designed to reduce segregation during hauling and provide easy cleaning and dumping. When using nonagitator trucks for hauling concrete, specifications may limit the truck speed and maximum haul distance that may be used. Temperature, road condition, truck body type, and mix design are the major factors that influence the maximum safe hauling distance. Railway cars designed for hauling concrete are also available but are not widely used.

Placing and Consolidating

The movement of plastic concrete into its final position (usually within forms) is called *placing.* Before placing concrete, the underlying surface and the interior of all concrete forms must be properly prepared. Concrete forms must be clean and tight and their interior surfaces coated with form oil or a parting agent to allow removal of the form from the hardened concrete without damaging the surface of the concrete. Additional considerations in placing concrete into forms are discussed in Chapter 10.

When concrete is poured directly onto a subgrade, the subgrade should be moistened or sealed by a moisture barrier to prevent the subgrade from absorbing water from the plastic concrete. When placing fresh concrete on top of hardened concrete, the surface of the hardened concrete should be roughened to provide an adequate bond between the two concrete layers. To improve bonding between the layers, the surface of the hardened concrete should also be coated with grout or a layer of mortar before the fresh concrete is placed. Concrete is usually placed in layers 6 to 24 in. (15 to 61 cm) thick except when pumping into the bottom of forms. When placing concrete in layers, care must be taken to ensure that the lower layer does not take its initial set before the next layer is poured.

Concrete may also be pneumatically placed by spraying it onto a surface. Concrete placed by this process is designated *shotcrete* by the American Concrete Institute but is also called *pneumatically applied concrete, gunned concrete,* or *gunite.* Since a relatively dry mix is used, shotcrete may be applied to overhead and vertical surfaces. As a result, shotcrete is often used for constructing tanks, swimming pools, and tunnel liners, as well as for repairing damaged concrete structures.

Concrete may be placed underwater by the use of a tremie or by pumping. A *tremie* (see Figure 9-20) is nothing more than a vertical tube with a gate at the bottom and a hopper on top. In operation the tremie tube must be long enough to permit the concrete hopper to remain above water when the lower end of the tremie is placed at the desired location. With the gate closed, the tremie is filled with concrete and lowered into position. The gate is then opened, allowing concrete to flow into place. The pressure of the plastic concrete inside the tremie prevents water from flowing into the tremie. The tremie is raised as concrete is poured, but care must be taken to keep the bottom end of the tremie immersed in the plastic concrete.

Consolidation is the process of removing air voids in concrete as it is placed. Concrete vibrators are normally used for consolidating concrete, but hand rodding or spading may be employed. Immersion-type electric, pneumatic, or hydraulic concrete vibrators are widely used. However, form vibrators or vibrators attached to the outside of the concrete forms are sometimes employed. Vibrators should not be used to move concrete horizontally, as this practice may produce segregation of the concrete mix. Vibrators should be inserted into the concrete vertically and allowed to penetrate several inches into the previously placed layer of concrete. The vibrator should be withdrawn and moved to another location when cement paste becomes visible at the top of the vibrator.

Finishing and Curing

Finishing is the process of bringing the surface of concrete to its final position and imparting the desired surface texture. Finishing operations include screeding, floating, troweling, and brooming. *Screeding* is the process of striking off the concrete in order to bring the concrete surface to the required grade. When the concrete has hardened enough so that a worker's foot makes only a small impression in the surface, the concrete is floated with a wood or metal float. *Floating* smooths and compacts the surface while embedding aggregate particles. *Troweling* with a steel trowel follows floating when a smooth dense surface is desired. Finally, the concrete may be *broomed* by drawing a stiff broom across the surface. This technique is used when a textured skid-resistant surface is desired.

The completion of cement hydration requires that adequate moisture and favorable temperatures be maintained after concrete is placed. The process of providing the required water and maintaining a favorable temperature for a period of time after placing concrete is referred to as *curing.* Methods for maintaining proper concrete temperatures in hot-weather and cold-weather concreting are described in Chapter 10. Methods used to retain adequate curing moisture include covering the concrete surface with wet straw or burlap, ponding water on the surface, covering the surface with paper or plastic sheets, and applying curing compounds. The use of sprayed-on curing compounds applied immediately after finishing has become widespread in recent years.

Vacuum dewatering may be employed to reduce the amount of free water present in plastic concrete after the concrete has been placed and screeded. The dewatering process involves placing a mat having a porous lower surface on top of the concrete and applying a vacuum to the mat. Vacuum within the mat causes excess water from the mix to flow into the mat and eventually to the vacuum source. Removal of excess water results in a lower water/cement ratio and a denser mix. Floating and troweling then follow as usual. In concept, vacuum dewatering permits placing concrete with a high water content (for good workability) while obtaining the strength and durability of concrete with a low water/cement ratio. Other advantages claimed for concrete placed by this method include high early strength, increased ultimate strength and wear resistance, reduced shrinkage, reduced permeability, and increased resistance to freeze/thaw damage. While the vacuum dewatering process was invented and patented in the United States in 1935, it has not been widely used in this country. Recent improvements in the equipment used for the process have led to increased use of the process in both Europe and the United States.

8-3 CONCRETE PAVING

Form-Riding Equipment

The frequent use of concrete for paving highways and airfields has led to the development of specialized concrete paving equipment. While slipform pavers that do not require the use of forms are becoming increasingly popular, paving is still accomplished using metal forms to retain the plastic concrete while it is placed and finished. Since much of this equipment is designed to ride on the concrete forms, the equipment is often referred to as *form-riding equipment.* The pieces of equipment used to perform the operations of mixing, placing, finishing, and curing are often referred to as a *paving train,* because they travel together in series along the roadway.

Standard metal paving forms are 10-ft (3-m) long and 8 to 12 in. (20 to 30 cm) in height. Metal pins are driven into the ground through holes in the form, and the form ends are locked together to hold them in alignment. Form-riding subgraders similar to the grade excavator described in Chapter 5 are available to bring the pavement subgrade or base to precise elevation before concrete is poured. Concrete is placed within the forms by a paving mixer or by truck mixers. A form-riding concrete spreader is used to spread, strike off, and consolidate the concrete. Combination placer/spreader units equipped with conveyor belts are available which are capable of operating with either form-riding or slipform paving equipment.

Finishing follows concrete placing and spreading. Form-riding equipment often includes both a transverse and longitudinal finisher. The transverse finisher is used to bring the surface to final elevation and provide initial finishing. The longitudinal finisher provides final machine finishing. Hand finishing may follow, employing a form-riding finishing bridge to permit workers to reach the entire surface of the pavement. Finishers may be followed by an automatic curing machine equipped with a power spray that applies curing compound.

When constructing large slabs and decks, concrete may be placed by chutes, buckets, or side discharge conveyors. Mechanical finishing may be supplied by roller finishers, oscillating strike-off finishers, large power floats, or other types of finishers. Figure 8-2 shows a large slab being poured directly from a truck mixer and finished by a roller finisher.

Slipform Paving

A *slipform paver* is capable of spreading, consolidating, and finishing a concrete slab without the use of conventional forms. The concrete develops sufficient strength to be self-supporting by the time it leaves the paver, as shown in Figure 8-3. Since the paver's tracks completely span the pavement slab, reinforcing steel may be placed

Figure 8-2 Roller finisher being used on large slab pour. (Courtesy of CMI Corp.)

Figure 8-3 Large slipform paver in operation. (Courtesy of CMI Corp.)

ahead of the paver. Typical slipform pavers are capable of placing slabs up to 10 in. (25.4 cm) thick and 24 ft (7.3 m) wide at speeds up to 20 ft/min (6 m/min). Other paving equipment, such as tube finishers and curing machines, may be used in conjunction with a slipform paver.

Small slipform pavers such as the one shown in Figure 8-4 are widely used for pouring curbs and gutters. Some machines are combination grade trimmers and pavers, capable of both preparing the subgrade and placing the curb and gutter. Slipform pavers have placed over 1 mi (1.6 km) of curb and gutter per day, although typical production is about one-half of that amount. Small slipform pavers are also capable of constructing sidewalks, highway median barriers, and similar structures.

Concrete saws equipped with diamond or abrasive blades are often used to cut joints in concrete slabs to control shrinkage cracking. The depth of control joints should be about one-fourth of the slab thickness, but not less than the maximum size of the aggregate used. Sawing should be done when the concrete is still green but has hardened sufficiently to produce a clean cut. This is usually 6 to 30 h after the concrete has been placed.

Figure 8-4 Slipform paver for curbs and gutters. (Courtesy of GOMACO Corporation)

8-4
BITUMINOUS MATERIALS

Bituminous materials include both asphalt and tar. Although asphalt is the type of bituminous material most frequently used in surfacing roads and airfields, road tars are sometimes used. Most properties of asphalt and tar are similar except that tars are not soluble in petroleum products. As a result, tar is often used when the pavement is likely to be subjected to spills of petroleum fuels. A major disadvantage of tar is its tendency to change consistency with small variations in temperature. Since asphalt predominates in construction, the words "bituminous" and "asphalt" are often used interchangeably in construction practice.

Bituminous surfaces (pavements and surface treatments) are used to provide a roadway wearing surface and to protect the underlying material from moisture. Because of their plastic nature, bituminous surfaces are often referred to as flexible pavements, in contrast to concrete pavements, which are identified as rigid pavements. Bituminous surfaces are produced by mixing solid particles (aggregates) and a bituminous material. Since the bituminous material serves to bond the aggregate particles together, it is referred to as *binder.*

The aggregate in a bituminous surface actually provides the load-carrying ability of the surface. The aggregate also resists the abrasion of traffic and provides skid resistance to the travel surface. In addition to the coarse aggregate (gravel) and fine aggregate (sand) used in concrete mixes, asphalt mixes often contain a third size of aggregate called *fines.* Fines, also called *mineral filler* or mineral dust, consist of any inert, nonplastic material passing the No. 200 sieve. Material used as fines includes rock dust, portland cement, and hydrated lime. Aggregates used in asphalt mixes should be angular, hard, durable, well graded, clean, and dry, in order to provide the required strength to the mix and to bond with the binder.

Asphalt cement, the solid form of asphalt, must be heated to a liquid state for use in bituminous mixes. Asphalt cements are viscosity-graded and range from AC-2.5 (soft) to AC-40 (hard). When petroleum distillates are mixed with asphalt cement, an asphalt *cutback* is created, which is liquid at room temperature. Asphalt cutbacks are classified as medium-curing (MC) or rapid-curing (RC), depending on the type of solvent used in their production. Road oils or slow-curing (SC) asphalt may be residual asphalt oils or may be produced by blending asphalt cement with residual oils. The classification symbol used for road oils and cutbacks includes a number that indicates the viscosity of the mixture. Viscosity grades range from 30 (viscosity similar to water) to 3000 (barely deforms under its own weight).

Asphalt *emulsions* contain particles of asphalt dispersed in water by means of emulsifying agents. Asphalt emulsions have several important advantages: they can be applied to wet aggregates and they are not flammable or toxic. Asphalt emulsions are classified as rapid setting (RS), medium setting (MS), or slow setting (SS).

Road tars are designed by the symbol RT plus a number indicating viscosity. Twelve grades are available, ranging from RT-1 (low viscosity) to RT-12 (solid at room temperature). Two tar cutback grades, RTCB-5 and RTCB-6, are also available.

Handling Bituminous Materials

When cutbacks are heated for mixing or spraying, they are usually above their flash point. The *flash point* of a liquid is the temperature at which it produces sufficient vapor to ignite in the presence of air and an open flame. Since the flash point is reached at a temperature below that at which the liquid would normally burn, extreme care must be taken when heating cutbacks or when handling the heated material. No open flame or spark-producing equipment should be allowed near the hot liquid. Use only equipment specifically designed for the purpose when heating, storing, mixing, or spraying cutbacks. Adequate fire-extinguishing equipment must be readily available together with personnel properly trained in their use. Proper precautions must also be taken to prevent burns when working with hot materials. Hot surfaces must be conspicuously marked or guarded to protect workers against contacting them. Gloves and other protective clothing must be used by workers handling hot equipment.

The Bituminous Distributor

The bituminous or asphalt distributor illustrated in Figure 8-5 is used to apply liquid bituminous materials. It is utilized in almost all types of bituminous construction. The rate of liquid bituminous application is expressed in gallons per square yard (liters per square meter). The rate at which the bituminous material is applied by a distributor depends on spray bar length, travel speed, and pump output. Spray bar length may range from 4 ft (1.2 m) to 24 ft (7.3 m). Travel speed is measured by a bitumeter calibrated in ft/min (m/min). Pump output is measured by a pump tachometer calibrated in gal/min (ℓ/min). Since standard asphalt volume is measured at

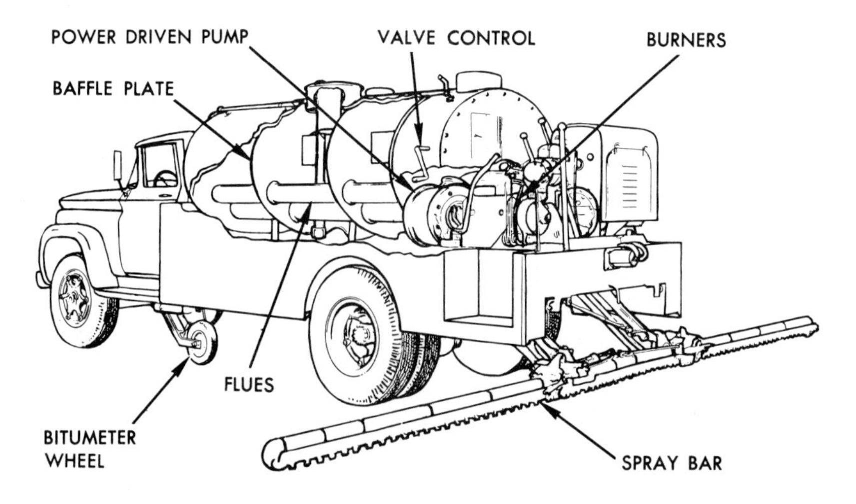

Figure 8-5 Bituminous distributor components. (Courtesy of The Asphalt Institute)

a temperature of 60°F (15.5°C), a volumetric correction factor must be applied to convert asphalt volume at other temperatures to the standard volume (see Table 8-1).

$$S = \frac{9 \times P}{W \times R} \text{ ft/min} \tag{8-3A}$$

$$S = \frac{P}{W \times R} \text{ m/min} \tag{8-3B}$$

where S = road speed (ft/min or m/min)
P = pump output (gal/min or ℓ/min)
W = spray bar width (ft or m)
R = application rate (gal/sq yd or ℓ/m²)

8-5 BITUMINOUS SURFACE TREATMENTS

Types of Surface Treatments

Bituminous surface treatments are used to bond old and new surfaces, to seal and rejuvenate old pavements, or to provide a fresh waterproofing and wearing surface. A wide variety of bituminous surface treatments are available, including prime coats, tack coats, dust palliatives, seal coats, single-pass surface treatments, and multiple-pass surface treatments.

A *prime coat* is a coating of light bituminous material applied to a porous unpaved surface. The purpose of the prime coat is to seal the existing surface and to

Table 8-1 Volumetric correction factor for asphalt*

Temperature °F	Temperature °C	To Obtain Standard Volume, Multiply Measured Volume by:
60	16	1.0000
80	27	0.9931
100	38	0.9862
120	49	0.9792
140	60	0.9724
160	71	0.9657
180	82	0.9590
200	93	0.9523
220	104	0.9458
240	116	0.9392
260	127	0.9328
280	138	0.9264
300	149	0.9201
320	160	0.9138
340	171	0.9076

*Specific gravity above 0.966. Applicable to all grades of asphalt cement and liquid asphalt grades 250, 800, and 3000.

provide a bond between the existing surface and the new bituminous surface. Bituminous materials commonly used for prime coats include RT-1, RT-2, RT-3, RC-70, RC-250, MC-30, MC-70, MC-250, SC-20, and SC-250. The usual rate of bituminous application varies from 0.25 to 0.50 gal/sq yd (1.1 to 2.3 ℓ/m²). All liquid bituminous should be absorbed within 24 h and it should cure in about 48 h.

A *tack coat* is a thin coating of light bituminous material applied to a previously paved surface to act as a bonding agent. Bituminous materials commonly used for tack coats include RC-70, RC-250, RS-1, RS-2, RT-7, RT-8, and RT-9. The usual rate of application is 0.1 gal/sq yd (0.45 ℓ/m²) or less. The tack coat must be allowed to cure to a tacky condition before the new surfacing layer is placed.

A *dust palliative* is a light coating of bituminous material designed to penetrate and bond particles in the surface of unpaved roads in order to reduce dust and provide some waterproofing. Bituminous materials commonly employed include MC-30, MC-70, and diluted slow-setting emulsions.

SEAL COATS

A *fog seal* is a light application of a slow-setting asphalt emulsion diluted by 1 to 3 parts of water. It is used to seal small cracks and voids and to rejuvenate old asphalt surfaces. The usual application rate is 0.1 to 0.2 gal/sq yd (0.4 to 0.9 ℓ/m²).

An *emulsion slurry seal* is composed of a mixture of slow-setting asphalt emulsion, fine aggregate, mineral filler, and water. Usual mixtures contain by weight 20 to 25% asphalt emulsion, 50 to 65% fine aggregate, 3 to 10% mineral filler, and 10 to 15% water. The slurry is placed in a layer ¼ in. (0.6 cm) or less in thickness using hand-operated squeeges, spreader boxes, or slurry seal machines.

A *sand seal* is composed of a light application of a medium-viscosity liquid asphalt covered with fine aggregates. Bituminous materials commonly used include RT-7, RT-8, RT-9, RC-250, RC-800, MC-250, MC-800, RS-1, and SS-1. The rate of application varies from 0.10 to 0.15 gal/sq yd (0.45 to 0.68 ℓ/m²). Fine aggregate is applied at a rate of 10 to 15 lb/sq yd (5.4 to 8.1 kg/m²).

Single- and Multiple-Pass Surface Treatments

Single-pass and multiple-pass surface treatments, sometimes called *aggregate surface treatments,* are made up of alternate applications of asphalt and aggregate. Aggregate surface treatments are used to waterproof a roadway and to provide an improved wearing surface. Such surface treatments are widely used because they require a minimum of time, equipment, and material. They also lend themselves to stage construction; that is, successive applications are repeated over a period of time to produce a higher level of roadway surface.

A *single-pass surface treatment* is constructed by spraying on a layer of asphalt and covering it with a layer of aggregate approximately one stone in depth. Hence the thickness of the finished surface is approximately equal to the maximum diameter of the aggregate used. A typical single surface treatment consists of 25 to 30 lb/sq yd (13 to 16 kg/m²) of ½-in. (1.3-cm) or smaller aggregate covering 0.25 to

0.30 gal/sq yd (1.1 to 1.4 ℓ/m^2) of binder. The type and quantity of binder selected will depend on ambient temperature, aggregate absorbency, and aggregate size.

The sequence of operations involved in placing a single surface treatment is as follows:

1. Sweep the existing surface.
2. Apply prime coat and cure, if required.
3. Apply binder at the specified rate.
4. Apply aggregate at the specified rate.
5. Roll the surface.
6. Sweep again to remove loose stone.

Rotary power booms are most often used for cleaning the existing surface, but blowers or water sprays may be used. The prime coat and binder are applied with an asphalt distributor. Spreading of aggregate must follow immediately after binder application. Since binder temperature has been found to drop to ambient surface temperature in about 2 min, every effort must be made to apply aggregate within 2 min after binder application. Major types of aggregate spreaders, including whirl spreaders, vane spreaders, hopper spreaders, and self-propelled spreaders, operate in conjunction with dump trucks. Spreaders must apply aggregate uniformly and at the specified rate. After the application of aggregate the surface is rolled to embed the aggregate in the binder and to interlock aggregate particles. Either pneumatic or steel wheel rollers may be used for compaction, but pneumatic rollers are preferred because they produce less bridging action and their contact pressure can be easily varied to prevent aggregate crushing. After compaction, the surface is again swept to remove loose stone that might cause damage when thrown by fast vehicles.

Multiple-pass surface treatments consist of two or more single surface treatments placed on top of each other. The construction sequence is the same as that shown above except that steps 3 to 5 are repeated as required. Thus a double surface treatment consists of two binder/aggregate layers, a triple surface treatment consists of three binder/aggregate layers, and so on. The maximum size of aggregate used in each layer should be about one-half the size used in the underlying layer.

8-6 ASPHALT PAVING

The principal types of asphalt pavements include penetration macadam and pavements constructed from road mixes and plant mixes. Paving mixes may be either hot mixes or cold mixes. Hot mixes are used in producing high-type pavements for major highways and airfields. Cold mixes are employed primarily for roadway patching but may also be used for paving secondary roads.

Penetration macadam, while usually classified as a pavement, is constructed using equipment and procedures very similar to those employed for constructing aggregate surface treatments. Penetration macadam may be used as a base as well as a pavement. To construct penetration macadam, a single layer of coarse aggregate, which may be 4 in. (10 cm) or more in thickness, is placed. This layer is then compacted and interlocked by rolling with a pneumatic or steel wheel roller. Binder is then applied followed immediately by an application of an intermediate size aggregate ("key" aggregate). The pavement is then rolled again to compact the key stone and force it into the binder. Another application of binder and smaller key stone may follow. The surface is swept after completion of rolling.

Road mixes or mixed-in-place construction are produced by mixing binder with aggregate directly on the roadway. This mix is then spread and compacted to form a pavement. Road mixes may be produced by motor graders, rotary mixers, or travel plants. To produce a road mix using the motor grader, aggregate is spread along the roadway and binder is applied by a distributor. The materials are then mixed by moving them laterally with the grader, spread to the required depth, and compacted. Rotary mixers use a pulverizing rotor and a spray bar to mix aggregate and binder in one operation. Travel plants pick up aggregate from a windrow on the roadway, mix it with binder, and deposit the mix back on the roadway or into a finishing machine. Problems often encountered in mixed-in-place construction include difficulty in obtaining aggregate moisture control, lack of uniformity in the mix, and difficulty in obtaining uniform spreading of the mix. As a result, the quality of road mixes is generally substantially inferior to that of plant mixes.

Asphalt Plants

While cold mixes may be produced, asphalt plants are primarily used to produce hot mixes for constructing high-level asphalt pavements. Asphalt plant types include batch plants, continuous-mix plants, and drum-mix plants. In a *batch plant* hot aggregates are proportioned by weight and placed into a mixing chamber (pugmill). As the required quantity of asphalt is added, the materials are mixed and then discharged into haul units or storage tanks.

A *continuous-flow plant* uses a cold feed hopper and cold elevator to feed a continuous supply of cold aggregate into the dryer. From the dryer, hot aggregate is moved by a hot elevator into the gradation control unit (GCU). Here the hot aggregate is separated by screening and placed into bins by size. Calibrated feeder gates in the GCU provide a continuous supply of aggregate and mineral filler in the required proportions to a second hot elevator, which feeds the pugmill. Asphalt is injected into the pugmill at a controlled rate and a continuous stream of hot mix flows out of the plant.

In recent years the *drum-mix plant* illustrated in Figure 8-6 has begun to replace the conventional continuous-flow plant. As you can see, both drying and mixing takes place in the dryer drum. The process eliminates the gradation control unit, hot elevators, and pugmill of the conventional continuous-flow plant. The dust emitted by the dryer is also less than that emitted by conventional plants because the asphalt tends to trap the fines inside the drum. This reduces the amount of pollution control equipment needed. As the result of these differences, drum-mix plants cost less and have lower operating costs than do continuous flow plants of the same capacity.

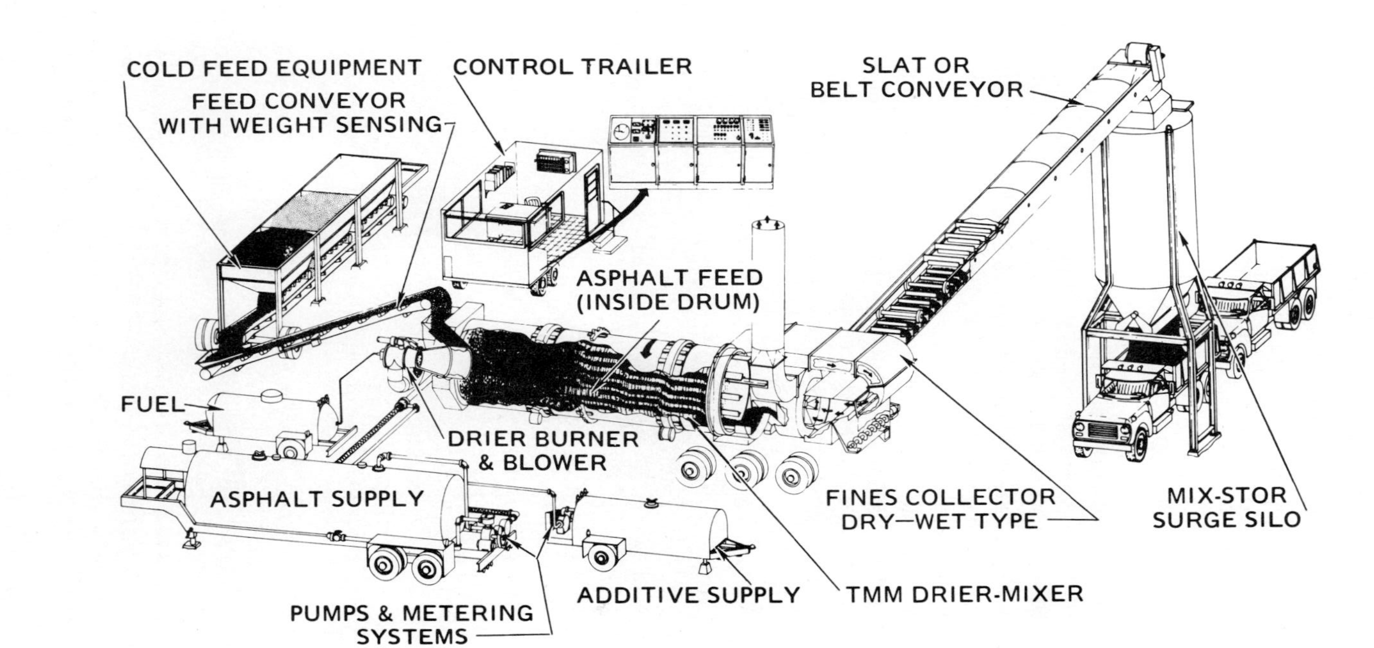

Figure 8-6 Components of a drum mix asphalt plant. (Courtesy of Iowa Manufacturing Co.)

Frequent sampling and testing of plant mixed bituminous material is required to assure adequate quality control. Insulated storage bins such as that shown in Figure 8-7 are available to store plant output when hauling capacity is limited or uncertain. Loading and hauling must be carefully conducted to prevent degradation of the mix. Trucks should be clean and dry before loading. Insulated or heated trucks may be required to ensure that the mix is delivered to the job site at the specified temperature.

Estimating Asphalt Plant Production

An asphalt mix is composed of asphalt, coarse aggregate (gravel), fine aggregate (sand), and mineral filler (or fines). The amount of asphalt in a mix is expressed as a percentage of total mix weight. Aggregate is heated in the dryer to permit bonding with the hot asphalt. Since fines are largely lost as the aggregate passes through the dryer, mineral filler is usually added directly to the pugmill along with the asphalt and hot aggregate.

Dryer capacity, which depends on aggregate moisture content, is normally the controlling factor in asphalt plant capacity. Thus, the maximum hourly plant capacity may be calculated from the dryer capacity and the percentage of asphalt and fines in the mix. The procedure is illustrated in Example 8-2. Notice that the calcu-

Figure 8-7 Insulated hot mix asphalt storage bin with skip hoist. (Courtesy of CMI Corp.)

lations are based on dry aggregate weights. The weight of coarse aggregate and sand must be corrected for moisture to obtain the actual field weight of these materials required to feed the dryer. Additional information on asphalt plant calibration is contained in reference 9.

Example 8-2

PROBLEM (a) Calculate the maximum hourly production of an asphalt plant based on the data below.

(b) Find the required feed rate (ton/h) for each mix component to achieve this production.

Mix composition:

Asphalt = 6%
Aggregate composition:
Coarse A = 42%
Coarse B = 35%
Sand = 18%
Mineral filler = 5%
Aggregate moisture = 8%
Dryer capacity at 8% moisture removal = 110 ton/h

SOLUTION

(a) $$\text{Plant capacity} = \frac{\text{dryer capacity} \times 10^4}{(100 - \text{asphalt }\%)(100 - \text{fines }\%)} \tag{8-4}$$

$$= \frac{110}{(100 - 6)(100 - 5)} = 123 \text{ ton/h}$$

(b) Feed rate (ton/h):

Component	*Fraction*	*Total*	*Rate*
Asphalt	0.06	123.0	7.4
Aggregate (dry)	0.94	123.0	115.6
	1.00	123.0	123.0
Aggregate components (dry weight)			
Coarse A	0.42	115.6	48.5
Coarse B	0.35	115.6	40.5
Sand	0.18	115.6	20.8
Mineral filler	0.05	115.6	5.8
	1.00	115.6	115.6

Paving Operations

Paving operations involve the delivery of bituminous mix, spreading of the mix, and compacting the mix. Spreading and initial compacting of the mix is accomplished by the asphalt paver or finishing machine shown in Figure 8-8. In operation the paver engages the material supply truck, couples the two units together, and pushes the truck as the mix is unloaded and the mat placed. Paver automatic control units

Figure 8-8 Asphalt paving machine in operation. (Courtesy of Barber-Greene Company)

which operate off stringlines or other reference marks are available to control the elevation of the pavement being placed. Pavers require a high volume of material to keep them supplied with mix. For example, a paver moving at 50 ft/min (15.2 m/min) while laying a mat 3 in. (7.6 cm) thick and 12 ft (3.7 m) wide requires 600 tons/h (544 t/h) of hot mix.

Compacting or rolling of the mix should begin immediately after it is placed by the paver. The usual sequence of rolling involves breakdown rolling with a steel wheel roller, intermediate rolling with a pneumatic roller, and final or finish rolling with a steel wheel roller. Vibratory rollers are now frequently used for rolling asphalt pavements, as shown in Figure 8-9. Joints (transverse and longitudinal) and the outside pavement edge should be rolled before the remainder of the pavement is rolled.

8-7 REPAIR AND REHABILITATION

Concern over the declining condition of the U.S. highway system has caused the U.S. Federal-Aid Highway Act to expand the definition of highway construction to include resurfacing, restoration, rehabilitation, and reconstruction. Within the transportation industry, these categories of work are often identified as *4R* construction. The use of *pavement management systems* to maintain pavements in satis-

Figure 8-9 Rolling an asphalt pavement with vibratory roller. (Courtesy of BOMAG(USA))

factory condition at the lowest possible cost is becoming widespread. Such computer-based systems require continuing data collection and evaluation to permit a timely decision on the maintenance strategy to be employed.

Resurfacing may involve surface treatments or overlays of asphalt or concrete. *Restoration* and *rehabilitation* are broad terms that include any of the work required to return the highway to an acceptable condition. One technique growing in popularity is the mechanical removal of the upper portion of the pavement by *planing* or *milling* followed by a new pavement overlay. Often, the material removed is recycled and used as a portion of the aggregate for the new overlay. In addition to reducing cost, recycling reduces the demand for new aggregate sources as well as the problems associated with disposal of the old material. Restoration or rehabilitation may also require that subgrades or base courses (Section 5-3) be strengthened by soil stabilization or drainage improvements. *Reconstruction* refers to complete removal of the old pavement structure and construction of a new pavement.

Bridge management systems are also being developed to improve bridge life and lower costs by optimum bridge maintenance. Bridge decks often require resurfacing or reconstruction as a result of the corrosion of the concrete reinforcing steel due to salt penetration into the concrete. The use of epoxy-coated reinforcing steel, chemical sealing of the pavement surface, and the use of chemical additives in the concrete mix all show promise in reducing deterioration of bridge decks.

Work zone safety, or the prevention of accidents while traffic is maintained during highway repair, is receiving increasing attention from contractors, highway officials, and the Occupational Health and Safety Administration (OSHA). See Chapter 18 for additional information on construction safety.

PROBLEMS

1. A one-sack trial concrete mix that meets specification requirements has the proportions given in the accompanying table. Determine the quantity of each ingredient by weight required to batch a 16S mixer using an integer number of sacks of cement. Assume that the aggregate is saturated, surface-dry, and allow a 10% mixer overload.

	Weight		
Component	*lb*	*kg*	*Specific Gravity*
Cement	94	42.6	3.15
Sand	235	106.6	2.65
Gravel	415	188.2	2.66
Water	54	24.5	1.00

2. Determine the actual field weight (lb or kg) required to charge a 16-cu ft (0.45-m^3) mixer without overload using the mix proportions given in the accompanying table. The field excess moisture content of the sand is 4%. All other ingredients are saturated, surface-dry.

	Weight		
Component	*lb*	*kg*	*Specific Gravity*
Cement	94	42.6	3.15
Sand	200	90.7	2.65
Gravel	400	181.4	2.66
Water	50	22.7	1.00

3. Estimate the hourly production of a 16S concrete mixer using a 10% overload, a cycle time of 2 min, and a job efficiency of 0.83.

4. Using the information of Problem 2, estimate the mixer production if job efficiency is equal to a 50-min hour and the average mixer cycle time is 2.0 min.

5. Explain the purpose of consolidating concrete and the proper employment of a concrete vibrator.

6. Find the asphalt application rate obtained if the volume of asphalt used (standard conditions) was 1000 gal (3785 ℓ), spray bar length was 16 ft (4.9 m), and the length of the road section sprayed was 2000 ft (610 m).

7. Calculate the asphalt application rate (standard volume) actually achieved by a bituminous distributor under the following conditions: tank gauge reading before spread, 2000 gal (7570 ℓ); tank gauge reading after spread, 200 gal (757 ℓ); temperature, 220° F (104° C); width of spray bar, 18 ft (5.5 m); length of spread, 2500 ft (762 m).

8. Why have asphalt emulsions largely replaced asphalt cutbacks in road construction and maintenance work in recent years?

9. Calculate the feed rate (ton/h) for a drum-mix asphalt plant under the following conditions.

    Asphalt content = 5%
    Aggregate composition:
        Coarse A = 45%
        Coarse B = 30%
        Sand = 20%
        Mineral filler = 5%
    Gravel and sand moisture = 6%
    Dryer capacity at 6% moisture removal = 138 ton/h

10. Write a computer program to calculate the field batch weight of each component for a specified concrete mix proportion and mixer capacity. Input should include rated mixer capacity, percent overload, and mix proportions. For each mix component include SSD weight for the specified mix, specific gravity, and percent excess moisture. Solve Problem 2 using your computer program.

11. Write a computer program to calculate the feed rate (ton/h) for an asphalt plant. For drum-mix plants, output the moist weight of coarse and fine aggregate. For batch and continuous-flow plants, output the dry weight of all aggregates. Input should include type of plant, aggregate composition, asphalt content, aggregate moisture, and dryer capacity at the specified moisture removal. Provide for at least two components of coarse aggregate. Solve Problem 9 using your computer program.

REFERENCES

1. *The Asphalt Handbook*. The Asphalt Institute, Lexington, Ky., 1989.
2. *Asphalt Paving Manual* (MS-8), 3rd ed. The Asphalt Institute, Lexington, Ky., 1983.
3. *Asphalt Plant Manual* (MS-3), 5th ed. The Asphalt Institute, Lexington, Ky., 1983.
4. *Asphalt Surface Treatments—Construction Techniques* (ES-12). The Asphalt Institute, Lexington, Ky., 1982.
5. *A Basic Asphalt Emulsion Manual* (MS-19), 2nd ed. The Asphalt Institute, Lexington, Ky., 1979.
6. *Bituminous Construction Handbook,* 5th ed. Barber-Greene Company, Aurora, Ill., 1976.
7. *Compaction Handbook*. Hyster Company, Kewanee, Ill., 1978.
8. *Design and Control of Concrete Mixes,* 13th ed. Portland Cement Association, Skokie, Ill., 1988.
9. Nunnally, S.W. *Managing Construction Equipment*. Englewood Cliffs, N.J.: Prentice Hall, 1977.

PART TWO
BUILDING CONSTRUCTION

9
Foundations

9-1
FOUNDATION SYSTEMS

The *foundation* of a structure supports the weight of the structure and its applied loads. In a broad sense the term "foundation" includes the soil or rock upon which a structure rests, as well as the structural system designed to transmit building loads to the supporting soil or rock. Hence the term *foundation failure* usually refers to collapse or excessive settlement of a building's supporting structure resulting from soil movement or consolidation rather than from a failure of the foundation structure itself. In this chapter the term "foundation" will be used in its more limited sense to designate those structural components that transfer building loads to the supporting soil or rock. A foundation is a part of a building's substructure—that portion of the building which is located below the surrounding ground surface. The principal types of foundation systems include spread footings, piles, and piers. These are illustrated in Figure 9-1 and described in the following sections.

9-2
SPREAD FOOTINGS

A *spread footing* is the simplest and probably the most common type of building foundation. It usually consists of a square or rectangular reinforced concrete pad that serves to distribute building loads over an area large enough so that the result-

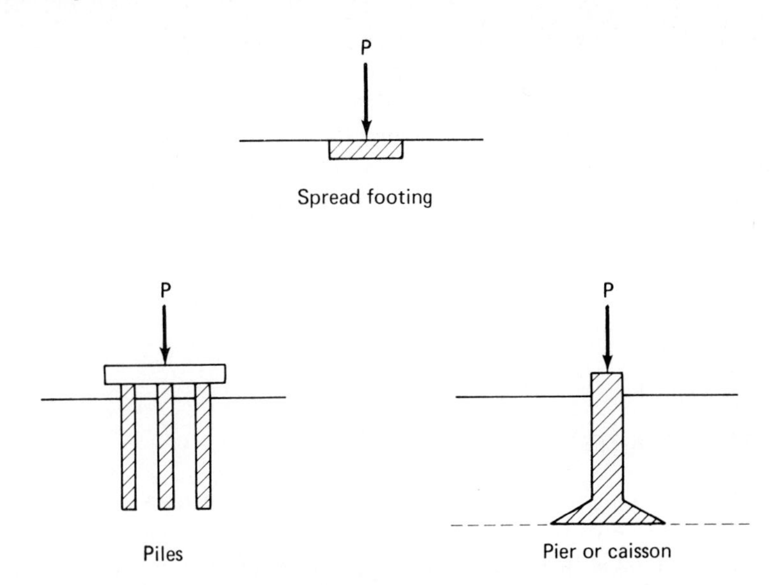

Figure 9-1 Foundation systems.

ing pressure on the supporting soil does not exceed the soil's allowable bearing strength. The principal types of spread footings are illustrated in Figure 9-2. They include individual footings, combined footings, and mat foundations. *Individual footings* include isolated (or single) footings, which support a single column (Figure 9-2a), and wall footings (Figure 9-2b), which support a wall. *Combined footings* support a wall and one or more columns, or several columns (Figure 9-2c).

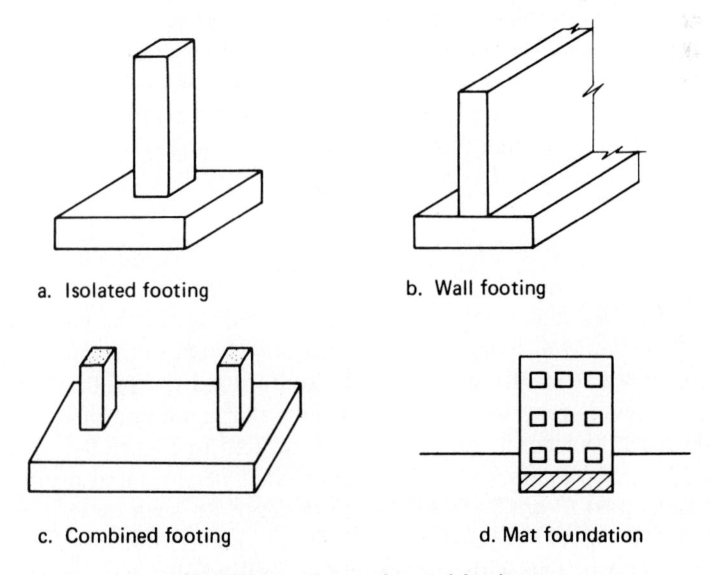

Figure 9-2 Types of spread footings.

Mat or *raft foundations* (Figure 9-2d) consist of a heavily reinforced concrete slab extending under the entire structure, in order to spread the structure's load over a large area. Because such foundations are usually employed for large buildings, they generally involve deep excavation and large-scale concrete pours. A *floating foundation* is a type of mat foundation in which the weight of the soil excavated approximately equals the weight of the structure being erected. Thus, in theory, the erection of the building would not result in any change in the load applied to the soil and hence there would be no settlement of the structure. In practice, however, some soil movement does occur, because the soil swells (or rebounds) during excavation and then recompresses as the building is erected.

9-3 SOIL IMPROVEMENT

If the underlying soil can be strengthened, the allowable bearing pressure on the soil surface will be increased. As a result, it may be possible to use spread footings for foundation loads that would normally require piles or other deep foundation methods. The process of improving soils in place is called *soil improvement* or *soil modification.* In addition to improving bearing capacity, soil improvement may also reduce foundation settlement, groundwater flow, and the subsidence resulting from seismic action. Some of the available soil improvement methods include dynamic compaction, vibratory compaction, vibratory replacement, surcharging with drains, and grouting.

Dynamic Compaction

Probably the simplest form of soil improvement, *dynamic compaction,* involves dropping a heavy weight from a crane onto the ground surface to achieve soil densification. Typically, weights of 10 to 40 tons (9 to 36 t) are used with a drop height of 50 to 100 ft (15 to 30 m) to produce soil densification to a depth of about 30 ft (9 m). The horizontal spacing of drop points usually ranges from 7 to 25 ft (2 to 8 m).

Vibratory Methods

Vibratory compaction, also called *vibroflotation* and *vibrocompaction,* is the process of densifying cohesionless soils by inserting a vibratory probe into the soil. After the probe is jetted and/or vibrated to the required depth, the vibrator is turned on and the device is slowly withdrawn while the soil is kept saturated. Clean, granular material is added from the surface as the soil around the probe densifies and subsides. The process is repeated in a pattern such that a column of densified soil is created under each footing or other load as illustrated in Figure 9-3. This process is quite effective on granular soils having less than 15% fines and often allows bearing capacities up to 5 tons/sq ft (479 kPa) or more. In such cases, vibratory compaction will usually be less expensive than installing piles. However, the process can also be used in conjunction with pile foundations to increase pile capacity.

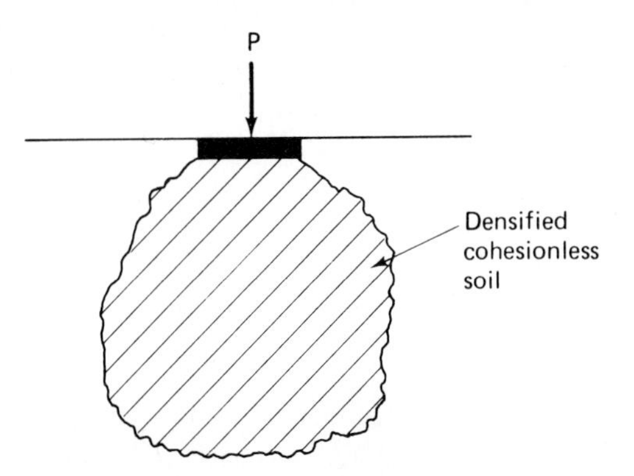

Figure 9-3 Soil densification under footing.

A related technique for strengthening cohesive soils is called *vibratory replacement, vibro-replacement,* or *stone column* construction. The process is similar to vibratory compaction except that the fill added as the probe is withdrawn consists of crushed stone or gravel rather than sand. The resulting stone column is vibrated to increase its density and interaction with the surrounding soil. Stone column capacities of 10 to 40 tons (9 to 36 t) are typically developed.

Surcharging and Drains

Saturated cohesive soils are particularly difficult to densify since the soil grains cannot be forced closer together unless water is drained from the soil's void spaces (see Section 5-1). *Surcharging,* or placing additional weight on the soil surface, has long been used to densify cohesive soils. However, this is a very long term process (months to years) unless natural soil drainage can be increased. *Sand columns* consisting of vertical drilled holes filled with sand have often been used for this purpose. A newer technique that provides faster drainage at lower cost involves forcing *wicks,* or plastic drain tubes, into the soil at intervals of a few feet.

Other Techniques

Grouting (Section 9-9), soil reinforcement (Section 9-6), and electroosmosis (Section 9-8) may also be employed to strengthen soil and reduce groundwater flow. These techniques are described in the indicated sections.

9-4 PILES

A *pile* is nothing more than a column driven into the soil to support a structure by transferring building loads to a deeper and stronger layer of soil or rock. Piles may be classified as either end-bearing or friction piles, according to the manner in which the pile loads are resisted. However, in actual practice, virtually all piles are supported by a combination of skin friction and end bearing.

Pile Types

The principal types of piles include timber, precast concrete, cast-in-place concrete, steel, composite, and bulb piles. *Timber piles* are inexpensive, easy to cut and splice, and require no special handling. However, maximum pile length is limited to about 100 ft, load-carrying ability is limited, and pile ends may splinter under driving loads. Timber piles are also subject to insect attack and decay. However, the availability of pressure-treated wood described in Chapter 14 has greatly reduced the vulnerability of timber piles to such damage.

Precast concrete piles may be manufactured in almost any desired size or shape. Commonly used section shapes include round, square, and octagonal shapes. Advantages of concrete piles include high strength and resistance to decay. However, a precast concrete pile is usually the heaviest type of pile available for a given pile size. Because of their brittleness and lack of tensile strength, they require care in handling and driving to prevent pile damage. Since they have little strength in bending, they may be broken by improper lifting procedures. Cutting requires the use of pneumatic hammers and cutting torches or special saws. Splicing is relatively difficult and requires the use of special cements.

Cast-in place concrete piles (or shell piles) are constructed by driving a steel shell into the ground and then filling it with concrete. Usually, a steel mandrel or core attached to the pile driver is placed inside the shell to reduce shell damage during driving. Although straight shells may be pulled as they are filled with concrete, shells are usually left in place and serve as additional reinforcement for the concrete. The principal types of shell pile include uniform taper, step-taper, and straight (or monotube) piles. The shells for cast-in-place piles are light, easy to handle, and easy to cut and splice. Since shells may be damaged during driving, they should be visually inspected before filling with concrete. Shells driven into expansive soils should be filled with concrete as soon as possible after driving to reduce the possibility of shell damage due to lateral soil pressure.

Steel piles are capable of supporting heavy loads, can be driven to great depth without damage, and are easily cut and spliced. Common types of steel piles include H-piles and pipe piles, where the name indicates the shape of the pile section. Pipe piles are usually filled with concrete after driving to obtain additional strength. The principal disadvantage of steel pile is its high cost.

Composite piles are piles made up of two or more different materials. For example, the lower section of pile might be timber while the upper section might be a shell pile. This would be an economical pile for use where the lower section would be continuously submerged (hence not subject to decay) while the upper section would be exposed to decay.

Bulb piles are also known as *compacted concrete piles, Franki piles,* and *pressure-injected footings.* They are a special form of cast-in-place concrete pile in which an enlarged base (or bulb) is formed during driving. The enlarged base increases the effectiveness of the pile as an end bearing pile. The driving procedure is illustrated in Figure 9-4. A drive tube is first driven to the desired depth of the base either by

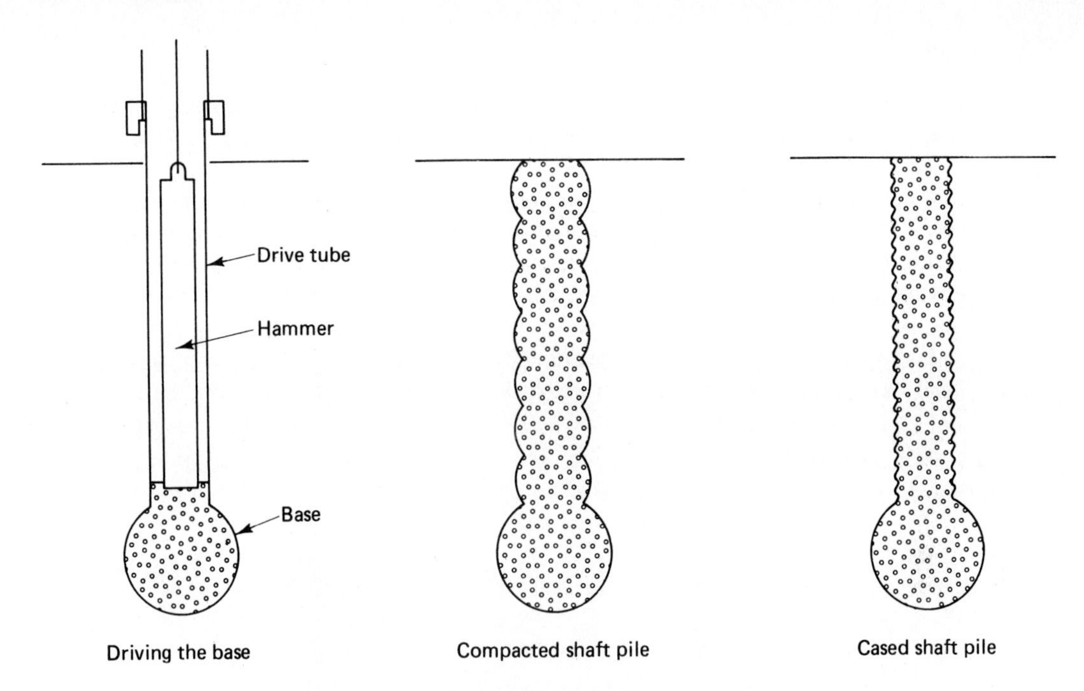

Figure 9-4 Bulb piles.

a powered hammer operating on the top of the drive tube (called *top driving*) or by placing a plug of zero-slump concrete [concrete having a slump of 1 in. (25 mm) or less] into the drive tube and driving both the concrete plug and the drive tube simultaneously using a drop hammer operating inside the drive tube (called *bottom driving*). The drive tube is then held in place and more zero-slump concrete added and hammered out of the end of the drive tube to form the base. Finally, the body or shaft of the pile is constructed by either of two methods. A compacted concrete shaft is formed by hammering zero-slump concrete into the ground as the drive tube is raised. A cased shaft is constructed by placing a steel shell inside the drive tube and then hammering a plug of zero-slump concrete into place to form a bond between the base and the shell. The shell is then filled in the same manner as a conventional cast-in-place concrete pile. Compacted shaft piles usually have a higher load capacity than do cased shaft piles due to the increased pressure between the shaft and the surrounding soil.

Minipiles or *micro piles* are small-diameter [2 to 8 in. (5 to 20 cm)], high-capacity [to 60 tons (54 t)] piles. They are most often employed in areas with restricted access or limited headroom to underpin (provide temporary or additional support to) building foundations. Some other applications include strengthening bridge piers and abutments, anchoring or supporting retaining walls, and stabilizing slopes. While they may be driven in place, minipiles are often installed by drilling a steel-cased hole 2 to 8 in. (5 to 20 cm) in diameter, placing reinforcing in the casing, and then bonding the soil, casing, and reinforcement together by grouting.

Pile Driving

In ancient times, piles were driven by raising and dropping a weight such as a large stone onto the pile. The drop hammer, the modern version of this type of pile driver, is illustrated in Figure 9-5. As you see, the pile-driving assembly is attached to a mobile crane, which provides the support and the power for the pile driver. The *leads* act as guides for the drop weight and the pile. Driving operations consist of lifting the pile, placing it into the leads, lowering the pile until it no longer penetrates the soil under its own weight, and then operating the drop hammer until the pile is driven to the required resistance. Safety requirements for drop hammers include the use of stop blocks to prevent the hammer from being raised against the head block (which could result in collapse of the boom), the use of a guard across

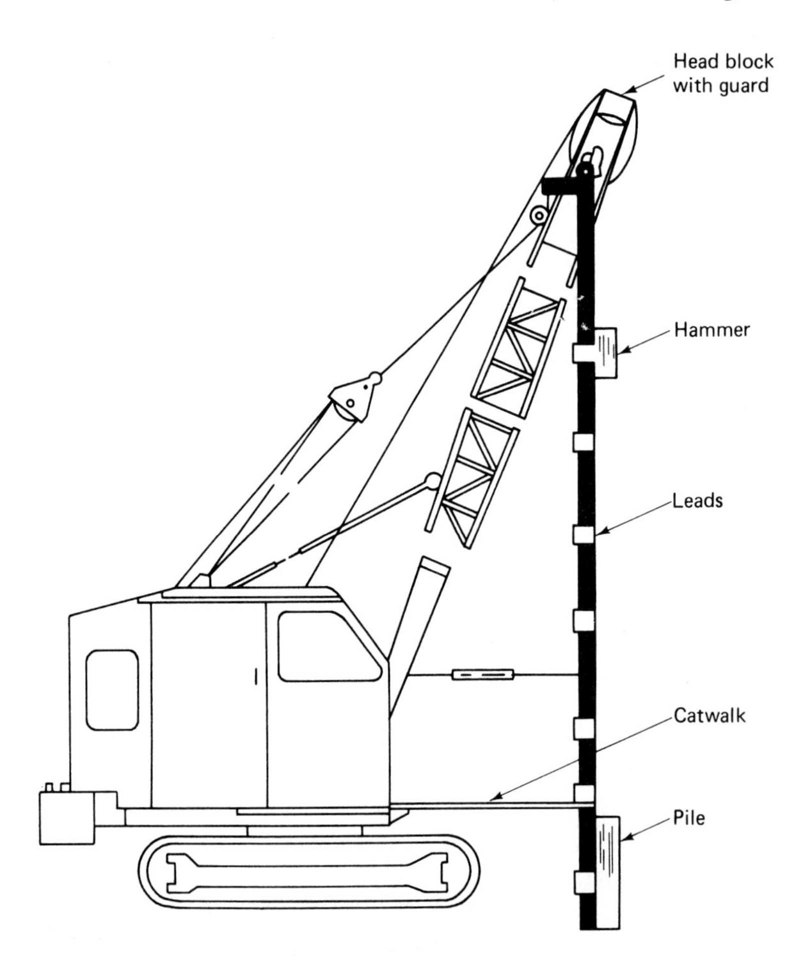

Figure 9-5 Drop hammer pile driver.

the head block to prevent the drop cable from jumping out of the sheaves, and placing a blocking device under the hammer whenever workers are under the hammer.

The remaining types of pile drivers are all powered hammers. That is, they use a working fluid rather than a cable to propel the ram (driving weight). Early powered hammers used steam as a working fluid. Steam power has now been largely replaced by compressed air power. Hydraulic power is replacing compressed air in many newer units. Single-acting hammers use fluid power to lift the ram, which then falls under the force of gravity. Double-acting and differential hammers use fluid power to both lift the ram and then drive the ram down against the pile. Thus double-acting and differential hammers can be lighter than a single-acting hammer of equal capacity. Typical operating frequencies are about 60 blows/min for single-acting hammers and 120 blows/min for double-acting hammers. Differential hammers usually operate at frequencies between these two values.

A diesel hammer contains a free-floating ram-piston that operates in a manner similar to that of a one-cylinder diesel engine. The principle of operation is illustrated in Figure 9-6. The hammer is started by lifting the ram (B) with the crane hoist line (A). The trip mechanism (C) automatically releases the ram at the top of the cylinder. As the ram falls, it actuates the fuel pump cam (D), causing fuel to be injected into the fuel cup in the anvil (E) at the bottom of the cylinder. As the ram continues to fall, it blocks the exhaust ports (F), compressing the fuel-air mixture. When the ram strikes the anvil, it imparts an impact blow to the pile top and also

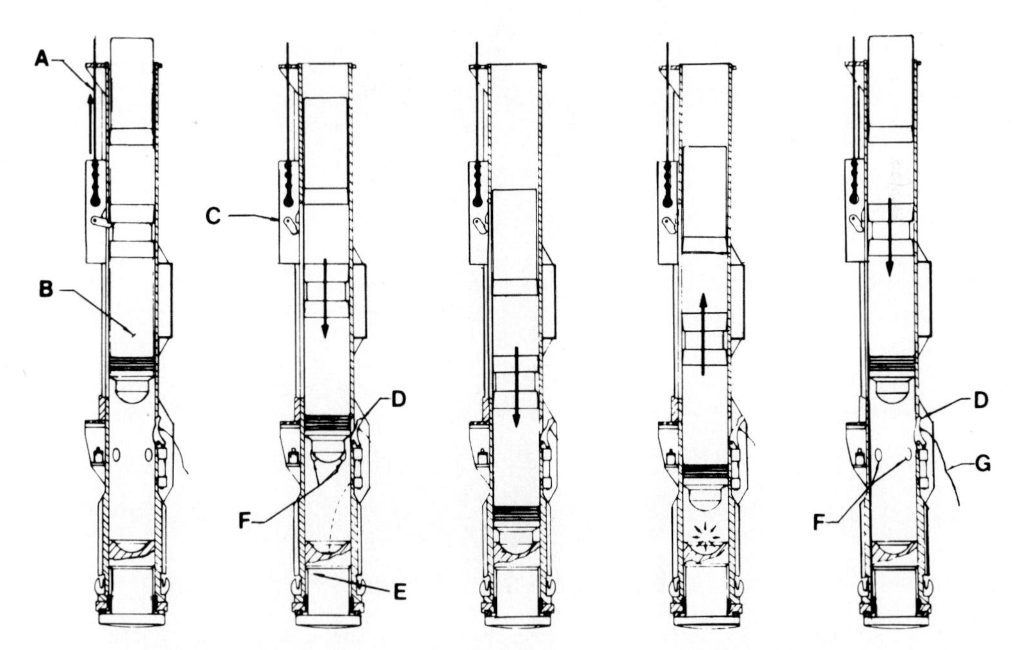

Figure 9-6 Operation of a diesel pile hammer. (Courtesy of MKT Geotechnical Systems)

fires the fuel-air mixture. As the cylinder fires, it forces the body of the hammer down against the pile top and drives the ram upward to start a new cycle. Operation of the hammer is stopped by pulling the rope (G), which disengages the fuel pump cam (D). Diesel hammers are compact, light, and economical and can operate in freezing weather. However, they may fail to operate in soft soil, where hammer impact may be too weak to fire the fuel-air mixture.

Vibratory hammers drive piles by a combination of vibration and static weight. As you might expect, they are most effective in driving piles into clean granular soils. Sonic hammers are vibratory hammers that operate at very high frequencies. Figure 9-7 shows a hydraulically powered vibratory driver/extractor in operation.

PILE-DRIVING PROCEDURES

A typical pile-driving operation for a straight shell pile is illustrated in Figure 9-8. Figure 9-8a shows the piles stockpiled at the job site. Notice the depth marks that have been painted on the pile. These will be used during driving to facilitate counting the number of blows required to obtain a foot of penetration. In Figure 9-8b the pile has been hooked to the hoist cable and is being swung into position for lowering into a hollow casing previously driven into the ground. After the shell has been lowered into the casing, the pile driver's mandrel (Figure 9-8c) is lowered into the shell. The shell and mandrel are then raised from the casing and swung into position for driving. The hammer (in this case a single-acting compressed air hammer) then

Figure 9-7 Hydraulically powered vibratory driver/extractor. (Courtesy of MKT Geotechnical Systems)

Figure 9-8 Driving a shell pile.

drives the pile (Figure 9-8d) until the required depth or driving resistance is obtained. After the mandrel is raised, the shell is cut off at the required elevation with a cutting torch (Figure 9-8e). When the reinforcing steel for the pile cap has been placed (Figure 9-8f), the shell is ready to be filled with concrete.

For driving piles with an impact-type pile driver it is recommended that a hammer be selected that will yield the required driving resistance at a final penetration of 8 to 12 blows/in. (reference 5). For fluid-powered hammers, it is also recommended that the weight of the ram be at least one-half the pile weight. For diesel hammers, ram weight should be at least one-fourth of the pile weight. When selecting a vibratory driver/extractor, a machine should be used that will yield a driving amplitude of ¼ to ½ in. (0.6 to 1.2 cm).

$$\text{Driving amplitude (in.)} = 2 \times \frac{\text{Eccentric moment (in.-lb)}}{\text{Vibrating mass (lb)}} \qquad (9\text{-}1)$$

In solving Equation 9-1 for driving amplitude, pile weight should be added to the weight of the driver's vibrating mass to obtain the value of the vibrating mass.

Powered hammers with leads should be used for driving piles at an angle (batter piles), because drop hammers lose significant energy to friction when the leads are inclined. Powered hammers without leads may be used in vertical driving, but the use of leads assists in maintaining pile alignment during driving. Double-acting, differential, and vibratory hammers may be used to extract piles as well as to drive them.

Determining Pile Load Capacity

The problem of determining pile load capacity is a complex one since it involves pile–soil–hammer interaction during driving, pile–soil interaction after the pile is in place, and the structural strength of the pile itself. The geotechnical engineer who designs the foundation must provide a pile design that is adequate to withstand driving stresses as well as to support the design load of the structure without excessive settlement. The best measure of in-place pile capacity is obtained by performing pile load tests as described later in this section.

A number of dynamic driving equations have been developed in attempting to predict the safe load capacity of piles based on behavior during driving. The traditional basis for such equations is to equate resisting energy to driving energy with adjustments for energy lost during driving. These equations treat the pile as a rigid body. A number of modifications to basic driving equations have been proposed in an attempt to provide better agreement with measured pile capacity. Equation 9-2, for determining the safe capacity of piles driven by powered hammers, has been incorporated in some U.S. building codes. Minimum hammer energy may also be specified by the building code.

$$R = \frac{2E}{S + 0.1}\,\frac{W_r + KW_p}{W_r + W_p} \qquad (9\text{-}2)$$

where R = safe load (lb)

S = average penetration per blow, last six blows (in.)

E = energy of hammer (ft-lb)

K = coefficient of restitution $\begin{cases} 0.2 \text{ for piles weighing 50 lb/ft or less} \\ 0.4 \text{ for piles weighing 50 to 100 lb/ft} \\ 0.6 \text{ for piles weighing over 100 lb/ft} \end{cases}$

W_r = weight of hammer ram (lb)

W_p = weight of pile, including driving appurtenances (lb)

Example 9-1

PROBLEM Using Equation 9-2 and the driving data below, determine the safe load capacity of a 6-in.-square concrete pile 60 ft long. Assume that the unit weight of the pile is 150 lb/cu ft.

$$\begin{aligned} \text{Pile driver energy} &= 14{,}000 \text{ ft-lb} \\ \text{Ram weight} &= 4000 \text{ lb} \\ \text{Weight of driving appurtenances} &= 1000 \text{ lb} \\ \text{Average penetration last six blows} &= \frac{1}{5} \text{ in./blow} \end{aligned}$$

SOLUTION

$$\begin{aligned} \text{Weight of pile} &= \frac{6 \times 6}{144} \times 60 \times 150 = 2250 \text{ lb} \\ W_p &= 2250 + 1000 = 3250 \text{ lb} \\ \text{Weight per foot of pile} &= \frac{2250}{60} = 37.5 \text{ lb/ft} \\ K &= 0.2 \\ S &= 0.2 \text{ in./blow} \\ R &= \frac{2E}{S + 0.1} \frac{W_r + KW_p}{W_r + W_p} \\ &= \frac{(2)(14{,}000)}{0.2 + 0.1} \left[\frac{4000 + (0.2)(3250)}{4000 + 3250} \right] \\ &= \frac{(28{,}000)(4650)}{(0.3)(7250)} = 59{,}862 \text{ lb} \end{aligned}$$

Equation 9-3 is used in several building codes and construction agency specifications for predicting the safe load capacity of bulb piles.

$$L = \frac{W \times H \times B \times V^{2/3}}{K} \tag{9-3}$$

where L = safe load capacity (tons)
W = weight of hammer (tons)
H = height of drop (ft)
B = number of blows per cubic foot of concrete used in driving final batch into base
V = uncompacted volume of concrete in base and plug (cu ft)
K = dimensionless constant depending on soil type and type of pile shaft

Nordlund (reference 7) has presented recommended K values which range from 9 for a compacted shaft pile in gravel to 40 for a cased shaft pile in very fine sand.

Example 9-2

PROBLEM Calculate the safe load capacity of a bulb pile based on the following driving data.

Hammer weight = 3 tons
Height of drop = 20 ft
Volume in last batch driven = 5 cu ft
Number of blows to drive last batch = 40
Volume of base and plug = 25 cu ft
Selected K value = 25

SOLUTION

$$B = \frac{40}{5} = 8 \text{ blows/cu ft}$$

$$R = \frac{W \times H \times B \times V^{2/3}}{K}$$

$$= \frac{(3)(20)(8)(25)^{2/3}}{25} = 164 \text{ tons}$$

A newer and better approach to predicting pile capacity during driving is provided by the use of wave equations to analyze dynamic forces developed during driving. In this method, force and velocity waves are measured near the top of the pile during driving and plotted against time. Computer programs are available to use these wave data to provide an estimate of pile capacity and stress in the pile during driving. In contrast to energy-based driving equations, wave equations treat the pile as an elastic body.

Pile capacity may be determined by performing pile load tests. One such test procedure (ASTM D-1143) involves loading the pile to 200% of design load at increments of 25% of the design load. Each load increment is maintained until the rate of settlement is not greater than 0.01 in./h or until 2 h have elapsed. The final load (200% of design load) is maintained for 24 h. Quick load tests utilizing a constant rate of penetration test or a maintained load test are also used. Quick load tests can usually be performed in 3 h or less.

With all of the foregoing methods of load testing, pile settlement is plotted against load to select a safe capacity. A number of methods have been proposed for identifying the failure load on a load–settlement curve (reference 6). One procedure for interpreting the load–settlement curve to determine pile capacity involves drawing tangents to the initial and final segments of the curve, as illustrated in Figure 9-9. The load (A) corresponding to the intersection of these two tangents is divided by 2 to yield the safe capacity.

Pile capacity usually increases after a period of time following driving. This increase in capacity is referred to as *soil setup* or *soil freeze.* However, in some cases pile capacity decreases with time. This decrease in capacity is referred to as *soil relaxation.* Soil setup or soil relaxation can be measured by performing load tests or by restriking the pile several days after pile driving. Building codes may specify a minimum waiting period between driving and loading a test pile.

9-5 PIERS AND CAISSONS

A *pier* is simply a column, usually of reinforced concrete, constructed below the ground surface. It performs much the same function as a pile. That is, it transfers the load of a structure down to a stronger rock or soil layer. Piers may be constructed in an open excavation, a lined excavation (caisson), or a drilled excavation. Since piers are often constructed by filling a caisson with concrete, the terms pier foundation, caisson foundation, and drilled pier foundation are often used interchangeably.

A *caisson* is a structure used to provide all-around lateral support to an excavation. Caissons may be either open or pneumatic. Pneumatic caissons are air- and watertight structures open on the bottom to permit the excavation of soil beneath the caisson. The caisson is filled with air under pressure to prevent water and soil from flowing in as excavation proceeds. To prevent workers from suffering from the

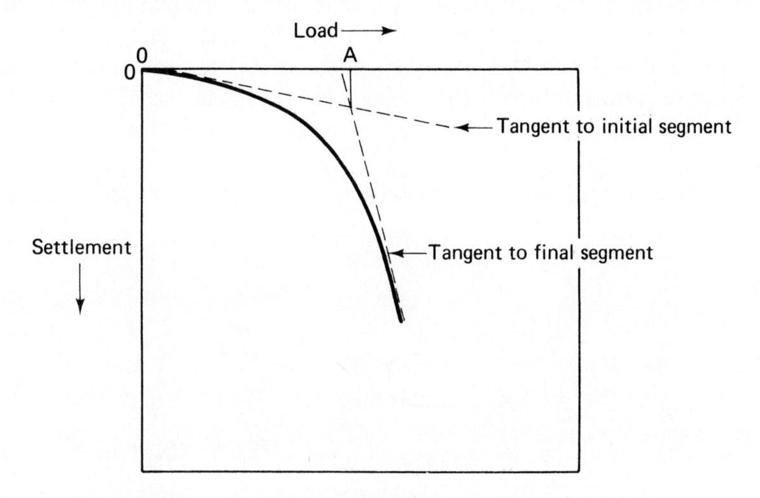

Figure 9-9 Determination of pile capacity from load test.

bends upon leaving pneumatic caissons, they must go through a decompression procedure like that employed for divers. Because of the health hazards and expense of this procedure, pneumatic caissons are rarely used today.

Drilled piers are piers placed in holes drilled into the soil. Holes drilled into cohesive soils are not usually lined. If necessary, the holes may be filled with a slurry of clay and water (such as bentonite slurry) during drilling to prevent caving of the sides. Concrete is then placed in the hole through a tremie, displacing the slurry. This procedure is similar to the slurry trench excavation method described in Section 9-6.

Holes drilled in cohesionless soils must be lined to prevent caving. Metal or fiber tubes are commonly used as liners. Linings may be left in place or they may be pulled as the concrete is placed. Holes for drilled piers placed in cohesive soil are often widened (or belled) at the bottom, as shown in Figure 9-1, to increase the bearing area of the pier on the supporting soil. Although this increases allowable pier load, such holes are more difficult to drill, inspect, and properly fill with concrete than are straight pier holes.

9-6 STABILITY OF EXCAVATIONS

Slope Stability

To understand the principal modes of slope failure, it is necessary to understand the basic concepts of soil strength. The soil identification procedures discussed in Chapter 2 included the classification of soil into cohesionless and cohesive types. As you recall, cohesionless soil is one whose grains do not show any tendency to stick together. The shear strength of a cohesionless soil is thus due solely to the friction developed between soil grains. A normal force (or force perpendicular to the sliding surface) is required to develop this strength. When an embankment composed of a cohesionless soil fails, it fails as shown in Figure 9-10. That is, material from the upper part of the slope breaks away and falls to the toe of the slope until the face of the embankment reaches the natural angle of repose for the soil.

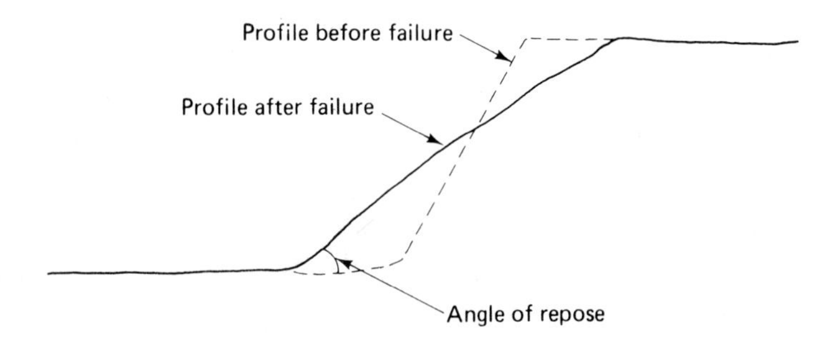

Figure 9-10 Slope failure of cohesionless soil.

In a cohesive soil, on the other hand, shear strength is provided primarily by the attraction between soil grains (which we call *cohesion*). Theoretically, a completely cohesive soil would exhibit no friction between soil grains. Failure of a highly cohesive soil typically occurs as shown in Figure 9-11. Notice that a large mass of soil has moved along a surface which we call a *slip plane*. The natural shape of this failure surface resembles the arc of an ellipse but is usually considered to be circular in soil stability analyses.

Embankment Failure During Construction

Most soils encountered in construction exhibit a combination of the two soil extremes just described. That is, their shear strength is due to a combination of intergranular friction and cohesion. However, the behavior of a highly plastic clay will closely approximate that of a completely cohesive soil.

Theoretically, a vertical excavation in a cohesive soil can be safely made to a depth that is a function of the soil's cohesive strength and its angle of internal friction. This depth can range from under 5 ft for a soft clay to 18 ft or so for a medium clay. The safe depth is actually less for a stiff clay than for a medium clay, because stiff clays commonly contain weakening cracks or fissures. In practice, however, the theoretically safe depth of unsupported excavation in clay can be sustained for only a limited time. As the clay is excavated, the weight of the soil on the sides of the cut causes the sides of the cut to bulge (or move inward at the bottom) with an accompanying settlement (or subsidence) of the soil at the top of the cut, as shown in Figure 9-12. Subsidence of the soil at the top of the cut usually results in the formation of tension cracks on the ground surface, as shown in Figure 9-13. Such cracks usually occur at a distance from the face of the cut equal to one-half to two-thirds of the depth of the cut. If lateral support is not provided, tension cracks will continue to deepen until failure of the embankment occurs. Failure may occur by sliding of the soil face into the cut (Figure 9-14a) or by toppling of the upper part of the face into the cut (Figure 9-14b).

The stability of an embankment or excavation is also affected by external factors. These include weather conditions, ground water level, the presence of loads such as material and equipment near the top of the embankment/excavation, and the presence of vibration from equipment or other sources (see also Section 18-4).

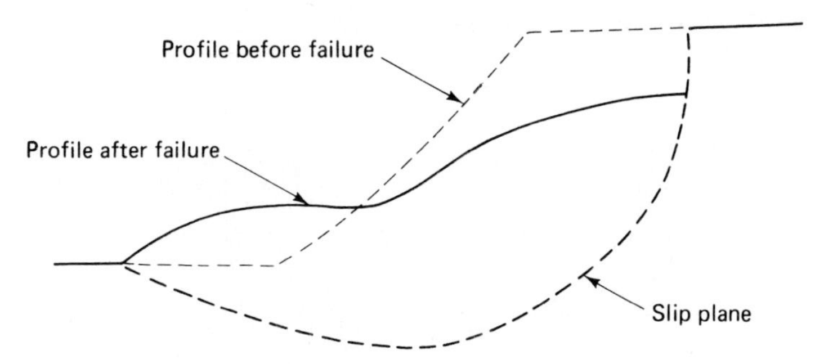

Figure 9-11 Slope failure of cohesive soil.

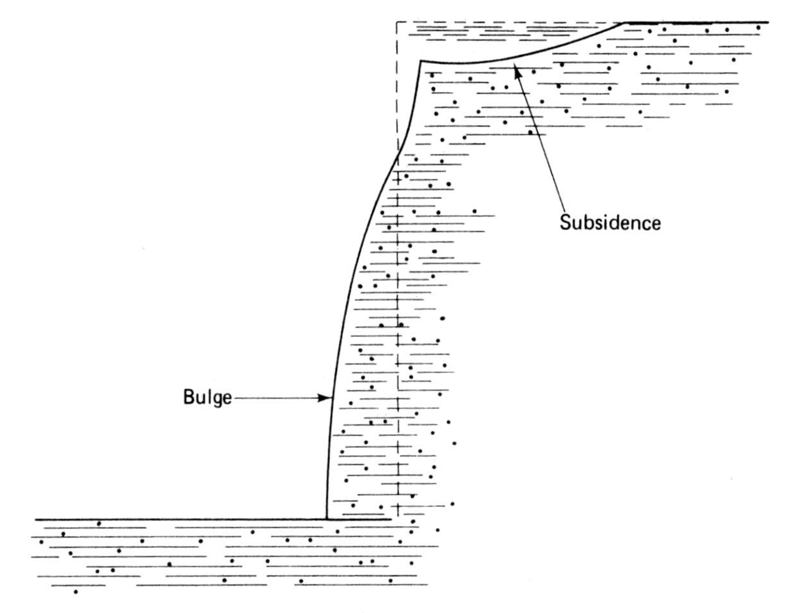

Figure 9-12 Subsidence and bulging.

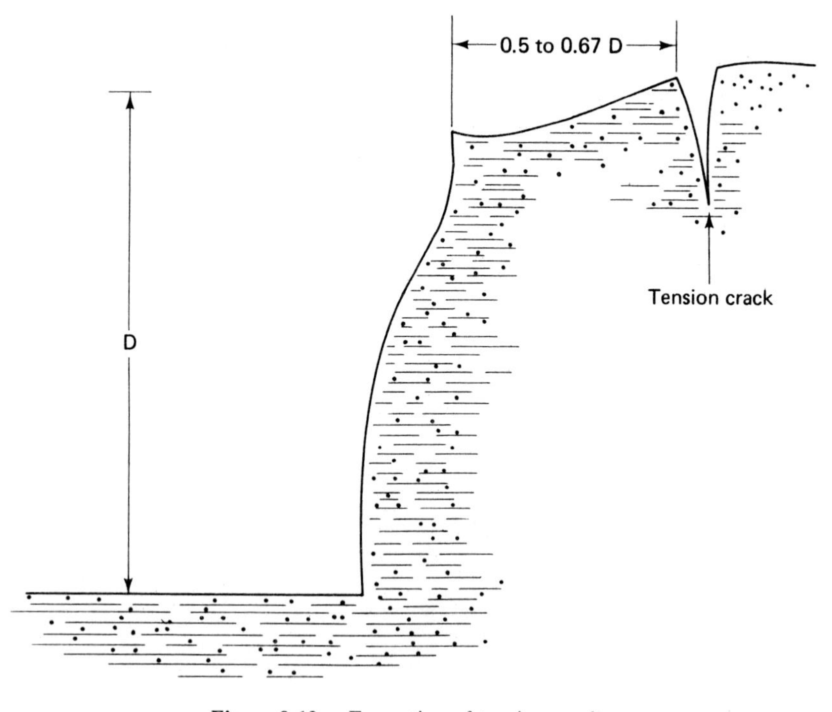

Figure 9-13 Formation of tension crack.

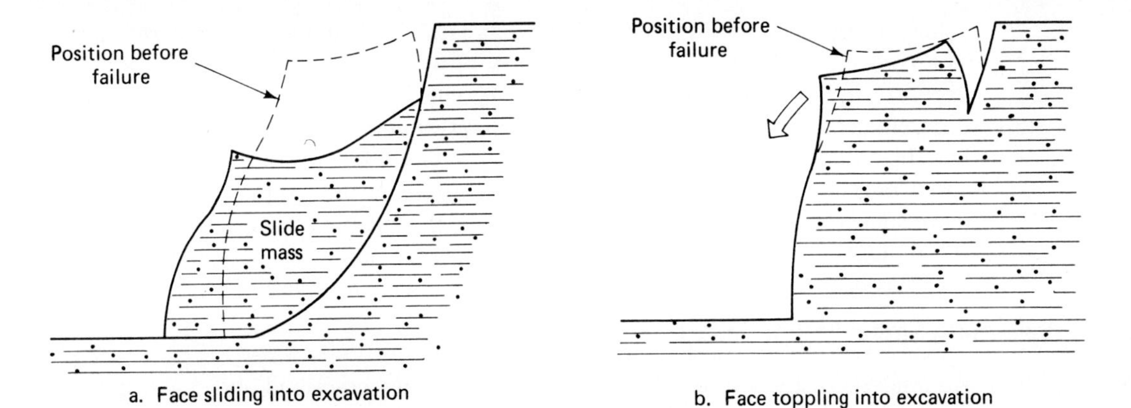

Figure 9-14 Modes of embankment failure.

Stability of Cut Bottom

Whenever cohesive soil is excavated, heaving (or rising) of the bottom of the cut will occur due to the weight of the soil on the sides of the cut. Heaving is most noticeable when the sides of the cut have been restrained, as shown in Figure 9-15. A more serious case of bottom instability may occur in cohesionless soils when a supply of water is present. If the sides of the cut are restrained and the bottom of the cut is

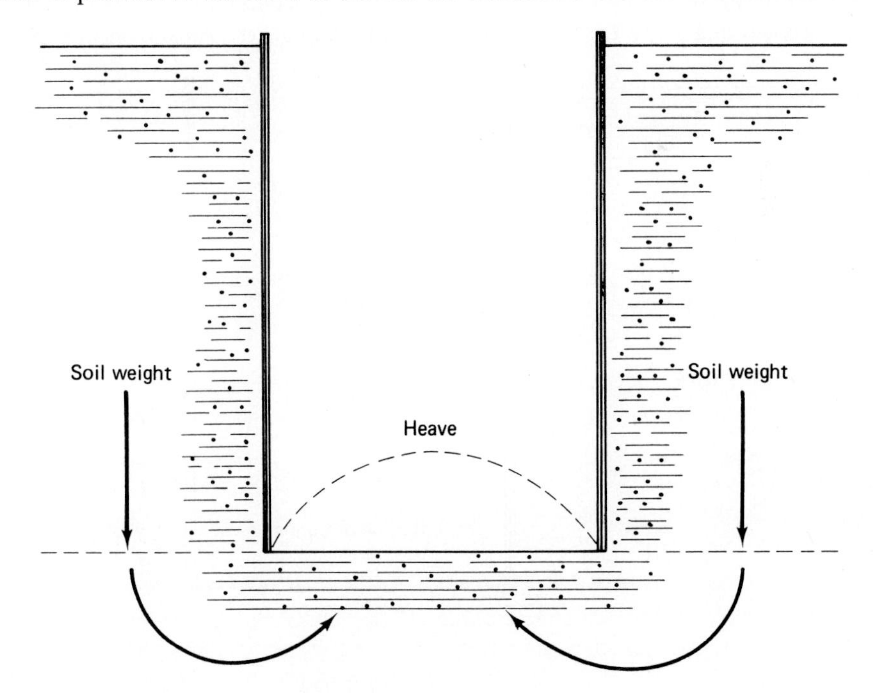

Figure 9-15 Heaving of cut bottom.

below the groundwater level, water will flow up through the bottom of the excavation, as shown in Figure 9-16. The upward flow of water reduces the effective pressure between the soil grains in the bottom of the cut. This may result in one of several different conditions. If the water pressure exactly equals soil weight, the soil will behave like a liquid and we have a condition called *liquefaction* (or quicksand). Such a soil is unable to support any applied load. If the water pressure is strong enough to move subsurface soil up through the bottom of the cut, this condition is called *boiling* or *piping*. Such a movement of soil often leads to failure of the surrounding soil. This has been the cause of the failure of some dams and levees.

Preventing Embankment Failure

An analysis of the causes of excavation slope failure described above will indicate methods that can be used to prevent such failures. Side slopes may be stabilized by cutting them back to an angle equal to or less than the angle of repose of the soil, or by providing lateral support for the excavation. Both side and bottom stability may be increased by dewatering the soil surrounding the excavation. Methods for dewatering and protecting excavations are described in the following sections.

To protect more permanent slopes, such as highway cuts, retaining walls are often used. Slopes of cohesive soil may be strengthened by increasing the shearing resistance along the potential slip plane. This may be done by driving piles or inserting stone columns into the soil across the potential slip plane. Another technique for reinforcing slopes is called *soil* (or earth) *reinforcement*. One form of this process is known under the trademark name Reinforced Earth. As shown in Figure 9-17, soil

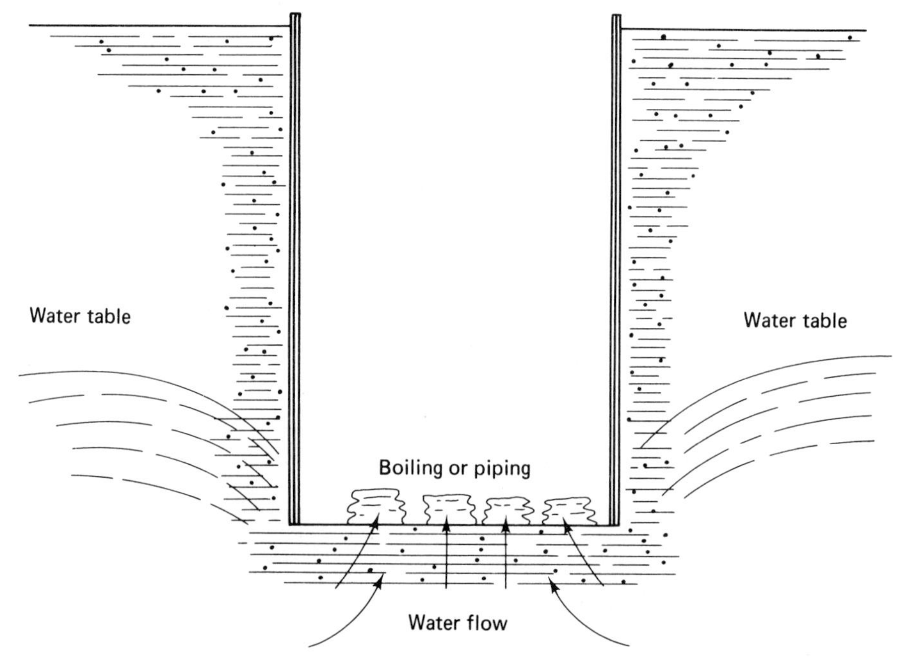

Figure 9-16 Boiling and piping of cut bottom.

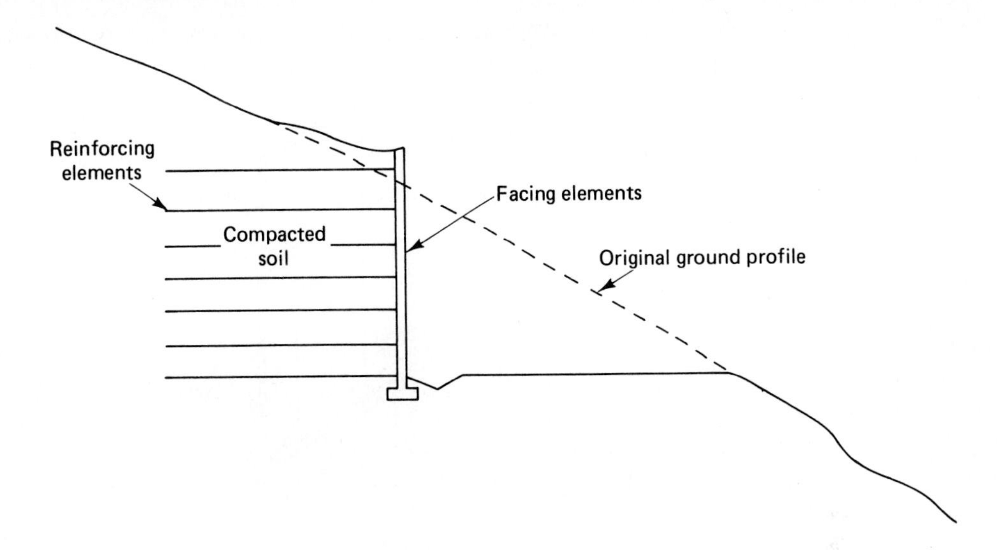

Figure 9-17 Soil reinforcement.

reinforcement involves embedding high-tensile-strength nonbiodegradable elements in a compacted soil mass. The embedded tensile elements are attached to facing material, usually of concrete or timber, to prevent erosion or raveling of soil at the cut surface. Soil reinforcement is often a less expensive method for stabilizing slopes than is the construction of conventional retaining walls.

9-7 PROTECTING EXCAVATIONS

Shoring Systems

Lateral support for the sides of an excavation is usually provided by *shoring.* A shoring system that completely encloses an excavation is essentially a cofferdam, which is a structure designed to keep water and/or soil out of an excavation area. A caisson is also a form of cofferdam, as we have seen. There are several types of shoring systems, of which the most common are sheeting, lagging, and sheet piling.

Sheeting involves the use of vertical members (usually timber) placed against the sides of the excavation and supported by horizontal beams called wales or stringers. Wales are braced by horizontal braces (called struts, cross braces, trench jacks, or trench braces), by inclined braces, or by anchors embedded in the sides of the excavation. A typical sheeting system is illustrated in Figure 9-18.

Lagging is nothing more than sheeting placed horizontally. However, in this case vertical supports (called soldier beams) are required between the lagging and-wales. Another lagging system uses soldier piles (such as H-piles) with the lagging placed between the open sides of the piling. Wales and struts or tiebacks are used to provide lateral support.

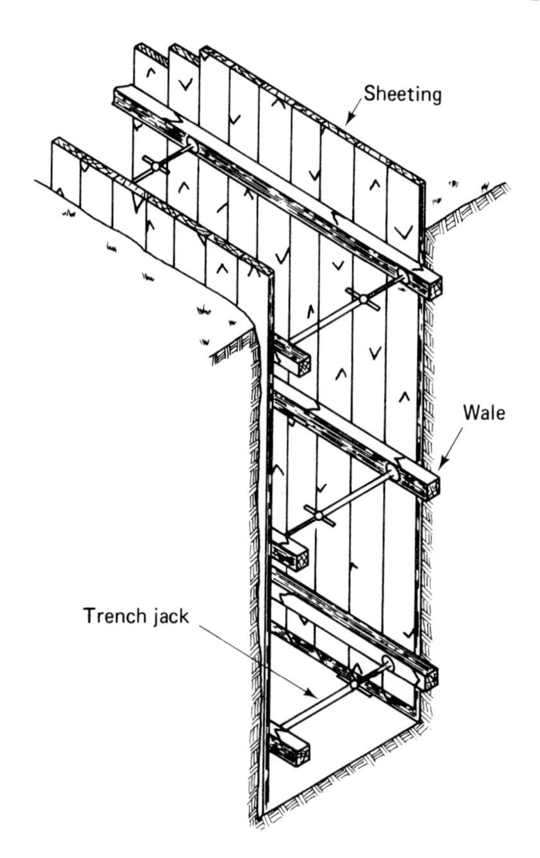

Figure 9-18 Trench shoring with sheeting.

Sheet piling is sheeting of concrete, steel, or timber that is designed to be driven by a pile driver. Sheet piling is used for constructing retaining walls, shoring, and cofferdams. Two sheet pile walls may be constructed parallel to each other, cross-braced, and filled with earth to form a cofferdam. When tight sheeting or sheet piling is used, the shoring system must be designed to withstand the full hydrostatic pressure of the groundwater level unless weep holes or other drains are provided in the shoring system.

Trench shields or trench boxes are used in place of shoring to protect workers during trenching operations. Figure 9-19 illustrates such a movable trench shield. The top of the shield should extend above the sides of the trench to provide protection for workers against objects falling from the sides of the trench. The trench shield is pulled ahead by the excavator as work progresses.

Slurry Trenches

A relatively new development in excavating and trenching is the construction of *slurry trenches.* In this technique, illustrated in Figure 9-20, a slurry (such as clay and water) is used to fill the excavation as soil is removed. The slurry serves to keep

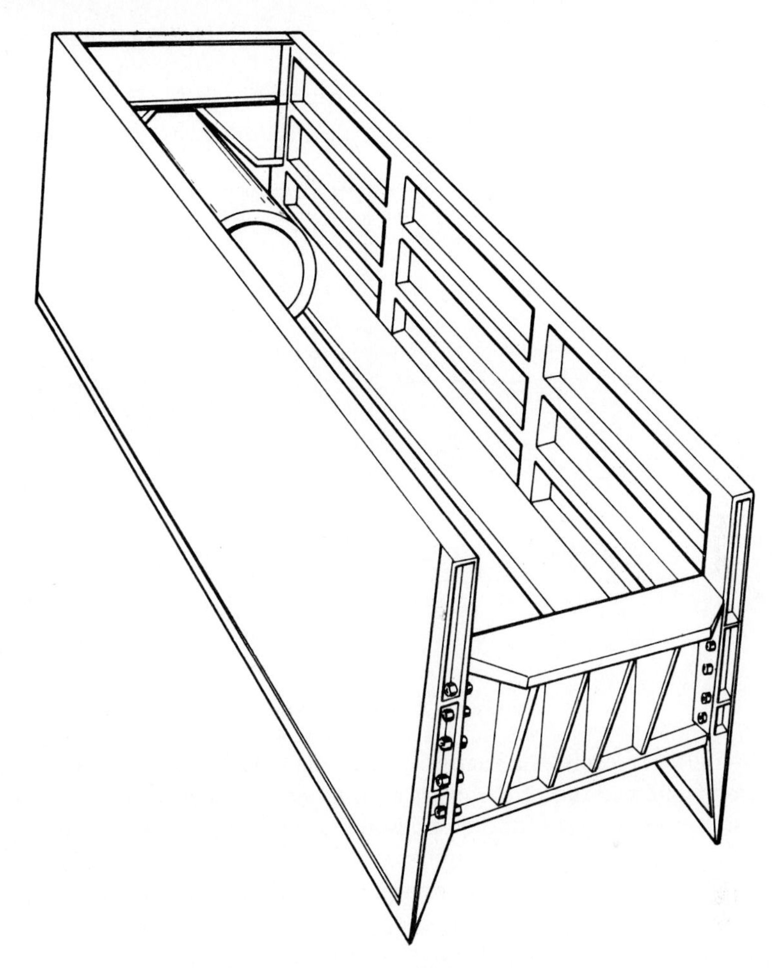

Figure 9-19 Trench shield.

the sides of the trench from collapsing during excavation. No lowering of the water table is required with this method. After completion of the trench, the slurry is displaced by concrete placed through the slurry by use of a tremie. The slurry is pumped away as it is displaced. The slurry trench technique eliminates the necessity for shoring and dewatering excavations. The soil between two rows of completed slurry trenches may be excavated to form a large opening such as a subway tunnel.

9-8 DEWATERING EXCAVATIONS

Dewatering is the process of removing water from an excavation. Dewatering may be accomplished by lowering the groundwater table before the excavation is begun. This method of dewatering is often used for placing pipelines in areas with high

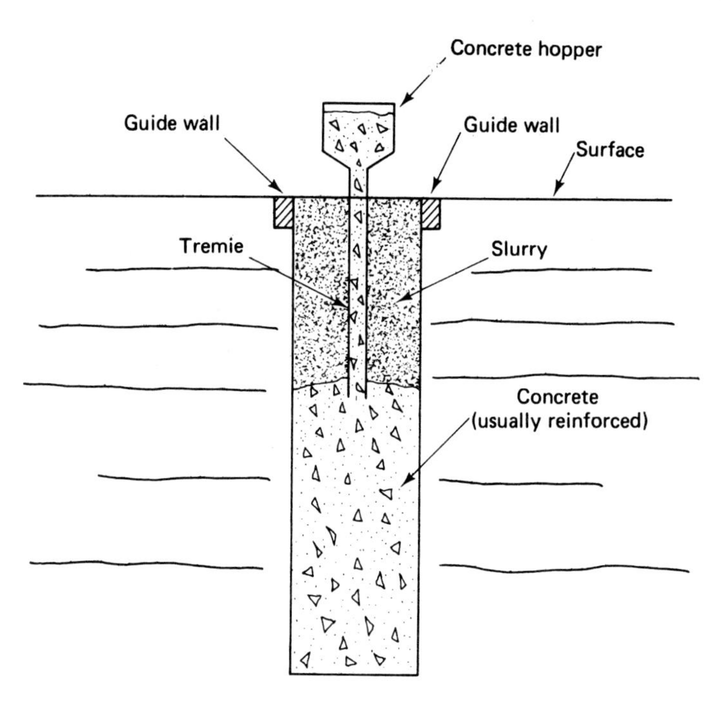

Figure 9-20 Slurry trench construction.

groundwater levels. Alternatively, excavation may be accomplished first and the water simply pumped out of the excavation as work proceeds. With either procedure, the result is a lowering of the groundwater level in the excavation area. Hence all dewatering methods involve pumping of water from the ground.

The selection of an appropriate dewatering method depends on the nature of the excavation and the permeability of the soil. *Soil permeability,* or the ease with which water flows through the soil, is primarily a function of a soil's grain size distribution. It has been found that the diameter of the soil particle which is smaller than 90% of the soil's grains (i.e., 10% of total soil grains are smaller than the designated grain size) is an effective measure of soil permeability. This soil grain size is referred to as the soil's *effective grain size* and is represented by the symbol D_{10}. Table 9-1 indicates appropriate dewatering methods as a function of effective soil grain size. Note that gravity drainage (use of pumps and well-points) is effective for soils whose effective grain size is about 0.1 mm (corresponding to a No. 150 sieve size) or larger.

Table 9-1 Appropriate dewatering methods

Effective Grain Size (D_{10})	*Dewatering Method*
Larger than 0.1 mm*	Sumps, ordinary wellpoints
0.1–0.004 mm	Vacuum wells or wellpoints
0.004–0.0017 mm	Electroosmosis

*No. 150 sieve size corresponds to an opening of 0.1 mm.

Wellpoint Systems

Figure 9-21 illustrates the use of a standard wellpoint system to dewater an area prior to excavation. Technically, a *wellpoint* is the perforated assembly placed on the bottom of the inlet pipe for a well. It derives its name from the point on its bottom used to facilitate driving the inlet pipe for a well. In practice, the term "wellpoint" is commonly used to identify each well in a dewatering system, consisting of a number of closely spaced wells. In sandy soils, the usual procedure is to jet the well point and riser into position. This is accomplished by pumping water down through the riser and wellpoint to loosen and liquefy the sand around the wellpoint. Under these conditions, the wellpoint sinks under its own weight to the desired depth. Additional wellpoints are sunk in a line surrounding the excavation area, then connected to a header pipe. Header pipes used for such systems are essentially manifolds consisting of a series of connection points with valves. After all well points are in place and connected to the header, the header pipes are connected to a self-priming centrifugal pump equipped with an air ejector. Since water from the wellpoints is drawn off by creating a partial vacuum at the pump inlet, the maximum height that water can be lifted by the pump is something less than 32 ft. In practice, the maximum effective dewatering depth is about 20 ft below the ground surface. Wellpoints are typically spaced 2 to 10 ft apart and yield flows ranging from 3 to 30 gal/min per wellpoint. Wellpoints placed in very fine sands may require the use of a coarse sand filter around the wellpoint to prevent an excessive flow of fine sand into the system. If the groundwater table must be lowered more than 20 ft a single stage of wellpoints will not be effective. In this situation two or more levels of wellpoints (called *stages*) may be used. The major disadvantage of such a system is the large area required for terracing the stages. For example, to lower the water table 36 ft using two stages with embankment side slopes of 1 on 2 and allowing a 5-ft-wide bench for each pump requires a total width of 82 ft on each side of the excavation. Alternatives to the use of staged

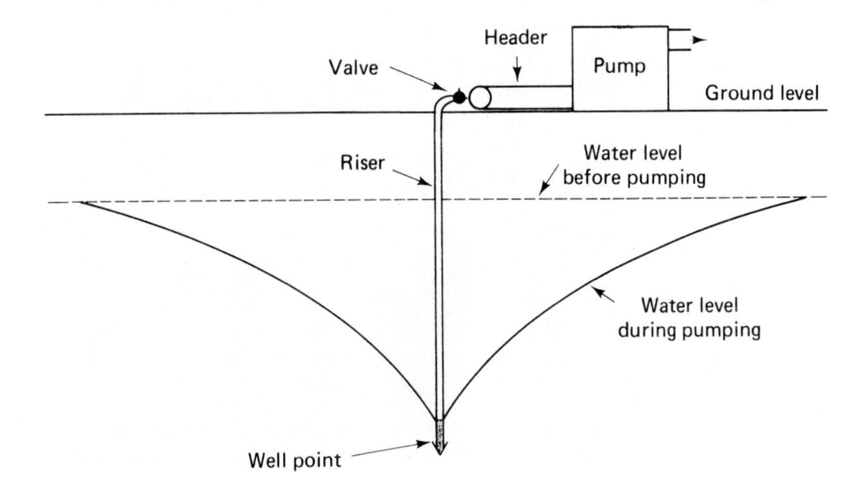

Figure 9-21 Wellpoint dewatering system.

wellpoints include the use of jet pumps and submersible pumps to lift water from the wells. Figure 9-22 shows an electrically powered submersible pump being placed into a dewatering well.

VACUUM WELLS

Vacuum wells are wellpoints that are sealed at the surface by placing a ring of bentonite or clay around the well casing. A vacuum pump is then connected to the header pipe. The resulting differential pressure between the well and the surrounding groundwater will accelerate the flow of water into the well. In fine-grained soils it may also be necessary to place a sand filter around the wellpoint and riser pipe.

ELECTROOSMOSIS

Electroosmosis is the process of accelerating the flow of water through a soil by the application of a direct current. Although the phenomenon of electroosmosis was discovered in the laboratory early in the nineteenth century, it was not applied to construction dewatering until 1939. As shown in Table 9-1, the method is applicable to relatively impervious soils such as silts and clays having an effective grain size as small as 0.0017 mm.

The usual procedure for employing electroosmosis in dewatering is to space wells at intervals of about 35 ft and drive grounding rods between each pair of wells. Each well is then connected to the negative terminal of a dc voltage source and each ground rod is connected to a positive terminal. A voltage of 1.5 to 4 V/ft of distance between the well and anode is then applied, resulting in an increased flow of water to the

Figure 9-22 Electrically powered submersible pump being placed into dewatering well. (Courtesy of Prosser Industries)

well (cathode). The applied voltage should not exceed 12 V/ft of distance between the well and anode to avoid excessive power loss due to heating. Typical current requirements of 15 to 30 A per well result in power demands of 0.5 to 2.5 kW per well.

A measure of the effectiveness of electroosmosis can be gained by comparing the flow developed by an electrical voltage with the flow produced by conventional hydraulic forces. Such a calculation for a clay of average permeability indicates that an electrical potential of 3 V/ft is equivalent to a hydraulic gradient of 50 ft/ft. To obtain a hydraulic gradient of 50 ft/ft by the use of vacuum wells would require wells to be spaced about 1 ft apart. This calculation provides a measure of the tremendous increase in water flow that electroosmosis provides in soils of low permeability over the flow produced by conventional hydraulic methods.

9-9 PRESSURE GROUTING

Grouting or *pressure grouting* is the process of injecting a grouting agent into soil or rock to increase its strength or stability, protect foundations, or reduce groundwater flow. Grouting of rock is widely employed in dam construction and tunneling. The need for such grouting is determined by explatory methods such as core drilling and visual observation in test holes. Pressure tests that measure the flow of water through injector pipes which have been placed and sealed into test holes may also be employed as a measure of the need for grouting and for measuring the effectiveness of grouting. Recent developments in grouting agents and injection methods have led to an increasing use of grouting in soils.

Common grouting patterns include blanket grouting, curtain grouting, and special grouting. *Blanket grouting* covers a large horizontal area, usually to a depth of 50 ft (15 m) or less. *Curtain grouting* produces a linear deep, narrow zone of grout that may extend to a depth of 100 ft (30 m) or more. It is commonly employed to form a deep barrier to water flow under a dam. *Special grouting* is grouting employed for a specific purpose, such as to consolidate rock or soil around a tunnel, fill individual rock cavities, or provide additional foundation support.

Grouting Methods

Major types of grouting include slurry grouting, chemical grouting, compaction grouting, and jet grouting (Figure 9-23).

Slurry grouting involves the injection of a slurry consisting of water and a grouting agent into soil or rock. Common grouting materials include portland cement, clay (bentonite), fly ash, sand, lime, and additives. In soil, regular portland cement grouts are only able to effectively penetrate gravel and coarse sand. Newer *microfine cement* (or fine-grind cement) grouts are able to penetrate medium and fine sands. Injection of *lime slurry* grout can be used to control the swelling of expansive clays. It can also be used to stabilize low-strength soils such as silts, dredge spoil, and saturated soils.

Chemical grouting involves the injection of a chemical into soil. It is used primarily in sands and fine gravel to cement the soil particles together for structural

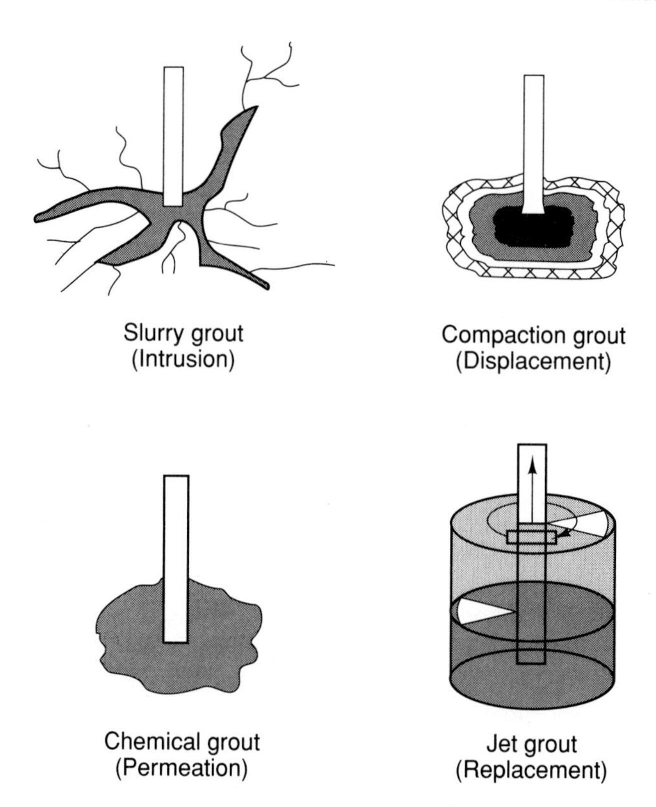

Figure 9-23 Types of grouting. (Courtesy of Hayward Baker Inc., A Keller Company)

support or to control water flow. The proper selection of a chemical grout and additives permit rather precise control of grout hardening (setting) time.

Compaction grouting is the process of injecting a very stiff mortar grout into a soil to compact and strengthen the soil. Grouting materials include silty sand, cement, fly ash, additives, and water. Compaction grouting is able to create grout bulbs or grout piles in the soil which serve to densify the soil and provide foundation support. Compaction grouting can also be used to raise (jack) foundations that have settled back to their original elevation.

Jet grouting employs a rotating jet pipe to remove soil around the grout pipe and replace the soil with grout. As a result, the technique is effective over a wide range of soil types to include silts and some clays. The unconfined compressive strength of the grouted soil structure may run as high as 2500 psi (17 MPa).

Injection Methods

The principal method for injecting grout into rock involves drilling a hole and then inserting an injector pipe equipped with expandable seals (*packers*) into the hole. Grout is then injected at the desired depths. Methods for injecting grout into soil

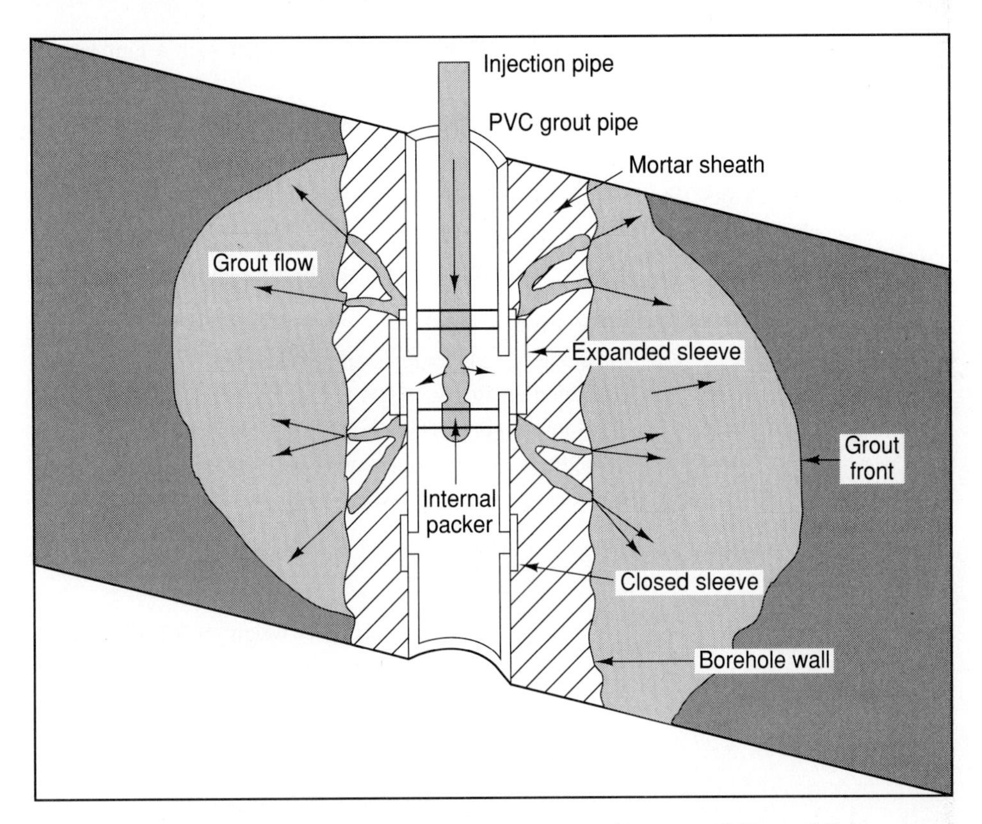

Figure 9-24 Grouting utilizing a sleeve port pipe. (Courtesy of Hayward Baker Inc., A Keller Company)

include driving an injector pipe into the soil, placing a sleeve port tube into the soil, and jet grouting. Grouting utilizing a sleeve port pipe is illustrated in Figure 9-24. Notice that the grout pipe is equipped with sleeves that cover ports spaced at intervals along the pipe. The sleeves serve as check valves to allow grout to flow out of the ports but prevent return flow. Packers serve to direct the flow of grout through the desired ports.

Selection of an optimum grouting agent and grouting system should be accomplished by experienced grouting specialists. Trial grouting and testing will usually be required before selecting the grouting system to be employed. Care must be taken to avoid the use of injection pressures that lift the ground surface, unless a jacking action is desired.

PROBLEMS

1. Briefly explain the term *soil improvement* and its purpose. List the five major methods used.

2. Briefly describe stone column construction. When might such a foundation system be used?

3. How might a saturated cohesive soil be densified? Describe the advantages and disadvantages of each major method.

4. Briefly describe the procedure for driving a shell pile.

5. Using Equation 9-2 and the driving data below, determine the safe load of an 8-in.-square concrete pile 40 ft long. Assume that the unit weight of the pile is 150 lb/cu ft.

   Pile driver:
   Single-acting compressed air hammer
   Rated energy = 15,000 ft-lb
   Ram weight = 5000 lb
   Weight of driving appurtenances = 2000 lb
   Average penetration last six blows = ¼ in./blow

6. Using Equation 9-3 and the driving data below, determine the safe load capacity of a bulb pile.

   Hammer weight = 3.5 tons
   Height of hammer drop = 20 ft
   Volume in last batch driven = 5 cu ft
   Number of blows to drive last batch = 45
   Volume of base and plug = 25 cu ft
   Soil K value = 26

7. What are minipiles? Identify their principal applications.

8. Briefly discuss the major causes for excavation slope failure and methods that can be used to prevent such failures.

9. What is jet grouting, and when might it be employed?

10. Write a computer program to predict the safe capacity of a pile driven by a powered hammer using Equation 9-2. Solve Problem 5 using your computer program.

11. Write a computer program to predict the safe capacity of a bulb pile using Equation 9-3. Solve Problem 6 using your computer program.

REFERENCES

1. AHLVIN, ROBERT G., AND VERNON A. SMOOTS. *Construction Guide for Soils and Foundations,* 2nd ed. New York: Wiley, 1988.
2. BAKER, W. H. "Grouting in Geotechnical Engineering," *Proceedings of ASCE Specialty Conference, New Orleans.* American Society of Civil Engineers, New York, 1982.
3. BOWLES, JOSEPH E. *Foundation Analysis and Design,* 3rd ed. New York: McGraw-Hill, 1982.

4. Butler, H. D., and Horace E. Hoy. *Users Manual for the Texas Quick-Load Method for Foundation Load Testing.* Federal Highway Administration, Washington, D.C., 1977.

5. Compton, G. Robert, Jr. *Selecting Pile Installation Equipment,* 3rd ed. MKT Geotechnical Systems, Dover, N.J., 1982.

6. Fellenius, B. H. "The Analysis of Results from Routine Pile Load Tests," *Ground Engineering,* vol. 13, no. 6 (1980), pp. 19–31.

7. Nordlund, Reymond L. "Dynamic Formula for Pressure Injected Footings," *Journal of the Geotechnical Engineering Division, ASCE,* vol. 108, no. GT3 (1982), pp. 419–437.

8. *OSHA Safety and Health Standards Digest: Construction Industry* (OSHA 2202). U.S. Department of Labor, Washington, D.C., 1990.

9. Welsh, J. P. "Soil Improvement—A Ten Year Update," *ASCE Geotechnical Special Publication No. 12.* American Society of Civil Engineers, New York, 1987.

10
CONCRETE CONSTRUCTION

10-1
CONSTRUCTION APPLICATIONS OF CONCRETE

Concrete, or more properly portland cement concrete, is one of the world's most versatile and widely used construction materials. Its use in the paving of highways and airfields is described in Chapter 8. Other construction applications range from its use in foundations for small structures, through structural components such as beams, columns, and wall panels, to massive concrete dams. Because concrete has little strength in tension, virtually all concrete used for structural purposes is *reinforced concrete*. That is, it contains reinforcing steel embedded in the concrete to increase the concrete member's tensile strength. The use of concrete reinforcing steel is described in Section 10-4.

Types of Concrete

Concrete is classified into several categories according to its application and density. *Normal-weight concrete* usually weighs from 140 to 160 lb/cu ft (2243 to 2563 kg/m^3), depending on the mix design and type of aggregate used. A unit weight of 150 lb/cu ft (2403 kg/m^3) is usually assumed for design purposes. Typical 28-day compressive strength ranges from 2000 to 4000 psi (13 790 to 27 580 kPa). *Structural lightweight concrete* has a unit weight less than 120 lb/cu ft (1922 kg/m^3) with a 28-day compressive strength greater than 2500 psi (17 237 kPa). Its light weight is obtained by using lightweight aggregates such as expanded shale, clay, slate, and slag. *Lightweight in-*

sulating concrete may weigh from 15 to 90 lb/cu ft (240 to 1442 kg/m^3) and have a 28-day compressive strength from about 100 to 1000 psi (690 to 6895 kPa). As the name implies, such concrete is primarily utilized for its thermal insulating properties. Aggregates frequently used for such concrete include perlite and vermiculite. In some cases, air voids introduced into the concrete mix in foam replace some or all of the aggregate particles.

Mass concrete is concrete used in a structure such as a dam in which the weight of the concrete provides most of the strength of the structure. Thus little or no reinforcing steel is used. Its unit weight is usually similar to that of regular concrete. *Heavyweight* is concrete made with heavy aggregates such as barite, magnetite, and steel punchings and used primarily for nuclear radiation shielding. Unit weights may range from 180 to about 400 lb/cu ft (2884 to 6408 kg/m^3). *No-slump concrete* is concrete having a slump of 1 in. (2.5 cm) or less. Slump is a measure of concrete consistency obtained by placing concrete into a test cone following a standard test procedure (ASTM C143) and measuring the decrease in height (slump) of the sample when the cone is removed. Applications of no-slump concrete include bedding for pipelines and concrete placed on inclined surfaces.

Refractory concrete is concrete that is suitable for high-temperature applications such as boilers and furnaces. The maximum allowable temperature for refractory concrete depends on the type of refractory aggregate used. *Precast concrete* is concrete that has been cast into the desired shape prior to placement in a structure. *Architectural concrete* is concrete that will be exposed to view and therefore utilizes special shapes, designs, or surface finishes to achieve the desired architectural effect. White or colored cement may be used in these applications. Surface textures may include exposed aggregates, raised patterns produced by form liners, sandblasted surfaces, and hammered surfaces. Architectural concrete panels are often precast and used for curtain walls and screens.

10-2 CONCRETE MATERIALS AND PROPERTIES

Desired Characteristics of Concrete

To meet design requirements while facilitating construction, it is important that concrete possess certain properties. Hardened concrete must meet design strength requirements and be uniform, watertight, durable, and wear-resistant. Desirable properties of plastic concrete include workability and economy. All of these properties are influenced by the concrete components and mix design used as well as by the construction techniques employed.

Concrete Components

The essential components of concrete are portland cement, aggregate, and water. Another component, an admixture or additive, is often added to impart certain desirable properties to the concrete mix. The characteristics and effects of each of these components on the concrete are discussed in the following paragraphs.

CEMENT

There are five principal types of portland cement, classified by the American Society for Testing and Materials (ASTM) as Types I to V, used in construction. Type I (normal) portland cement is a general-purpose cement suitable for all normal applications. Type II (modified) portland cement provides better resistance to alkali attack and produces less heat of hydration than does Type I cement. It is suitable for use in structures such as large piers and drainage systems, where groundwater contains a moderate level of sulfate. Type III (high early strength) cement provides 190% of Type I strength after 1 day of curing. It also produces about 150% of the heat of hydration of normal cement during the first 7 days. It is used to permit early removal of forms and in cold-weather concreting. Type IV (low heat) cement produces only 40 to 60% of the heat produced by Type I cement during the first 7 days. However, its strength is only 55% of that of normal cement after 7 days. It is produced for use in massive structures such as dams. Type V (sulfate-resistant) cement provides maximum resistance to alkali attack. However, its 7-day strength is only 75% of normal cement. It should be used where the concrete will be in contact with soil or water that contains a high sulfate concentration.

In addition to these five major types of cement, ASTM has established standards for a number of special cement types. Types IA, IIA, and IIIA are the same as Types I, II, and III, with the addition of an air-entraining agent. Type IS is similar to Type I except that it is produced from a mixture of blast-furnace slag and portland cement. Type IS-A contains an air-entraining agent. Types IP, IP-A, P, and P-A contain a pozzolan in addition to portland cement. Because of their reduced heat of hydration, pozzolan cements are often used in large hydraulic structures such as dams. Types IP-A and P-A cements also contain an air-entraining agent. White portland cement (ASTM C150 and C175) is also available and is used primarily for architectural purposes.

AGGREGATES

Aggregate is used in concrete to reduce the cost of the mix and to reduce shrinkage. Because aggregates make up 60 to 80% of concrete volume, their properties strongly influence the properties of the finished concrete. To produce quality concrete, each aggregate particle must be completely coated with cement paste and paste must fill all void spaces between aggregate particles. The quantity of cement paste required is reduced if the aggregate particle sizes are well distributed and the aggregate particles are rounded or cubical. Aggregates must be strong, resistant to freezing and thawing, chemically stable, and free of fine material that would affect the bonding of the cement paste to the aggregate.

WATER

Water is required in the concrete mix for several purposes. Principal among these is to provide the moisture required for hydration of the cement to take place. Hydration is the chemical reaction between cement and water which produces hardened cement. The heat that is produced by this reaction is referred to as *heat of hydration.* If aggregates are not in a saturated, surface-dry (SSD) condition when added to a concrete mix, they will either add or subtract water from the mix. Methods for correcting the amount of water added to a concrete batch to compensate for aggregate moisture are covered in Chapter 8. The amount of water in a mix also affects the plasticity or workability of the plastic concrete.

It has been found that the strength, watertightness, durability, and wear resistance of concrete are related to the water/cement ratio of the concrete mix. The lower the water/cement ratio, the higher the concrete strength and durability, provided that the mix has adequate workability. Thus the water/cement ratio is selected by the mix designer to meet the requirements of the hardened concrete. Water/cement ratios normally used range from about 0.40 to 0.70 by weight. In terms of water quality, almost any water suitable for drinking will be satisfactory as mix water. However, organic material in mix water tends to prevent the cement paste from bonding properly to aggregate surfaces. Alkalies or acids in mix water may react with the cement and interfere with hydration. Seawater may be used for mixing concrete, but its use will usually result in concrete compressive strengths 10 to 20% lower than normal. The use of a lower water/cement ratio can compensate for this strength reduction. However, seawater should not be used for prestressed concrete where the prestressing steel will be in contact with the concrete. When water quality is in doubt, it is recommended that trial mixes be tested for setting time and 28-day strength.

ADDITIVES

There are a number of types of additives or admixtures used in concrete. Some of the principal types of additives used are air-entraining agents, water-reducing agents, retarders, accelerators, pozzolans, and workability agents. *Air-entrained concrete* has significantly increased resistance to freezing and thawing as well as to scaling caused by the use of deicing chemicals. Entrained air also increases the workability of plastic concrete and the watertightness of hardened concrete. For these reasons air-entrained concrete is widely used for pavements and other structures exposed to freezing and thawing.

Water-reducing agents increase the slump or workability of a concrete mix. Thus with a water-reducing agent the amount of water in the mix may be reduced without changing the concrete's consistency. However, note that some water-reducing agents also act as retarders. *Retarders* slow the rate of hardening of concrete. Retarders are often used to offset the effect of high temperatures on setting time. They are also used to delay the setting of concrete when pumping concrete over long distances. The use of retarders to produce exposed-aggregate surfaces is discussed in Section 10-3. *Accelerators* act in the opposite manner to retarders. That is, they decrease setting time and increase the early strength of concrete. Since the most common accelerator, calcium chloride, is corrosive to metal it should not be used in concrete with embedded prestressing steel, aluminum, or galvanized steel.

Pozzolans are finely divided materials such as fly ash, diatomaceous earth, volcanic ash, and calcined shale, which are used to replace some of the cement in a concrete mix. Pozzolans are used to reduce the heat of hydration, increase the workability, and reduce the segregation of a mix. *Workability agents* or plasticizers increase the workability of a mix. However, air-entraining agents, water-reducing agents, pozzolans, and retarders will also increase the workability of a mix.

Mix Design

The concrete mix designer is faced with the problem of selecting the most economical concrete mix that meets the requirements of the hardened concrete while providing acceptable workability. The most economical mix will usually be the mix that uses the highest ratio of aggregate to cement while providing acceptable workability at the required water/cement ratio.

A suggested mix design procedure is to first select a water/cement ratio that satisfies requirements for concrete strength, durability, and watertightness. (Table 10-1 gives maximum water/cement ratio recommended by the American Concrete Institute for various applications.) Next, select the workability or slump required (see Table 10-2). The third step is to mix a trial batch using a convenient quantity of cement at the selected water/cement ratio. Quantities of saturated, surface-dry fine and coarse aggregate are then added until the desired slump is obtained. After weighing each trial mix component, the yield of the mix and the amount of each component required for a full-scale batch may be calculated by the method presented in Chapter 8.

10-3 CONCRETE CONSTRUCTION PRACTICES

Concreting Operations

Like concrete paving, structural concrete construction involves concrete batching, mixing, transporting, placing, consolidating, finishing, and curing. The equipment, methods, and recommended practices for each of these phases of concrete construction are explained in Chapter 8. Special considerations for performing structural concrete operations for pouring concrete during extremely hot or cold weather are described in the remainder of this section.

Hot-Weather Concreting

The rate of hardening of concrete is greatly accelerated when concrete temperature is appreciably higher than the optimum temperature of 50 to 60°F (10 to 15.5°C). Ninety degrees Fahrenheit (32°C) is considered a reasonable upper limit for concreting operations. In addition to reducing setting time, higher temperatures reduce the amount of slump for a given mix. If additional water is added to obtain the desired slump, additional cement must also be added or the water-cement ratio will be increased with corresponding strength reduction. High temperatures, especially when accompanied by winds and low humidity, greatly increase the shrinkage of concrete and often lead to surface cracking of the concrete. Several steps may be taken to reduce the effect of high temperature on concreting operation. The temperature of the plastic concrete may be lowered by cooling the mixing water and/or aggregates before mixing. Heat gain during hydration may be reduced by using Type IV (low-

Table 10-1 ACI recommended maximum permissible water/cement ratios for different types of structures and degrees of exposure* (Courtesy of Portland Cement Association)

	*Exposure Conditions***					
	Severe Wide Range in Temperature, or Frequent Alternations of Freezing and Thawing (Air-Entrained Concrete Only)			*Mild Temperature Rarely Below Freezing, or Rainy, or Arid*		
		At Water Line or Within Range of Fluctuating Water Level or Spray			*At Water Line or Within Range of Fluctuating Water Level or Spray*	
Type of Structure	*In Air*	*In Fresh Water*	*In Seawater or in Contact with Sulfates*†	*In Air*	*In Fresh Water*	*In Seawater or in Contact with Sulfates*†
A. Thin sections such as reinforced piles and pipe	0.49	0.44	0.40	0.53	0.49	0.40
B. Bridge decks	0.44	0.44	0.40	0.49	0.49	0.44
C. Thin sections such as railings, curbs, sills, ledges, ornamental or architectural concrete, and all sections with less than 1-in. concrete cover over reinforcement	0.49	—	—	0.53	0.49	—
D. Moderate sections, such as retaining walls, abutments, piers, girders, beams	0.53	0.49	0.44	††	0.53	0.44
E. Exterior portions of heavy (mass) sections	0.58	0.49	0.44	††	0.53	0.44
F. Concrete deposited by tremie under water	—	0.44	0.44	—	0.44	0.44
G. Concrete slabs laid on the ground	0.53	—	—	††	—	—
H. Pavements	0.49	—	—	0.53	—	—
I. Concrete protected from the weather, interiors of buildings, concrete below ground	††	—	—	††	—	—
J. Concrete which will later be protected by enclosure or backfill but which may be exposed to freezing and thawing for several years before such protection is offered	0.53	—	—	††	—	—

*Adapted from Recommended Practice for Selecting Proportions for Concrete (ACI 613-54).

**Air-entrained concrete should be used under all conditions involving severe exposure and may be used under mild exposure conditions to improve workability of the mixture.

†Soil or groundwater containing sulfate concentrations of more than 0.2%. For moderate sulfate resistance, the tricalcium aluminate content of the cement should be limited to 8%, and for high sulfate resistance to 5%; CSA Sulfate-Resisting cement limits tricalcium aluminate to 4%. At equal cement contents, air-entrained concrete is significantly more resistant to sulfate attack than non-air-entrained concrete.

††Water/cement ratio should be selected on basis of strength and workability requirements, but minimum cement content should not be less than 470 lb/cu yd.

Table 10-2 Typical slump ranges for various types of construction (Courtesy of Portland Cement Association)

Type of Construction	*Slump (in.)**	
	Maximum	*Minimum*
Reinforced foundation walls and footings	3	1
Unreinforced footings, caissons, and substructure walls	3	1
Reinforced slabs, beams, and walls	4	1
Building columns	4	1
Bridge decks	3	2
Pavements	2	1
Sidewalks, driveways, and slabs on ground	4	2
Heavy mass construction	2	1

*When high-frequency vibrators are *not* used, the values may be increased by about 50%, but in no case should the slump exceed 6 in.

heat) cement or by adding a retarder. Air-entraining agents, water-reducing agents, or workability agents may be used to increase the workability of the mix without changing water/cement ratios. It is also advisable to reduce the maximum time before discharge of ready-mixed concrete from the normal 1½ to 1 h or less. The use of shades or covers will be helpful in controlling the temperature of concrete after placement. Moist curing should start immediately after finishing and continue for at least 24 h.

Cold-Weather Concreting

The problems of cold-weather concreting are essentially opposite to those of hot-weather concreting. Concrete must not be allowed to freeze during the first 24 h after placing to avoid permanent damage and loss of strength. Specifications frequently require that when air temperature is 40 °F (5 °C) or less, concrete be placed at a minimum temperature of 50 °F (10 °C) and that this temperature be maintained for at least 3 days after placing. Type III (high early strength) cement or an accelerator may be used to reduce concrete setting time during low temperatures. Mix water and/or aggregates may be heated prior to mixing to raise the temperature of the plastic concrete. The use of unvented heaters inside an enclosure during the first 36 h after placing concrete may cause the concrete surface to dust after hardening. To avoid this problem, any fuel-burning heaters used during this period must be properly vented. When heat is used for curing, the concrete must be allowed to cool gradually at the end of the heating period or cracking may result.

Cast-in-Place Concrete

Concrete structural members have traditionally been built in-place by placing the plastic concrete into forms and allowing it to harden. The forms are removed after the concrete has developed sufficient strength to support its own weight and the weight of any construction loads. Typical shapes and types of concrete structural members are described in the following paragraphs. The construction and use of concrete forms are described in Section 10-4.

WALLS AND WALL FOOTINGS

Although almost any type of concrete wall may be cast in-place, this method of construction is now used primarily for foundation walls, retaining walls, tank walls, and walls for special-purpose structures such as nuclear reactor containment structures. High-rise concrete structures often use a concrete column and beam framework with curtain wall panels inserted between these members to form the exterior walls. Columns are normally of either circular or rectangular cross section. Some typical cast-in-place wall and column shapes are illustrated in Figure 10-1. In placing concrete into wall and column forms, care must be taken to avoid segregation of aggregate and paste that may result from excessive free-fall distances. Another problem frequently encountered in wall construction is the formation of void spaces in the concrete under blockouts for windows, pipe chases, and so on. This can be prevented by using concrete with adequate workability accompanied by careful tamping or vibration of the concrete in these areas during placing.

The relatively new technique of pumping concrete into vertical forms through the bottom of the form may also be used to eliminate the formation of voids in the concrete. Figure 10-2 shows a 2-ft (0.6-m)-square column form 18 ft (5.5 m) high being prepared for pumping. Notice the method of attachment of the pumping hose to the form shown in Figure 10-2b. When the form is filled to the required height, the gate on the form fixture is closed and the pumping hose is removed.

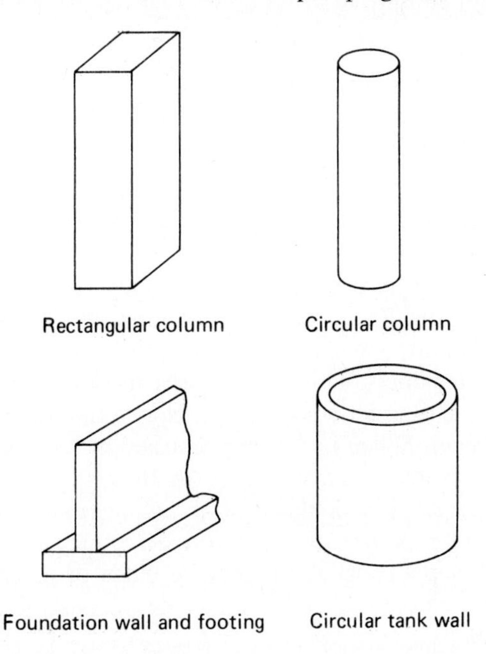

Figure 10-1 Typical cast-in-place column and wall shapes.

a. Preparation of form b. Pumping hose in place on fixture at bottom of form

Figure 10-2 Pumping concrete into bottom of column form. (Courtesy of Gates & Sons, Inc.)

FLOORS AND ROOFS

There are a number of different types of structural systems used for concrete floors and roofs. Such systems may be classified as one-way or two-way slabs. When the floor slab is principally supported in one direction (i.e., at each end) this is referred to as a *one-way slab. Two-way slabs* provide support in two perpendicular directions. *Flat slabs* are supported directly by columns without edge support.

One-Way Slabs

Supporting beams, girders, and slabs may be cast at one time (monolithically), as illustrated in Figure 10-3a. However, columns are usually constructed prior to casting the girders, beams, and slab to eliminate the effect of shrinkage of column concrete on the other members. This type of construction is referred to as *beam-and-slab* or as slab-beam-and-girder construction. Notice that the outside beam is referred to as a *spandrel beam.* When beams are replaced by more closely spaced joists, the type of construction illustrated in Figure 10-3b results. Joists may be either straight or tapered, as shown. The double joist in the illustration is used to carry the additional load imposed by the partition above it. Slabs may also be supported by nonintegral beams. Such supporting beams may be made of precast or cast-in-place concrete, timber, steel, or other materials. This type of construction is referred to as *solid slab construction.*

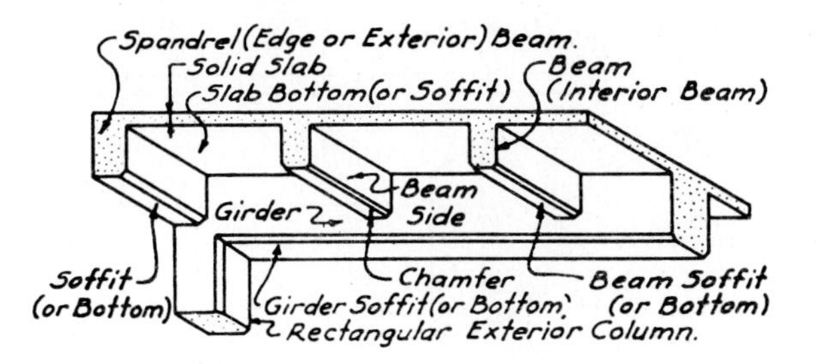

a. Slab-beam-and-girder floor

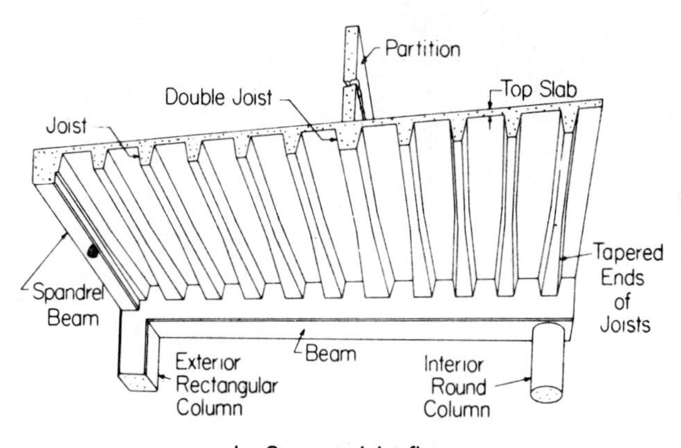

b. Concrete joist floor

Figure 10-3 Floor slab construction. (Courtesy of Concrete Reinforcing Steel Institute)

Two-Way Slabs

The principal type of two-way slab is the *waffle slab,* illustrated in Figure 10-4. Notice that this is basically a joist slab with joists running in two perpendicular directions.

Flat Slabs

Slabs may be supported directly by columns without the use of beams or joists. Such slabs are referred to as *flat slabs* or *flat plate slabs.* A flat plate slab is illustrated in Figure 10-5a. A flat slab is illustrated in Figure 10-5b. Note that the flat slab uses

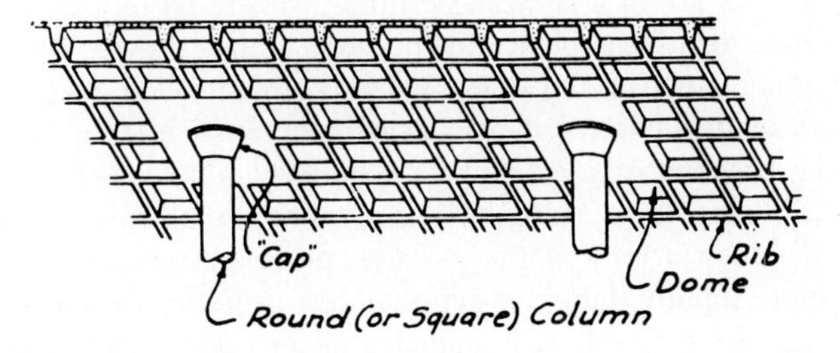

Figure 10-4 Waffle slab. (Courtesy of Concrete Reinforcing Steel Institute)

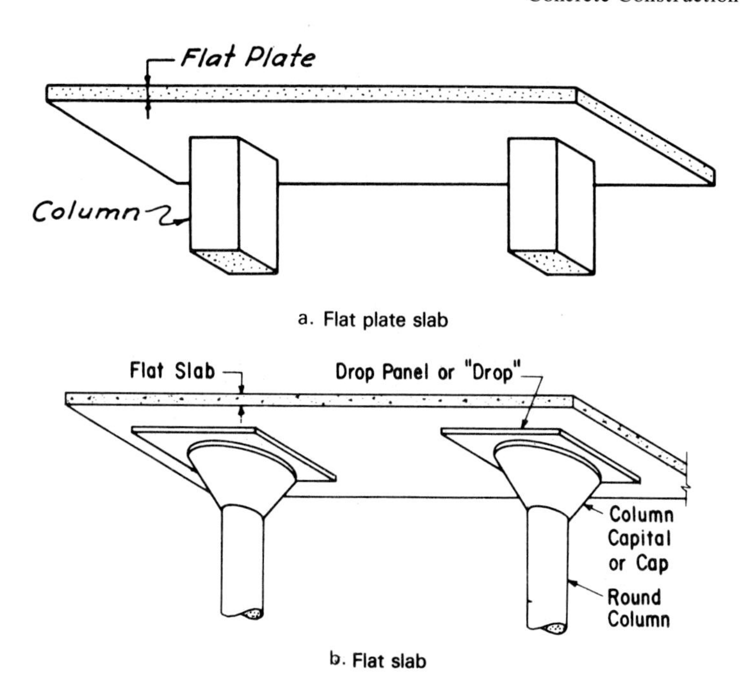

Figure 10-5 Flat slab and flat plate slab. (Courtesy of Concrete Reinforcing Steel Institute)

column capitals to distribute the column reaction over a larger area of slab, while the drop panels serve to strengthen the slab in this area of increased stress. Both of these measures reduce the danger of the column punching through the slab when the slab is loaded.

Precast Concrete

Precast concrete is concrete that has been cast into the desired shape prior to placement in a structure. There are a number of advantages obtained by removing the concrete forming, placing, finishing, and curing operations from the construction environment. Precasting operations usually take place in a central plant where industrial production techniques may be used. Since standard shapes are commonly used, the repetitive use of formwork permits forms to be of high quality at a low cost per unit. These forms and plant finishing procedures provide better surface quality than is usually obtained in the field. Because of controlled environment and procedures, concrete quality control is also usually superior to that of cast-in-place concrete. Forming procedures used make it relatively simple to incorporate prestressing in structural members. Many of the common members described below are prestressed. Upon arrival at the job site, precast structural members may be erected much more rapidly than conventional cast-in-place components.

There are a number of standard shapes commonly used for precast concrete structural members. Figure 10-6 illustrates some common beam and girder sections.

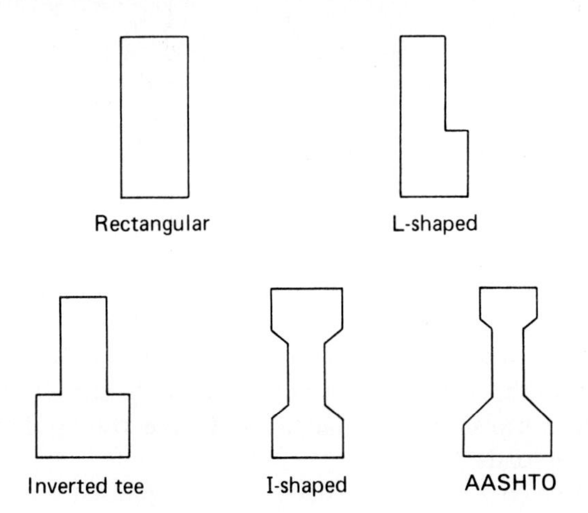

Figure 10-6 Precast beam and girder shapes.

The inverted tee shape is normally used with a cast-in-place concrete slab which forms the upper flange of the section.

Precast concrete joists and purlins (roof supports spanning between trusses or arches) are most often of the I- or T-section shape. Sizes commonly available provide depths of 8 to 12 in. (20 to 30 cm) and length of 10 to 20 ft (3 to 6 m). Precast roof and floor panels (often integral slabs and beams) include flat, hollowcore, tee, double-tee, and channel slabs. These shapes are illustrated in Figure 10-7. Concrete

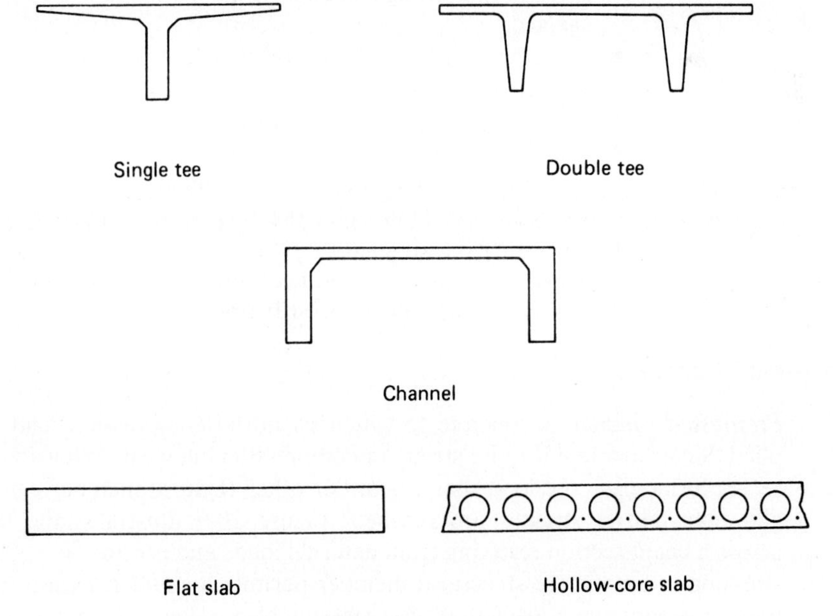

Figure 10-7 Precast slab shapes.

planks are commonly available in thicknesses of 1 to 4 in. (2.5 to 10.1 cm), widths of 15 to 32 in. (38 to 81 cm), and lengths of 4 to 10 ft (1.2 to 3 m). Hollow-core planks range from 4 to 12 in. (10 to 30 cm) in thickness, are usually 4 or 8 ft (1.2 or 2.4 m) wide, and range from 15 to 50 ft (4.6 to 15.3 m) in length. Channel slabs range from 2 to 5 ft (0.6 to 1.5 m) wide and from 15 to 50 ft (4.6 to 15.3 m) in length. Tee and double-tee slabs are available in widths of 4 to 12 ft (1.2 to 3.7 m) and spans of 12 ft (3.7 m) up to 100 ft (30.5 m).

Wall panels may also be precast, hauled to the site, and erected. However, the major use of precast wall panels (excluding tilt-up construction) is for curtain-wall construction, in which the panels fit between the structural framework to form exterior walls. In this type of construction, the walls serve to provide a weatherproof enclosure and transmit any wind loads to the frame. The building frame actually supports all loads.

Tilt-up construction is a special form of precast wall construction in which wall panels are cast horizontally at the job site and then erected. The wall panels are usually cast on the previously placed building floor slab using only edge forms to provide the panel shape. The floor slab thus serves as the bottom form for the panel. Panels may also be cast one on top of another where slab space is limited. A bond-breaker compound is applied to the slab to prevent the tilt-up panel from sticking to the slab. Figure 10-8 illustrates the major steps in a tilt-up construction project.

Some suggestions for obtaining the best results with tilt-up construction procedures include:

DO
- Pour a high-quality slab.
- Keep all plumbing and electrical conduits at least 1 in. under floor surface.
- Let cranes operate on the floor slab.
- Vibrate the slab thoroughly.
- Pour wall panels with their exterior face down.
- Use load spreading frames when lifting panels that have been weakened by windows and other cutouts.

DON'T
- Erect steel framework before raising wall panels.
- Fail to cure floor slab properly.
- Move crane farther than necessary when raising wall panels.
- Lay wall panels down after lifting.

Prestressed Concrete

Prestressed concrete is concrete to which an initial compression load has been applied. Since concrete is quite strong in compression but weak in tension, prestressing serves to increase the load that a beam or other flexural member can carry before allowable tensile stresses are reached. Figure 10-9 illustrates the stress pattern across a beam section resulting from external loads and prestressing. The use of prestressing in a concrete structural member permits a smaller, lighter member to be used in supporting a given load. Prestressing also reduces the amount of deflection

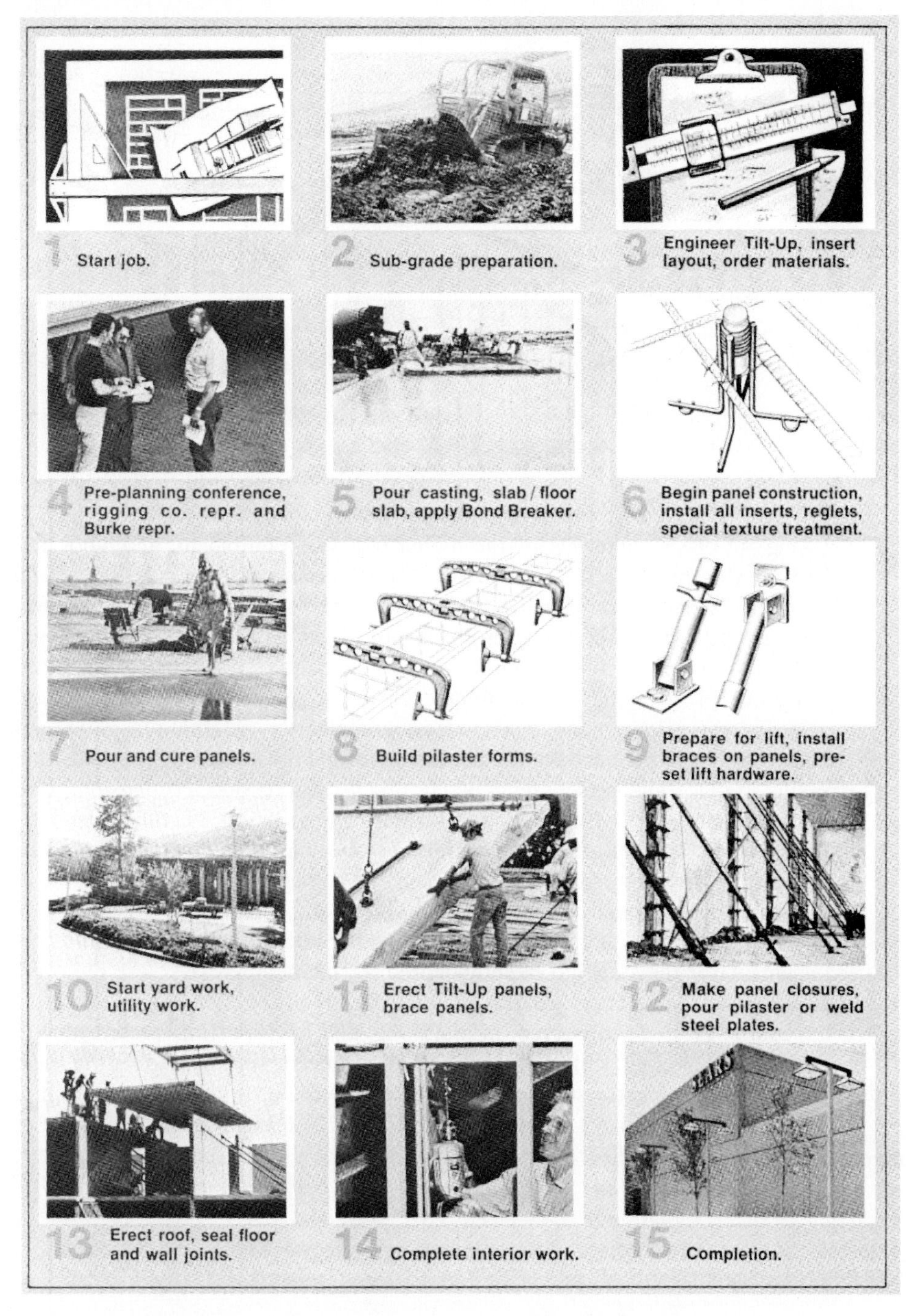

Figure 10-8 Steps in tilt-up construction. (Courtesy of The Burke Company)

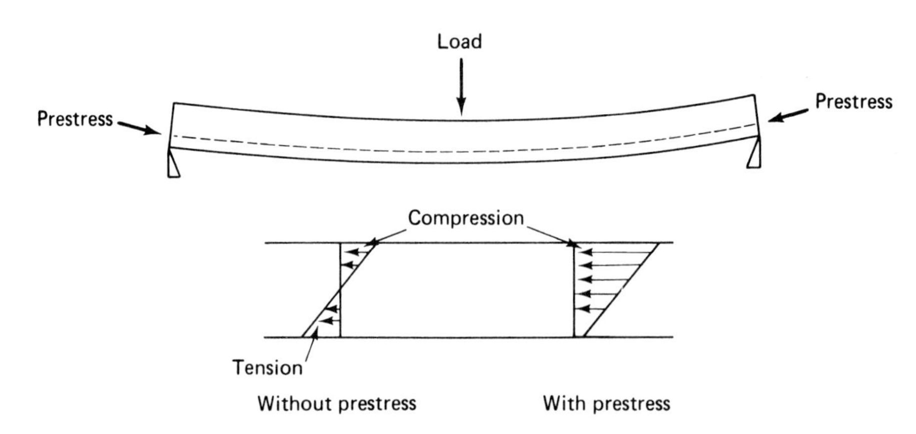

Figure 10-9 Stresses in a prestressed simple beam.

in a beam. Since the member is always kept under compression, any cracking that does occur will remain closed up and not be apparent. These advantages of prestressing are offset somewhat by the higher material, equipment, and labor cost involved in the production of prestressed components. Nevertheless, the use of prestressing, particularly in precast structural members, has become widespread.

There are two methods for producing prestress in concrete members, pretensioning and posttensioning. *Pretensioning* places the prestressing material (reinforcing steel or prestressing cables) under tension in the concrete form before the member is poured. After the concrete has hardened, the external tensioning devices are removed. Bonding between the concrete and the prestressing steel holds the prestressing in place and places the concrete under compression. *Posttensioning* places the prestressing steel (usually placed inside a metal or plastic tube cast into the member) under tension after the concrete member has been erected. The prestressing is then tensioned by jacks placed at each end of the member. After the prestressing load has been applied, the prestressing steel is anchored to the concrete member by mechanical devices at each end or by filling the prestressing tubes with a cementing agent. After the steel has been anchored to the member, jacks are removed and the prestressing steel is cut off flush with the ends of the member.

Caution must be observed in handling and transporting pretensioned prestressed members, particularly if they are unsymmetrically stressed. In the beam of Figure 10-10, the prestressing has been placed off center in the lower portion of the beam. This placement better offsets the tension that would normally occur in the

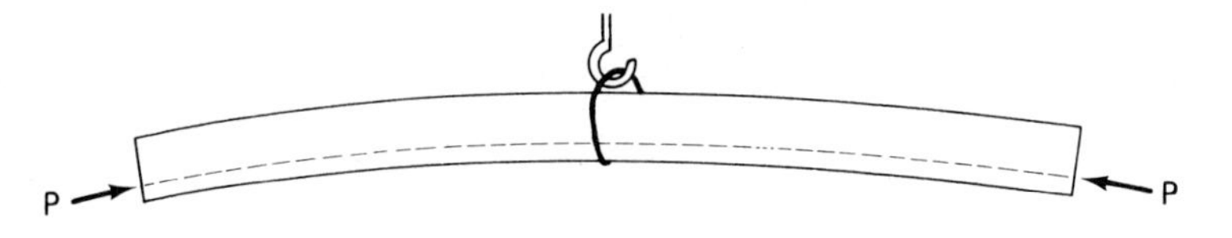

Figure 10-10 Lifting prestressed beam at the center.

lower chord when the beam is loaded. If this beam were to be lifted at the top center, it would tend to bend as shown, resulting in tension along the top chord. The presence of the off-center compression load provided by the prestressing would serve to increase the tension in the top chord and may cause failure of the member prior to erection. Hence this type of beam should not be raised using a center lift. It should be lifted by the ends or by using multiple lift points along the beam.

Architectural Concrete

The architectural use of concrete to provide appearance effects has greatly increased in recent years. Architectural effects are achieved by the shape, size, texture, and color used. An example of the use of shape and texture in a wall treatment is given in Figure 10-11. Here precast panels made from a mix of white quartz aggregate and white cement were applied to the exterior of the building frame to achieve the desired color and three-dimensional surface. The availability of plastic forms and form liners has made it possible to impart special shapes to concrete at a relatively low cost.

Some of the major methods used for obtaining architectural concrete effects include exposed aggregate surfaces (Figure 10-12a), special surface designs and textures achieved by the use of form liners (Figure 10-12b), and mechanically produced surfaces (Figure 10-13). Exposed aggregate surfaces are produced by removing the cement paste from the exterior surface, exposing the underlying aggregate. A method frequently used is to coat the interior surface of the form with a retarder. After the concrete has cured enough to permit the removal of forms, the cement

Figure 10-11 Application of architectural concrete. (Courtesy of Portland Cement Association)

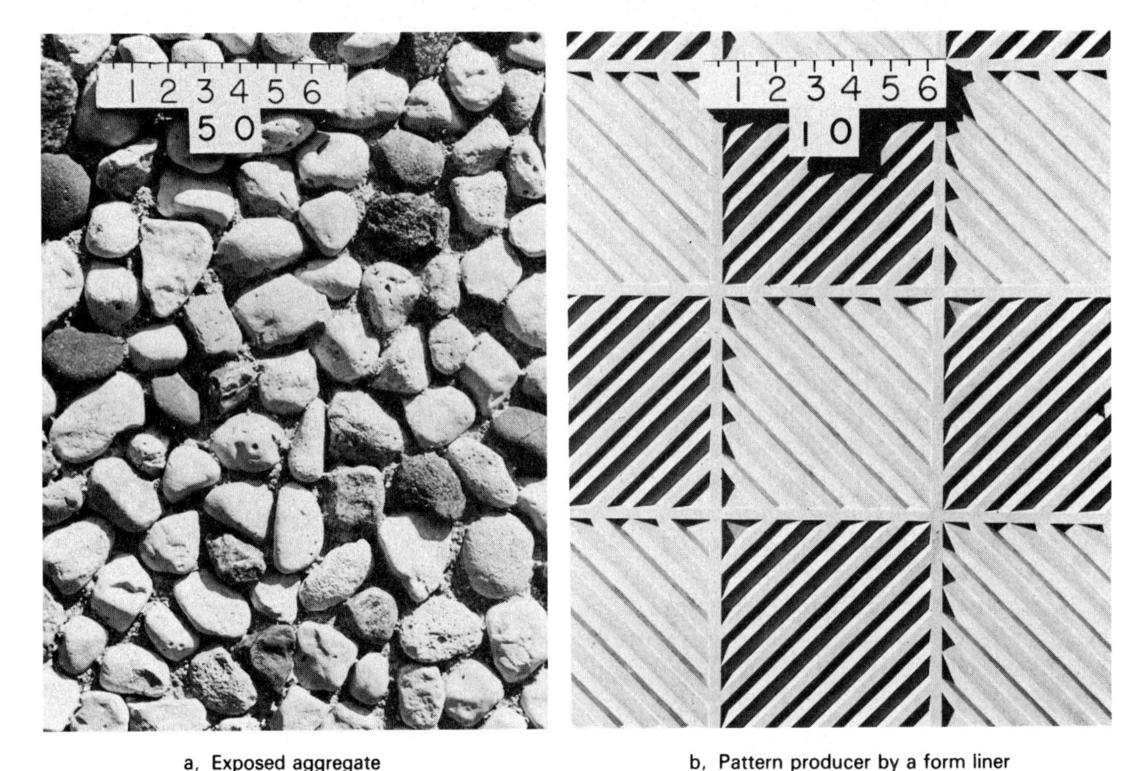

a, Exposed aggregate

b, Pattern producer by a form liner

Figure 10-12 Architectural concrete surfaces. (Courtesy of Portland Cement Association)

paste near the surface (whose curing has been retarded) is removed by brushing and washing or by sandblasting. In the case of horizontal surfaces, a retarder may be applied to the surface after final trowling. Surface textures and designs such as those illustrated in Figure 10-12b may be achieved by the use of form liners of plastic, rubber, or wood. Sandblasting or mechanical hammering may also be used to produce special surface effects. To achieve the surface texture of the building shown in Figure 10-13, a form liner was used to produce triangular surface ridges, which were then chipped with a hammer.

10-4 CONCRETE FORMWORK

General Requirements for Formwork

The principal requirements for concrete formwork are that it be safe, produce the desired shape and surface texture, and be economical. Procedures for designing formwork that will be safe under the loads imposed by plastic concrete, workers and

Figure 10-13 Mechanically produced concrete surface texture. (Courtesy of Portland Cement Association)

other live loads, and external forces (such as wind loads) are explained in Chapter 11. Construction procedures relating to formwork safety are discussed later in this section. Requirements for the shape (including deflection limitations) and surface texture of the finished concrete are normally contained in the construction plans and specifications. Since the cost of concrete formwork often exceeds the cost of the concrete itself, the necessity for economy in formwork is readily apparent.

Typical Formwork

A typical *wall form* with its components is illustrated in Figure 10-14. Sheathing may be either plywood or lumber. Double wales are often used as illustrated so that form ties may be inserted between the two wales. With a single wale it would be necessary to drill the wales for tie insertion. While the pressure of the plastic concrete is resisted by form ties, bracing must be used to prevent form movement and to provide support against wind loads or other lateral loads. Typical form ties are illustrated in Figure 10-15. Form ties may incorporate a spreader device to maintain proper spacing between form walls until the concrete is placed. Otherwise, a removable spreader bar must be used for this purpose. Ties are of two principal types, continuous single-member and internally disconnecting. *Continuous single-member ties* may be pulled out after the concrete has hardened or they may be broken off at a weakened point just below the surface after forms are removed. Common types of *internally disconnecting ties* include the coil tie and stud rod (or she-bolt) tie. With

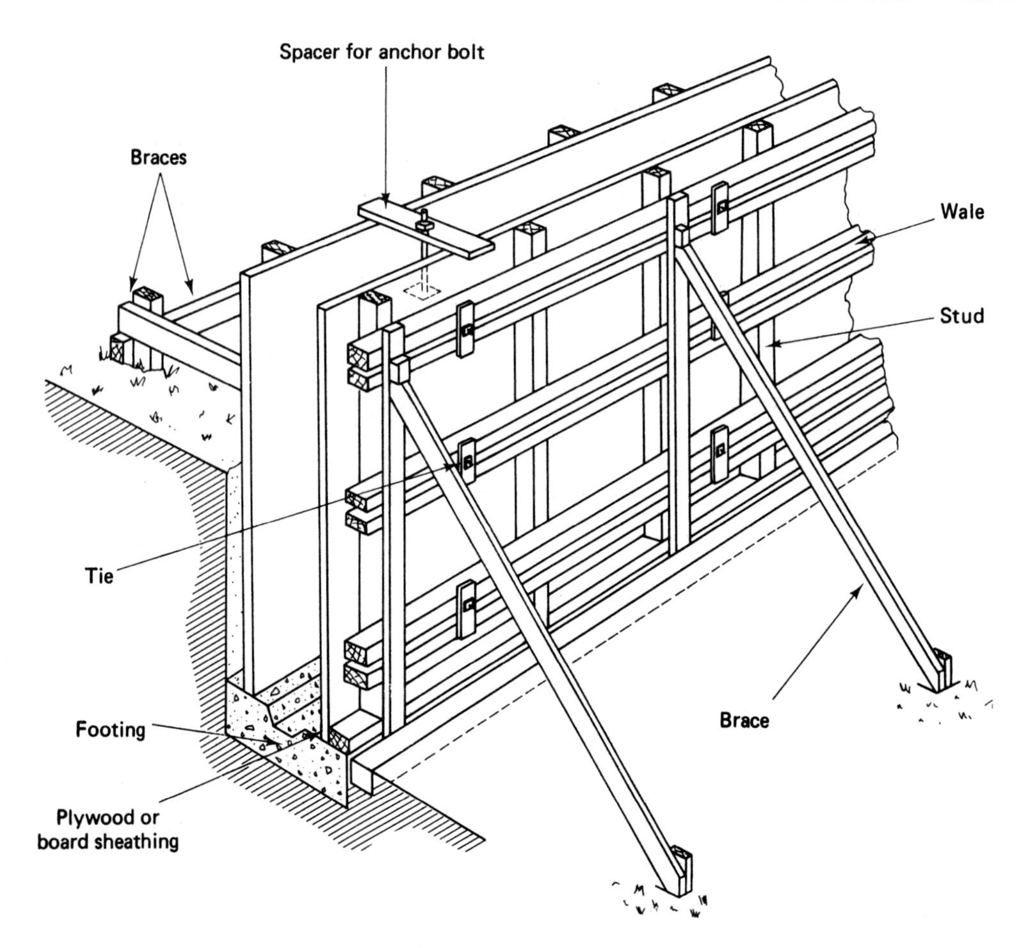

Figure 10-14 Typical wall form.

internally disconnecting ties, the ends are unscrewed to permit form removal with the internal section left embedded in the concrete. The holes remaining in the concrete surface after the ends of the ties are removed are later plugged or grouted.

Column forms are similar to wall forms except that studs and wales are replaced by column clamps or yokes that resist the internal concrete pressure. A typical column form is shown in Figure 10-16. Yokes may be fabricated of wood, wood and bolts (as shown), or of metal. Commercial column clamps (usually of metal) are available in a wide range of sizes (Figure 10-2). Round columns are formed with ready-made fiber tubes or steel reinforced fiberglass forms. Openings or "windows" may be provided at several elevations in high, narrow forms to facilitate placement of concrete. Special fittings may also be inserted near the bottom of vertical forms to permit pumping concrete into the form from the bottom.

Figure 10-17 illustrates a typical elevated floor or desk slab form with its components identified. Forming for a slab with an integral beam is illustrated in

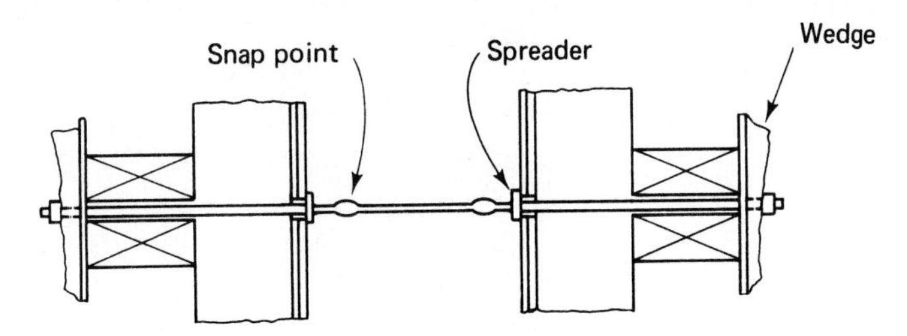

Snap tie with washer spreader

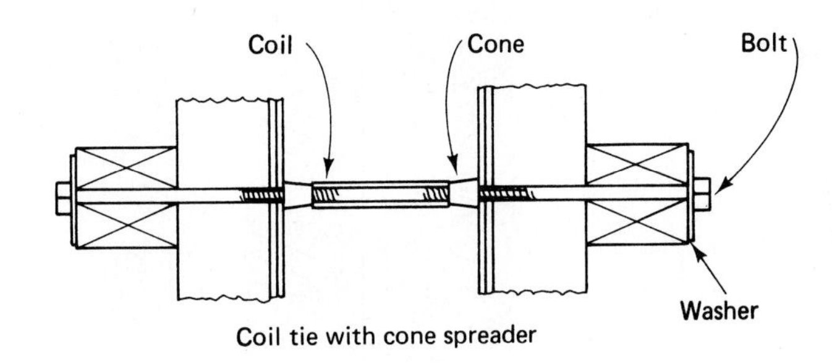

Coil tie with cone spreader

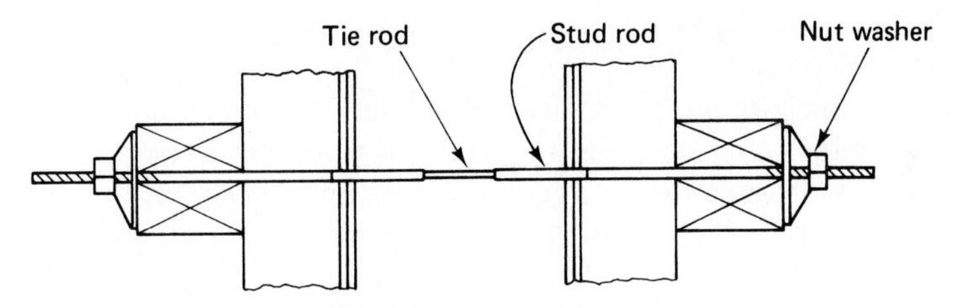

Stud rod (she-bolt) tie

Figure 10-15 Typical form ties.

Figure 10-18. Forming for the one-way and two-way slabs described in Section 10-3 is usually accomplished using commercial pan forms. Figure 10-19 illustrates the use of long pans for a one-way joist slab. Figure 10-20 shows a waffle slab formed with dome pans. Such pan forms may be made of metal or plastic. Wooden stairway forms suitable for constructing stairways up to 3 ft wide are illustrated in Figure 10-21.

Minimizing Cost of Formwork

Since formwork may account for 40 to 60% of the cost of concrete construction, it is essential that the formwork plan be carefully developed and thoroughly evaluated. A cost comparison should be made of all feasible forming systems and methods of op-

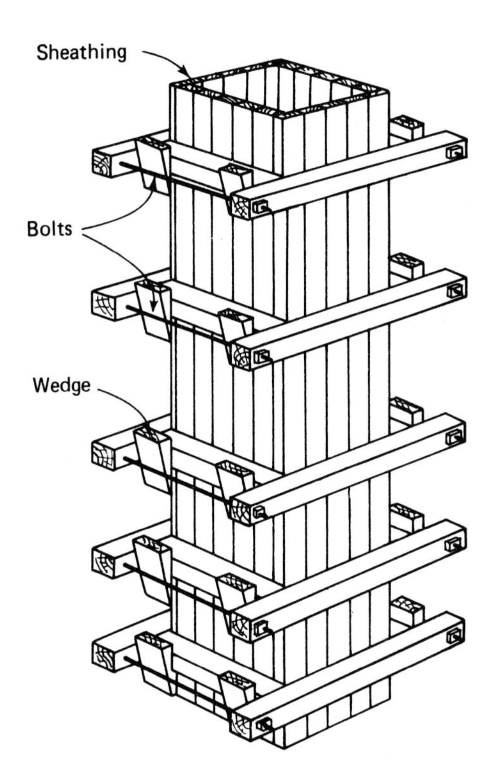

Figure 10-16 Typical column form. (U.S. Department of the Army)

eration. Such an analysis must include the cost of equipment and labor required to install reinforcing steel and to place and finish the concrete, as well as the cost of formwork, its erection, and removal. The formwork plan that provides the required safety and construction quality at the minimum overall cost should be selected for implementation.

In general, lower formwork cost will result from repetitive use of forms. Multiple-use forms may be either standard commercial types or custom-made by the contractor. Contractor-fabricated forms should be constructed using assembly-line techniques whenever possible. *Flying forms,* large sections of formwork moved by crane from one position to another, are often economical in repetitive types of concrete construction. Where appropriate, the use of slip forms and the tilt-up construction techniques described earlier can greatly reduce forming costs. A flying form is pictured in Figure 10-22.

Construction Practices

Forms must be constructed with tight joints to prevent the loss of cement paste, which may result in honeycombing. Before concrete is placed, forms must be aligned both horizontally and vertically and braced to remain in alignment. Form alignment should be continuously monitored during concrete placement and adjustments made if necessary. When a vertical form is wider at the bottom than at the top, an uplift

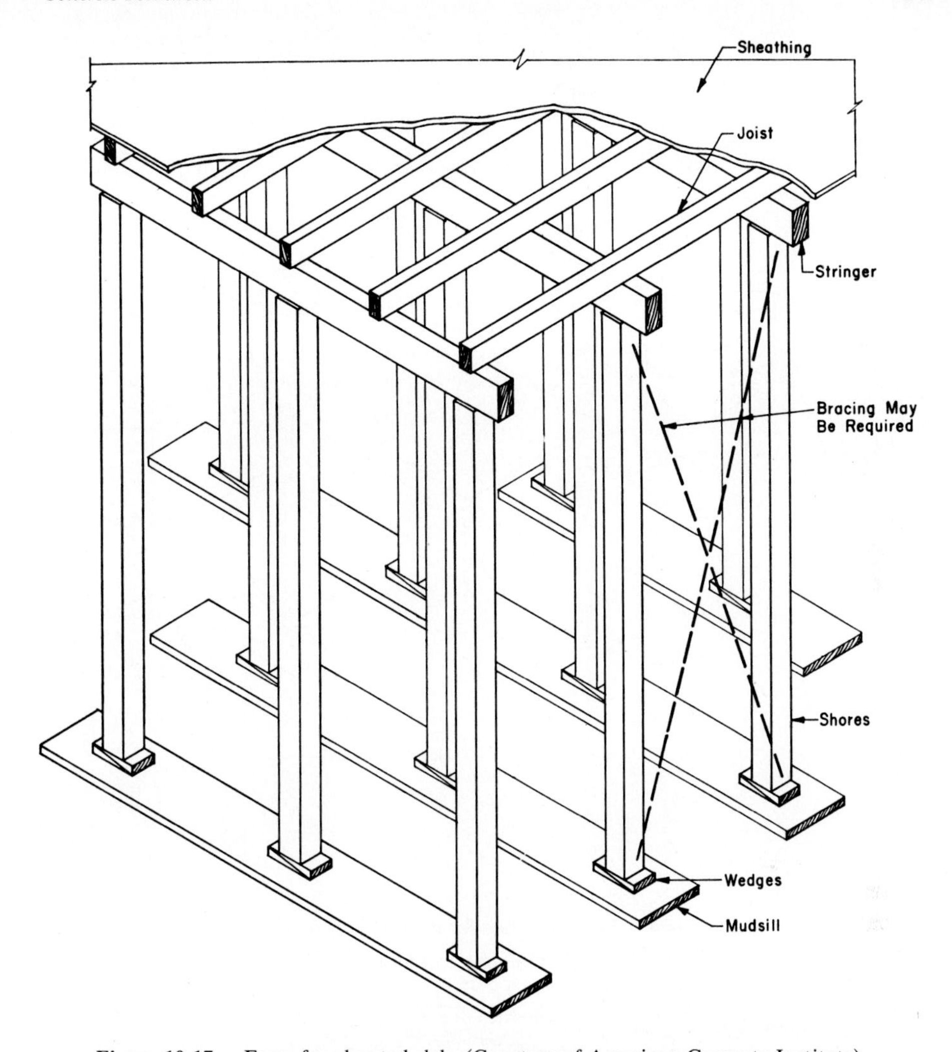

Figure 10-17 Form for elevated slab. (Courtesy of American Concrete Institute)

force will be created as the form is filled. Such forms must be anchored against uplift. Inspect the interior of all forms and remove any debris before placing concrete. Use drop chutes or rubber elephant trunks to avoid segregation of aggregate and paste when placing concrete into high vertical forms. Free-fall distance should be limited to 5 ft or less. When vibrating concrete in vertical forms, allow the vibrator head to penetrate through the freshly placed concrete about 1 in. (2.5 cm) [but not more than 8 in. (20 cm)] into the previously placed layer of concrete. It is possible to bulge or rupture any wall or column form by inserting a large vibrator deep into previously placed, partially set concrete. However, revibration of previously compacted concrete is not harmful to the concrete as long as it becomes plastic when vibrated. When pumping forms from the bottom, it is important to fill the forms rapidly so that the concrete does not start to set up before filling is completed. If the pump rate

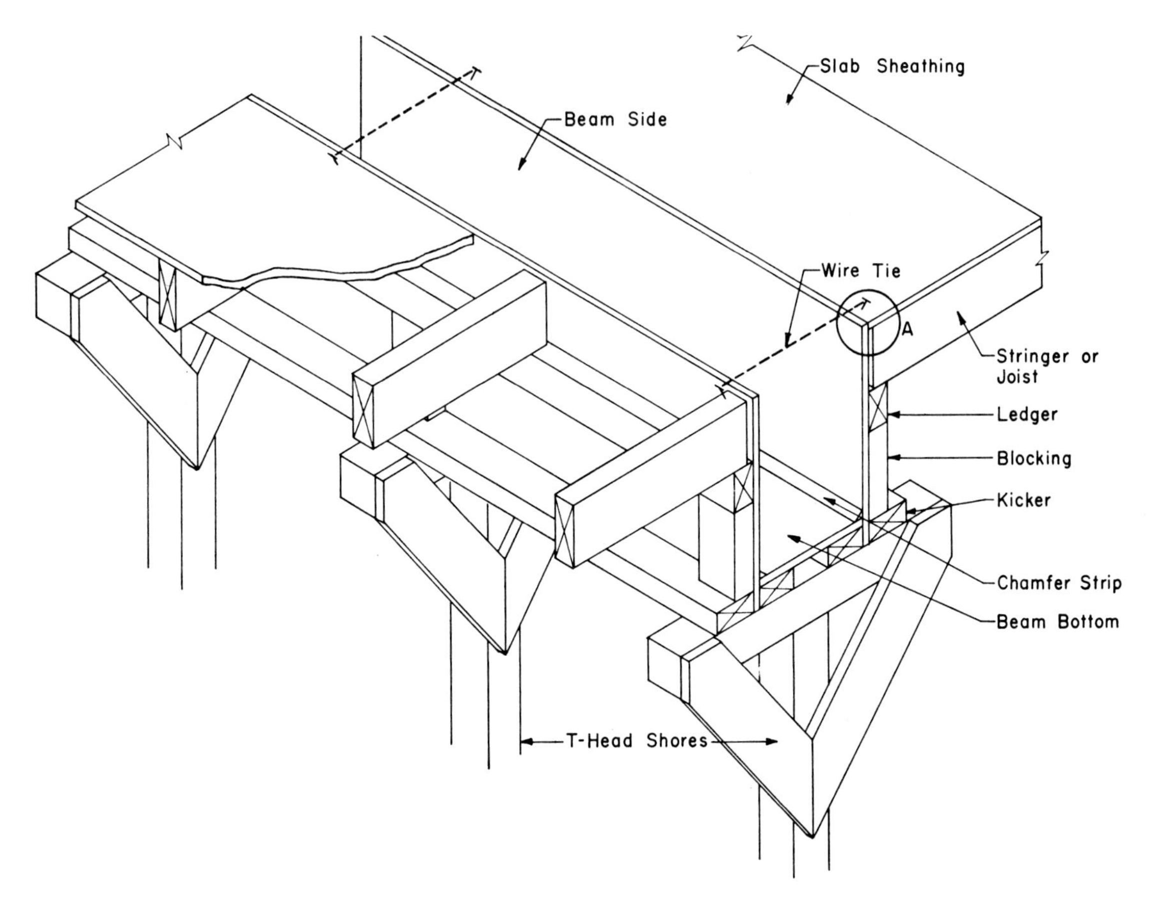

Figure 10-18 Beam and slab form. (Courtesy of American Concrete Institute)

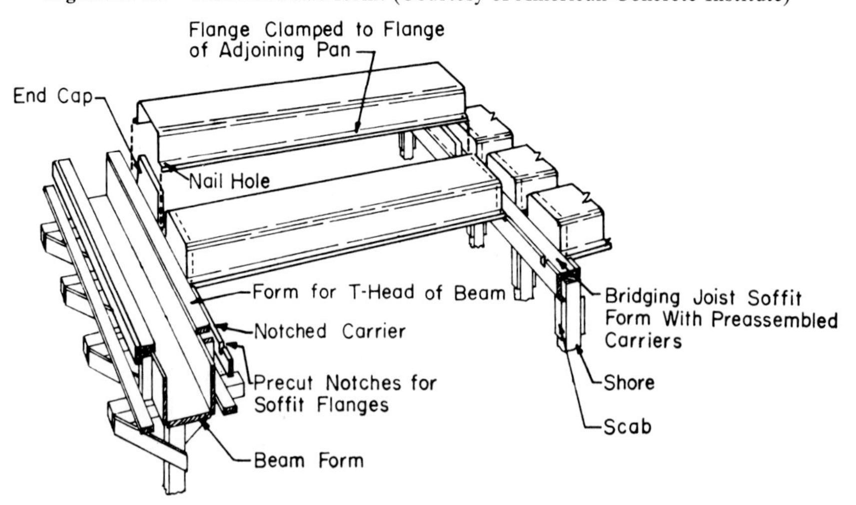

Figure 10-19 One-way slab form. (Courtesy of American Concrete Institute)

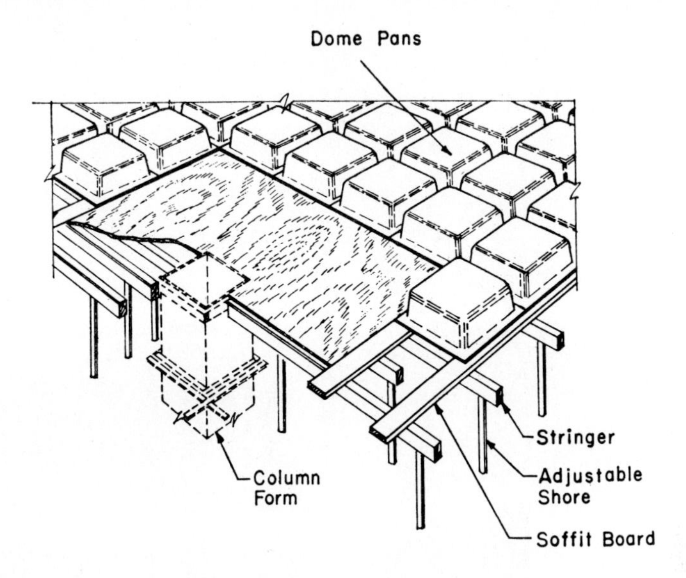

Figure 10-20 Two-way slab form. (Courtesy of American Concrete Institute)

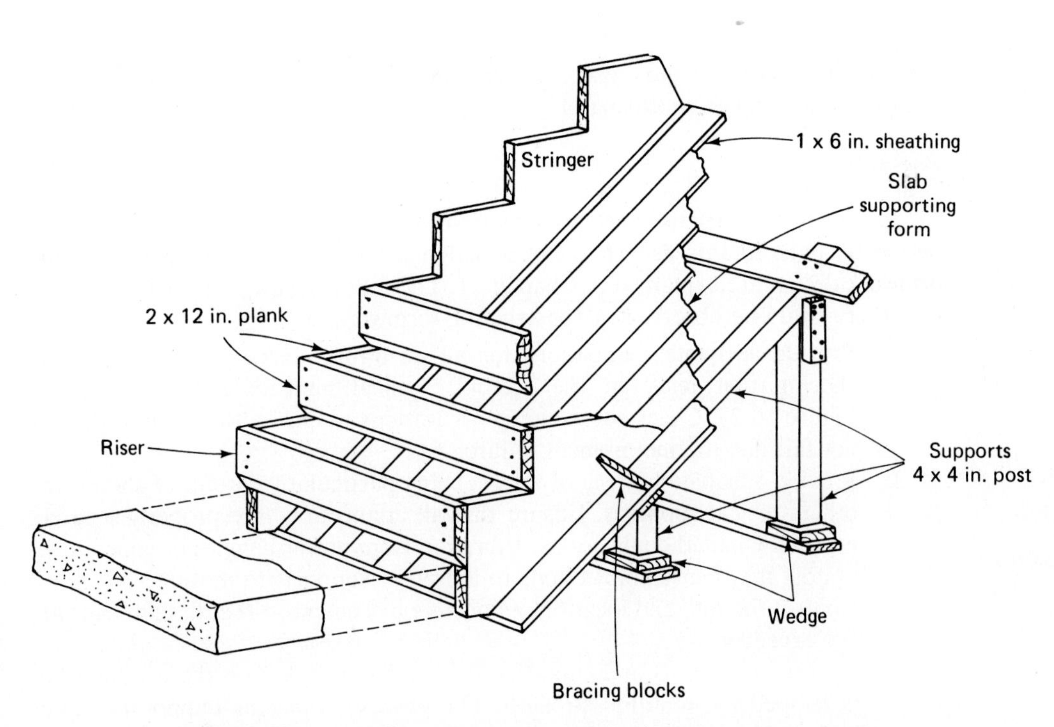

Figure 10-21 Wood form for stairway. (U.S. Department of the Army)

Figure 10-22 Repositioning flying form. (Courtesy of Lorain Division, Koehring Co.)

is so low that setting begins, excessive pressure will be produced inside the form, resulting in bulging or rupturing of the form.

Formwork Safety

The frequency and serious consequences of formwork failure require that special attention be paid to this aspect of construction safety. The requirements for safe formwork design are explained in Chapter 11. The following are some safety precautions that should be observed in constructing formwork.

1. Provide adequate foundations for all formwork. Place mudsills under all shoring that rests on the ground. Typical mudsills are illustrated in Figure 10-23. Check surrounding excavations to ensure that formwork does not fail due to embankment failure.
2. Provide adequate bracing of forms, being particularly careful of shores and other vertical supports. Ensure that all connections are properly secured, especially nailed connections. Vibration from power buggies or concrete vibrators may cause connections to loosen or supports to move.
3. Control the rate and location of concrete placement so that design loads are not exceeded.
4. Ensure that forms and supports are not removed before the concrete has developed the required strength. The process of placing temporary shores under slabs or structural members after forms have been stripped is called

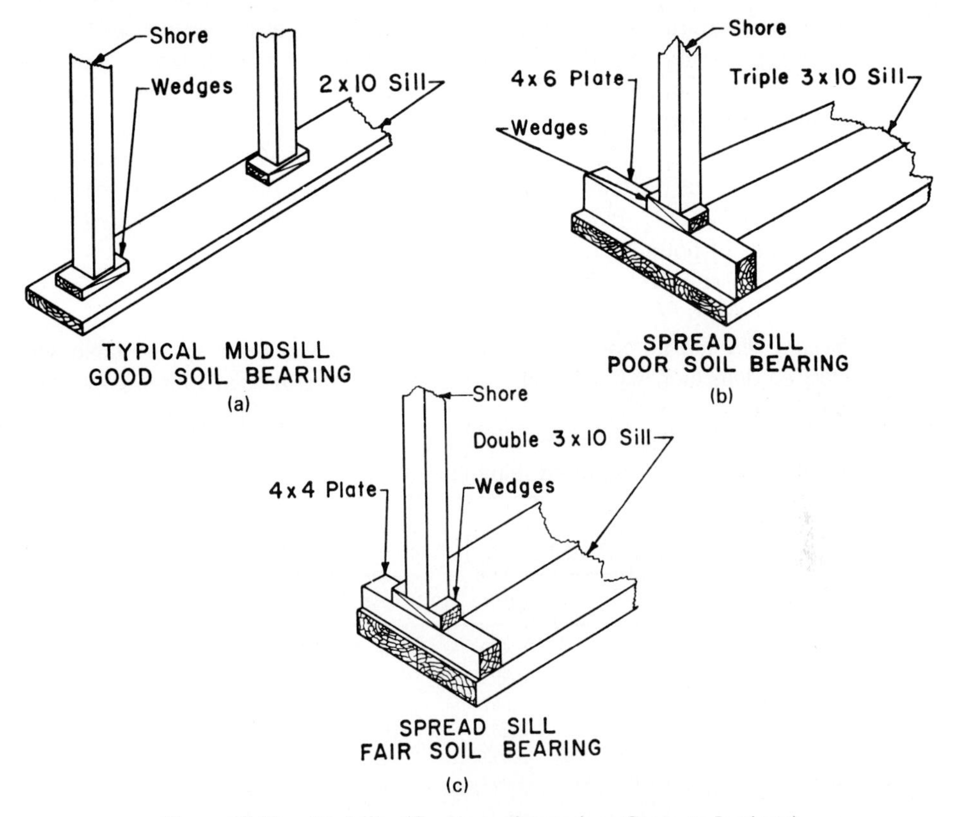

Figure 10-23 Mudsills. (Courtesy of American Concrete Institute)

reshoring. Reshoring is a critical operation that must be carried out exactly as specified by the designer. Only a limited area should be stripped and reshored at one time. No construction loads should be allowed on the partially hardened concrete while reshoring is under way. Adequate bracing must be provided for reshoring.

5. When placing prefabricated form sections in windy weather, care must be taken to avoid injury due to swinging of the form caused by wind forces.
6. Protruding nails are a major source of injury on concrete construction sites. As forms are stripped, form lumber must be promptly removed to a safe location and nails pulled.

10-5 REINFORCING STEEL

Concrete Reinforcing Steel

Concrete reinforcing steel is available as standard reinforcing bars, spirals (for column reinforcing), and welded wire fabric (WWF). *Reinforcing bars* are usually deformed; that is, they are manufactured with ridges that provide an interlocking bond with the surrounding concrete. Deformed bars are available in the 11 American Society for Testing and Materials (ASTM) standard sizes listed in Table 10-3. Note that the size number of the bar indicates the approximate diameter of the bar in eighths of an inch.

Two marking systems are used to identify ASTM standard reinforcing bars, the continuous line system and the number system. The systems are illustrated in Figure 10-24. The grade of reinforcing steel corresponds to its rated yield point in thousands of pounds per square inch.

WELDED WIRE FABRIC

Welded wire fabric, commonly used for slab reinforcement, is available with smooth wire or deformed wire. Fabric is made from bright wire unless galvanized wire is specified.

Table 10-3 ASTM standard reinforcing bar sizes

Size Number	Weight lb/ft	Weight kg/m	Diameter in.	Diameter mm	Section Area sq in.	Section Area mm^2
3	0.376	0.560	0.375	9.52	0.11	71
4	0.668	0.994	0.500	12.70	0.20	129
5	1.043	1.552	0.625	15.88	0.31	200
6	1.502	2.235	0.750	19.05	0.44	284
7	2.044	3.042	0.875	22.22	0.60	387
8	2.670	3.973	1.000	25.40	0.79	510
9	3.400	5.059	1.128	28.65	1.00	645
10	4.303	6.403	1.270	32.26	1.27	819
11	5.313	7.906	1.410	35.81	1.56	1006
14	7.650	11.384	1.693	43.00	2.25	1452
18	13.600	20.238	2.257	57.33	4.00	2581

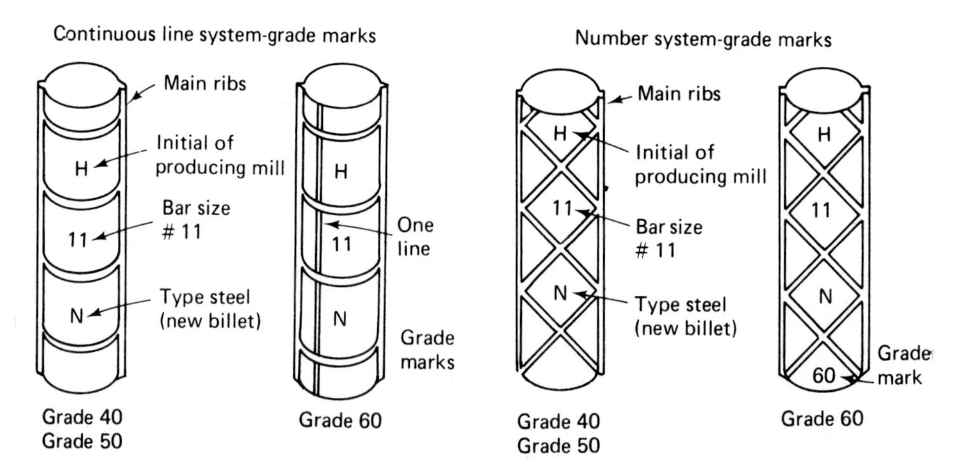

Figure 10-24 Reinforcing bar identification marks. (Courtesy of Concrete Reinforcing Steel Institute)

Welded wire fabric is identified by the letters WWF followed by the spacing of longitudinal wires [in. (mm)], the spacing of transverse wires [in. (mm)], the size of longitudinal wires [sq in. × 100 (mm^2)], the size of transverse wires [sq in. × 100 (mm^2)]. Metric sizes are identified by the letter M preceding the wire sizes. Standard wire sizes are given in Table 10-4. Deformed wire is indicated by the letter D preceding the wire size. For *example,* "WWF 6 × 6-4.0 × 4.0 [152 × 152 MW 25.8 × MW 25.8]" denotes a square wire pattern with both transverse and longitudinal wires spaced 6 in. (152 mm) on center. Both wires are size W4 [0.04 sq in. (25.8-mm^2) section area]. Requirements for welded wire fabric are given in ASTM A185 and A497.

SPIRALS

Spirals are available in three standard rod sizes: ⅜ in. (0.95 cm), ½ in. (1.27 cm), and ⅝ in. (1.59 cm) in diameter. Standard spiral diameters (outside to outside) range from 12 in. (30 cm) to 33 in. (84 cm). Pitch (distance between centers of adjacent spirals) ranges from 1¾ in. (4.4 cm) to 3¼ in. (8.3 cm) by ¼-in. (0.64-cm) increments. Steel grades available include grades 40, 60, and 70.

Table 10-4 Steel wire data for welded wire fabric

Wire Size Number		*Diameter*		*Area*		*Weight*	
Smooth	*Deformed*	*in.*	*mm*	*sq in.*	*mm²*	*lb/ft*	*kg/m*
W31	D31	0.628	16.0	0.31	200	1.054	1.568
W28	D28	0.597	15.2	0.28	181	0.952	1.417
W26	D26	0.575	14.6	0.26	168	0.934	1.390
W24	D24	0.553	14.1	0.24	155	0.816	1.214
W22	D22	0.529	13.4	0.22	142	0.748	1.113
W20	D20	0.505	12.8	0.20	129	0.680	1.012
W18	D18	0.479	12.2	0.18	116	0.612	0.911
W16	D16	0.451	11.5	0.16	103	0.544	0.810
W14	D14	0.422	10.7	0.14	90	0.476	0.708
W12	D12	0.391	9.9	0.12	77	0.408	0.607
W11	D11	0.374	9.5	0.11	71	0.374	0.557
W10	D10	0.357	9.1	0.10	65	0.340	0.506
W9.5		0.348	8.8	0.095	61	0.323	0.481
W9	D9	0.338	8.6	0.09	58	0.306	0.455
W8.5		0.329	8.4	0.085	55	0.289	0.430
W8	D8	0.319	8.1	0.08	52	0.272	0.405
W7.5		0.309	7.8	0.075	48	0.255	0.379
W7	D7	0.299	7.6	0.07	45	0.238	0.354
W6.5		0.288	7.3	0.065	42	0.221	0.329
W6	D6	0.276	7.0	0.06	39	0.204	0.304
W5.5		0.265	6.7	0.055	35	0.187	0.278
W5	D5	0.252	6.4	0.05	32	0.170	0.253
W4.5		0.239	6.1	0.045	29	0.153	0.228
W4	D4	0.226	5.7	0.04	26	0.136	0.202
W3.5		0.211	5.4	0.035	23	0.119	0.177
W2.9		0.192	4.9	0.029	19	0.099	0.147
W2.5		0.178	4.5	0.025	16	0.085	0.126
W2		0.160	4.1	0.02	13	0.068	0.101
W1.4		0.134	3.4	0.014	9	0.048	0.071

Placement of Reinforcing

Since concrete is weak in resistance to tensile forces, reinforcing steel is used primarily to resist tension and thus prevent cracking or failure of the concrete member under tension. Tension may be induced by shrinkage of concrete as it hardens and by temperature changes as well as by bending and shear forces. Typical placement of reinforcing steel in concrete structural members is illustrated in Figure 10-25.

To provide protection of reinforcing steel against corrosion and fire, a minimum cover of concrete must be furnished. Building codes usually specify minimum cover requirements. The American Concrete Institute (ACI) recommends the following minimum cover when not otherwise specified:

- Slabs, joists, and walls not exposed to weather or ground: ¾ in. (1.9 cm).
- Beams, girders, and columns not exposed to weather or ground: 1½ in. (3.8 cm).
- Concrete placed in forms but exposed to weather or ground: 1½ in. (3.8 cm). for No. 5 bars or smaller; 2 in. (5.1 cm) for bars larger than No. 5.
- Concrete placed without forms directly on the ground: 3 in. (7.6 cm).
- At least one bar diameter of cover should be used in any case.

Reinforcing steel must be placed within the tolerances specified by the designer. General placement tolerances suggested by the Concrete Reinforcing Steel Institute (CRSI) include:

- Spacing of outside top, bottom, and side bars in beams, joists, and slabs: ±¼ in. (0.64 cm).
- Lengthwise position of bar ends:
  Sheared bars ±2 in. (5.1 cm)
  Bars with hooked ends ±½ in. (1.3 cm).
- Horizontal spacing of bars in slabs and walls: ±1 in. (2.5 cm).
- Stirrup spacing (distance between adjacent stirrups): ±1 in. (2.5 cm).

The minimum clear distance between parallel bars in columns should be the greater of 1½ bar diameters, 1½ in. (3.8 cm) or 1½ times the maximum aggregate size. For other than columns, the minimum clear distance between parallel bars should be the greater of one bar diameter, 1 in. (2.5 cm) or 1⅓ times the maximum aggregate size. Bars are maintained in their specified position by tying to adjacent bars or by the use of bar supports. Standard types and sizes of wire bar supports are illustrated in Figure 10-26. Figure 10-27 illustrates the CRSI-suggested sequence for placing reinforcing steel in a deep, heavily reinforced concrete beam when a preassembled reinforcing cage cannot be used.

10-6 QUALITY CONTROL

Common Deficiencies in Concrete Construction

Adequate quality control must be exercised over concrete operations if concrete of the required strength, durability, and appearance is to be obtained. Quality control

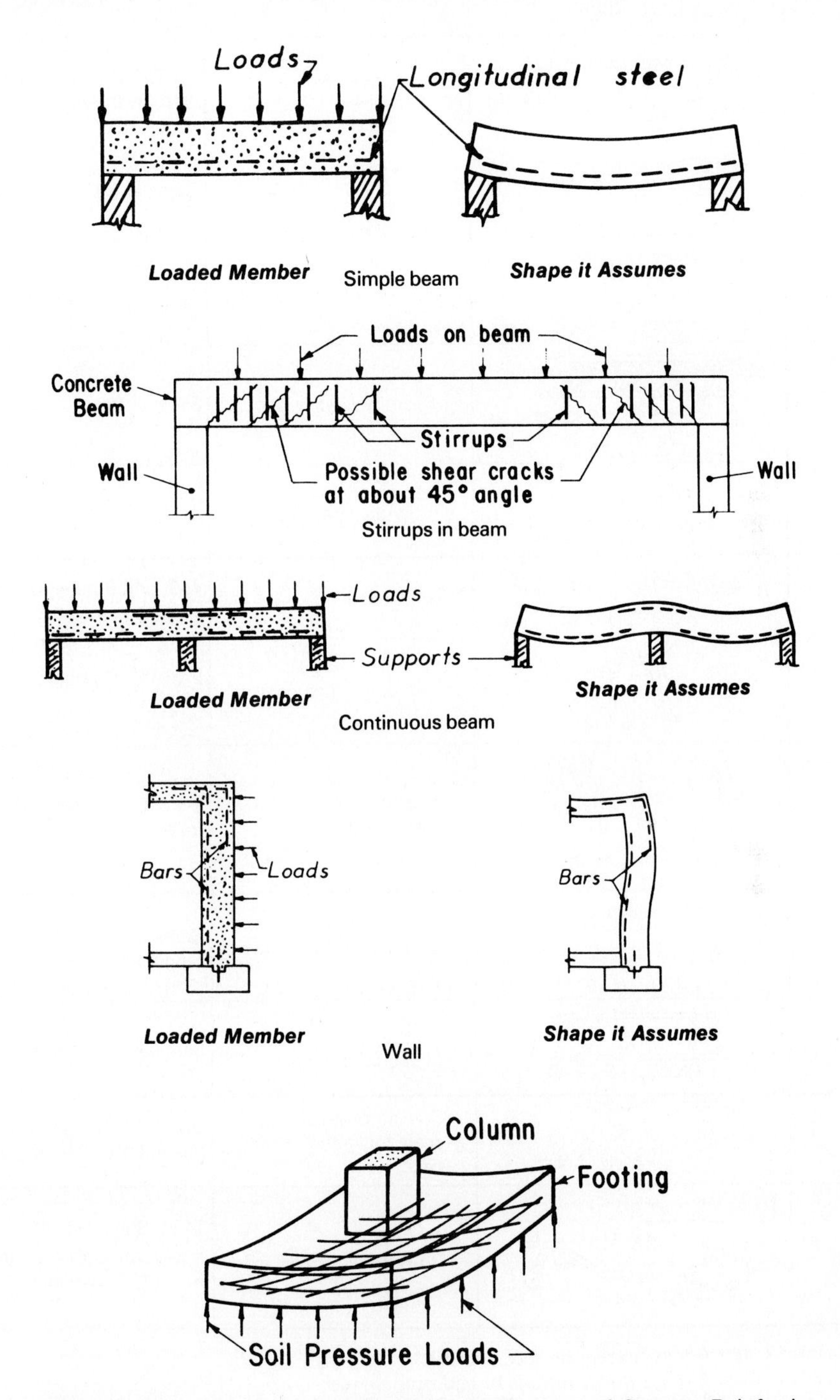

Figure 10-25 Placement of reinforcing steel. (Courtesy of Concrete Reinforcing Steel Institute)

Symbol	Bar support illustration	Type of support	Standard sizes
SB	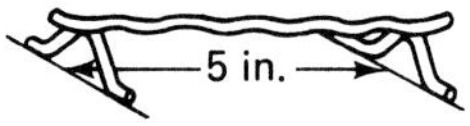 5 in.	Slab bolster	$\frac{3}{4}$, 1, $1\frac{1}{2}$, and 2 in. heights in 5 ft and 10 ft lengths
SBU*	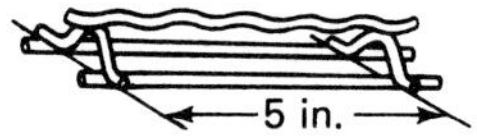 5 in.	Slab bolster upper	Same as SB
BB	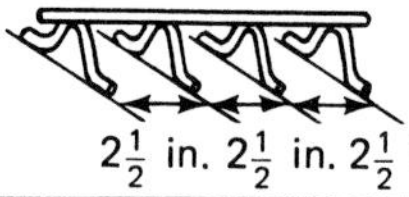 $2\frac{1}{2}$ in. $2\frac{1}{2}$ in. $2\frac{1}{2}$ in.	Beam bolster	1, $1\frac{1}{2}$, 2; over 2 in. to 5 in. height in increments of $\frac{1}{4}$ in. in lengths of 5 ft
BBU*	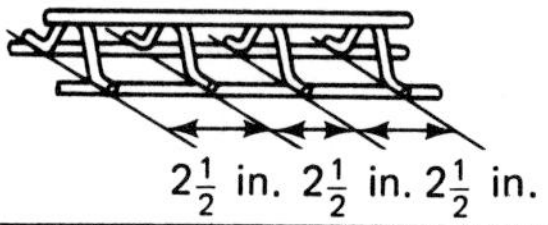 $2\frac{1}{2}$ in. $2\frac{1}{2}$ in. $2\frac{1}{2}$ in.	Beam bolster upper	Same as BB
BC		Individual bar chair	$\frac{3}{4}$, 1, $1\frac{1}{2}$, and $1\frac{3}{4}$ in. heights
JC		Joist chair	4, 5, and 6 in. widths and $\frac{3}{4}$, 1, and $1\frac{1}{2}$ in. heights
HC		Individual high chair	2 to 15 in. heights in increments of $\frac{1}{4}$ in.
HCM*	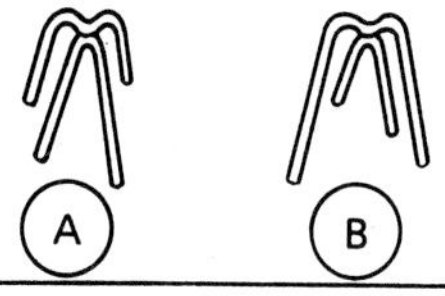 A B	High chair for metal deck	2 to 15 in. heights in increments of $\frac{1}{4}$ in.
CHC	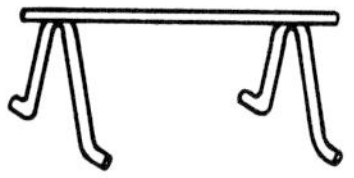	Continuous high chair	Same as HC in 5 foot and 10 foot lengths
CHCU*		Continuous high chair upper	Same as CHC
CHCM*	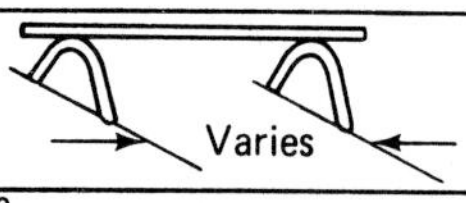 Varies	Continuous high chair for metal deck	Up to 5 in. heights in increments of $\frac{1}{4}$ in.
JCU**	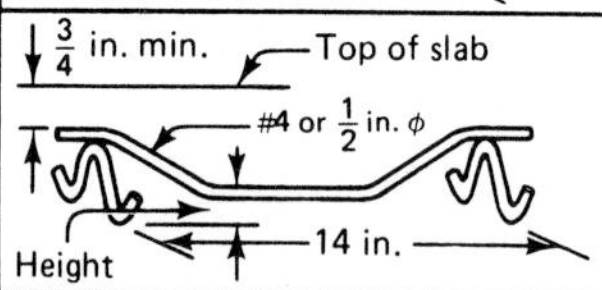 $\frac{3}{4}$ in. min.; Top of slab; #4 or $\frac{1}{2}$ in. ϕ; 14 in.; Height	Joist chair upper	14 in. Span. Heights −1 in. through $+3\frac{1}{2}$ in. vary in $\frac{1}{4}$ in. increments

* Available in Class A only, except on special order.

** Available in Class A only, with upturned or end bearing legs.

Figure 10-26 Wire bar supports. (Courtesy of Concrete Reinforcing Steel Institute)

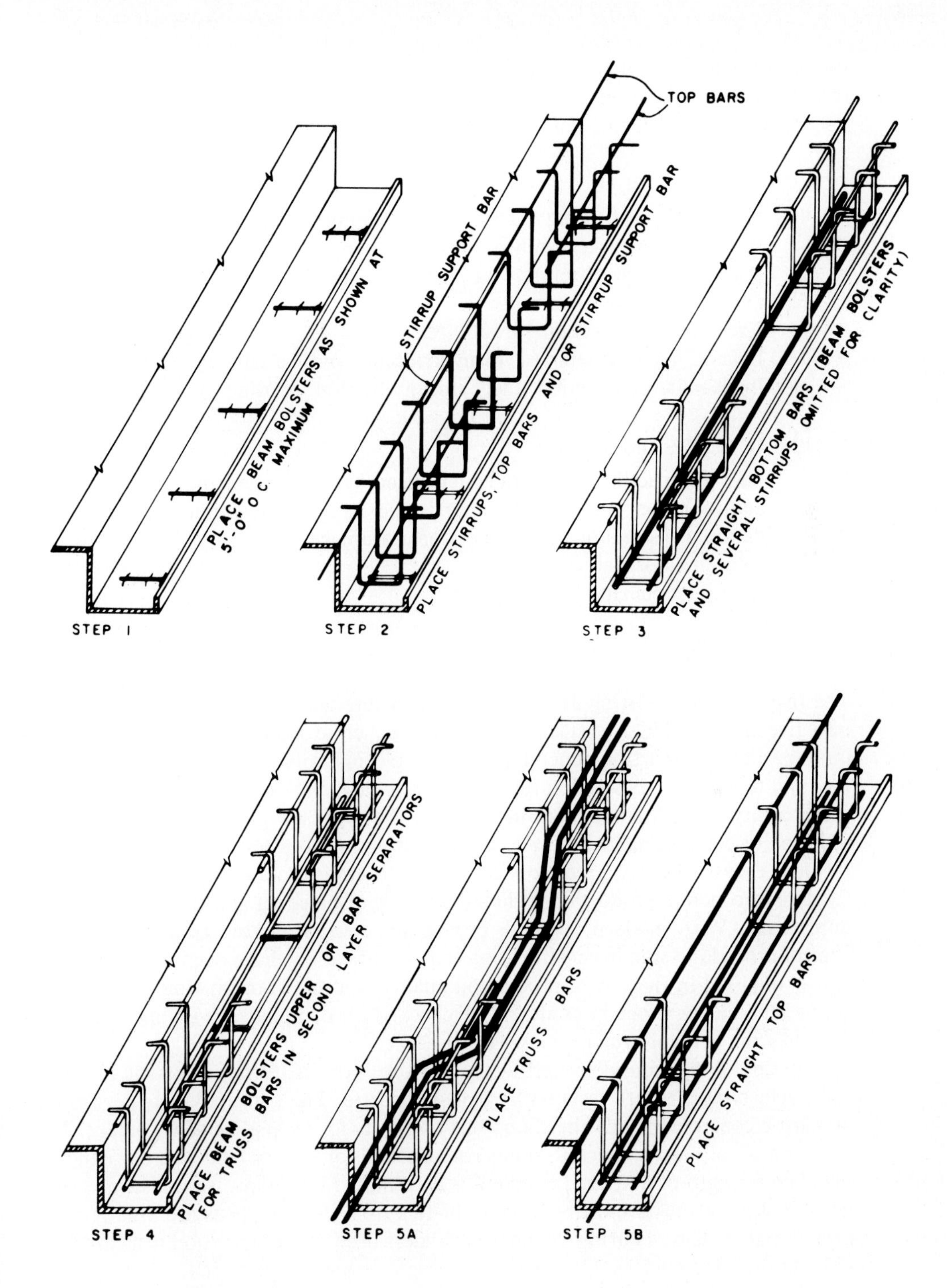

Figure 10-27 Placing reinforcing steel in a beam. (Courtesy of Concrete Reinforcing Steel Institute)

cies in concrete construction practice may usually be traced to inadequate supervision of construction operations. A review by the U.S. Army Corps of Engineers has produced the following list of repetitive deficiencies observed in concrete construction.

STRUCTURAL CONCRETE

1. Unstable form bracing and poor form alignment evidenced by form bulging, spreading, or inaccurately aligned members.
2. Poor alignment of reinforcing steel and exceeding prescribed tolerances.
3. Obvious cold joints in walls.
4. Excessively honeycombed wall areas.
5. Belated form tie removal, form stripping, and patching.
6. Inadequate compaction (mechanical vibration, rodding, or spading).

CONCRETE SLABS ON GRADE

1. Poor compaction of subgrade evidenced by slab settlement.
2. Saturation and damage to subgrade caused by water standing around foundation walls and/or inadequate storm drainage.
3. Uneven floor slab finishes.
4. Inadequate curing of floor slabs.

Inspection and Testing

The inspection and testing associated with concrete quality control may be grouped into five phases. These include mix design; concrete materials quality; batching, mixing, and transporting concrete; concrete placing, vibrating, finishing, and curing; and testing of fresh and hardened concrete at the job site. Mix design includes the quantity of each component in the mix, the type and gradation of aggregates, the type of cement, and so on. Aggregate testing includes tests for organic impurities and excessive fines, gradation, resistance to abrasion, and aggregate moisture. Control of concrete production includes accuracy of batching and the mixing procedures used. With modern concrete production equipment, the producer's quality control procedures and his certification that specifications have been met may be all that is required in the way of production quality control. Transporting, placing, finishing, and curing procedures should be checked for compliance with specifications and with the general principles explained earlier.

Testing of concrete delivered to the job site involves testing of plastic concrete and performing strength tests on hardened concrete. The principal tests performed on plastic concrete include the slump test and tests for air and cement content. The temperature of plastic concrete should be checked for hot- or cold-weather concreting. The strength of hardened concrete is determined by compression tests on cylinder samples, by tensile splitting tests, or by flexure tests. Such tests are usually made after 7 and 28 days of curing. Standard cylinders used for compression tests are 6 in. (15.2 cm) in diameter by 12 in. (30.5 cm) high. Beam samples for flexure tests are usually 6 in. (15.2 cm) square by 20 in. (50.8 cm) long. A procedure for evaluating compression tests results which is recommended by the American Concrete Institute is contained in ACI 214.

PROBLEMS

1. What effect does the water/cement ratio of a concrete mix have on the characteristics of the resulting concrete?
2. Explain the purpose and use of a bond-breaker compound or parting agent in concrete construction.
3. Briefly discuss the advantages and disadvantages of precast, prestressed concrete compared to cast-in-place concrete.
4. Why is the cost of concrete formwork an important factor in the cost of concrete construction?
5. A steel reinforcing bar contains the marking "C 18 N 60." What is the size and strength of the bar?
6. Design drawings show that the following types of welded wire fabric are to be used in a structure. Explain the meaning of these symbols.
   (a) WWF 6 × 12—W16 × W8 [152 × 305 MW103 × MW52]
   (b) WWF 6 × 6—W2 × W2 [152 × 152 MW13.3 × MW13.3]
7. When placing parallel No. 8 reinforcing bars in a beam form, what minimum clear distance between bars should be obtained if the maximum concrete aggregate size is 2 in. (5.1 cm)?
8. Give three major safety precautions that should be observed in constructing and utilizing concrete formwork.
9. What minimum concrete cover should be provided for a No. 6 reinforcing bar in a basement wall below grade with earth backfill?
10. Write a computer program to calculate the total weight of reinforcing steel required for a concrete construction project. Input should include bar size, bar length, and number of pieces of each bar dimension.

REFERENCES

1. *APA Design/Construction Guide: Concrete Forming.* American Plywood Association, Tacoma, Wash., 1990.
2. COLLINS, MICHAEL P., AND DENIS MITCHELL. *Prestressed Concrete Structures.* Englewood Cliffs, N.J.: Prentice Hall, 1991.
3. *Color and Texture in Architectural Concrete* (SP021A). Portland Cement Association, Skokie, Ill., 1980.
4. *Concrete Construction* (Compilation No. 2). American Concrete Institute, Detroit, Mich., 1968.
5. *CRSI Handbook,* 6th ed. Concrete Reinforcing Steel Institute, Chicago, 1984.

6. *Design and Control of Concrete Mixtures,* 13th ed. Portland Cement Association, Skokie, Ill., 1988.
7. "Guide to Troubleshooting Concrete Forming and Shoring Problems," *Concrete Construction,* vol. 24, no. 8 (1979).
8. HURD, M. K. *Formwork for Concrete,* 6th ed. American Concrete Institute, Detroit, Mich., 1989.
9. *Manual of Standard Practice,* 25th ed. Concrete Reinforcing Steel Institute, Chicago, 1990.
10. *Placing Reinforcing Bars,* 5th ed. Concrete Reinforcing Steel Institute, Chicago, 1989.

11
Concrete Form Design

11-1
DESIGN PRINCIPLES

The design of concrete formwork that has adequate strength to resist failure and will not deflect excessively when the forms are filled is a problem in structural design. Unless commercial forms are used, this will usually involve the design of wall, column, or slab forms constructed of wood or plywood. In such cases, after the design loads have been established, each of the primary form components may be analyzed as a beam to determine the maximum bending and shear stresses and the maximum deflection that will occur. Vertical supports and lateral bracing are then analyzed for compression and tension loads. The procedures and applicable equations utilizing customary units are presented in this chapter. Refer to Appendix A for metric conversion factors.

11-2
DESIGN LOADS

Wall and Column Forms

For vertical forms (wall and column forms), design load consists of the lateral pressure of the concrete against the forms. The maximum lateral pressure that concrete exerts against a form has been found to be a function of the type of concrete, temperature of the concrete, vertical rate of placing, and height of the form. For ordinary (150 lb/cu ft) internally vibrated concrete, the American Concrete Institute (ACI) recommends the use of the following formulas to determine the maximum lateral concrete pressure expressed in pounds per square foot.

- For all columns and for walls with a vertical rate of placement of 7 ft/h or less:

$$p = 150 + \frac{9000R}{T} \qquad (11\text{-}1)$$

where p = lateral pressure (lb/sq ft)
R = rate of vertical placement (ft/h)
T = temperature (°F)
h = height of form (ft)

Maximum pressure = 3000 lb/sq ft for columns, 2000 lb/sq ft for walls, or 150 h, whichever is less.

- For walls with a vertical rate of placement of 7 to 10 ft/h:

$$p = 150 + \frac{43{,}400}{T} + \frac{2800R}{T} \qquad (11\text{-}2)$$

Maximum pressure = 2000 lb/sq ft or 150 h, whichever is less.

- For walls with a vertical rate of placement greater than 10 ft/h:

$$p = 150h \qquad (11\text{-}3)$$

When forms are vibrated externally, it is recommended that a design load twice that given by Equations 11-1 and 11-2 be used. When concrete is pumped into vertical forms from the bottom (both column and wall forms), Equation 11-3 should always be used.

Floor and Roof Slab Forms

The design load to be used for elevated slabs consists of the weight of concrete and reinforcing steel, the weight of the forms themselves, and any live loads (equipment, workers, material, etc.). For normal reinforced concrete the design load for concrete and steel is based on a unit weight of 150 lb/cu ft. The American Concrete Institute recommends a minimum design load of 50 lb/sq ft be used for the weight of equipment, materials, and workers. When motorized concrete buggies are utilized, the live load should be increased to at least 75 lb/sq ft. Any unusual loads would be in addition to these values. A minimum total design load of 100 lb/sq ft should be used. This should be increased to 125 lb/sq ft when motorized buggies are used.

Lateral Loads

Formwork must be designed to resist lateral loads such as those imposed by wind, the movement of equipment on the forms, and the placing of concrete into the forms. Such forms are usually resisted by lateral bracing whose design is covered in

Table 11-1 Recommended minimum lateral design load for wall forms

Wall Height, h (ft)	*Design Lateral Force Applied at Top of Form (lb/ft)*
13.33 or less	100 but at least $\frac{h \times wf^*}{2}$
Over 13.33	$7.5h$ but at least $\frac{h \times wf^*}{2}$

$^*wf$ = wind force prescribed by local code (lb/sq ft).

Section 11-6. Minimum lateral design loads recommended for tied wall forms are given in Table 11-1. When form ties are not used, bracing must be designed to resist the internal concrete pressure as well as any external loads.

For slab forms, the minimum lateral design load is expressed as follows:

$$H = 0.02 \times dl \times ws \tag{11-4}$$

where H = lateral force applied along the edge of the slab (lb/ft); minimum value = 100 lb/ft
dl = design dead load (lb/sq ft)
ws = width of slab perpendicular to form edge (ft)

In using Equation 11-4, design dead load includes the weight of concrete plus formwork. In determining the value of ws, consider only that part of the slab being placed at one time.

11-3 METHOD OF ANALYSIS

Basis of Analysis

After appropriate design loads have been selected, the sheathing, joists or studs, and stringers or wales are analyzed in turn, considering each member to be a uniformly loaded beam supported in one of three conditions (single-span, two-span, or three-span or larger) and analyzed for bending, shear, and deflection. Vertical supports and lateral bracing must be checked for compression and tension stresses. Except for sheathing, bearing stresses must be checked at supports to ensure against crushing.

Using the methods of engineering mechanics, the maximum values of bending moment, shear, and deflection developed in a uniformly loaded, simply supported beam of uniform cross section are given in Table 11-2.

The maximum fiber stresses in bending, shear, and compression resulting from a specified load may be determined from the following equations.

Table 11-2 Maximum bending, shear, and deflection in a uniformly loaded beam

Type	Support Conditions		
	1 Span	*2 Spans*	*3 Spans*
Bending moment (in.-lb)	$M = \frac{wl^2}{96}$	$M = \frac{wl^2}{96}$	$M = \frac{wl^2}{120}$
Shear (lb)	$V = \frac{wl}{24}$	$V = \frac{5wl}{96}$	$V = \frac{wl}{20}$
Deflection (in.)	$\Delta = \frac{5wl^4}{4608EI}$	$\Delta = \frac{wl^4}{2220EI}$	$\Delta = \frac{wl^4}{1740EI}$

Notation:
l = length of span (in.)
w = uniform load per foot of span (lb/ft)
E = modulus of elasticity (psi)
I = moment of inertia (in.^{4})

Bending:

$$f_b = \frac{M}{S} \tag{11-5}$$

Shear:

$$f_v = \frac{1.5V}{A} \quad \text{for rectangular wood members} \tag{11-6}$$

$$f_v = \frac{V}{Ib/Q} \quad \text{for plywood} \tag{11-7}$$

Compression:

$$f_c \text{ or } f_{c\perp} = \frac{P}{A} \tag{11-8}$$

Tension:

$$f_t = \frac{P}{A} \tag{11-9}$$

where f_b = actual unit stress for extreme fiber in bending (psi)
f_c = actual unit stress in compression parallel to grain (psi)
$f_{c\perp}$ = actual unit stress in compression perpendicular to grain (psi)
f_t = actual unit stress in tension (psi)
f_v = actual unit stress in horizontal shear (psi)
A = section area (sq in.)
M = maximum moment (in.-lb)

P = concentrated load (lb)
S = section modulus (cu in.)
V = maximum shear (lb)
lb/Q = rolling shear constant (sq in./ft)

Since the grain of a piece of timber runs parallel to its length, axial compressive forces result in unit compressive stresses parallel to the grain. Thus, a compression force in a formwork brace (Figure 11-4) will result in unit compressive stresses parallel to the grain (f_c) in the member. Loads applied to the top or sides of a beam, such as a joist resting on a stringer (Figure 11-1b), will result in unit compressive stress perpendicular to the grain ($f_{c\perp}$) in the beam. Equating allowable unit stresses to the maximum unit stresses developed in a beam subjected to a uniform load of w pounds per linear foot yields the design equations of Table 11-3. When design load and beam section properties have been specified, these equations may be solved directly for the maximum allowable span. Given a design load and span length, the equations may be solved for the required size of the member. Design properties for Plyform are given in Table 11-4 and section properties for dimensioned lumber and timber are given in Table 11-5. Typical allowable unit stress values for lumber are given in Table 11-6. The allowable unit stress values in Table 11-6 (but not modulus of elasticity values) may be multiplied by a load duration factor of 1.25 (7-day load) when designing formwork for light construction and single use or very limited reuse of forms. However, allowable stresses for lumber sheathing (not Plyform) should be reduced by the factors given in Table 11-6 for wet conditions. The values for Plyform properties presented in Table 11-4 are based on wet strength and 7-day load duration, so no further adjustment in these values is required.

11-4 SLAB FORM DESIGN

Method of Analysis

The procedure for applying the equations of Table 11-3 to the design of a deck or slab form is to first consider a strip of sheathing of the specified thickness and 1 ft wide (see Figure 11-1a). Determine in turn the maximum allowable span based on the allowable values of bending stress, shear stress, and deflection. The lower of these values will, of course, determine the maximum spacing of the supports (joists). For simplicity and economy of design, this maximum span value is usually rounded down to the next lower integer or modular value when selecting joist spacing.

Based on the selected joist spacing, the joist itself is analyzed to determine its maximum allowable span. The load conditions for the joist are illustrated in Figure 11-1b. The joist span selected will be the spacing of the stringers. Again, an integer or modular value is selected for stringer spacing.

Based on the selected stringer spacing, the process is repeated to determine the maximum stringer span (distance between vertical supports or shores). Notice in

Table 11-3 Concrete form design equations

Design Condition	Support Conditions: 1 Span	2 Spans	3 or More Spans
Bending	$l = 4.0d\left(\frac{F_b b}{w}\right)^{1/2}$	$l = 4.0d\left(\frac{F_b b}{w}\right)^{1/2}$	$l = 4.46d\left(\frac{F_b b}{w}\right)^{1/2}$
	$l = 9.8\left(\frac{F_b S}{w}\right)^{1/2}$	$l = 9.8\left(\frac{F_b S}{w}\right)^{1/2}$	$l = 10.95\left(\frac{F_b S}{w}\right)^{1/2}$
Shear			
Wood	$l = 16\frac{F_v A}{w} + 2d$	$l = 12.8\frac{F_v A}{w} + 2d$	$l = 13.3\frac{F_v A}{w} + 2d$
Plywood	$l = 24\frac{F_v (Ib/Q)}{w} + 2d$	$l = 19.2\frac{F_v (Ib/Q)}{w} + 2d$	$l = 20\frac{F_v (Ib/Q)}{w} + 2d$
Deflection	$l = 5.51\left(\frac{EI\Delta}{w}\right)^{1/4}$	$l = 6.86\left(\frac{EI\Delta}{w}\right)^{1/4}$	$l = 6.46\left(\frac{EI\Delta}{w}\right)^{1/4}$
If $\Delta = l/180$	$l = 1.72\left(\frac{EI}{w}\right)^{1/3}$	$l = 2.31\left(\frac{EI}{w}\right)^{1/3}$	$l = 2.13\left(\frac{EI}{w}\right)^{1/3}$
If $\Delta = l/240$	$l = 1.57\left(\frac{EI}{w}\right)^{1/3}$	$l = 2.10\left(\frac{EI}{w}\right)^{1/3}$	$l = 1.94\left(\frac{EI}{w}\right)^{1/3}$
If $\Delta = l/360$	$l = 1.37\left(\frac{EI}{w}\right)^{1/3}$	$l = 1.83\left(\frac{EI}{w}\right)^{1/3}$	$l = 1.69\left(\frac{EI}{w}\right)^{1/3}$
Compression	$f_c \text{ or } f_{c\perp} = \frac{P}{A}$		
Tension	$f_t = \frac{P}{A}$		

Notation:

l = length of span, center to center of supports (in.)
F_b = allowable unit stress in bending (psi)
F_c = allowable unit stress in compression parallel to grain (psi)
$F_{c\perp}$ = allowable unit stress perpendicular to grain (psi)
F_v = allowable unit stress in horizontal shear (psi)
f_c = actual unit stress in compression parallel to grain (psi)
$f_{c\perp}$ = actual unit stress in compression perpendicular to grain (psi)
f_t = actual unit stress in tension (psi)
A = area of section ($in.^2$)*
E = modulus of elasticity (psi)
I = moment of inertia ($in.^4$)*
Ib/Q = rolling shear constant (sq in./ft)
P = applied force (compression or tension) (lb)
S = section modulus ($in.^3$)*
Δ = deflection (in.)
b = width of member (in.)
d = depth of member (in.)
w = uniform load per foot of span (lb/ft)

*For a rectangular member: $A = bd$, $S = bd^2/6$, $I = bd^3/12$.

Table 11-4 Properties of Plyform* (Courtesy of American Plywood Association)*

Thickness (in.)	Approx. Weight (psf)	Properties for Face Grain Across Supports: Moment of Inertia, I (in.^{4}/ft)	Effective Section Modulus, KS (in.^{3}/ft)	Rolling Shear Constant, Ib/Q (in.^{2}/ft)	Properties for Face Grain Parallel to Supports: Moment of Inertia, I (in.^{4}/ft)	Effective Section Modulus, KS (in.^{3}/ft)	Rolling Shear Constant, Ib/Q (in.^{2}/ft)
Class I							
$\frac{1}{2}$	1.5	0.077	0.268	5.153	0.024	0.130	2.739
$\frac{5}{8}$	1.8	0.130	0.358	5.717	0.038	0.175	3.094
$\frac{3}{4}$	2.2	0.199	0.455	7.187	0.092	0.306	4.063
$\frac{7}{8}$	2.6	0.296	0.584	8.555	0.151	0.422	6.028
1	3.0	0.427	0.737	9.374	0.270	0.634	7.014
$1\frac{1}{8}$	3.3	0.554	0.849	10.430	0.398	0.799	8.419
Class II							
$\frac{1}{2}$	1.5	0.075	0.267	4.891	0.020	0.167	2.727
$\frac{5}{8}$	1.8	0.130	0.357	5.593	0.032	0.225	3.074
$\frac{3}{4}$	2.2	0.198	0.454	6.631	0.075	0.392	4.049
$\frac{7}{8}$	2.6	0.300	0.591	7.990	0.123	0.542	5.997
1	3.0	0.421	0.754	8.614	0.220	0.812	6.987
$1\frac{1}{8}$	3.3	0.566	0.869	9.571	0.323	1.023	8.388
Structural I							
$\frac{1}{2}$	1.5	0.078	0.271	4.908	0.029	0.178	2.725
$\frac{5}{8}$	1.8	0.131	0.361	5.258	0.045	0.238	3.073
$\frac{3}{4}$	2.2	0.202	0.464	6.189	0.108	0.418	4.047
$\frac{7}{8}$	2.6	0.317	0.626	7.539	0.179	0.579	5.991
1	3.0	0.479	0.827	7.978	0.321	0.870	6.981
$1\frac{1}{8}$	3.3	0.623	0.955	8.841	0.474	1.098	8.377

*All properties adjusted to account for reduced effectiveness of plies with grain perpendicular to applied stress. Stresses adjusted for wet condition, load duration, and experience factors.

	Plyform Class I	Plyform Class II	Structural I Plyform
Modulus of elasticity (psi)	1,650,000	1,430,000	1,650,000
Bending stress (psi)	1,930	1,330	1,930
Rolling shear stress (psi)	72	72	102

Table 11-5 Section properties of lumber and timber (reference 1)*, **

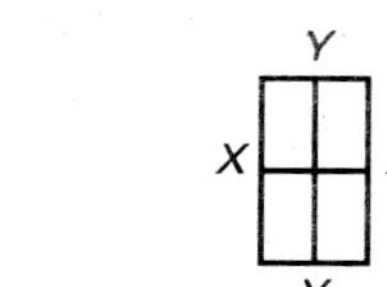

Nominal Size $b \times d$ (in.)	Standard Dressed Size (S4S), $b \times d$ (in.)	Area of Section, A ($in.^2$)	X-X Axis: Moment of Inertia, I ($in.^4$)	X-X Axis: Section Modulus, S ($in.^3$)	Y-Y Axis: Moment of Inertia, I ($in.^4$)	Y-Y Axis: Section Modulus, S ($in.^3$)	Board Measure per Lineal Foot	Weight (lb/lin ft) of Piece When Weight (lb) of Wood per Cubic Foot Equals: 25	30	35	40	45	50
1 × 3	$\frac{3}{4} \times 2\frac{1}{2}$	1.875	0.977	0.781	0.088	0.234	$\frac{1}{4}$	0.326	0.391	0.456	0.521	0.586	0.651
1 × 4	$\frac{3}{4} \times 3\frac{1}{2}$	2.625	2.680	1.531	0.123	0.328	$\frac{1}{3}$	0.456	0.547	0.638	0.729	0.820	0.911
1 × 6	$\frac{3}{4} \times 5\frac{1}{2}$	4.125	10.398	3.781	0.193	0.516	$\frac{1}{2}$	0.716	0.859	1.003	1.146	1.289	1.432
1 × 8	$\frac{3}{4} \times 7\frac{1}{4}$	5.438	23.817	6.570	0.255	0.680	$\frac{2}{3}$	0.944	1.133	1.322	1.510	1.699	1.888
1 × 10	$\frac{3}{4} \times 9\frac{1}{4}$	6.938	49.466	10.695	0.325	0.867	$\frac{5}{6}$	1.204	1.445	1.686	1.927	2.168	2.409
1 × 12	$\frac{3}{4} \times 11\frac{1}{4}$	8.438	88.989	15.820	0.396	1.055	1	1.465	1.758	2.051	2.344	2.637	2.930
2 × 3	$1\frac{1}{2} \times 2\frac{1}{2}$	3.750	1.953	1.563	0.703	0.938	$\frac{1}{2}$	0.651	0.781	0.911	1.042	1.172	1.302
2 × 4	$1\frac{1}{2} \times 3\frac{1}{2}$	5.250	5.359	3.063	0.984	1.313	$\frac{2}{3}$	0.911	1.094	1.276	1.458	1.641	1.823
2 × 6	$1\frac{1}{2} \times 5\frac{1}{2}$	8.250	20.797	7.563	1.547	2.063	1	1.432	1.719	2.005	2.292	2.578	2.865
2 × 8	$1\frac{1}{2} \times 7\frac{1}{4}$	10.875	47.635	13.141	2.039	2.719	$1\frac{1}{3}$	1.888	2.266	2.643	3.021	3.398	3.776
2 × 10	$1\frac{1}{2} \times 9\frac{1}{4}$	13.875	98.932	21.391	2.602	3.469	$1\frac{2}{3}$	2.409	2.891	3.372	3.854	4.336	4.818
2 × 12	$1\frac{1}{2} \times 11\frac{1}{4}$	16.875	177.979	31.641	3.164	4.219	2	2.930	3.516	4.102	4.688	5.273	5.859
2 × 14	$1\frac{1}{2} \times 13\frac{1}{4}$	19.875	290.775	43.891	3.727	4.969	$2\frac{1}{3}$	3.451	4.141	4.831	5.521	6.211	6.901

*For lumber surfaced $1\frac{5}{8}$ in. thick, instead of $1\frac{1}{2}$ in., the area (A), moment of inertia (I) and section modulus (S) about the X-X axis may be increased 8.33%.

**For members over 12 in. in dimension, check availability with local suppliers. Lengths are generally available in 2-ft increments with a limit on maximum length; check with local suppliers for lengths normally available.

Source: Copyright © 1974 by John Wiley & Sons, Inc. Reprinted by permission.

Table 11-5 *(continued)*

Nominal Size $b \times d$ (in.)	Standard Dressed Size (S4S), $b \times d$ (in.)	Area of Section, A (in.^{2})	X-X Axis: Moment of Inertia, I (in.^{4})	X-X Axis: Section Modulus, S (in.^{3})	Y-Y Axis: Moment of Inertia, I (in.^{4})	Y-Y Axis: Section Modulus, S (in.^{3})	Board Measure per Lineal Foot	Weight (lb/lin ft) of Piece When Weight (lb) of Wood per Cubic Foot Equals: 25	30	35	40	45	50
3 × 4	$2\frac{1}{2} \times 3\frac{1}{2}$	8.750	8.932	5.104	4.557	3.646	1	1.519	1.823	2.127	2.431	2.734	3.038
3 × 6	$2\frac{1}{2} \times 5\frac{1}{2}$	13.750	34.661	12.604	7.161	5.729	$1\frac{1}{2}$	2.387	2.865	3.342	3.819	4.297	4.774
3 × 8	$2\frac{1}{2} \times 7\frac{1}{4}$	18.125	79.391	21.901	9.440	7.552	2	3.147	3.776	4.405	5.035	5.664	6.293
3 × 10	$2\frac{1}{2} \times 9\frac{1}{4}$	23.125	164.886	35.651	12.044	9.635	$2\frac{1}{2}$	4.015	4.818	5.621	6.424	7.227	8.030
3 × 12	$2\frac{1}{2} \times 11\frac{1}{4}$	28.125	296.631	52.734	14.648	11.719	3	4.883	5.859	6.836	7.813	8.789	9.766
3 × 14	$2\frac{1}{2} \times 13\frac{1}{4}$	33.125	484.625	73.151	17.253	13.802	$3\frac{1}{2}$	5.751	6.901	8.051	9.201	10.352	11.502
3 × 16	$2\frac{1}{2} \times 15\frac{1}{4}$	38.125	738.870	96.901	19.857	15.885	4	6.619	7.943	9.266	10.590	11.914	13.238
4 × 4	$3\frac{1}{2} \times 3\frac{1}{2}$	12.250	12.505	7.146	12.505	7.146	$1\frac{1}{3}$	2.127	2.552	2.977	3.403	3.828	4.253
4 × 6	$3\frac{1}{2} \times 5\frac{1}{2}$	19.250	48.526	17.646	19.651	11.229	2	3.342	4.010	4.679	5.347	6.016	6.684
4 × 8	$3\frac{1}{2} \times 7\frac{1}{4}$	25.375	111.148	30.661	25.904	14.802	$2\frac{2}{3}$	4.405	5.286	6.168	7.049	7.930	8.811
4 × 10	$3\frac{1}{2} \times 9\frac{1}{4}$	32.375	230.840	49.911	33.049	18.885	$3\frac{1}{3}$	5.621	6.745	7.869	8.933	10.117	11.241
4 × 12	$3\frac{1}{2} \times 11\frac{1}{4}$	39.375	415.283	73.828	40.195	22.969	4	6.836	8.203	9.570	10.938	12.305	13.672
4 × 14	$3\frac{1}{2} \times 13\frac{1}{4}$	46.375	678.475	102.411	47.340	27.052	$4\frac{2}{3}$	8.047	9.657	11.266	12.877	14.485	16.094
4 × 16	$3\frac{1}{2} \times 15\frac{1}{4}$	53.375	1,034.418	135.661	54.487	31.135	$5\frac{1}{3}$	9.267	11.121	12.975	14.828	16.682	18.536
6 × 6	$5\frac{1}{2} \times 5\frac{1}{2}$	30.250	76.255	27.729	76.255	27.729	3	5.252	6.302	7.352	8.403	9.453	10.503
6 × 8	$5\frac{1}{2} \times 7\frac{1}{2}$	41.250	193.359	51.563	103.984	37.813	4	7.161	8.594	10.026	11.458	12.891	14.323
6 × 10	$5\frac{1}{2} \times 9\frac{1}{2}$	52.250	392.963	82.729	131.714	47.896	5	9.071	10.885	12.700	14.514	16.328	18.142
6 × 12	$5\frac{1}{2} \times 11\frac{1}{2}$	63.250	697.068	121.229	159.443	57.979	6	10.981	13.177	15.373	17.569	19.766	21.962

Table 11-5 (continued)

Nominal Size $b \times d$ (in.)	Standard Dressed Size (S4S), $b \times d$ (in.)	Area of Section, A (in.^{2})	X-X Axis: Moment of Inertia, I (in.^{4})	X-X Axis: Section Modulus, S (in.^{3})	Y-Y Axis: Moment of Inertia, I (in.^{4})	Y-Y Axis: Section Modulus, S (in.^{3})	Board Measure per Lineal Foot	Weight (lb/lin ft) of Piece When Weight (lb) of Wood per Cubic Foot Equals: 25	30	35	40	45	50
6 × 14	$5\frac{1}{2} \times 13\frac{1}{2}$	74.250	1,127.672	167.063	187.172	68.063	7	12.891	15.469	18.047	20.625	23.203	25.781
6 × 16	$5\frac{1}{2} \times 15\frac{1}{2}$	85.250	1,706.776	220.229	214.901	78.146	8	14.800	17.760	20.720	23.681	26.641	29.601
6 × 18	$5\frac{1}{2} \times 17\frac{1}{2}$	96.250	2,456.380	280.729	242.630	88.229	9	16.710	20.052	23.394	26.736	30.078	33.420
6 × 20	$5\frac{1}{2} \times 19\frac{1}{2}$	107.250	3,398.484	348.563	270.359	98.313	10	18.620	22.344	26.068	29.792	33.516	37.240
6 × 22	$5\frac{1}{2} \times 21\frac{1}{2}$	118.250	4,555.086	423.729	298.088	108.396	11	20.530	24.635	28.741	32.847	36.953	41.059
6 × 24	$5\frac{1}{2} \times 23\frac{1}{2}$	129.250	5,948.191	506.229	325.818	118.479	12	22.439	26.927	31.415	35.903	40.391	44.878
8 × 8	$7\frac{1}{2} \times 7\frac{1}{2}$	56.250	263.672	70.313	263.672	70.313	$5\frac{1}{3}$	9.766	11.719	13.672	15.625	17.578	19.531
8 × 10	$7\frac{1}{2} \times 9\frac{1}{2}$	71.250	535.859	112.813	333.984	89.063	$6\frac{2}{3}$	12.370	14.844	17.318	19.792	22.266	24.740
8 × 12	$7\frac{1}{2} \times 11\frac{1}{2}$	86.250	950.547	165.313	404.297	107.813	8	14.974	17.969	20.964	23.958	26.953	29.948
8 × 14	$7\frac{1}{2} \times 13\frac{1}{2}$	101.250	1,537.734	227.813	474.609	126.563	$9\frac{1}{3}$	17.578	21.094	24.609	28.125	31.641	35.156
8 × 16	$7\frac{1}{2} \times 15\frac{1}{2}$	116.250	2,327.422	300.313	544.922	143.313	$10\frac{2}{3}$	20.182	24.219	28.255	32.292	36.328	40.365
8 × 18	$7\frac{1}{2} \times 17\frac{1}{2}$	131.250	3,349.609	382.813	615.234	164.063	12	22.786	27.344	31.901	36.458	41.016	45.573
8 × 20	$7\frac{1}{2} \times 19\frac{1}{2}$	146.250	4,634.297	475.313	684.547	182.813	$13\frac{1}{3}$	25.391	30.469	35.547	40.625	45.703	50.781
8 × 22	$7\frac{1}{2} \times 21\frac{1}{2}$	161.250	6,211.484	577.813	755.859	201.563	$14\frac{2}{3}$	27.995	33.594	39.193	44.792	50.391	55.990
8 × 24	$7\frac{1}{2} \times 23\frac{1}{2}$	176.250	8,111.172	690.313	826.172	220.313	16	30.599	36.719	42.839	48.958	55.078	61.198
10 × 10	$9\frac{1}{2} \times 9\frac{1}{2}$	90.250	678.755	142.896	678.755	142.896	$8\frac{1}{3}$	15.668	18.802	21.936	25.069	28.203	31.337
10 × 12	$9\frac{1}{2} \times 11\frac{1}{2}$	109.250	1,204.026	209.396	821.651	172.979	10	18.967	22.760	26.554	30.347	34.141	37.934
10 × 14	$9\frac{1}{2} \times 13\frac{1}{2}$	128.250	1,947.797	288.563	964.547	203.063	$11\frac{2}{3}$	22.266	26.719	31.172	35.625	40.078	44.531

Table 11-5 *(continued)*

Nominal Size $b \times d$ (in.)	Standard Dressed Size (S4S), $b \times d$ (in.)	Area of Section, A (in.^{2})	X-X Axis: Moment of Inertia, I (in.^{4})	X-X Axis: Section Modulus, S (in.^{3})	Y-Y Axis: Moment of Inertia, I (in.^{4})	Y-Y Axis: Section Modulus, S (in.^{3})	Board Measure per Lineal Foot	Weight (lb/lin ft) of Piece When Weight (lb) of Wood per Cubic Foot Equals: 25	30	35	40	45	50
10 × 16	$9\frac{1}{2} \times 15\frac{1}{2}$	147.250	2,948.068	380.396	1,107.443	233.146	$13\frac{1}{3}$	25.564	30.677	35.790	40.903	46.016	51.128
10 × 18	$9\frac{1}{2} \times 17\frac{1}{2}$	166.250	4,242.836	484.896	1,250.338	263.229	15	28.863	34.635	40.408	46.181	51.953	57.726
10 × 20	$9\frac{1}{2} \times 19\frac{1}{2}$	185.250	5,870.109	602.063	1,393.234	293.313	$16\frac{2}{3}$	32.161	38.594	45.026	51.458	57.891	64.323
10 × 22	$9\frac{1}{2} \times 21\frac{1}{2}$	204.250	7,867.879	731.896	1,536.130	323.396	$18\frac{1}{3}$	35.460	42.552	49.644	56.736	63.828	70.920
10 × 24	$9\frac{1}{2} \times 23\frac{1}{2}$	223.250	10,274.148	874.396	1,679.026	353.479	20	38.759	46.510	54.262	62.014	69.766	77.517
12 × 12	$11\frac{1}{2} \times 11\frac{1}{2}$	132.250	1,457.505	253.479	1,457.505	253.479	12	22.960	27.552	32.144	36.736	41.328	45.920
12 × 14	$11\frac{1}{2} \times 13\frac{1}{2}$	155.250	2,357.859	349.313	1,710.984	297.563	14	26.953	32.344	37.734	43.125	48.516	53.906
12 × 16	$11\frac{1}{2} \times 15\frac{1}{2}$	178.250	3,568.713	460.479	1,964.463	341.646	16	30.946	37.135	43.325	49.514	55.703	61.892
12 × 18	$11\frac{1}{2} \times 17\frac{1}{2}$	201.250	5,136.066	586.979	2,217.943	385.729	18	34.939	41.927	48.915	55.903	62.891	69.878
12 × 20	$11\frac{1}{2} \times 19\frac{1}{2}$	224.250	7,105.922	728.813	2,471.422	429.813	20	38.932	46.719	54.505	62.292	70.078	77.865
12 × 22	$11\frac{1}{2} \times 21\frac{1}{2}$	247.250	9,524.273	885.979	2,724.901	473.896	22	42.925	51.510	60.095	68.681	77.266	85.851
12 × 24	$11\frac{1}{2} \times 23\frac{1}{2}$	270.250	10,274.148	1,058.479	2,978.380	517.979	24	46.918	56.302	65.686	75.069	84.453	93.837
14 × 16	$13\frac{1}{2} \times 15\frac{1}{2}$	209.250	4,189.359	540.563	3,177.984	470.813	$18\frac{2}{3}$	36.328	43.594	50.859	58.125	65.391	72.656
14 × 18	$13\frac{1}{2} \times 17\frac{1}{2}$	236.250	6,029.297	689.063	3,588.047	531.563	21	41.016	49.219	57.422	65.625	73.828	82.031
14 × 20	$13\frac{1}{2} \times 19\frac{1}{2}$	263.250	8,341.734	855.563	3,998.109	592.313	$23\frac{1}{3}$	45.703	54.844	63.984	73.125	82.266	91.406
14 × 22	$13\frac{1}{2} \times 21\frac{1}{2}$	290.250	11,180.672	1,040.063	4,408.172	653.063	$25\frac{2}{3}$	50.391	60.469	70.547	80.625	90.703	100.781
14 × 24	$13\frac{1}{2} \times 23\frac{1}{2}$	317.250	14,600.109	1,242.563	4,818.234	713.813	28	55.078	66.094	77.109	88.125	99.141	110.156

Table 11-6 Typical values of allowable stress for lumber

Species (No. 2 Grade, 4 × 4 or Smaller)	Allowable Unit Stress (psi) (Moisture Content = 19%)					
	F_b	F_v	$F_{c\perp}$	F_c	F_t	E
Douglas fir-larch	1450	185	385	1000	850	1.7×10^6
Hemlock-fir	1150	150	245	800	675	1.4×10^6
Southern pine	1400	180	405	975	825	1.6×10^6
California redwood	1400	160	425	1000	800	1.3×10^6
Eastern spruce	1050	140	255	700	625	1.2×10^6
Reduction factor for wet conditions	0.86	0.97	0.67	0.70	0.84	0.97

the design of stringers that the joist loads are actually applied to the stringer as a series of concentrated loads at the points where the joists rest on the stringer. However, it is simpler and sufficiently accurate to treat the load on the stringer as a uniform load. The width of the uniform design load applied to the stringer is equal to the stringer spacing as shown in Figure 11-1c. The calculated stringer span must next be checked against the capacity of the shores used to support the stringers. The load on each shore is equal to the shore spacing multiplied by the load per foot of stringer. Thus the maximum shore spacing (or stringer span) is limited to the lower of these two maximum values.

Although the effect of intermediate form members was ignored in determining allowable stringer span, it is necessary to check for crushing at the point where each joist rests on the stringer. This is done by dividing the load at this point by the bearing area and comparing the resulting stress to the allowable unit stress in compression perpendicular to the grain. A similar procedure is applied at the point where each stringer rests on a vertical support.

To preclude buckling, the maximum allowable load on a rectangular wood column is a function of its unsupported length and least dimension (or l/d ratio). The l/d ratio must not exceed 50 for a simple solid wood column. For l/d ratios less than 50, the following equation applies:

$$F'_c = \frac{0.3E}{(l/d)^2} \leqslant F_c \tag{11-10}$$

where F_c = allowable unit stress in compression parallel to the grain (psi)
F'_c = allowable unit stress in compression parallel to the grain, adjusted for l/d ratio (psi)
E = modulus of elasticity (psi)
l/d = ratio of member length to least dimension

In using this equation, note that the maximum value used for F'_c may not exceed the value of F_c.

These design procedures are illustrated in the following example. Sheathing design employing plywood is illustrated in Example 11-2.

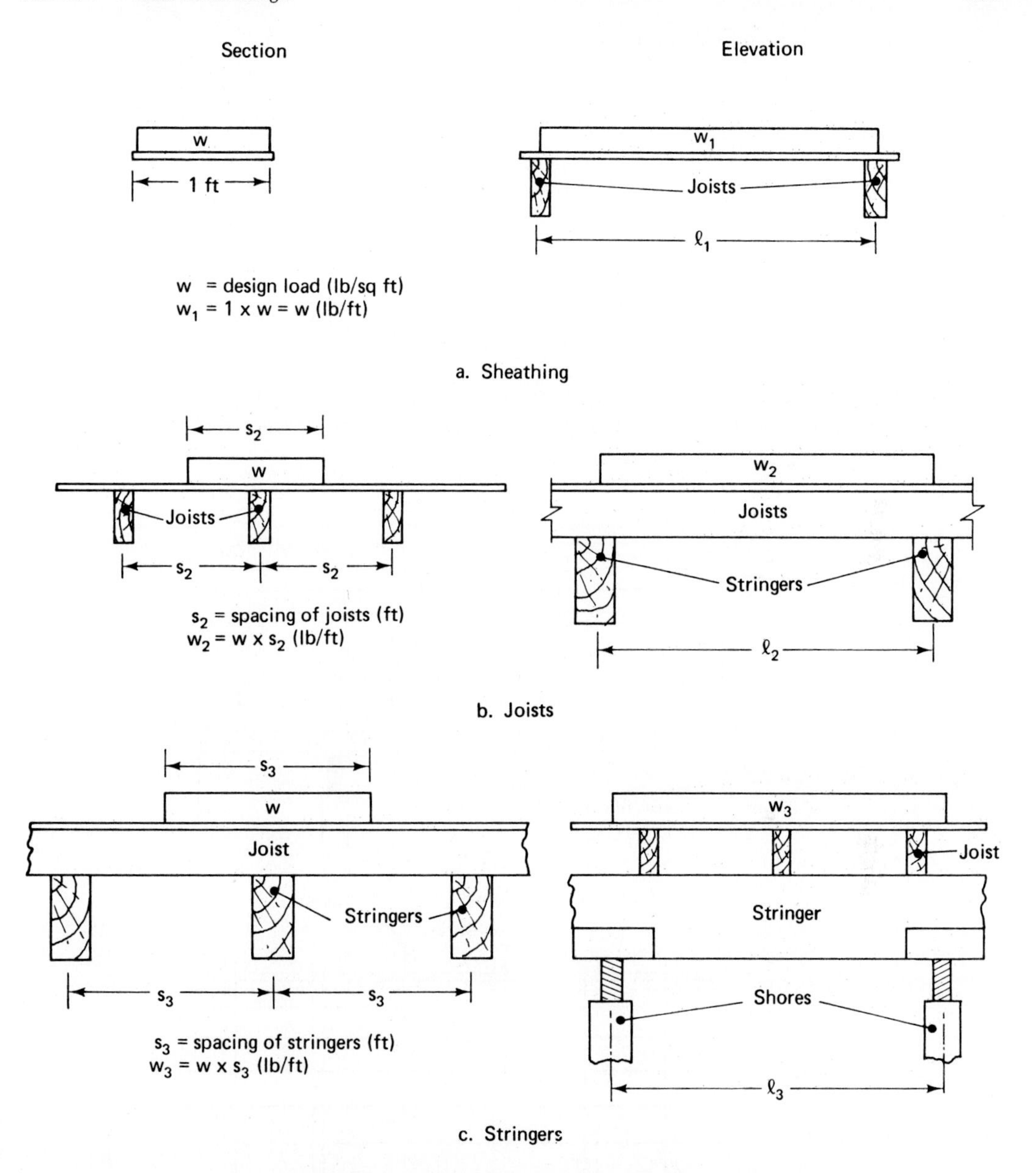

Figure 11-1 Design analysis of form members.

Example Problem—Slab Form Design

Example 11-1

PROBLEM Design the formwork (Figure 11-2) for an elevated concrete floor slab 6 in. thick. Sheathing will be 1-in. (nominal)-thick lumber, while 2 × 8 in. lumber will be used for joists. Stringers will be 4 × 8 in. lumber. Assume that all members are continuous over three or more spans. Commercial 4000-lb shores will be used. It is estimated that the weight of the formwork will be 5 lb/sq ft. The adjusted allowable stresses for the lumber being used are as follows:

	Sheathing (psi)	*Other Members (psi)*
F_b	1075	1250
F_v	174	180
$F_{c\perp}$		405
F_c		850
E	1.36×10^6	1.4×10^6

Maximum deflection of form members will be limited to $l/360$. Use the minimum value of live load permitted by ACI. Determine appropriate joist spacing, stringer spacing, and shore spacing.

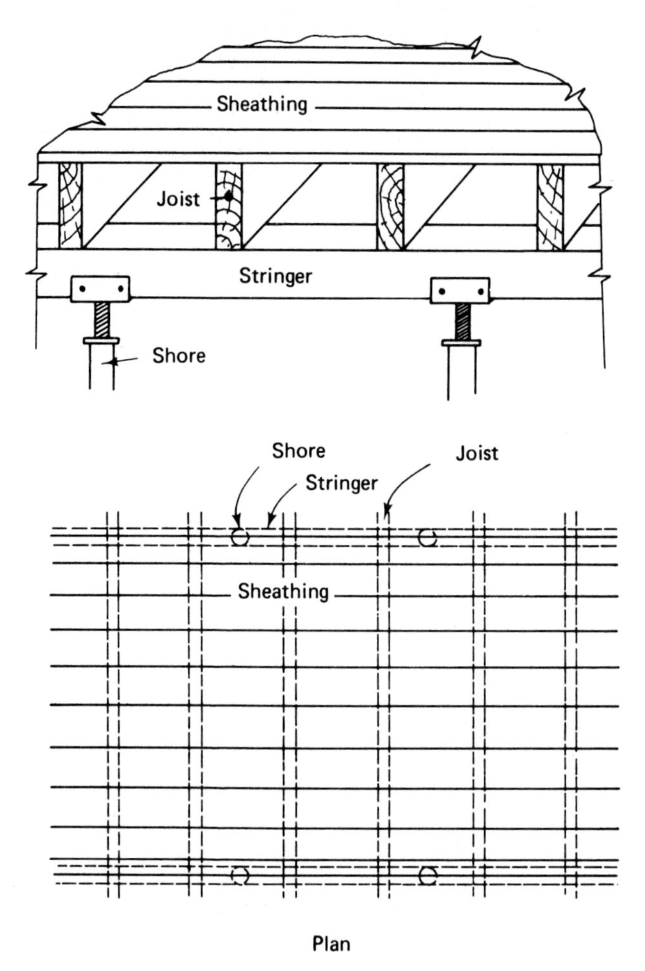

Figure 11-2 Slab form, Example 11-1.

SOLUTION

Design Load

Concrete = 1 sq ft × 6/12 ft × 150 lb/cu ft = 75 lb/sq ft
Form work = 5 lb/sq ft
Live load = 50 lb/sq ft
Design load = 130 lb/sq ft

Deck Design. Consider a uniformly loaded strip of decking (sheathing) 12 in. wide perpendicular to the joists (Figure 11-1a) and analyze it as a beam. Assume that a deck is continuous over three or more spans and use equations for three or more spans in Table 11-3.

$$w = (1 \text{ sq ft/lin ft}) \times (130 \text{ lb/sq ft}) = 130 \text{ lb/ft}$$

(a) Bending:

$$l = 4.46d\left(\frac{F_b b}{w}\right)^{1/2}$$
$$= (4.46)(0.75)\left[\frac{(1075)(12)}{130}\right]^{1/2} = 33.3 \text{ in.}$$

(b) Shear:

$$l = 13.3\frac{F_v A}{w} + 2d = 13.3\frac{F_v bd}{w} + 2d$$
$$= \frac{(13.3)(174)(12)(0.75)}{130} + (2)(0.75) = 161.7 \text{ in.}$$

(c) Deflection:

$$l = 1.69\left(\frac{EI}{w}\right)^{1/3} = 1.69\left(\frac{Ebd^3}{w12}\right)^{1/3}$$
$$= 1.69\left[\frac{(1.36 \times 10^6)(12)(0.75)^3}{(130)(12)}\right]^{1/3} = 27.7 \text{ in.}$$

Deflection governs in this design and the maximum allowable span is 27.7 in. We will select a 24-in. joist spacing for the design.

Joist Design. Consider the joist as a uniformly loaded beam supporting a strip of design load 24 in. wide (same as joist spacing; see Figure 11-1b). Joists are 2 × 8 in. lumber. Assume that joists are continuous over three spans.

$$w = (2 \text{ ft} \times 1 \text{ ft}) \times (130 \text{ lb/sq ft}) = 260 \text{ lb/ft}$$

(a) Bending:

$$l = 10.95\left(\frac{F_b S}{w}\right)^{1/2}$$
$$= 10.95\left[\frac{(1250)(13.141)}{260}\right]^{1/2} = 87.0 \text{ in.}$$

(b) Shear:

$$l = 13.3\frac{F_v A}{w} + 2d$$

$$= \frac{(13.3)(180)(10.875)}{260} + (2)(7.25) = 114.6 \text{ in.}$$

(c) Deflection:

$$l = 1.69\left(\frac{EI}{w}\right)^{1/3}$$

$$= 1.69\left[\frac{(1.4 \times 10^6)(47.635)}{260}\right]^{1/3} = 107.4 \text{ in.}$$

Thus bending governs and the maximum joist span is 87 in. Select a stringer spacing (joist span) of 84 in.

Stringer Design. To analyze the stringer design, consider a strip of design load 7 ft wide (equal to stringer spacing) as resting directly on the stringer (Figure 11-1c). Assume the stringer to be continuous over three spans. Stringers are 4 × 8 lumber.

$$w = (7)(130) = 910 \text{ lb/ft}$$

Now analyze the stringer as a beam and determine the maximum allowable span.

(a) Bending:

$$l = 10.95\left(\frac{F_b S}{w}\right)^{1/2}$$

$$= 10.95\left[\frac{(1250)(30.661)}{910}\right]^{1/2} = 71.1 \text{ in.}$$

(b) Shear:

$$l = 13.3\frac{F_v A}{w} + 2d$$

$$= \frac{(13.3)(180)(25.375)}{910} + (2)(7.25) = 81.3 \text{ in.}$$

(c) Deflection:

$$l = 1.69\left(\frac{EI}{w}\right)^{1/3}$$

$$= 1.69\left[\frac{(1.4 \times 10^6)(111.148)}{910}\right]^{1/3} = 93.8 \text{ in.}$$

Bending governs and the maximum span is 71.1 in.

Now we must check shore strength before selecting the stringer span (shore spacing). The maximum stringer span in feet based on shore strength is equal to the shore strength divided by the load per feet of stringer.

$$l = \frac{4000}{910} \times 12 = 52.7 \text{ in.}$$

Thus the maximum stringer span is limited by shore strength to 52.7 in. Select a shore spacing of 4 ft (48 in.).

Before completing our design, we should check for crushing at the point where each joist rests on a stringer. The load at this point is the load per foot of joist multiplied by the joist span in feet.

$$P = (260)\left(\frac{84}{12}\right) = 1820 \text{ lb}$$

$$\text{Bearing area } (A) = (1.5)(3.5) = 5.25 \text{ sq in.}$$

$$f_{c\perp} = \frac{P}{A} = \frac{1820}{5.25} = 347 \text{ psi} < 405 \text{ psi} \quad (F_{c\perp}) \quad ok$$

Final Design

Decking: 1-in. lumber.

Joists: 2 × 8's at 24-in. spacing.

Stringers: 4 × 8's at 84-in. spacing.

Shores: 4000-lb commercial shores at 48-in. intervals.

11-5 WALL AND COLUMN FORM DESIGN

Design Procedures

The design procedure for wall and column forms is similar to that used for slab forms substituting studs for joists, wales for stringers, and ties for shores. First, the maximum lateral pressure against the sheathing is determined from the appropriate equation (Equation 11-1, 11-2, or 11-3). If the sheathing thickness has been specified, the maximum allowable span for the sheathing based on bending, shear, and deflection is the maximum stud spacing. If the stud spacing is fixed, calculate the required thickness of sheathing.

Next, calculate the maximum allowable stud span (wale spacing) based on stud size and design load, again considering bending, shear, and deflection. If the stud span has already been determined, calculate the required size of the stud. After stud size and wale spacing have been determined, determine the maximum allowable spacing of wale supports (tie spacing) based on wale size and load. If tie spacing has been preselected, determine the minimum wale size. Double wales are commonly used (see Figure 11-3) to avoid the necessity of drilling wales for tie insertion.

Next check the tie's ability to carry the load imposed by wale and tie spacing. The load (lb) on each tie is calculated as the design load (lb/sq ft) multiplied by the product of tie spacing (ft) and wale spacing (ft). If the load exceeds tie strength, a stronger tie must be used or the tie spacing must be reduced.

The next step is to check bearing stresses (or compression perpendicular to the grain) where the studs rest on wales and where tie ends bear on wales. Maximum bearing stress must not exceed the allowable compression stress perpendicular to the grain or crushing will result. Finally, design lateral bracing to resist any expected lateral loads, such as wind loads.

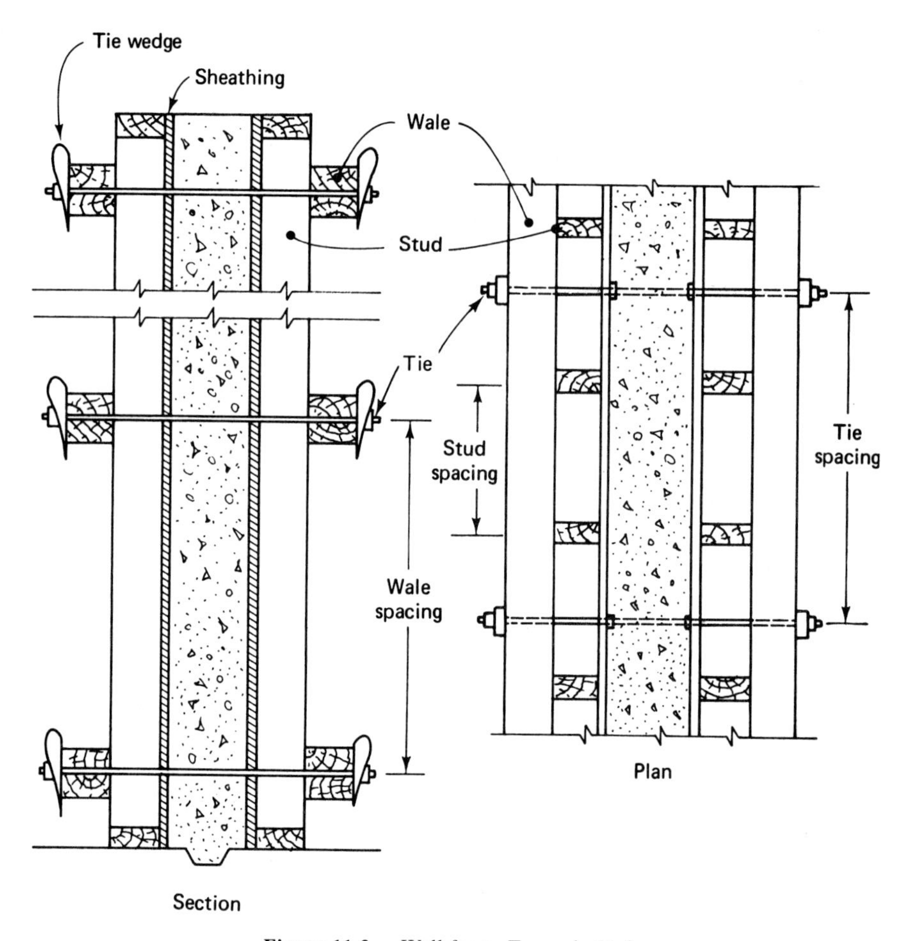

Figure 11-3 Wall form, Example 11-2.

Example Problem — Wall Form Design

Example 11-2

PROBLEM Forms are being designed for an 8-ft-high concrete wall to be poured at the rate of 4 ft/h, internally vibrated, at 90°F temperature (see Figure 11-3). Sheathing will be 4 × 8 ft sheets of ⅝-in.-thick class I Plyform with face grain perpendicular to studs. Studs and double wales will be 2 × 4 in. lumber. Snap ties are 3000-lb capacity with 1½-in.-wide wedges bearing on wales. Deflection must not exceed $l/360$. Determine stud, wale, and tie spacing. Use Plyform section properties and allowable stresses from Table 11-4 and lumber section properties from Table 11-5. Allowable stresses for studs and wales are

$$f_b = 1810 \text{ psi}$$
$$f_v = 120 \text{ psi}$$
$$f_{c\perp} = 485 \text{ psi}$$
$$E = 1.7 \times 10^6 \text{ psi}$$

SOLUTION

Design Load

$$p = 150 + \frac{9000R}{T} = 150 + \frac{(9000)(4)}{90} = 550 \text{ lb/sq ft} \qquad (11\text{-}1)$$

Check: $550 < 150h$ (= 1200 lb/sq ft) < 2000 lb/sq ft *ok*

Select Stud Spacing (Three or More Spans)

Material: ⅝-in. class I Plyform (Table 11-4)

Consider a strip 12 in. wide.

$$w = 1 \times 1 \times 550 = 550 \text{ lb/ft}$$

(a) Bending:

$$l = 10.95\left(\frac{F_b S}{w}\right)^{1/2}$$

$$= 10.95\left[\frac{(1930)(0.358)}{550}\right]^{1/2} = 12.27 \text{ in.}$$

(b) Shear:

$$l = 20\frac{F_v(Ib/Q)}{w} + 2d$$

$$= (20)\frac{(72)(5.717)}{550} + (2)(0.625) = 16.22 \text{ in.}$$

(c) Deflection:

$$l = 1.69\left(\frac{EI}{w}\right)^{1/3}$$

$$= 1.69\left(\frac{(1.65 \times 10^6)(0.130)}{550}\right)^{1/3} = 12.35 \text{ in.}$$

Bending governs. Maximum span = 12.27 in. Use a 12-in. stud spacing.

Select Wale Spacing (Three or More Spans). Since the stud spacing is 12 in., consider a uniform design load 1 ft wide resting on each stud.

$$w = 1 \times 1 \times 550 = 550 \text{ lb/ft}$$

(a) Bending:

$$l = 10.95\left(\frac{F_b S}{w}\right)^{1/2}$$

$$= 10.95\left[\frac{(1810)(3.063)}{550}\right]^{1/2} = 34.76 \text{ in.}$$

(b) Shear:

$$l = 13.3\frac{F_v A}{w} + 2d$$
$$= \frac{(13.3)(120)(5.25)}{550} + (2)(3.5) = 22.23 \text{ in.}$$

(c) Deflection:

$$l = 1.69\left(\frac{EI}{w}\right)^{1/3}$$
$$= 1.69\left[\frac{(1.7 \times 10^6)(5.359)}{550}\right]^{1/3} = 43.08 \text{ in.}$$

Shear governs, so maximum span (wale spacing) is 22.23 in. Use a 16-in. wale spacing for modular units.

Select Tie Spacing (Three or More Spans, Double Wales). Based on a wale spacing of 16 in.,

$$w = \frac{16}{12} \times 550 = 733.3 \text{ lb/ft}$$

(a) Bending:

$$l = 10.95\left(\frac{F_b S}{w}\right)^{1/2}$$
$$= 10.95\left[\frac{(1810)(2 \times 3.063)}{733.3}\right]^{1/2} = 42.58 \text{ in.}$$

(b) Shear:

$$l = 13.3\frac{F_v A}{w} + 2d$$
$$= 13.3\frac{(120)(2 \times 5.25)}{733.3} + (2)(3.5) = 29.85 \text{ in.}$$

(c) Deflection:

$$l = 1.69\left(\frac{EI}{w}\right)^{1/3}$$
$$= 1.69\left[\frac{(1.7 \times 10^6)(2 \times 5.359)}{733.3}\right]^{1/3} = 49.31 \text{ in.}$$

Shear governs. Maximum span is 29.85 in. Select a 24-in. tie spacing.

(d) Check tie load:

$$P_{tie} = \text{wale spacing} \times \text{tie spacing} \times w$$
$$= \left(\frac{16}{12}\right)\left(\frac{24}{12}\right)(550) = 1467 \text{ lb/tie} < 3000 \text{ lb} \quad ok$$

Check Bearing

(a) Stud on wales:

Bearing area (double wales) = (2) (1.5) (1.5) = 4.5 sq in.

Load at each panel point (P) = (load/ft of stud) × [wale spacing (ft)]

$$P = (550)\left(\frac{16}{12}\right) = 733 \text{ lb}$$

$$f_{c\perp} = \frac{P}{A} = \frac{733}{4.5} = 163 \text{ psi} < 485 \text{ psi} \quad (F_{c\perp})\ ok$$

(b) Tie wedges on wales:

Tie load (P) = 1467 lb

Bearing area (A) = (1.5) (2) (1.5) = 4.5 sq in.

$$f_{c\perp} = \frac{P}{A} = \frac{1467}{4.5} = 326 \text{ psi} < 485 \text{ psi} \quad (F_{c\perp})\ ok$$

Final Design

Sheathing: 4 × 8 ft sheets of ⅝-in. class I Plyform placed horizontally.

Studs: 2 × 4's at 12 in. on center.

Wales: Double 2 × 4's at 16 in. on center.

Ties: 3000-lb snap ties at 24 in. on center.

11-6 DESIGN OF LATERAL BRACING

Many failures of formwork have been traced to omitted or inadequately designed lateral bracing. Minimum lateral design load values were given in Section 11-2. Design procedures for lateral bracing are described and illustrated in the following paragraphs.

Lateral Braces for Wall and Column Forms

For wall and column forms, lateral bracing is usually provided by inclined rigid braces or guy-wire bracing. Since wind loads, and lateral loads in general, may be applied in either direction perpendicular to the face of the form, guy-wire bracing must be placed on both sides of the forms. When rigid braces are used they may be placed on only one side of the form if designed to resist both tension and compression forces. When forms are placed on only one side of a wall with the excavation serving as the second form, lateral bracing must be designed to resist the lateral pressure of the concrete as well as other lateral forces.

Inclined bracing will usually resist any wind uplift forces on vertical forms. However, uplift forces on inclined forms may require additional consideration and the use of special anchors or tiedowns. The strut load per foot of form developed by the design lateral load can be calculated by the use of Equation 11-11. The total load per strut is then P' multiplied by strut spacing.

$$P' = \frac{H \times h \times l}{h' \times l'} \tag{11-11}$$

$$l = (h'^2 + l'^2)^{1/2} \tag{11-12}$$

where P' = strut load per foot of form (lb/ft)
H = lateral load at top of form (lb/ft)
h = height of form (ft)
h' = height of top of strut (ft)
l = length of strut (ft)
l' = horizontal distance from form to bottom of strut (ft)

If struts are used on only one side of the form, the allowable unit stress for strut design will be the lowest of the three possible allowable stress values (F_c, F_c', or F_t).

Example 11-3

PROBLEM Determine the maximum spacing of 2 × 4 in. lateral braces for the wall form of Example 11-2 placed as shown in Figure 11-4. Assume that local code wind load requirements are less stringent than Table 11-1. Allowable stress values for braces are $E = 1.4 \times 10^6$ psi, $F_c = 850$ psi, $F_t = 725$ psi.

SOLUTION Determine the design lateral force per linear foot of form:

$$H = 100 \text{ lb/lin ft (Table 11-1)}$$

Determine the length of the strut:

$$l = (h'^2 + l'^2)^{1/2}$$
$$= (6^2 + 5^2)^{1/2} = 7.81 \text{ ft} \times 12 = 93.7 \text{ in.} \tag{11-12}$$

The axial concentrated load on the strut produced by one foot of form may now be determined from Equation 11-11.

$$P' = \frac{H \times h \times l}{h' \times l'} = \frac{(100)(8)(7.81)}{(6)(5)} = 208.3 \text{ lb/ft of form}$$

Now determine the allowable compressive stress for the strut using Equation 11-10. First check the l/d ratio for the strut.

$$l/d = \frac{93.7}{1.5} = 62.5 > 50$$

Since this value exceeds 50, the strut itself must be provided lateral bracing to reduce its unsupported length. Try a single lateral support at the midpoint of the strut, reducing l to 46.9 in.

$$F_c' = \frac{0.3E}{(l/d)^2} = \frac{(0.3)(1.4 \times 10^6)}{(46.9/1.5)^2} = 430 \text{ psi}$$

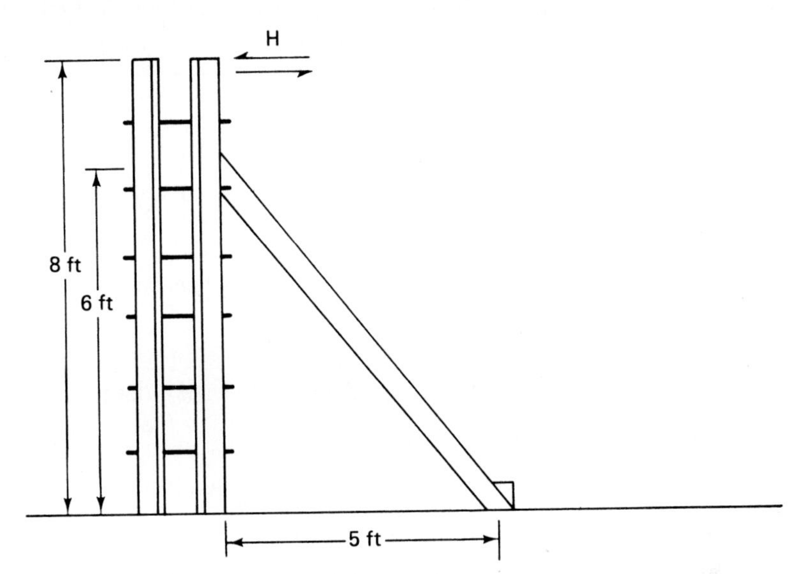

Figure 11-4 Wall form bracing, Example 11-3.

As $F_c' < F_t < F_c$, the value of F_c' governs.

$$\text{Maximum allowable compression per strut} = (1.5)(3.5)(430) = 2257 \text{ lb}$$

$$\text{Maximum strut spacing} = \frac{2257}{208} = 10.8 \text{ ft}$$

Remember that this design is based on providing lateral support to each strut at the midpoint of its length.

Lateral Braces for Slab Forms

For elevated floor or roof slab forms, lateral bracing may consist of cross braces between shores or inclined bracing along the outside edge of the form similar to that used for wall forms. The following example illustrates the method of determining the design lateral load for slab forms.

Example 11-4

PROBLEM Determine the design lateral force for a slab form 6 in. thick, 20 ft wide, and 100 ft long, shown in Figure 11-5. The slab is to be placed in one pour. Use design weight of formwork of 15 lb/sq ft.

SOLUTION

$$\text{Dead load} = \left(\frac{1}{2}\right)(1)(150) + 15 = 90 \text{ lb/sq ft}$$

$$H = 0.02 \times dl \times ws \tag{11-4}$$

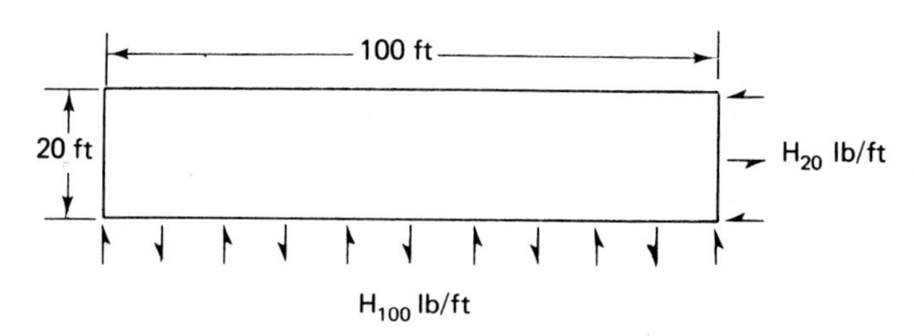

Figure 11-5 Slab form bracing design, Example 11-4.

For the 20-ft face, width of slab is 100 ft.

$$H_{20} = (0.02)(90)(100) = 180 \text{ lb/lin ft}$$

For the 100-ft face, width of slab is 20 ft.

$$H_{100} = (0.02)(90)(20) = 36 \text{ lb/ft} < 100 \text{ lb/ft}$$

Use $H_{100} = 100$ lb/lin ft (minimum load).

PROBLEMS

1. Calculate the design load in pounds per square foot for a wall form 8 ft high to be poured at the rate of 6 ft/h at a temperature of 50 °F.
2. What should be the design load for a column form 18 ft high that is to be filled by pumping concrete from the bottom?
3. What is the maximum allowable span for ¾-in. Class II Plyform decking with face grain across supports for a design load of 200 lb/sq ft? Assume that the decking is continuous over three or more spans. Limit the deflection to 1/360 span length.
4. Determine the maximum allowable spacing of 2 × 4 in. studs for a wall form sheathed with nominal 1-in. lumber. The design load is 600 lb/sq ft. Assume that the sheathing is continuous over three or more spans and is hem-fir. Limit deflection to 1/240 span length.
5. Determine the maximum allowable spacing of 3500-lb capacity shores used with a 4 × 6 in. stringer of southern pine. Stringer spacing is 5 ft on center. Design load is 150 lb/sq ft. Use allowable stresses of Table 11-6. Maximum allowable deflection is 1/240 span length. Assume that stringers are continuous over two spans.
6. Determine the maximum allowable span of 2 × 4 in. wall form studs carrying a design load of 950 lb/sq ft. Stud spacing is 18 in. on center. Use allowable stresses from Table 11-6 and 7-day load duration. Assume that studs are continuous over three or more spans. Based on the maximum allowable span, check for crushing of studs on double 2 × 4 in. wales. Limit deflection to 1/240 span length. All lumber is Douglas fir.
7. Design the formwork for a wall 6 ft high to be poured at the rate of 6 ft/h at a temperature of 80°F. Sheathing will be ⅞-in. Class I Plyform with face grain across supports. All lumber is No. 2 southern pine. Use 2 × 4 in. studs, double 2 × 4 in. wales, and 3000-lb

snap ties. Lateral bracing of 2 × 4's will be attached at a height of 5 ft above the form bottom and anchored 4 ft away from the base of the form. Use allowable lumber stresses from Table 11-6 adjusted for 7-day load and wet conditions. Limit deflection to 1/360 of the span length. Wind load equals 20 lb/sq ft.

8. Design the formwork for a concrete slab 6 in. thick whose net width between beam faces is 15.5 ft. Hand concrete buggies will be used to place concrete. Formwork is estimated to weigh 5 lb/sq ft. Decking will be ¾-in. class I Plyform with face grain across supports. Joists will be nominal 2-in.-wide Douglas fir. One 4-in.-wide Douglas fir stringer will be used between beams. Limit deflection to 1/360 span length. Commercial shores of 4000-lb capacity will be used. Lateral support will be provided by beam forms.

9. Design the formwork for an 8-in.-thick concrete floor slab based on the following data. Hand concrete buggies will be used to place concrete. Formwork is estimated to weigh 10 lb/sq ft. Decking will be ¾-in. class I Plyform with face grain across supports. Joists will be 2 × 12's and stringers will be 4 × 10's. All lumber will be Douglas fir (use allowable stresses from Table 11-6 and 7-day load duration). Maximum deflection must be limited to 1/360 span length. Shores of 8000-lb capacity will be used. The slab will be 40 ft wide × 50 ft long, poured at one time. Guy-wire bracing capable of carrying a load of 2000 lb each will be used on all four sides of the form, attached at slab elevation and making a 45° angle with the ground.

10. Write a computer program to calculate maximum span length of a wood formwork member as determined by shear, bending, and deflection based on the equations of Table 11-3. Input should include support conditions, allowable deflection, allowable stresses, load per foot of span, and dimensions of the member.

11. Write a computer program to calculate the maximum span of a plywood deck (or sheathing) based on the equations of Table 11-3 and the Plyform properties given in Table 11-4. Input should include design load, support conditions, Plyform type and thickness, and whether face grain is across supports or parallel to supports.

REFERENCES

1. American Institute of Timber Construction. *Timber Construction Manual,* 3rd ed. New York: Wiley, 1985.
2. *APA Design/Construction Guide: Concrete Forming.* American Plywood Association, Tacoma, Wash., 1990.
3. *Design of Wood Formwork for Concrete Structures* (Wood Construction Data No. 3). National Forest Products Association, Washington, D.C., 1961.
4. Hurd, M. K. *Formwork for Concrete,* 6th ed. American Concrete Institute, Detroit, Mich., 1989.
5. *National Design Specification for Wood Construction.* National Forest Products Association, Washington, D.C., 1982.
6. Peurifoy, R. L. *Formwork for Concrete Structures,* 2nd ed. New York: McGraw-Hill, 1976.
7. *Plywood Design Specifications.* American Plywood Association, Tacoma, Wash., 1985.

12
Wood Construction

12-1
INTRODUCTION

Wood is one of man's oldest construction materials. Today it is still widely used to construct residential, commercial, and industrial buildings, as well as such varied structures as piers, bridges, retaining walls, and power transmission towers. In the United States, for example, 90% of all houses are constructed of wood. In this chapter we will consider the properties of wood that influence its use in construction, together with the principles and practices of both frame and timber construction.

12-2
WOOD MATERIALS AND PROPERTIES

Types

Wood is divided into two major classes, hardwood and softwood, according to its origin. *Hardwood* is produced from deciduous (leaf-shedding) trees. *Softwood* comes from conifers (trees having needlelike or scalelike leaves), which are primarily evergreens. The terms "hardwood" and "softwood" indicate only the wood species and may be misleading, because some softwoods are actually harder than some hardwoods. In the United States, lumber is grouped into several grading types, which have similar properties. Most of the lumber used in the United States for structural purposes is softwood.

Moisture Content

The moisture content of lumber (which is defined as the weight of moisture in the wood divided by the oven-dry weight of the wood and then expressed as a percentage) has a great influence on its strength properties. At moisture contents above 30%, wood is essentially in its natural state and no changes in size or strength properties occur. At moisture contents below 30%, wood shrinks and its strength properties increase. For example, the bending strength of common softwood at moisture contents below 19% is approximately 2½ times its bending strength at moisture contents above 30%. Warping of lumber often occurs as it shrinks.

Structural Wood

Lumber is any wood that is cut into a size and shape suitable for use as a building material. *Timber* is broadly classified as lumber having a smallest dimension of at least 5 in. (12.7 cm). Structural lumber is further divided into *board, dimension, beam and stringer,* and *post and timber* classifications. The board classification applies to lumber less than 2 in. (5 cm) thick and at least 2 in. (5 cm) wide. Dimension applies to lumber at least 2 in. (5 cm) but less than 5 in. (12.7 cm) thick and 2 in. (5 cm) or more wide. The beam and stringer classification applies to lumber at least 5 in. (12.7 cm) thick and 8 in. (20 cm) wide, graded for its strength in bending with the load applied to the narrow face (thickness). The post and beam classification applies to lumber that is approximately square in cross section, at least 5 in. (12.7 cm) in thickness and width, and intended for use where bending strength is not important. Lumber may be either rough or dressed. *Rough lumber* has been sawn on all four sides but not surfaced (planed smooth or dressed). *Dressed lumber* has been surfaced on one or more sides. Possible classifications include surfaced one side (S1S), surfaced two sides (S2S), surfaced one edge (S1E), surfaced two edges (S2E), and combinations of sides and edges (S1S1E, S1S2E, and S4S).

Structural lumber is usually available in lengths from 10 ft (3 m) to 20 ft (6 m) in 2-ft (0.6-m) increments. Studs are available in 8-ft (2.4-m) lengths. Longer lengths may be available on special order. Section dimensions and properties for common sizes of dimension lumber are given in Table 11-5. Warping can be minimized by shaping lumber after it has been dried to within a few percent of the moisture content at which it will be used. Grading rules define *green lumber* as having a moisture content greater than 19%, *dry lumber* as having a moisture content of 19% or less, and *kiln dried lumber* as having a moisture content of 15% or less.

Strength

In the United States, lumber grading rules are set by a national Grading Rule Committee established by the U.S. Department of Commerce. The allowable stresses for dimensioned lumber are determined by the wood species, moisture content, and grade. Allowable stresses for common species are set forth in Table 1 of reference 6. Allowable stresses should be adjusted for duration of load and wet conditions as explained in the notes to Table 1 of reference 6.

Wood Preservation

Wood is subject to damage by decay and by wood-boring insects. Mechanical shields may be used to reduce exposure to insect damage. However, wood preservation by chemical treatment is the principal method used today to provide protection against decay and insect damage. Surface treatment to wood has been largely replaced by pressure treatment which forces the preservatives deep into the wood cells. The principal wood preservatives now used include creosote, waterborne salts, and pentachlorophenol in a solvent. Creosote turns the wood surface black and the surface cannot be painted. Waterborne preservatives normally leave a clean, odorless, and paintable surface. Pentachlorophenol preservatives usually leave a brown surface, which gradually turns silver. Depending on the solvent used, the surface may be paintable.

Glued Laminated Timber

Glulam, glued laminated timber (Figure 12-1), is composed of layers of wood 2 in. (5 cm) or less in thickness which are glued together to form a solid structural member. Glued laminated timber has several advantages over sawn timber. It provides a way to manufacture wood structural members of great size, curved as well as straight. Since the individual wood pieces used for lamination are rather thin, they can be readily dried to a moisture content that produces a dimensionally stable

Figure 12-1 Glued laminated timber beam. (Courtesy of American Institute of Timber Construction)

member of high strength. The strength of a glued laminated timber member can be closely controlled by placing high-strength lumber in areas of high stress and lower strength lumber in areas of lower stress. This practice reduces the cost of the structural member. The production of glued laminated timber under carefully controlled conditions results in precisely dimensioned structural members of high strength at a minimum cost.

Glued laminated members are widely used in large buildings such as churches, auditoriums, shopping centers and sports arenas, as well as in industrial plants. The radial arch structure shown in Figure 12-2 has a clear span of 240 ft (73 m). Other structural applications range from bridge beams to power transmission towers. Reference 1 provides data on standard sizes and allowable stresses for glued laminated timber.

Plywood

Plywood is a wood structural material formed by gluing three or more thin layers of wood (veneers) together with the grain of alternate layers running perpendicular to each other. This process results in a material having a high strength/weight ratio which can be produced in a wide range of strength and appearance grades. Grading

Figure 12-2 Erecting large glued laminated timber arches. (Courtesy of American Institute of Timber Construction)

rules established by the American Plywood Association divide wood veneers into groups 1 through 5, based on strength and stiffness, with group 1 having the highest strength characteristics. In addition to the basic Exterior and Interior type classification, Engineered Grades and an Identification Index indicating the maximum allowable span of the member under standard loads are incorporated in the grading rules. Exterior-type plywood is manufactured with waterproof glue and higher-grade veneers than those used for Interior-type plywood.

Surface appearance grades of N, A, B, C, and D are available, with grade N being of the highest appearance quality. Plyform, a grade of plywood intended for use in concrete formwork, can be manufactured in two grades, class I and class II. Class II is most readily available. High-Density Overlay (HDO) and Medium-Density Overlay (MDO) plywoods have an abrasion resistant resin-fiber overlay on one or both faces. Plywood with rough-sawn, grooved, or other special faces is available for siding and other uses where appearance is a major consideration. The usual size of plywood sheets in the United States is 4 ft by 8 ft (1.2 m by 2.4 m). Thickness of ⅜ in. (1.0 cm) to 1⅛ in. (2.9 cm) are available. Plywood is available with tongue-and-groove edges for use in floor construction. Plywood design specifications are given in reference 9. Plywood applications and construction guidelines are described in reference 3.

Other Wood Products

There are a growing number of other wood products that can be used in place of structural wood and plywood. Some of these include laminated veneer lumber, Com-Ply, built-up I-beams, particleboard, waferboard, and oriented strand board. Most of these products consist wholly or partly of wood fibers or chips bonded together with resins. Since they are able to utilize portions of trees that would otherwise be wasted, these products are relatively economical to produce. In addition, they resist warping and swelling and are often stronger than structural wood members of the same size.

Laminated veneer lumber is similar to plywood but is produced in sizes that can replace structural lumber. MICRO=LAM, for example, is produced as a billet 2 ft (61 cm) wide, 80 ft (24 m) long, and ¾ to 2½ in. (1.9 to 6.4 cm) thick. Billets are then cut to form lumber of the desired width and length.

Com-Ply studs and beams consist of wood veneer facings bonded to two sides of a wood chip core.

Wood I-beams consist of structural wood top and bottom flanges bonded to a plywood or laminated veneer lumber web.

Particleboard is produced in sheets by bonding wood chips together with resin. The usual panel size is 4 ft × 8 ft (1.2 m × 2.4 m). Usual thicknesses are ¼ in. to 1½ in. (6 to 38 mm).

Waferboard is similar to particleboard except that it is manufactured from larger wood chips.

Oriented strand board is built up in layers like plywood. However, each layer consists of wood strands bonded together by a resin.

12-3 FRAME CONSTRUCTION

Frame construction utilizes studs [typically spaced 16 or 24 in. (0.4 or 0.6 m) on center], joists, and rafters to form the building frame. Framing members are usually of 2 in. (5 cm) nominal thickness. This frame is then covered with siding and roof sheathing of plywood or lumber. Frame construction is widely used in the United States for single-family residences, as well as for small multiple-family residences, offices, and shops. Building codes frequently specify procedures or minimum dimensions to be used in frame construction. The procedures described in this section are those widely recommended in the absence of specific code requirements. The two principal forms of frame construction, platform frame construction and balloon frame construction, are described in the following paragraphs.

Platform Frame Construction

Platform frame construction is illustrated in Figure 12-3. In this type of construction, the subfloor of each story extends to the outside of the building and provides a platform for the construction of the building walls. This method of framing is widely used because it provides a good working platform at each level during construction and also permits preassembled wall sections to be quickly set in place once the subfloor is completed.

The principal framing members are identified in Figure 12-3. In this example, the first floor joists are supported by sills (placed on top of foundation walls) and ledger strips attached to the girder. Wall panels are composed of sole plates (or soles), studs, and top plates. Double top plates are used for bearing walls (walls that support a load from an upper level).

Balloon Frame Construction

In *balloon frame construction,* exterior wall studs extend all the way from the sill to the top of the second floor wall, as shown in Figure 12-4. The outside ends of second-floor joists are supported by ribbon strips notched (or let-in) into the studs. Balloon framing is especially well suited for use in two-story buildings that have exterior walls covered with masonry veneer, since this method of framing reduces the possibility of movement between the building frame and the exterior veneer.

Foundation and Floor Construction

Platform frame construction supported by foundation walls is illustrated in Figure 12-5. In this illustration the floor joists are lapped and rest on top of the girder rather than resting on the ledger strip used in Figure 12-3. Notice also the use of a header joist (or band) to close off the exterior end of joists. Lateral bracing (bridging or cross bracing) between joists may be either solid bridging or diagonal bridging as shown. Board or plywood subflooring may be used as shown in Figure 12-5. Notice that board subflooring should be placed at an angle of 45° to the joists to provide additional stiffness to the floor structure and to permit the finish

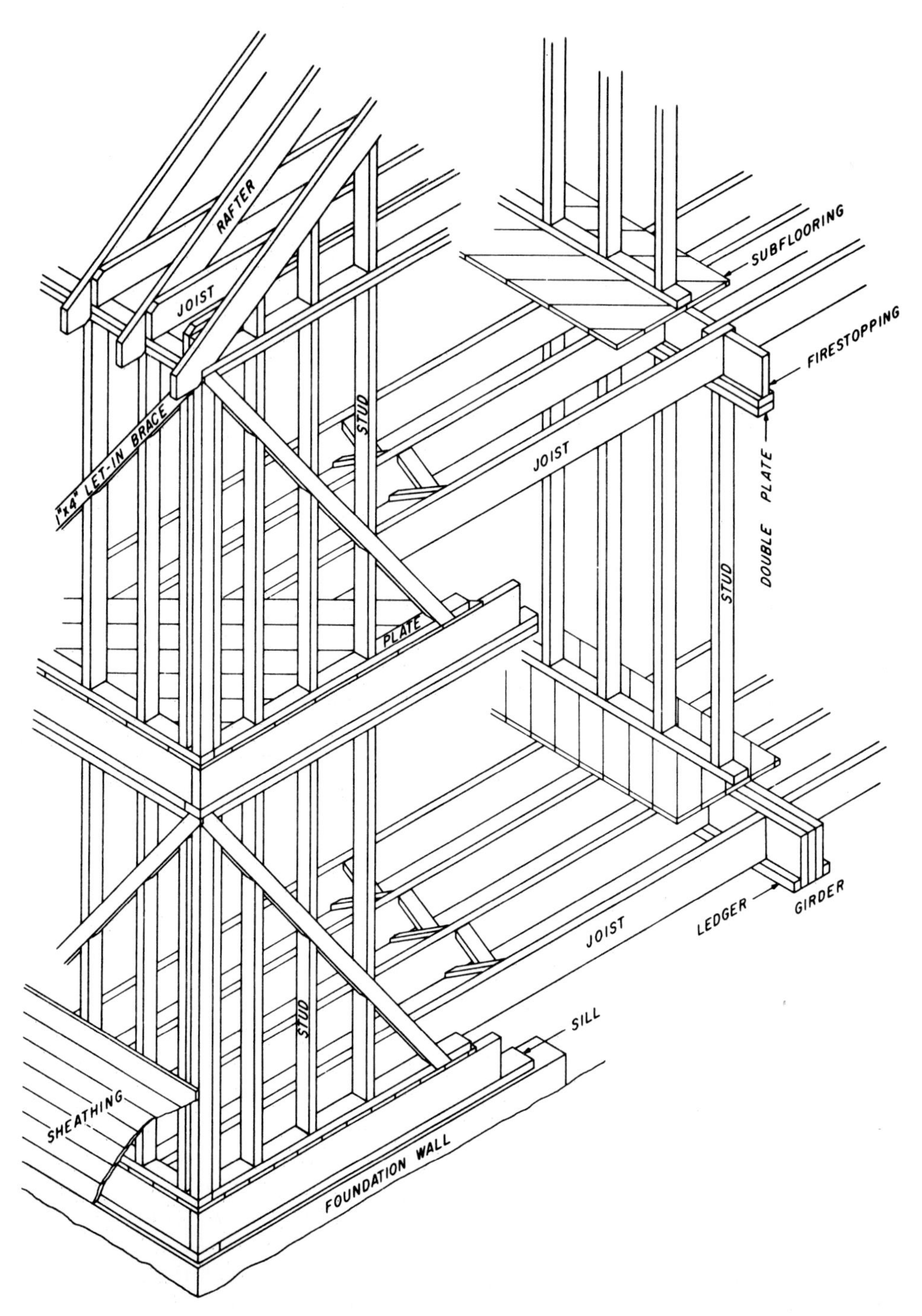

Figure 12-3 Platform frame construction. (Courtesy of National Forest Products Association)

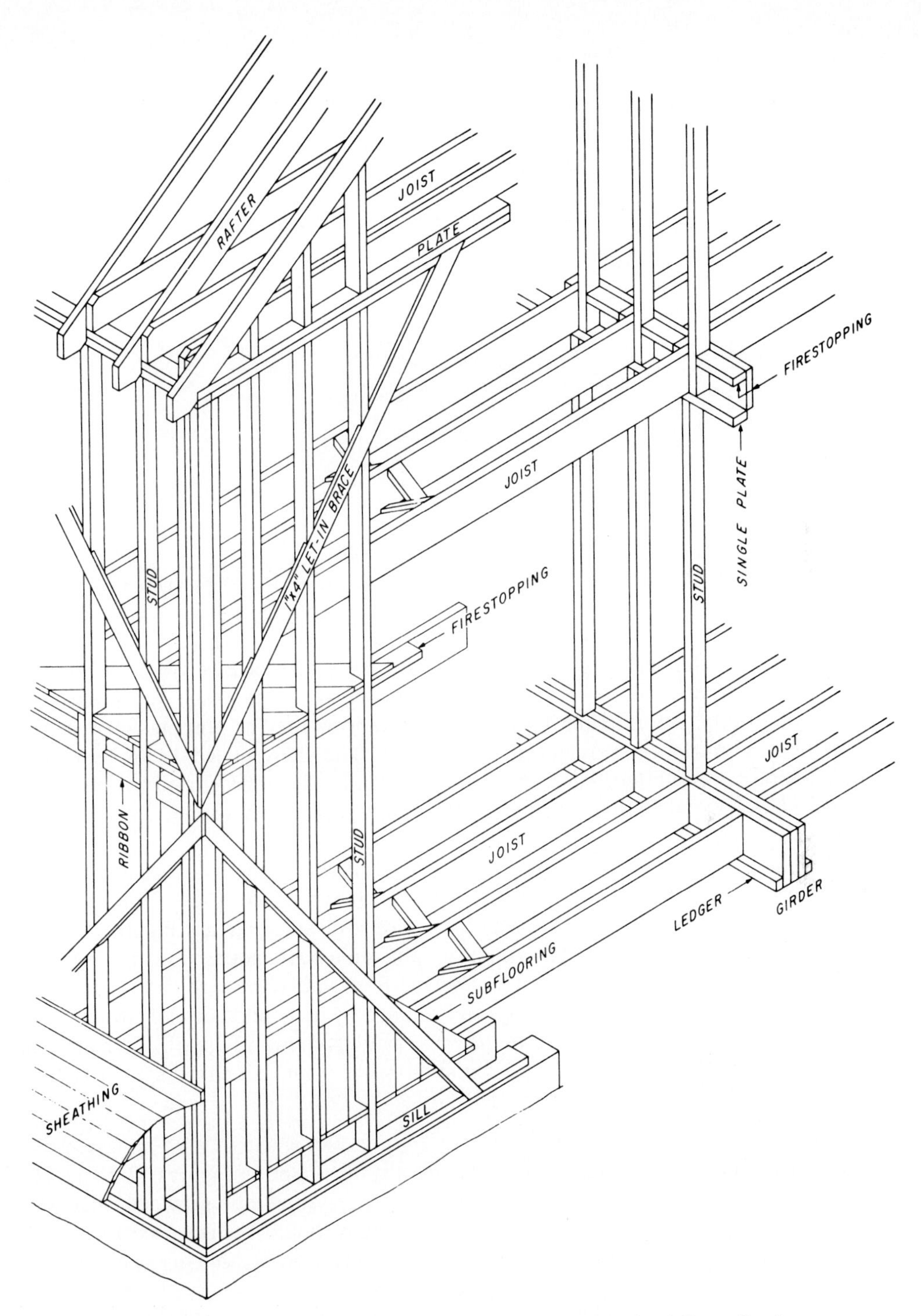

Figure 12-4 Balloon frame construction. (Courtesy of National Forest Products Association)

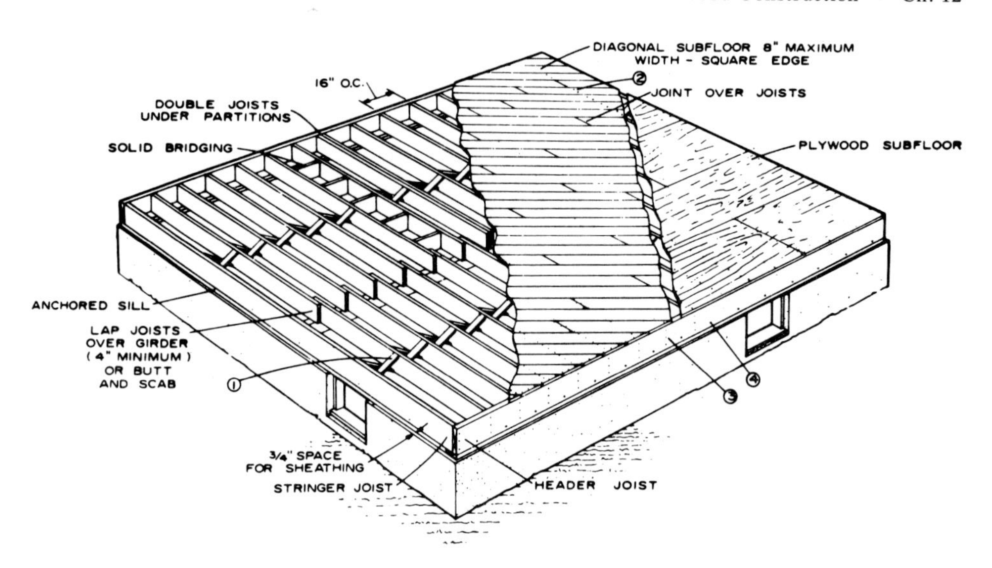

Figure 12-5 Floor framing for platform frame construction. (U.S. Department of Agriculture)

flooring to be laid either parallel or perpendicular to the joists. Plywood subflooring should be placed with the face grain perpendicular to the joists. When carpeting or other nonstructural flooring is used, subflooring may be eliminated by using a combined subfloor-underlayment of plywood. The APA Glued Floor System, in which plywood subflooring is glued to the joists, has been developed by the American Plywood Association to reduce subfloor cost and to increase the stiffness of the floor system.

Figure 12-6 illustrates slab on grade construction using a separate foundation wall. Notice the use of rigid insulation on the interior face of the foundation wall and under the edge of the floor slab to provide a thermal barrier. Figure 12-7 illustrates slab-on-grade construction using a foundation beam poured integrally with the floor slab. Such construction is also referred to as *thickened-edge slab construction.* Finish flooring of wood, carpeting, and so on, may be applied directly to the top of the slab or it may be supported on *sleepers,* as illustrated in Figure 12-6. Notice the use of a vapor barrier to prevent ground moisture from rising through the slab.

Framing Details

Two methods of supporting joists by beams are illustrated in Figures 12-3 and 12-5. A minimum bearing length along the joist of 1½ in. should be provided when joists rest on wood or metal beams. The bearing length should be increased to 3 in. for bearing on masonry. Framing anchors or joist hangers may be used in place of a ledger strip to support joists, as shown in Figure 12-8.

The use of both solid and diagonal bridging to provide lateral bracing of joists is illustrated in Figure 12-5. Prefabricated metal diagonal bridging with integral fasten-

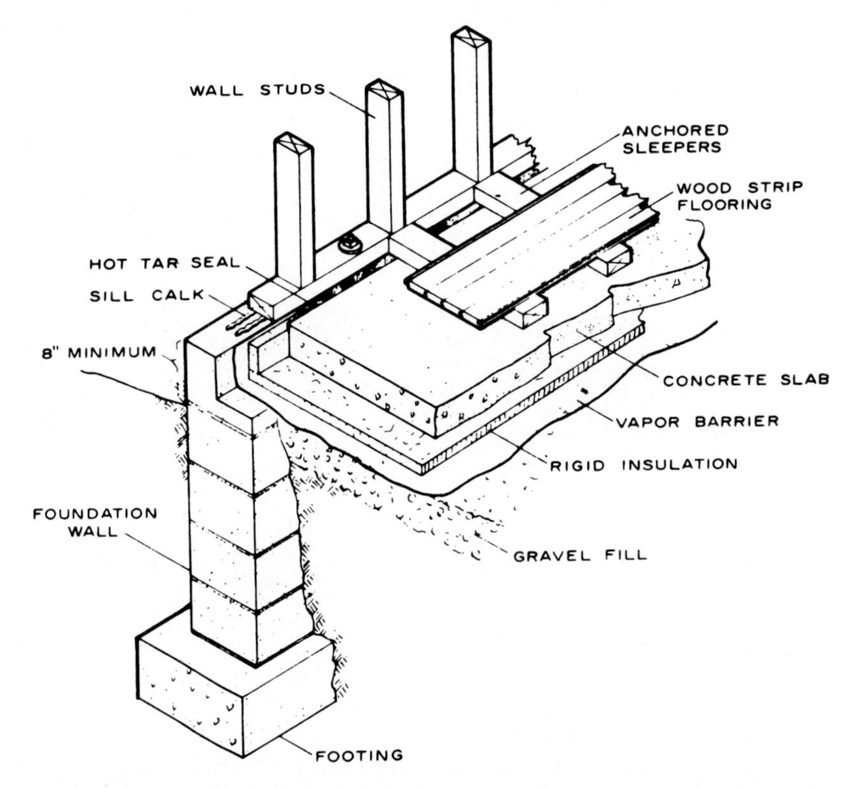

Figure 12-6 Slab on grade with foundation wall. (U.S. Department of Agriculture)

ing devices is also available for standard joist spacings. Bridging between joists is not required when the Glued Floor System is used.

Typical platform frame construction of an exterior wall including a window opening is illustrated in Figure 12-9. The load on the top of the window opening is carried by a header, which is in turn supported by double studs at the sides of the opening. Note the use of let-in braces (braces notched into the studs) to reinforce the wall at building corners. Plywood panels may be used as sheathing at building corners to replace corner braces. Corner braces are not required when full plywood sheathing is used.

Roof Construction

One method of roof construction, called *joist and rafter framing,* is illustrated in Figure 12-10. Rafters are notched where they rest on wall plates and are held in place by nailing them to the wall plates or by the use of metal framing anchors. The collar beam shown is used to assist in resisting wind loads on the roof.

A roof truss is shown in Figure 12-11. The use of the roof trusses permits interior walls to be nonbearing, because all roof loads are supported by the exterior

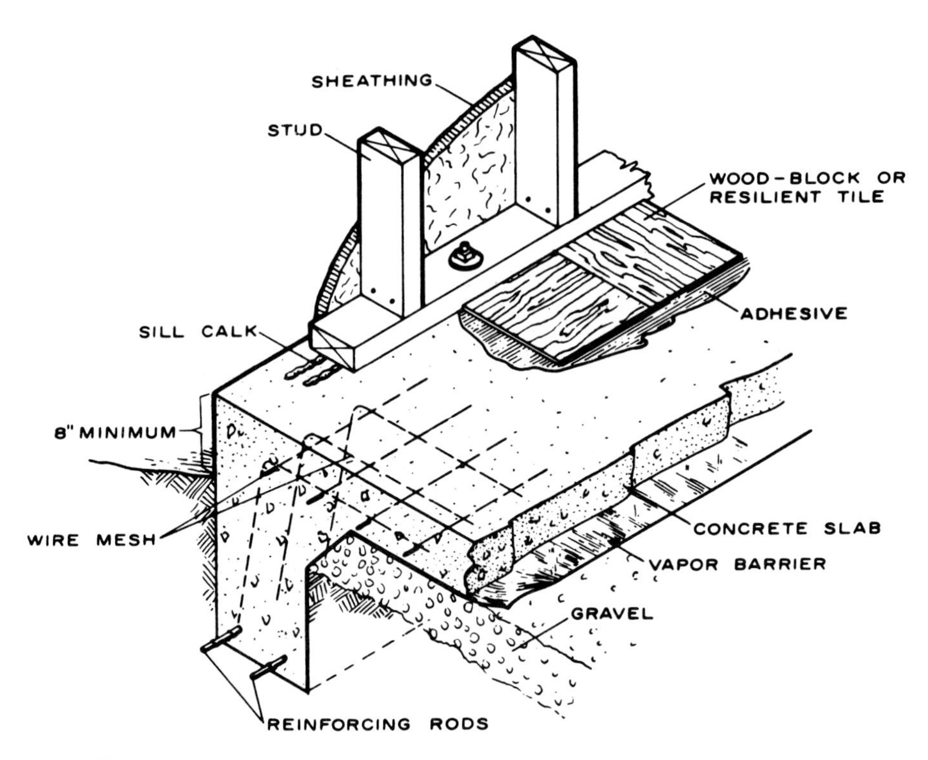

Figure 12-7 Combined slab and foundation. (U.S. Department of Agriculture)

walls. Trusses are usually prefabricated so that they may be rapidly erected as soon as exterior walls are in place.

Roof sheathing normally consists of plywood or nominal-1-in. (2.5-cm) boards applied perpendicular to the rafters or trusses. Roofing is applied over the roof sheathing to provide a waterproof enclosure.

Siding

Exterior frame walls are most often covered with wood or plywood siding or a masonry veneer applied over sheathing. Sheathing may consist of nominal-1-in. (2.5-cm) boards placed diagonally, plywood, or nonstructural sheathing. Sheathing may be omitted when plywood siding is applied in accordance with the recommendations of the American Plywood Association. Such construction is referred to as *single-wall construction.* To make a watertight enclosure, joints in the siding of single-wall construction must be caulked, lapped, battened, or backed with building paper.

Brick veneer siding over plywood sheathing is shown in Figure 12-12. Note that building paper is not required over plywood sheathing when an air space and weep holes are provided. However, building paper should be used over wood sheathing. A stucco exterior wall finish over plywood sheathing is illustrated in Figure 12-13.

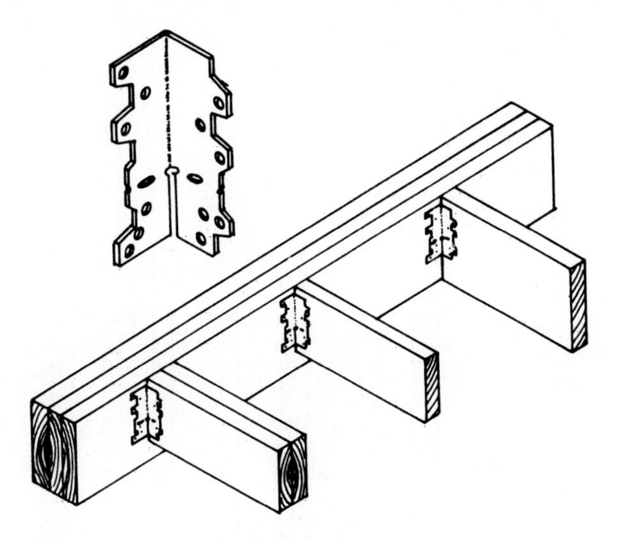

All purpose framing anchors.

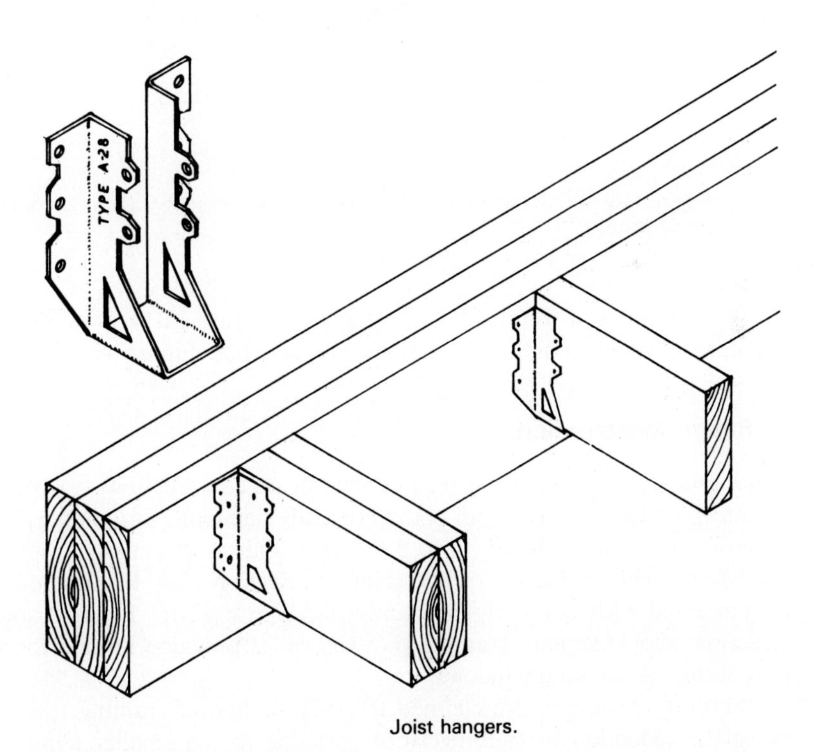

Joist hangers.

Figure 12-8 Joists supported by joist hangers and framing anchors. (Courtesy of TECO, Washington, DC 20015)

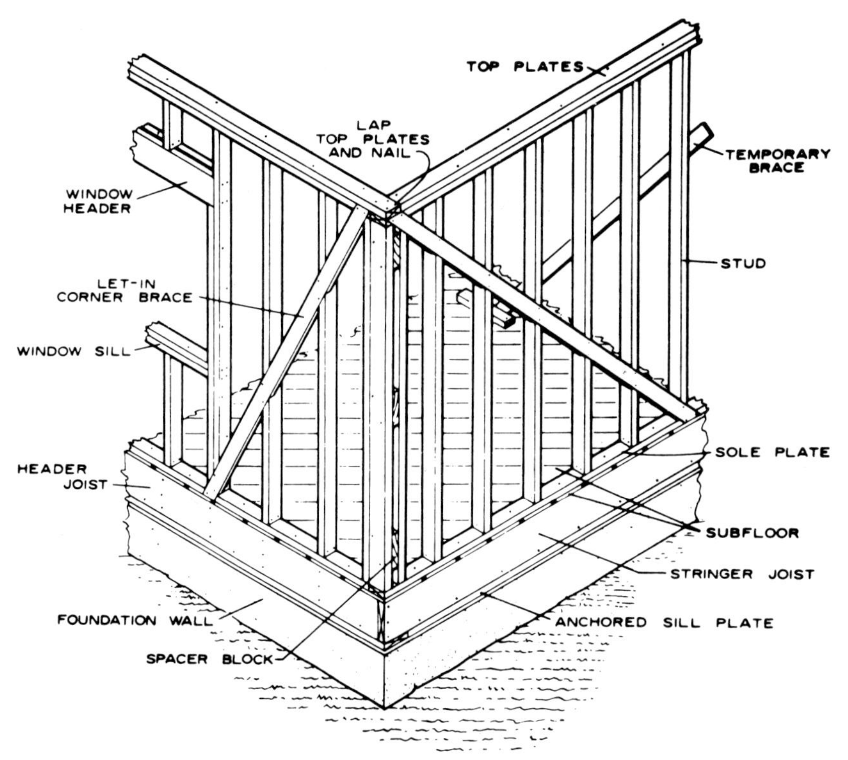

Figure 12-9 Exterior wall framing, platform construction. (U.S. Department of Agriculture)

Typical types of wood siding are shown in Figure 12-14. Note the nailing methods used. Plywood siding over sheathing is shown in Figure 12-15.

Plank-and-Beam Construction

Plank-and-beam construction (or post-and-beam construction) is a method of framing in which flooring and roof planks (usually nominal-2-in. lumber) are supported by posts and beams spaced up to 8 ft apart. This is essentially a lighter version of the heavy timber construction described in Section 12-4. Plank-and-beam framing is contrasted with conventional framing in Figure 12-16. In plank-and-beam construction, supplementary framing (not shown) is provided in exterior walls to support siding, doors, and windows.

Several advantages are claimed for this method of framing, the principal one being the reduction in framing labor cost due to the smaller number of framing members required. The system also produces a distinctive architectural effect that many people find attractive. Some construction details for plank-and-beam framing of a one-story residence are shown in Figure 12-17.

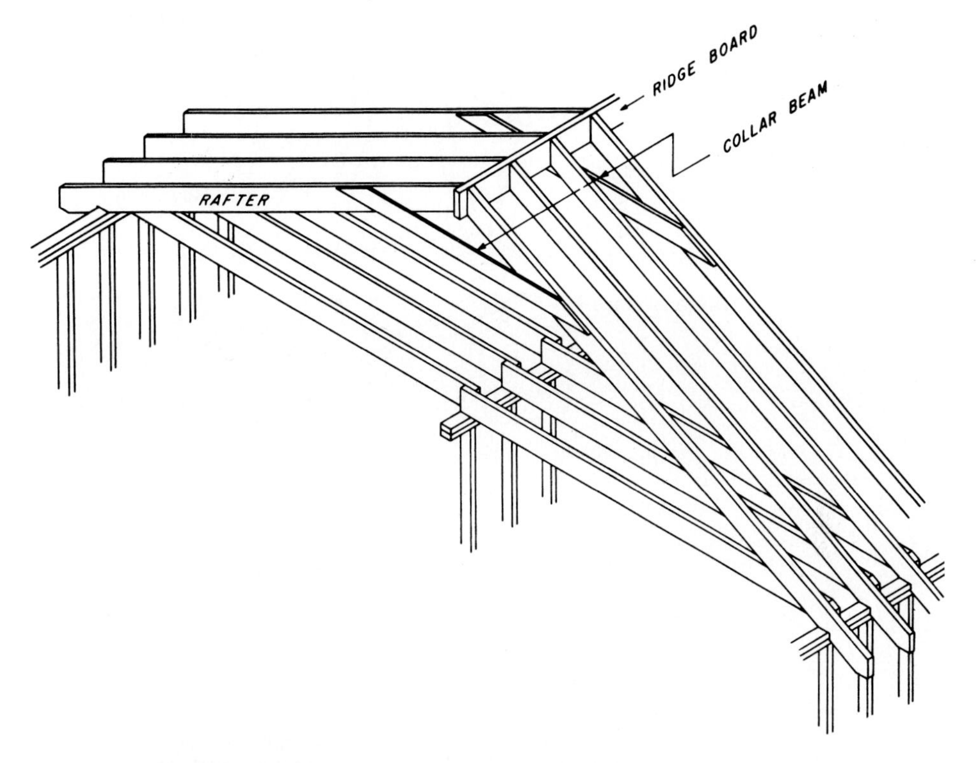

Figure 12-10 Roof framed with rafters and ceiling joists. (Courtesy of National Forest Products Association)

12-4
TIMBER CONSTRUCTION

Buildings

The term *heavy timber construction* originally identified a multistory structure whose structural members (except for exterior walls) were primarily composed of timber. Such structures were widely used for industrial and storage purposes. Today heavy timber construction indicates the type of wood building construction that carries the highest fire-resistance classification. Such high fire resistance is obtained by specifying construction details, the minimum sizes of wood structural members, the composition and minimum thickness of floors and roofs, the types of fasteners and adhesives used, and the fire resistance of walls, as well as by prohibiting concealed spaces under floors and roofs.

Both glued laminated and sawn timber are used in modern heavy timber construction. Modern structures are often only one story in height. Such construction is widely used for schools, churches, auditoriums, sports arenas, and stores, as well as for industrial and storage buildings.

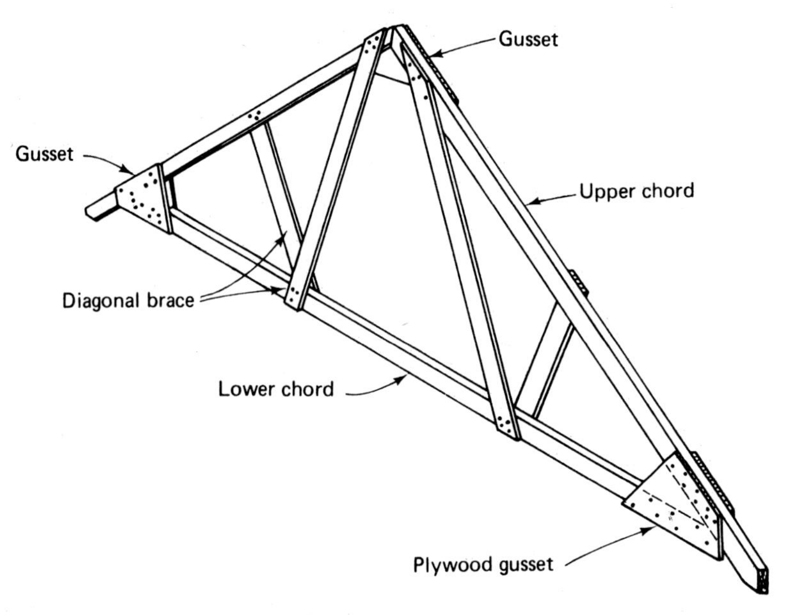

Figure 12-11 Roof truss.

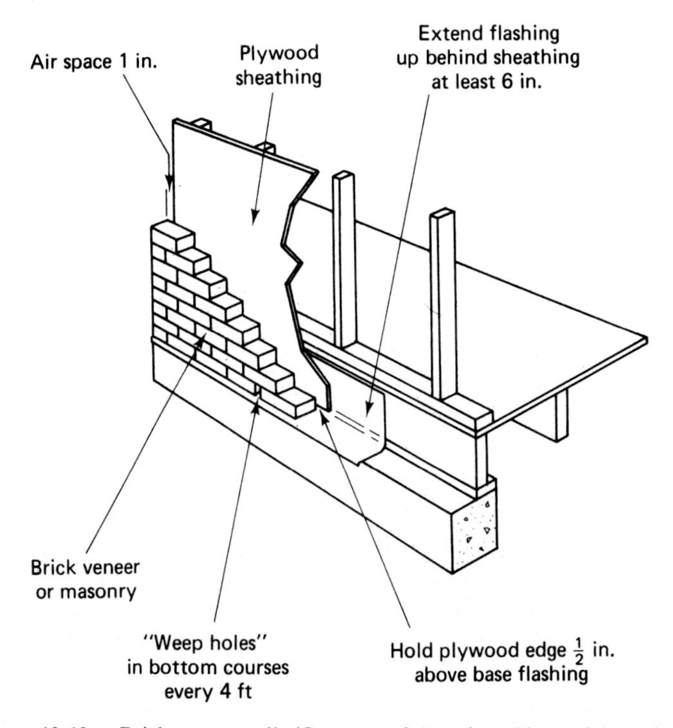

Figure 12-12 Brick veneer wall. (Courtesy of American Plywood Association)

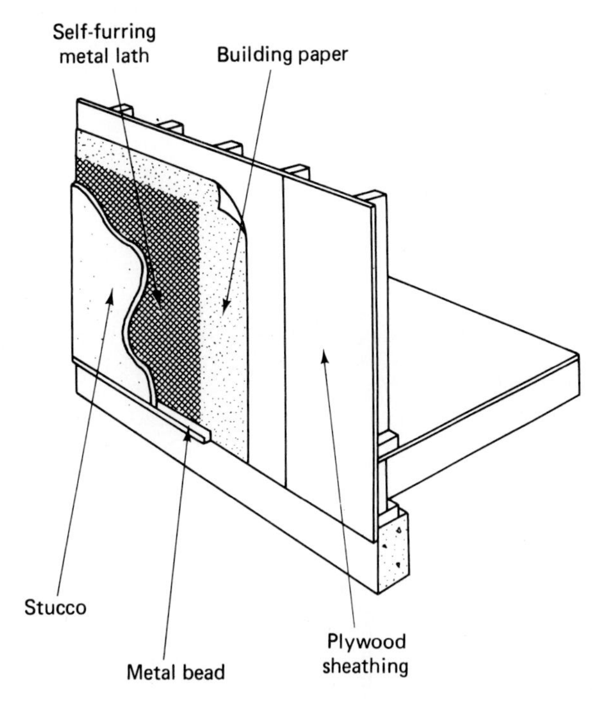

Figure 12-13 Stucco over sheathing. (Courtesy of American Plywood Association)

A typical multistory building of traditional heavy timber construction is illustrated in Figure 12-18. Some construction details recommended by the National Forest Products Association for roof beam and column connections are shown in Figure 12-19. Such details are typical of the practices that are specified to attain the high fire resistance of heavy timber construction. Some typical varieties of modern heavy timber buildings are illustrated in Figures 12-20 to 12-22. Figure 12-20 shows a rigid arch structure using glued laminated timber arches that are supported at ground level. A barrel arch roof using curved glued laminated timber arches supported by exterior piers is depicted in Figure 12-21. A bowstring roof truss supported by wood columns is used in the building of Figure 12-22. The knee brace shown in Figure 12-22 may be eliminated when plywood roof sheathing is used and the building's perimeter frame is designed to carry lateral loads.

Bridges

Timber bridges have been used throughout recorded history to span streams and valleys. Major types of timber bridge structures include trestle bridges, truss bridges, and arch bridges. *Trestle bridges* consist of stringers whose ends are supported by timber or pile bents, as illustrated in Figure 12-23. Loads are transferred to the stringer by decking laid across the stringers. Note the use of sway bracing (or cross bracing) on the bents of Figure 12-24. Tower bents consisting of several parallel bent frames and connected by bent caps and longitudinal bracing may also be used. Stringers may be fabricated of sawn or glued laminated timber or other materials.

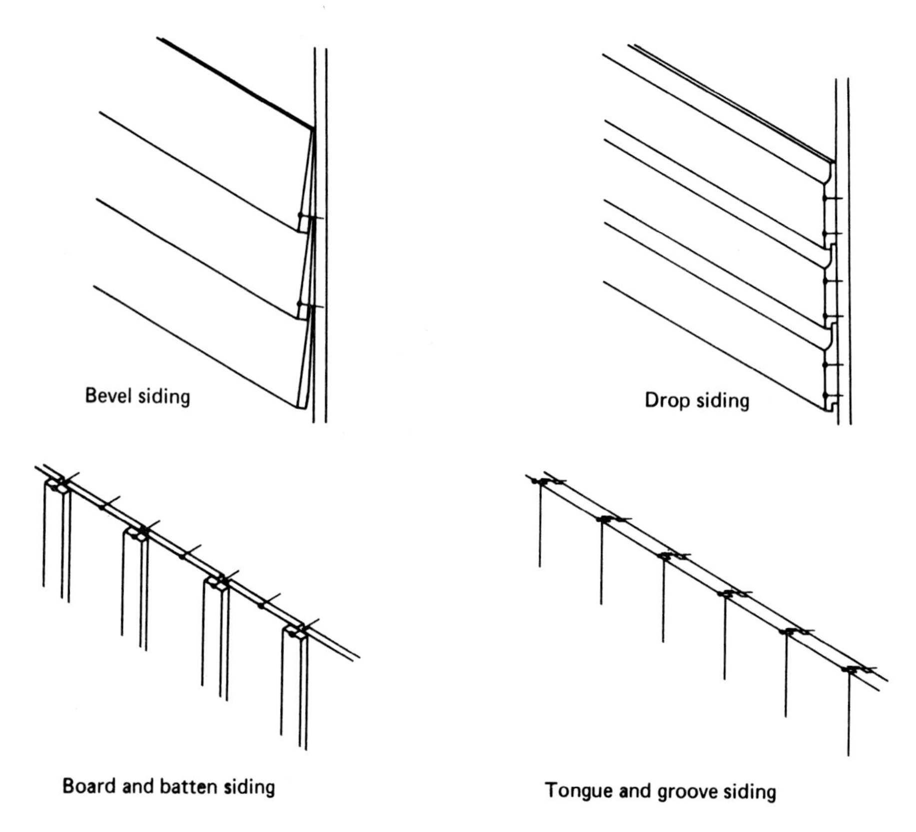

Figure 12-14 Common types of wood siding. (Courtesy of National Forest Products Association)

Girder, truss, and arch bridges are capable of spanning greater distances than can trestle bridges. Wood girders are usually fabricated of glued laminated timber. The trusses used in timber *truss bridges* are similar in design to those used for roof trusses. Truss designs frequently used include parallel chord trusses, triangular trusses, and bowstring trusses (Figure 12-25).

Timber *arch bridges* utilize arches built up from wood members. Arches are usually fabricated of glued laminated timber. A highway overpass whose glued laminated timber arches span 155 ft (47 m) is shown in Figure 12-26. Note the smaller bridge, that utilizes a curved continuous-span glued laminated timber girder 170 ft (52 m) long.

Other Structures

Timber construction is often used for many other types of structures, such as tanks, water towers, observation towers, and power transmissions towers. Timber crossarms are sometimes used on metal power transmission towers because of wood's good diaelectric properties.

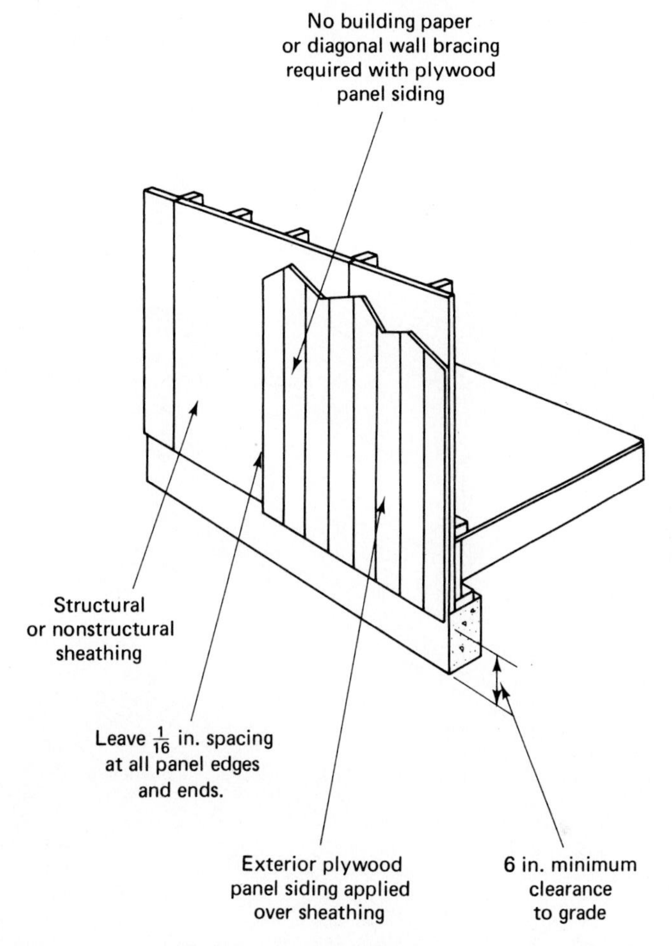

Figure 12-15 Plywood siding over sheathing. (Courtesy of American Plywood Association)

12-5 FASTENINGS, CONNECTIONS, AND NOTCHING

Fastenings

As in any mechanical system, a wood structure cannot develop the full strength of its members unless connections between members are at least as strong as the members. There are a number of types of fasteners used to join wood members, the most common being nails and wood screws. Sizes of common wire nails are presented in Table 12-1. Other major types of fasteners include bolts, lag-screws, spikes, and drift-bolts (or drift-pins). Major factors controlling the allowable strength of mechanical fasteners include the lumber species, the angle of the load with respect to

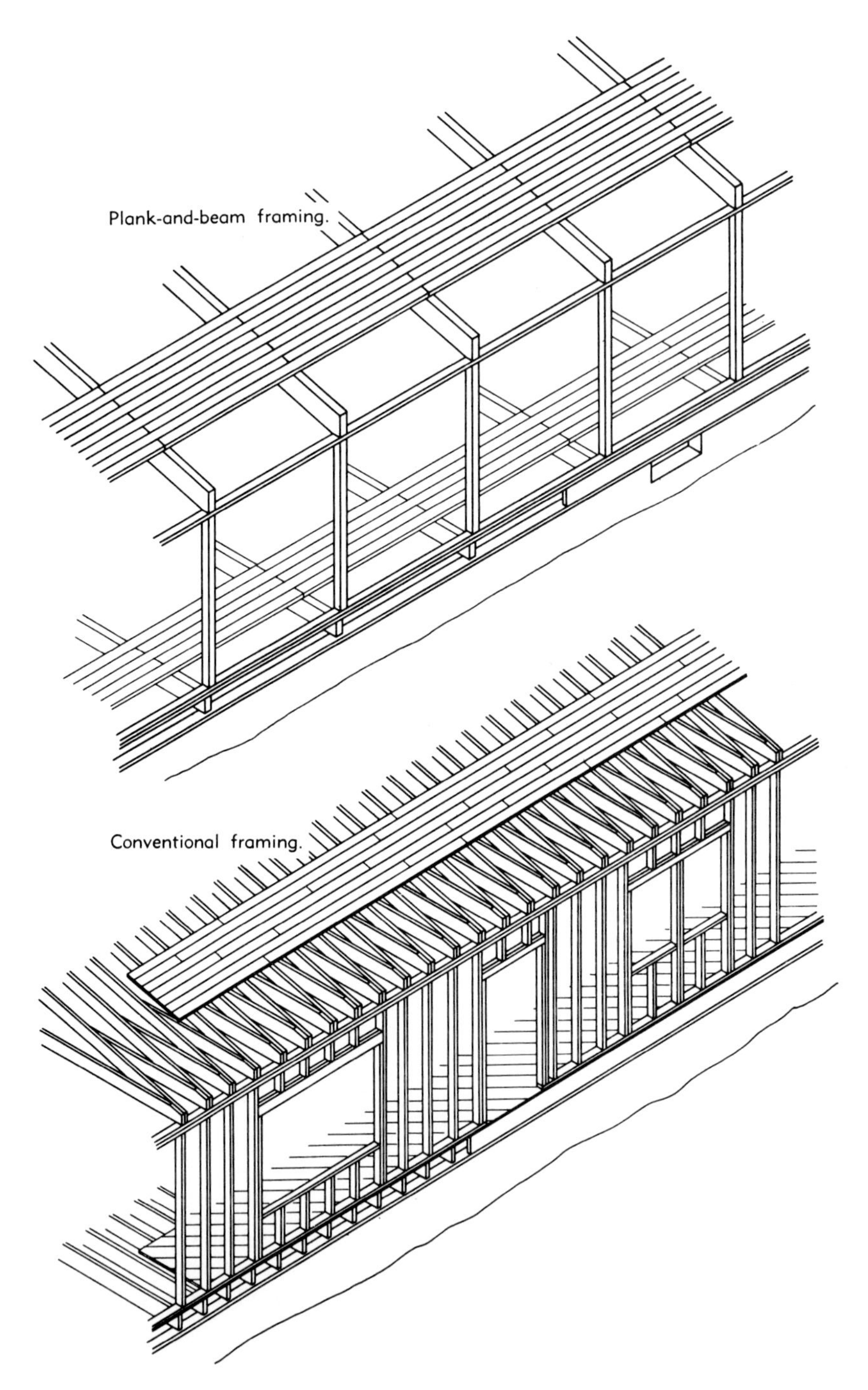

Figure 12-16 Comparison of plank-and-beam and conventional framing. (Courtesy of National Forest Products Association)

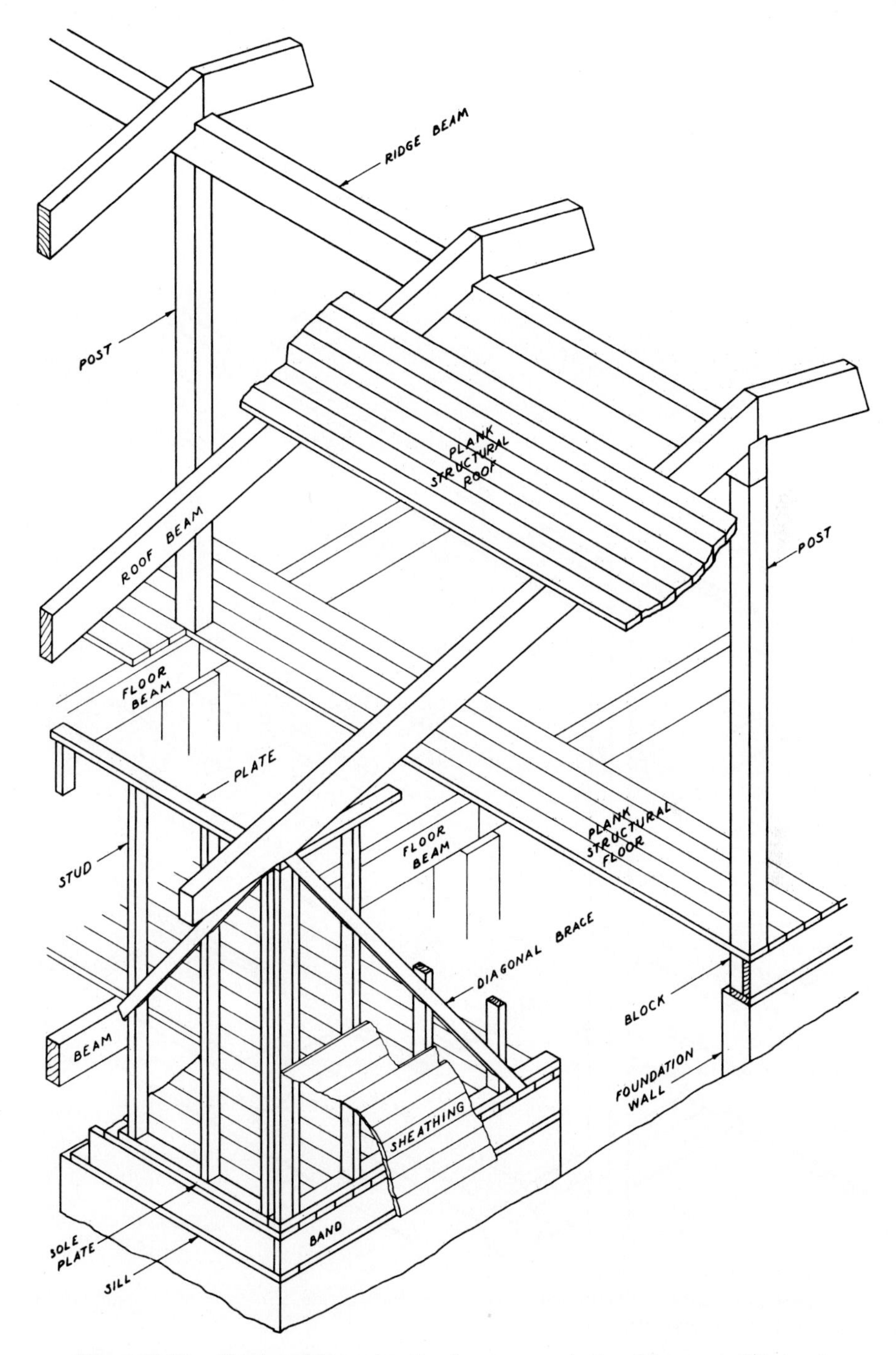

Figure 12-17 Plank-and-beam framing for one-story house. (Courtesy of National Forest Products Association)

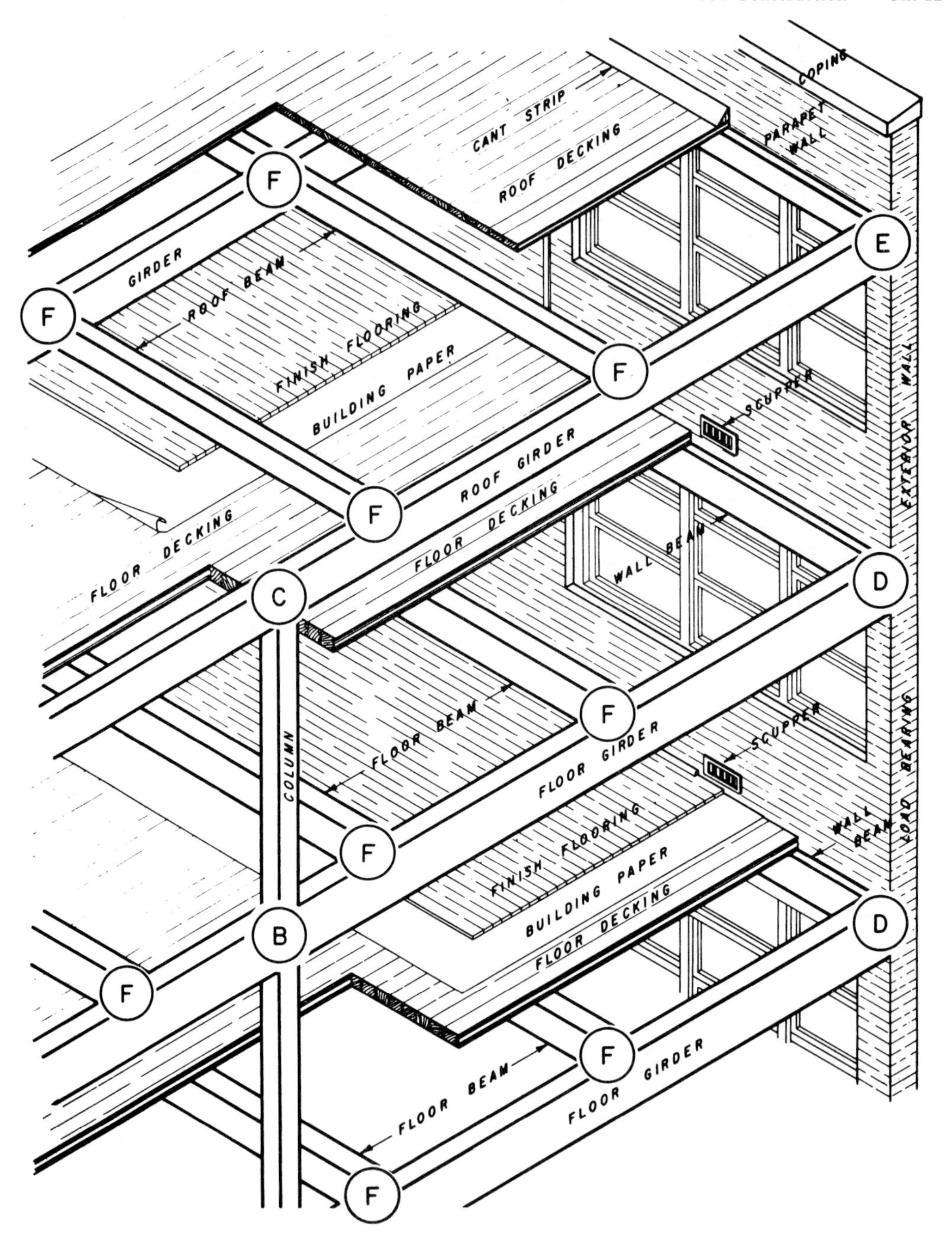

Figure 12-18 Traditional heavy timber construction. (Courtesy of National Forest Products Association)

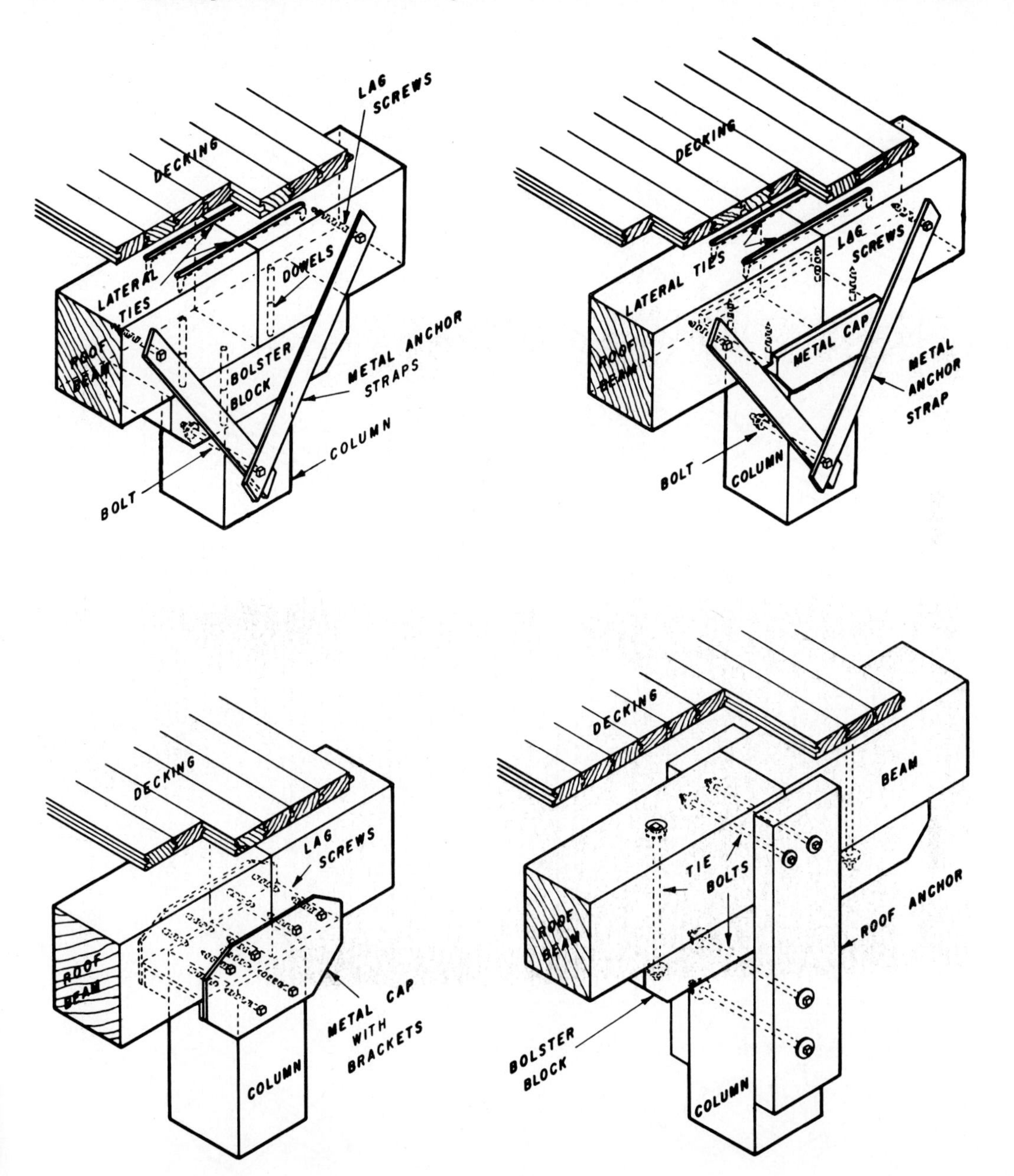

Figure 12-19 Typical roof beam and column connection details. (Courtesy of National Forest Products Association)

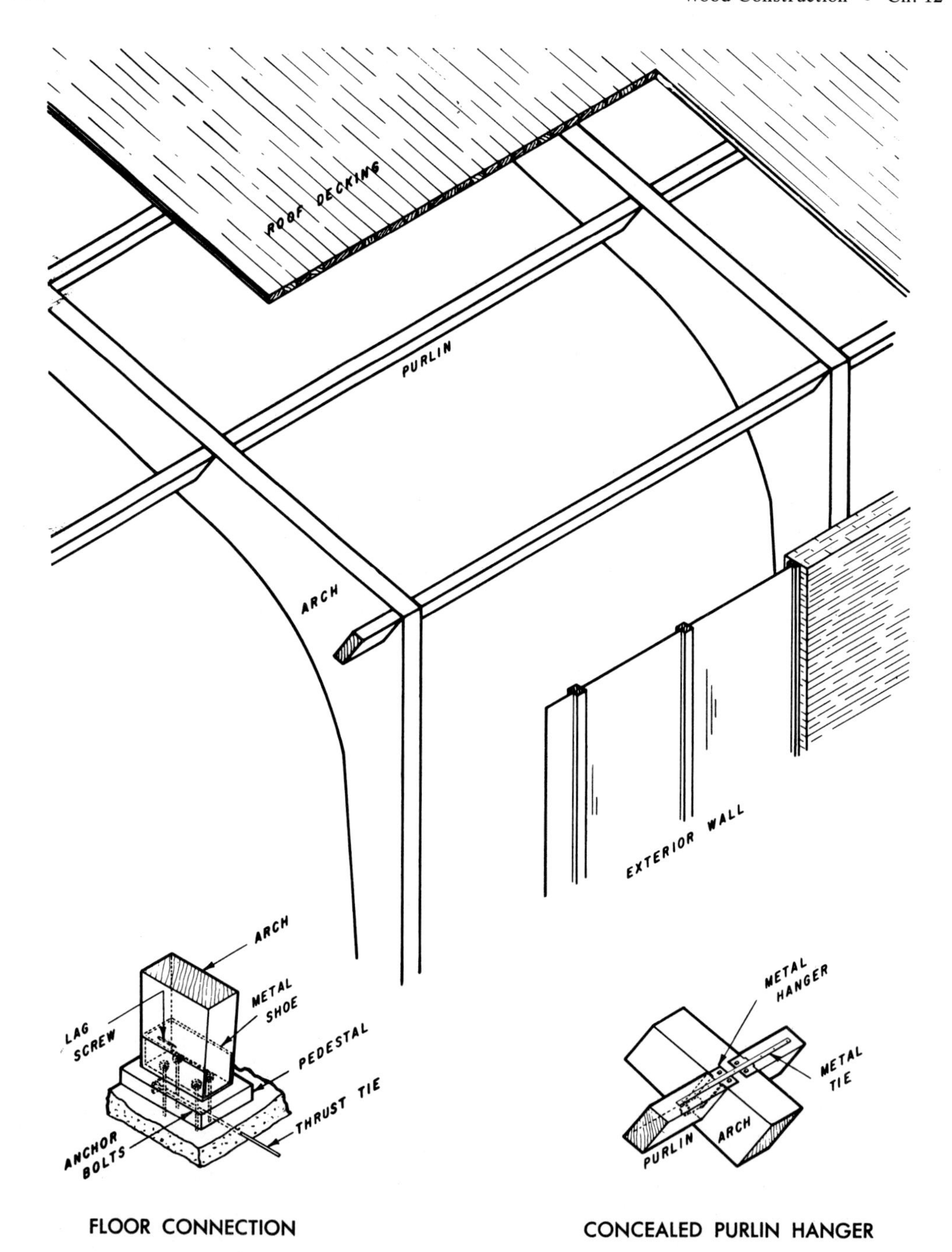

Figure 12-20 Rapid arch frame supported at floor. (Courtesy of National Forest Products Association)

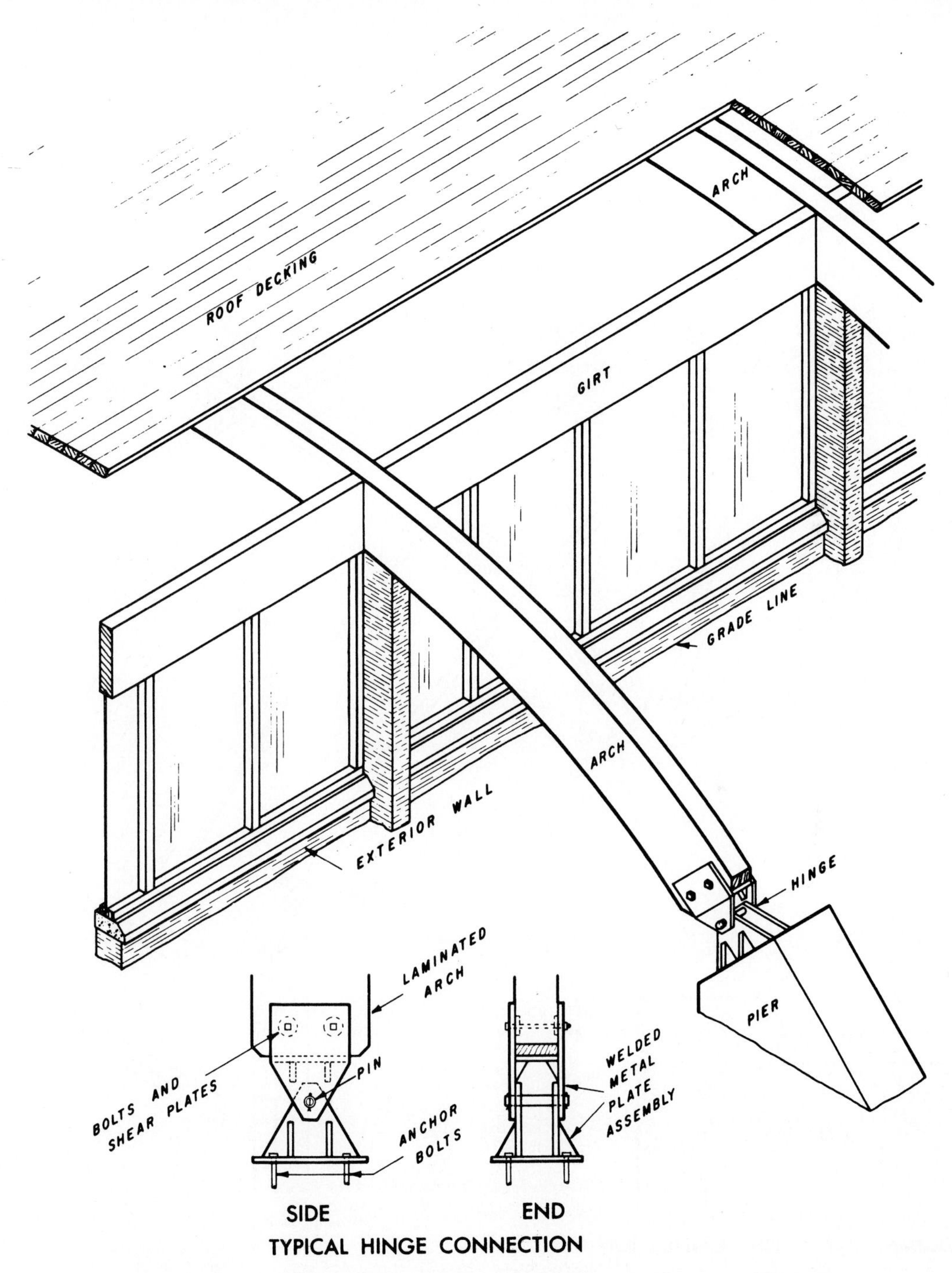

Figure 12-21 Barrel arch frame supported by exterior pier. (Courtesy of National Forest Products Association)

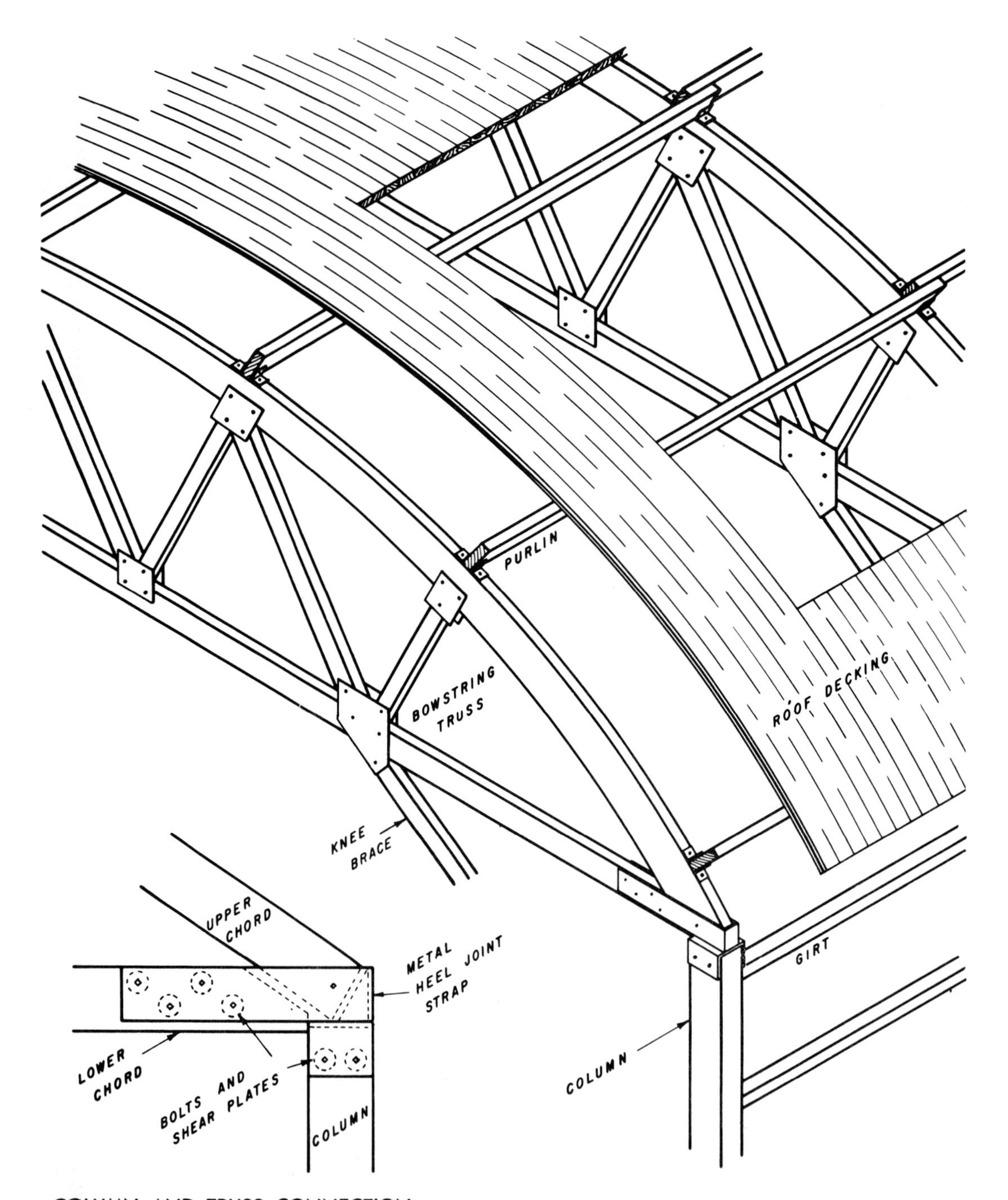

Figure 12-22 Bowstring roof truss supported by wood column. (Courtesy of National Forest Products Association)

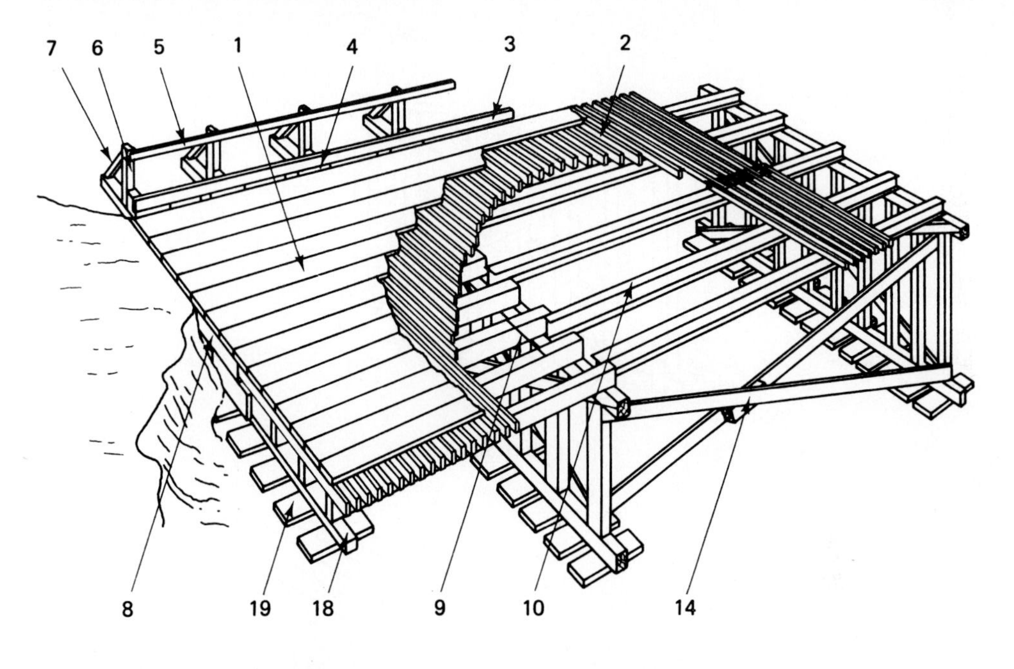

1 – Tread
2 – Open-laminated deck
3 – Curb
4 – Curb riser block
5 – Handrail
6 – Handrail post
7 – Handrail brace
8 – End dam
9 – Timber stringers
10 – Steel stringers
14 – Longitudinal bracing
18 – Abutment sill
19 – Abutment footing

Figure 12-23 Timber trestle bridge with frame bent. (U.S. Department of the Army)

Table 12-1 Common wire nail sizes

Size: Penny (d)	Wire Gauge	Length in.	Length cm
4	$12\frac{1}{2}$	1.50	3.8
6	$11\frac{1}{2}$	2.00	5.1
8	$10\frac{1}{4}$	2.50	6.4
10	9	3.00	7.6
12	9	3.25	8.3
16	8	3.50	8.9
20	6	4.00	10.2
30	5	4.50	11.4
40	4	5.00	12.7
50	3	5.50	14.0
60	2	6.00	15.2

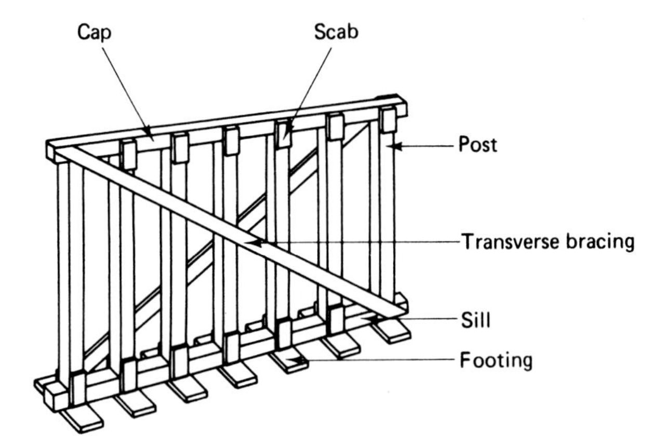

Timber trestle bent

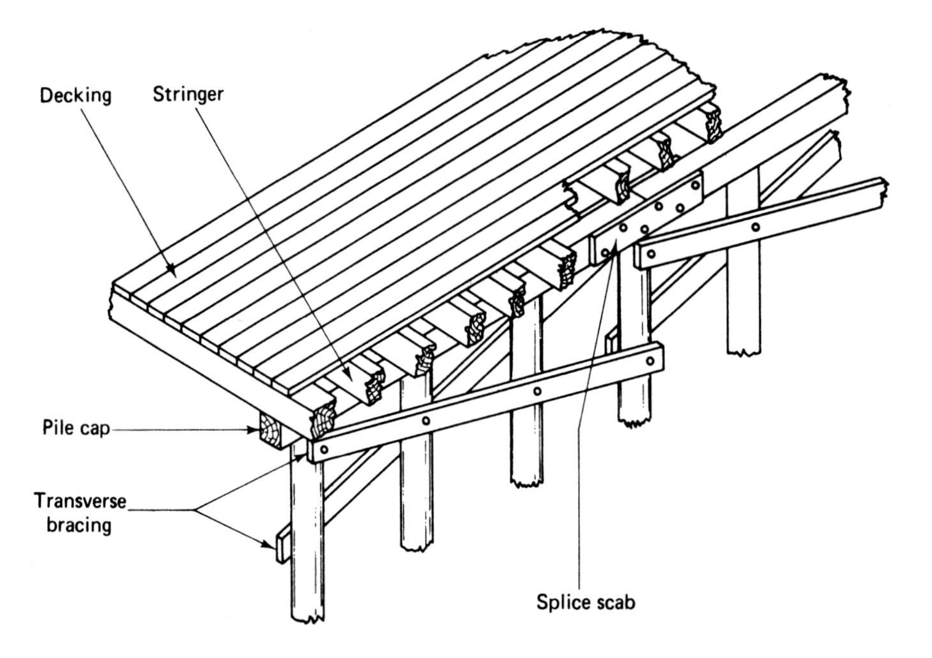

Pile bent

Figure 12-24 Typical timber trestle and pile bridge bent.

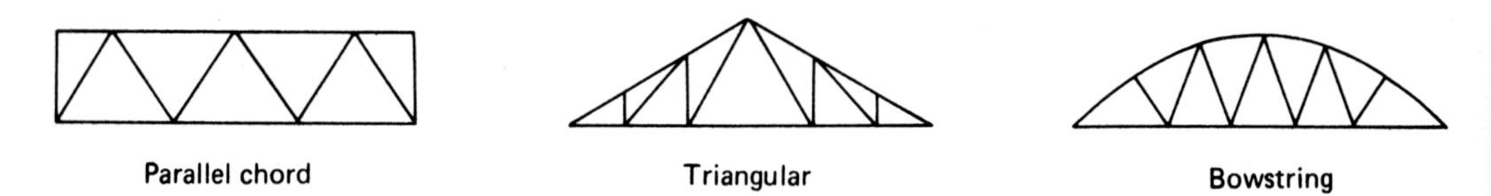

Figure 12-25 Truss types.

Figure 12-26 Highway bridges supported by glued laminated timber beams. (Courtesy of American Institute of Timber Construction)

the wood grain, the size of the member perpendicular to the load, the distance of the fastener from the edge of the wood, and the spacing of fasteners. Methods for determining the allowable load on common wood fasteners are given in references 1 and 6.

Connectors

To provide the most efficient use of materials and labor while providing the required strength, a number of special timber connectors have been developed. Major types of timber connectors include split-ring connectors, toothed-ring connectors, and shear plates. These are illustrated in Figure 12-27. These connectors use a bolt or lag screw to join the wood members and place the connector under compression. Split-ring connectors and shear plates fit into grooves precut into the wood members. Toothed-ring connectors are forced into the wood under the pressure of the bolt joining the members.

Light metal framing devices are available in a wide range of types and sizes, some of which are illustrated in Figure 12-28. Light metal connector plates may incorporate integral teeth or may use nails for load transfer. All-purpose framing anchors may be used for a variety of connections, such as rafters to wall plates and studs to top and sole plates.

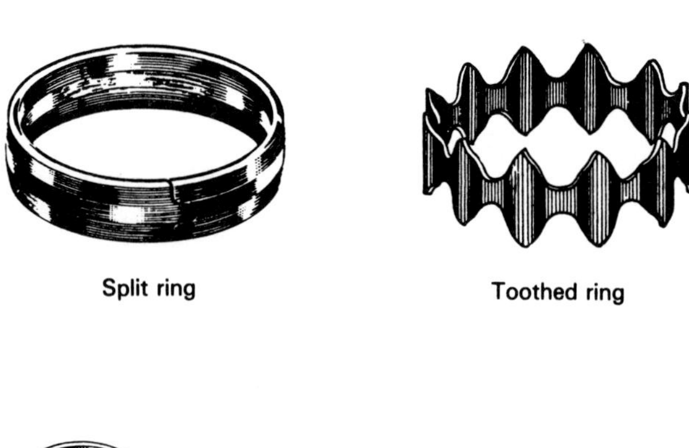

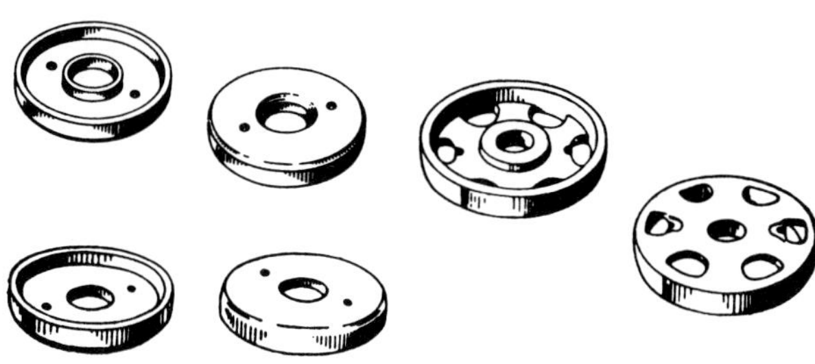

Figure 12-27 Typical timber connectors. (Courtesy of National Forest Products Association)

Notching and Boring of Beams

Notching the top or bottom of a beam will seriously reduce its bending strength. In short, heavily loaded beams, horizontal shear stress may be critical. The safe vertical reaction on an end-notched beam (such as a joist) which is notched on the tension side (as shown in Figure 12-29) may be calculated as follows:

$$R_v = \frac{2F_v b d_e^2}{3d} \tag{12-1}$$

where R_v = safe vertical end reaction (lb)
F_v = allowable shear stress (psi)
b = width of beam (in.)
d = depth of beam (in.)
d_e = depth of beam above notch (in.)

When the notch is curved, or is beveled over a distance greater than d_e, Equation 12-2 may be used in lieu of Equation 12-1 to calculate the maximum allowable end reaction for the beam.

$$R_v = \frac{2F_v b d_e}{3} \tag{12-2}$$

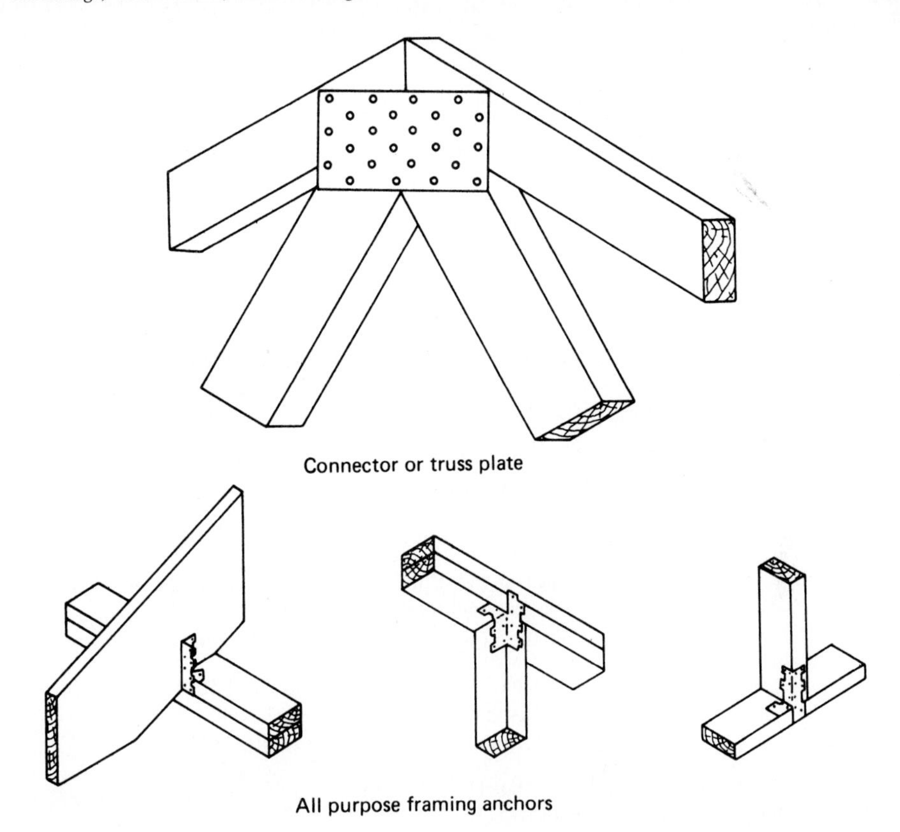

Figure 12-28 Typical light-metal framing devices. (Courtesy of TECO, Washington, DC 20015)

When it is necessary to notch joists to provide passage for piping or electrical cables, the following limits should not be exceeded without a design analysis of the joist. Notches in the top or bottom of joists should not exceed one-sixth of the joist depth and should be located less than one-third of the joist length from either end of the joist. The diameter of holes bored in joists should not exceed one-third of the depth of the joist and should not extend closer than 2 in. to the top or bottom edge of the joist.

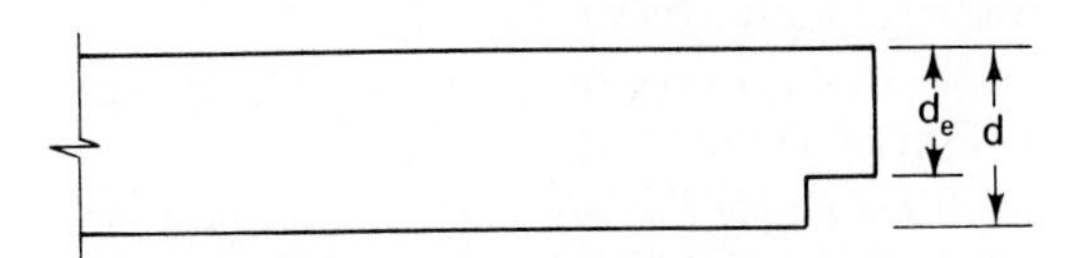

Figure 12-29 End notching of beam.

PROBLEMS

1. Explain the importance of moisture content relative to the strength of structural lumber.
2. Briefly explain the meaning of the term *glulam* and identify the characteristics of this material.
3. Briefly describe the following wood products and their use.
   (a) Plywood
   (b) Particleboard
   (c) Laminated veneer lumber
4. Give the actual dimensions of the following lumber (S4S).
   (a) 2×4 in.
   (b) 1×12 in.
   (c) 4×8 in.
5. What is platform frame construction?
6. In what direction should the face grain of a plywood subfloor be placed?
7. The vertical reaction at the end of a 2×6 in. floor joist is 550 lb. If the allowable shear stress in the joist is 150 psi, what is the maximum depth that the joist can be safely end-notched? Assume that the notch is a square notch on the tension side of the joist.
8. A 2×10 in. floor joist has the end notched on the bottom to a depth of 3.0 in. The notch is beveled over a distance of 6.0 in. If the allowable shear stress is 185 psi, what is the maximum safe vertical reaction at the end of the joist?
9. Explain the limitations that should be observed when notching joists for the passage of pipe or electrical conduit.
10. Write a computer program to calculate the safe vertical end reaction on an end-notched beam (Equations 12-1 and 12-2). Input should include allowable shear stress, beam width, beam depth, depth of notch, and distance over which the notch is beveled. Using your computer program, solve Problem 8.

REFERENCES

1. AMERICAN INSTITUTE OF TIMBER CONSTRUCTION. *Timber Construction Manual,* 3rd ed. New York: Wiley, 1985.
2. ANDERSON, L. O. *Wood-Frame House Construction* (Agriculture Handbook No. 73). U.S. Department of Agriculture, Washington, D.C., 1975.
3. *APA Design/Construction Guide: Residential and Commercial.* American Plywood Association, Tacoma, Wash., 1988.
4. *Heavy Timber Construction Details* (Wood Construction Data No. 5). National Forest Products Association, Washington, D.C., 1971.
5. *Manual for Wood Frame Construction* (Wood Construction Data No. 1). National Forest Products Association, Washington, D.C., 1988.
6. *National Design Specification for Wood Construction.* National Forest Products Association, Washington, D.C., 1986.

7. NUNNALLY, S.W., AND J. A. NUNNALLY. *Residential and Light Building Construction.* Englewood Cliffs, N.J.: Prentice Hall, 1990.
8. *Plank-and-Beam Framing for Residential Buildings* (Wood Construction Data No. 4). National Forest Products Association, Washington, D.C., 1970.
9. *Plywood Design Specification.* American Plywood Association, Tacoma, Wash., 1985.

13
Steel Construction

13-1
INTRODUCTION

Elements of Steel Construction

Structural steel construction is a specialized task that is usually performed by specialty subcontractors. However, construction managers and inspectors must understand the principles and procedures involved. The process of steel construction can be broken down into the three major elements of advanced planning, steel fabrication and delivery to the job site, and field operations. Each of these elements involve a number of operations which are described in this chapter.

For large or complex projects, advanced planning includes divisioning the steel and planning shipping and erection procedures. *Divisioning* is the process of dividing a structure into units (called *divisions*) which are used to schedule the fabrication and delivery of structural steel members to the job site. Since divisioning is determined by the order in which the structure will be erected, it must be performed as a joint effort of the steel fabricator and the erection manager. When planning shop fabrications procedures, the size and weight of large members must be checked against plant capacity, transportation size and weight limits, and the capacity of erection equipment. In planning erection procedures, the type of equipment to be utilized and the procedures to be followed are determined by the type of structure being erected and the anticipated site conditions. Lifting equipment, alignment requirements, and field connections are described in succeeding sections of this chapter.

Field Operations

Field operations include receiving and unloading, sorting (or "shaking out"), inspecting, storing, and erecting the steel. The process of unloading steel to a temporary storage area and then moving it from storage to the point of erection is called *yarding.* Structural steel members are often carelessly handled during unloading at the job site. They may be thrown off the truck or railcar and stacked up in a manner that will cause distortion in the member and damage to its paint. Such practices must be avoided. In unloading long flexible members and trusses, double slings should be used to avoid bending the member. If the steel has not been inspected at the fabrication shop, it must be inspected after unloading for conformance to the shop drawings and the tolerances specified in Table 13-1. Camber and sweep of beams are illustrated in Figure 13-1. In any case, members must be checked at the job site for possible shipping and unloading damage.

Shaking out steel is the process of sorting it out by identifying each member, and storing it in such a manner that it can be easily obtained during erection. Code numbers are often painted on the members to facilitate identification during erection. Steel should be stored off the ground on platforms, skids, or other supports, and protected from dirt, grease, and corrosion. Erection, the final element of field operations, is described in Section 13-3.

Table 13-1 Fabrication and mill tolerance for steel members

Dimensions	*Tolerance*
Depth	$\pm \frac{1}{8}$ in. (0.32 cm)
Width	$+ \frac{1}{4}$ in. (0.64 cm), $- \frac{3}{16}$ in. (0.48 cm)
Flanges out-of-square	
Depth 12 in. (30 cm) or less	$\frac{1}{4}$ in. (0.64 cm)
Depth over 12 in. (30 cm)	$\frac{5}{16}$ in. (0.79 cm)
Area and weight	$\pm$ 2.5%
Length	
End contact bearing	$\pm \frac{1}{32}$ in. (0.08 cm)
Other members	
Length 30 ft (9.2 m) or less	$\pm \frac{1}{16}$ in. (0.16 cm)
Length over 30 ft (9.2 m)	$\pm \frac{1}{8}$ in. (0.32 cm)
Ends out-of-square	$\frac{1}{64}$ in./in. (cm/cm) of depth or flange width, whichever is greater
Straightness	
General	$\frac{1}{8}$ in./10 ft (0.1 cm/m) of length
Compression members	Deviation from straightness of $\frac{1}{1000}$ of axial length between points of lateral support

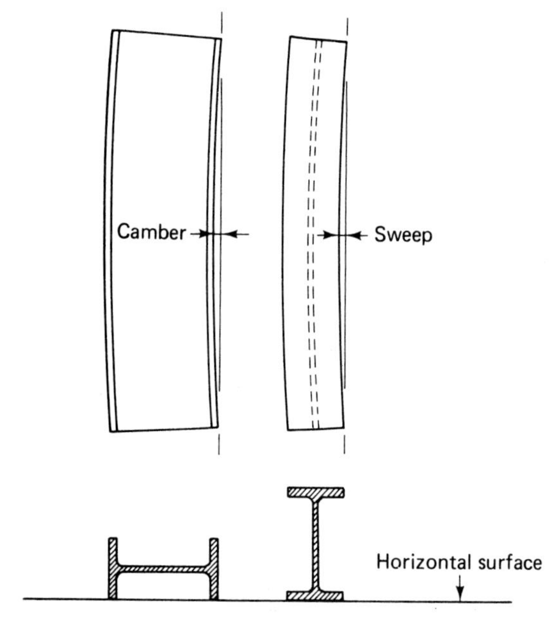

Figure 13-1 Camber and sweep of beams.

13-2 STRUCTURAL STEEL

Types of Steel

The type of steel contained in a structural steel member is designated by the letter A followed by the American Society for Testing and Materials (ASTM) designation number. The principal types of structural steel include:

- A36 Carbon Structural Steel.
- A572 High-Strength Low-Alloy Structural Steel.
- A588 Corrosion-Resistant High-Strength Low-Alloy Structural Steel.

Steel strength is designated by the symbol F_y, which indicates the minimum yield point of the steel expressed in thousands of pounds per square inch (ksi), pounds per square inch (psi), or megapascals (MPa). Type A36 steel has a yield strength of 36 ksi (36,000 psi or 248.2 MPa). The high-strength steels (types A572 and A588) are available in yield strengths of 42 ksi (289.6 MPa) to 65 ksi (448.2 MPa).

Weathering steel is a type of steel that develops a protective oxide coat on its surface upon exposure to the elements so that painting is not required for protection against most atmospheric corrosion. That natural brown color that develops with exposure blends well with natural settings. However, care must be taken to prevent staining of structural elements composed of other materials which are located in the vicinity of the weathering steel and thus exposed to the runoff or windblown water from the weathering steel.

Standard Rolled Shapes

There are a number of rolled steel shapes produced for construction which have been standardized by the American Society for Testing and Materials. Figure 13-2 illustrates five major section shapes. A list of standard shapes and their AISC designation is given in Table 13-2. Note that the usual designation code includes a letter symbol (identifying the section shape) followed by two numbers (indicating the section depth in inches and the weight per foot). Designations for angles, bars, and tubes are slightly different, in that the numbers used identify principal section

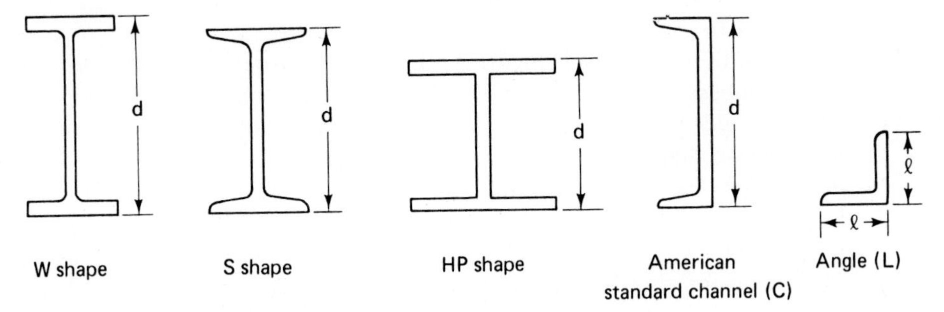

Figure 13-2 Rolled steel section shapes.

Table 13-2 Rolled-steel shape designations

Type of Shape	*Example Designation*
W shape	W27 × 114
S shape	S20 × 95
M shape	M8 × 25
American Standard Channel	C12 × 30
Miscellaneous Channel	MC12 × 50
HP (bearing pile) shape	HP14 × 89
Equal leg angle	L6 × 6 × $\frac{1}{2}$
Unequal leg angle	L8 × 4 × $\frac{1}{2}$
Structural tee cut from:	
W shape	WT8 × 18
S shape	ST6 × 25
M shape	MT4 × 16.3
Plate	PL $\frac{1}{2}$ × 12
Square bar	Bar 2 ◫
Round bar	Bar ϕ
Flat bar	Bar 2 × $\frac{1}{2}$
Pipe	Pipe 6 std.
Structural tubing	
Square	TS6 × 6 × .250
Rectangular	TS6 × 4 × .250
Circular	TS4 OD × .250

dimensions in inches rather than the section depth and weight. Detailed section properties as well as the weight of pipe, plates, and crane rails are given in reference 2.

Built-Up Members

Girders are used when regular rolled shapes are not deep enough or wide enough to provide the required section properties. Plate girders (Figure 13-3a) normally consist of a web and top and bottom flanges. Stiffners may be added if needed to prevent buckling of the web. Box girders are constructed using two webs as shown in Figure 13-3b.

Open web steel joists (Figure 13-4) and joist girders are other forms of built-up steel members. These are lightweight open trusses that are strong and economical. They are widely used for supporting floors and roofs of buildings. *Bar joists* are steel joists whose diagonal members consist of steel bars. Standard *open web steel joist* designations include K, LH, and DLH series. All are designed to support uniform loads. K series are parallel chord joists that span up to 60 ft (18.3 m) with a maximum depth of 30 in. (76 cm). Series K uses steel with a yield strength of 50 ksi (345 MPa) for chords and either 36 ksi (248 MPa) or 50 ksi (345 MPa) for webs. Series LH (longspan joists) and DLH (deep longspan joists) joists are available with parallel chords or with the top chord pitched one way or two ways (Figure 13-5).

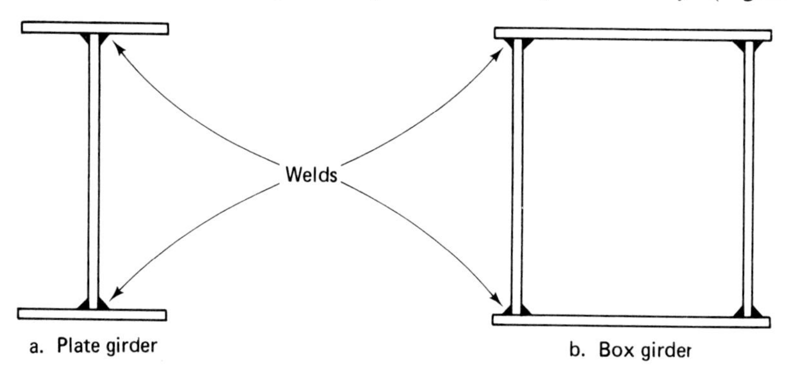

Figure 13-3 Built-up steel members.

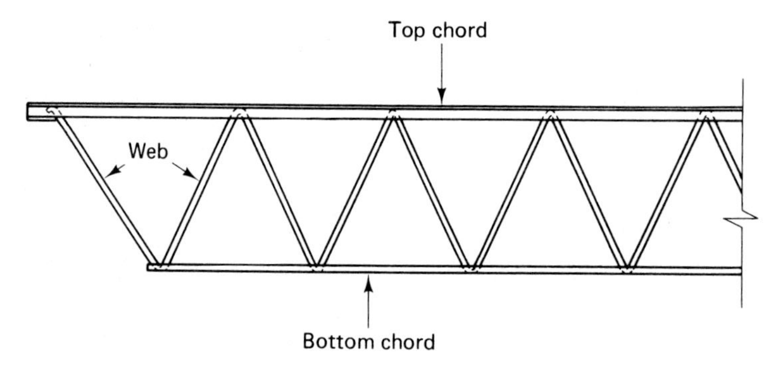

Figure 13-4 Open-web steel joist.

The standard pitch is ⅛ in./ft (1 cm/m) to provide drainage. Longspan and deep longspan joists are normally cambered to offset the deflection of the joist due to its own weight. They use steel with a yield strength of either 36 or 50 ksi (248 or 345 MPa). Series LH joists span up to 96 ft (29.3 m) with a maximum depth of 48 in. (122 cm). Series DLH joists span up to 144 ft (43.9 m) with depths to 72 in. (183 cm).

Joist girders, Series G, are similar to open web steel joists except that they are designed to support panel point loads. Series G girders use steel with a yield strength of 36 to 50 ksi (248 to 345 MPa), span up to 60 ft (18.3 m), and have a maximum depth of 72 in. (183 cm). Joist girders and open web steel joists are available with square ends, underslung ends, or extended ends, as shown in Figure 13-6.

Castellated steel beams are created from standard rolled shapes by shearing one side and then joining two sections together to create the shape shown in Figure 13-7. Such beams are deeper and have a higher strength/weight ratio than do standard rolled sections. The open portions of the web also facilitate the installation of building utilities.

13-3 STEEL ERECTION

Erection Procedure

The usual steel erection procedure employs three crews (a raising crew, a fitting crew, and a fastening crew) which operate in sequence as erection proceeds. The raising crew lifts the steel member into position and makes temporary bolted connections that will hold the member safely in place until the fitting crew takes over. OSHA safety regulations use the term *structural integrity* to indicate the ability of a

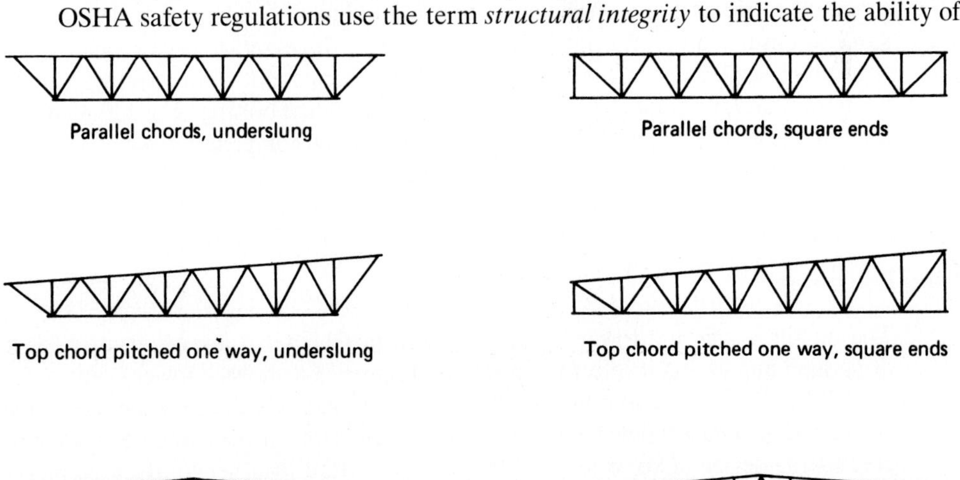

Figure 13-5 Steel joist types.

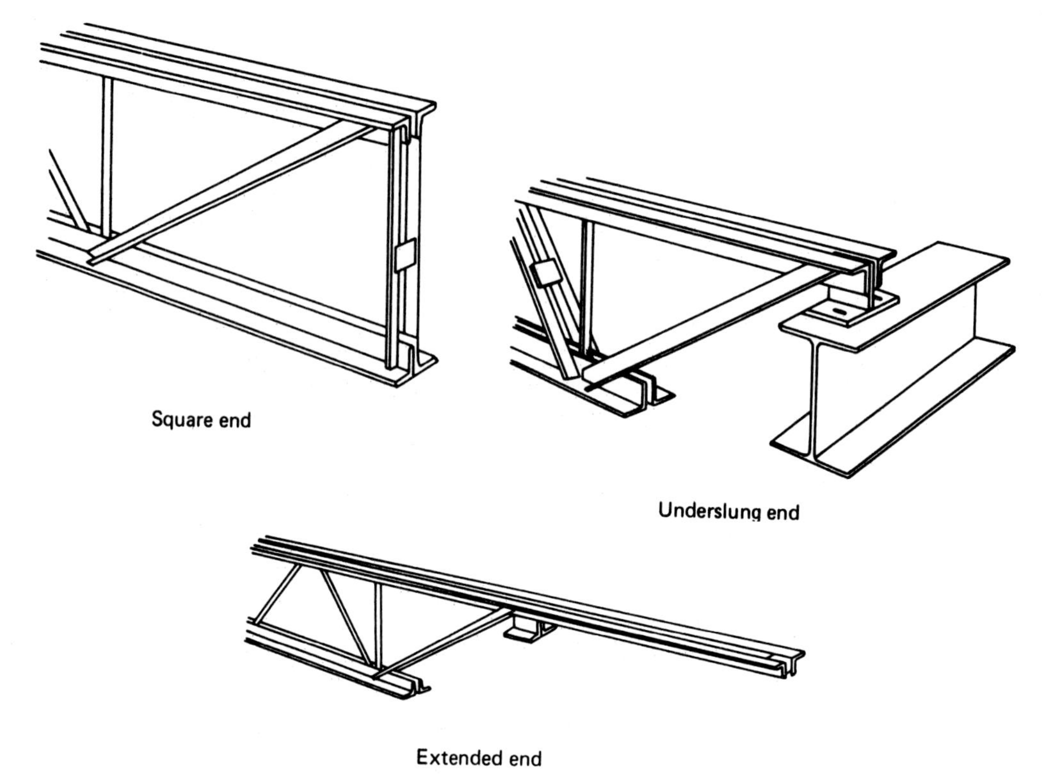

Figure 13-6 Types of joist ends.

structure to safely stand up during erection and has prescribed specific safety measures to ensure structural integrity. For example, the erection deck cannot be more than eight stories above the highest completed permanent floor. Neither can there be more than four floors or 48 ft (14.6 m) of unfinished bolting or welding above the highest permanently secured floor (not necessarily completed floor). The fitting crew brings the member into proper alignment and tightens enough bolts to hold the structure in alignment until final connections are made. The fastening crew makes the final connections (bolted or welded) to meet specification requirements.

Lifting Equipment

The mobile crane and tower crane described in Chapter 3 are often used for handling steel and lifting it into final position. Figure 13-8 shows a tower crane erecting steel. There are also a number of other lifting devices which are often used in steel construction. The gin pole shown in Figure 13-9a is one of the simplest types of powered lifting device. Two or more of these may be used together to lift large pieces of equipment such as boilers or tanks. A guy derrick is shown in Figure 13-9b. This is probably the most widely used lifting device in high-rise building construction. An advantage of the guy derrick is that it may easily be moved (or jumped) from one

Figure 13-7 Castellated steel beams. (Courtesy of American Institute of Steel Construction)

floor to the next as construction proceeds. Figure 13-9c illustrates a heavy-duty lifting device called a stiffleg derrick. Stiffleg derricks may be mounted on tracks to facilitate movement within a work area.

Alignment of Steel

Alignment of steel members must be accomplished within the tolerances of the AISC Code of Standard Practice (see reference 2). Under AISC standards the vertical (or plumb) error cannot exceed 1 unit in 500 units of height and the centerline of exterior columns cannot be more than 1 in. (2.5 cm) toward or 2 in. (5 cm) away from the building line in 20 stories.

The minimum clearance between steel members is also specified in reference 2. *Coping* or *blocking* is the name applied to notching beams to provide the necessary clearance when beams connect to columns or other beams. Electrical, plumbing, and other trades often find it convenient to make attachments to steel or to cut openings (inserts) in the steel to facilitate installation of their equipment. No attachments or inserting, including blocking and coping, should be permitted without the approval of the structural designer.

Guy ropes and supports are often used in the process of bringing steel into alignment, as illustrated in Figure 13-10. Erection planning should include the number, type, and location of all guys and supports to be used. Guys should be placed so as to minimize interference with travel ways and erection equipment, and must be kept taut. Care must be taken not to overstress guys during alignment.

Figure 13-8 Tower crane erecting steel. (Courtesy of FMC Corporation)

Erection of Steel Joists

The requirements for the lateral bracing of steel joists by bridging have been established by the Steel Joist Institute (SJI) (reference 1). The lateral bracing of longspan and deep longspan joists during erection is especially critical. For these joists the SJI requires that hoisting cables not be released until a minimum number of lines of bridging have been installed: one line for spans to 60 ft (18.3 m), two lines for spans of 60 to 100 ft (18.3 to 30.5 m), and all lines for spans over 100 ft (30.5 m). Joists should be completely braced before any loads are applied.

13-4 FIELD CONNECTIONS

Fastening Systems

The three principal systems used for connecting steel members include bolting, riveting, and welding. Riveting is now seldom used for making field connections or for shop fabrications. Riveting procedures will not be described here.

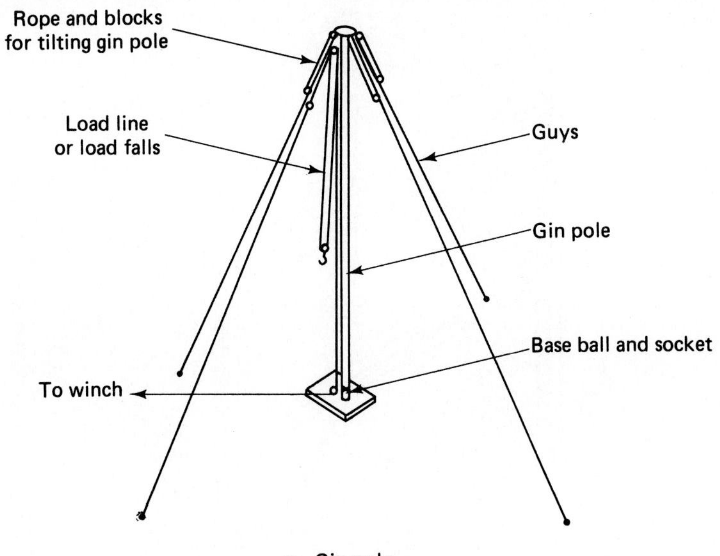

a. Gin pole

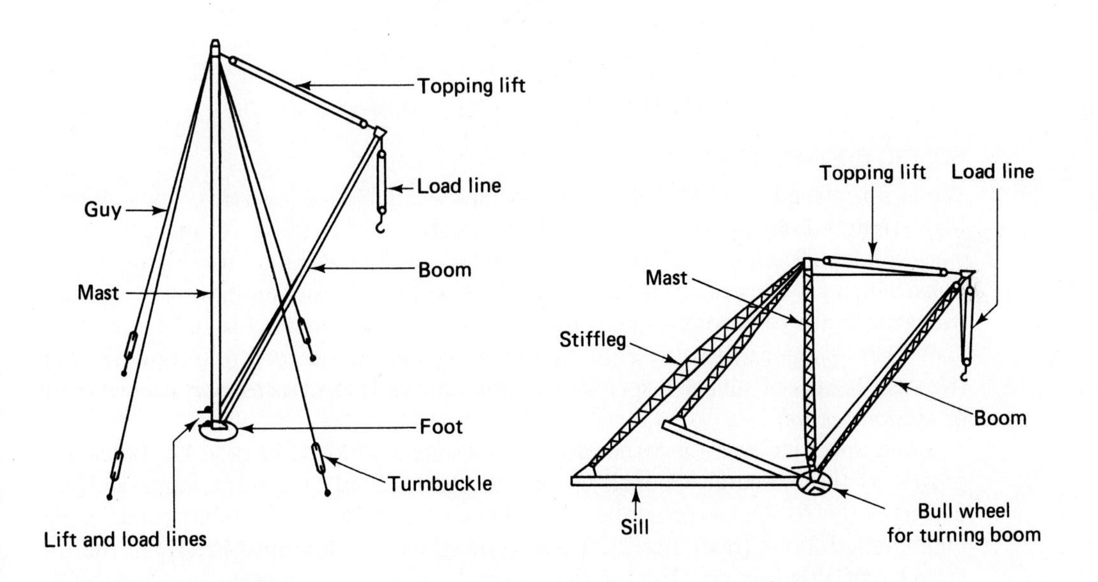

b. Guy derrick

c. Stiffleg derrick

Figure 13-9 Steel lifting equipment.

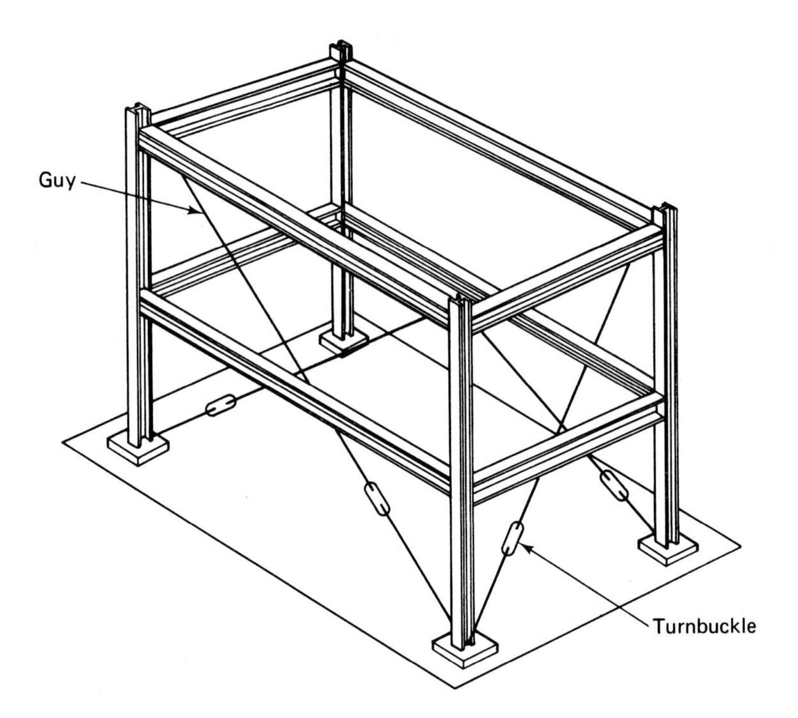

Figure 13-10 Plumbing a steel structure.

BOLTED CONNECTIONS

While unfinished (ASTM A307) bolts are still available for low stress applications, high-strength bolts are used in most of today's steel construction. To prevent confusion in identification, ASTM has prescribed special markings for high-strength bolts, which are illustrated in Figure 13-11. Bolts that are driven into place and use oversize shanks to prevent turning during tightening are referred to as *interference-body* or *interference-fit bolts*. Bolts that incorporate a torque control groove so that the stem breaks off under a specified torque are referred to as *tension control bolts* or *tension set bolts.*

The Specifications for Structural Joists Using ASTM A325 or A490 Bolts, approved by the Research Council on Riveted and Bolted Structural Joints and endorsed by the AISC, has prescribed acceptable procedures for assembling steel using high strength bolts (reference 2). These procedures are described briefly in the remainder of this section. Two of the methods used for tightening standard high-strength bolts to the specified tension are the turn-of-nut method and the calibrated wrench method.

Quality control procedures may require the use of a torque wrench to verify that the required bolt tension is being obtained. When used, torque wrenches should be calibrated with a bolt-tension calibrator at least once a day by tightening at least three bolts of each diameter being used. A *bolt-tension calibrator* is a device that can be used to calibrate both impact wrenches (used for bolt tightening) and hand-indicator torque wrenches (used by inspectors for checking the tension of bolts that

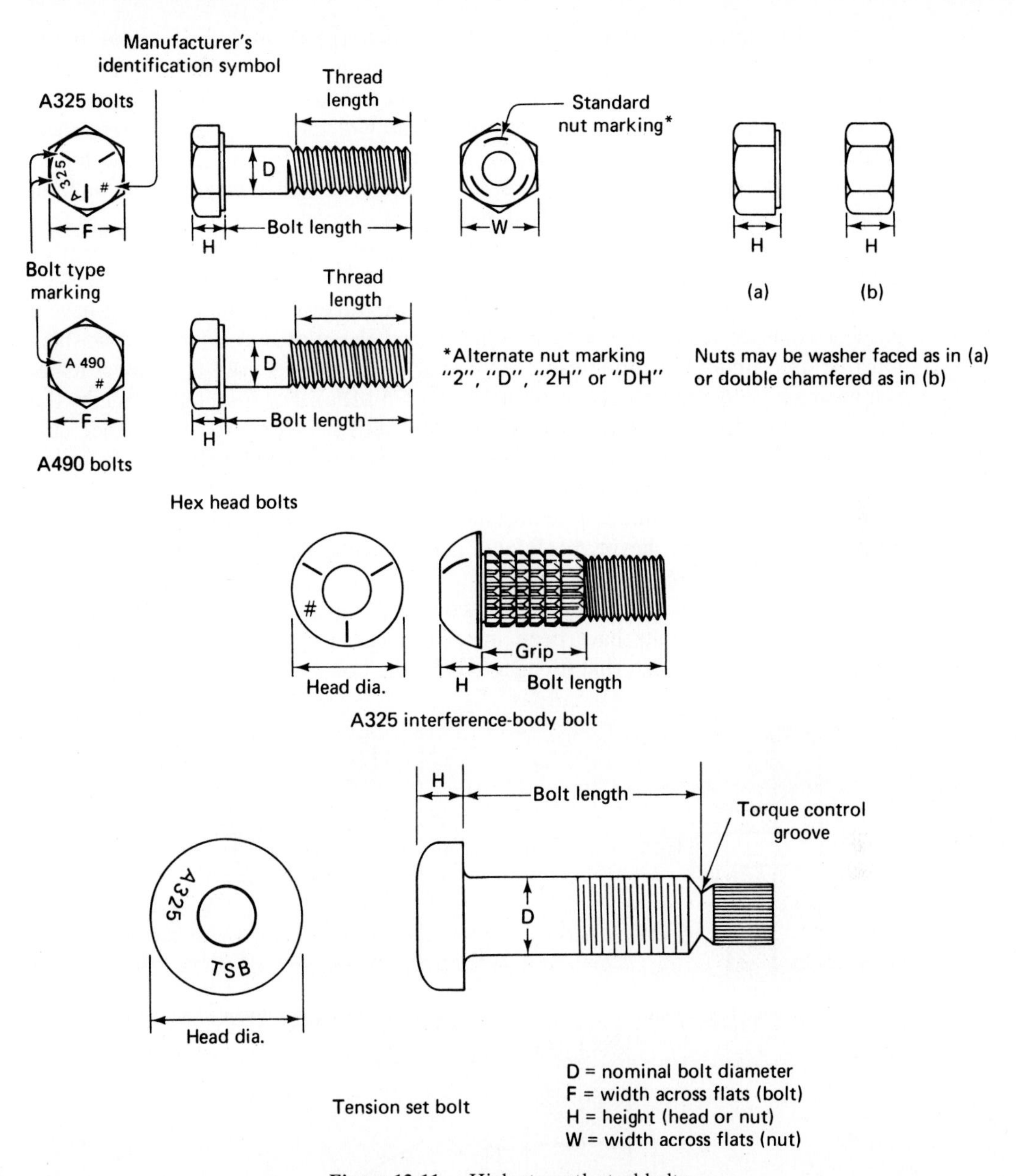

Figure 13-11 High-strength steel bolts.

have been tightened by either method). Torque-control devices on impact wrenches should be set to produce a bolt tension 5 to 10% greater than the specified minimum bolt tension. Air pressure at the impact wrench used for bolt tightening should be at least 100 psi (690 kPa), and the wrench should be capable of producing the required bolt tightening in about 10 seconds.

When tightening a high-strength bolt by either the turn-of-nut method or the calibrated wrench method, the bolt is first brought to a snug condition. (The snug condition is reached when an impact wrench begins to impact solidly or when a worker uses his full strength on an ordinary spud wrench.) Except for interference-body bolts, final bolt tightening may be accomplished by turning either the nut or the bolt head. Residual preservative oil on bolts may be left in place during tightening. When the surface to be bolted is inclined at a slope greater than 1 in 20 to a perpendicular to the bolt axis, a beveled washer must be used to provide full bearing for the nut or head. Both A325 and A490 bolts tightened by the calibrated wrench tightening method and A490 bolts installed by the turn-of-nut method must have a hardened washer under the element being turned (head or nut). Hardened washers must be used under both the head and the nut of A490 bolts used to connect material having a yield strength of less than 40 ksi (276 MPa). For final tightening by the calibrated wrench method, the bolt is impacted until the torque-control device cuts off. Using the turn-of-nut method, the specified rotation must be obtained from the snug condition while the stationary end (bolt head or nut) is held by hand wrench to prevent rotation. The tightening requirement for a bolt not more than 8 diameters or 8 in. (20 cm) in length, having both faces perpendicular to the bolt axis, is one-half turn from the snug condition.

The procedure for tightening tension control or tension set bolts is illustrated in Figure 13-12. After bolts have been installed finger-tight, the installation tool is placed over the bolt end so that it engages both the bolt spline and nut. The installa-

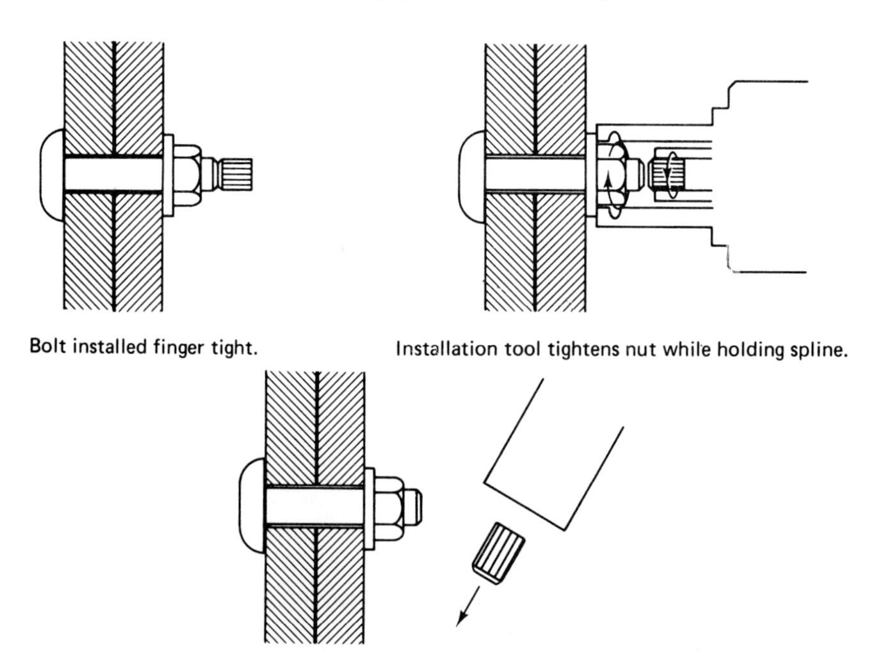

Figure 13-12 Installation of tension control bolts.

tion tool holds the bolt spline to prevent the bolt from rotating while torque is applied to the nut. When the torque on the nut reaches the required value, the bolt spline will shear off at the torque control groove. Visual inspection will indicate whether bolts have been properly tightened by determining that the spline end of the bolt has sheared off. If desired, bolt tension may be verified by the use of a calibrated torque wrench, as described in the previous paragraph.

WELDED CONNECTIONS

Welding (Figure 13-13) is another specialized procedure that must be accomplished properly if adequate connection strength is to be provided. Welding requirements for steel construction are contained in reference 2 and the publications of the American Welding Society (AWS). A few of the principal welding requirements are described in this section. In the United States, all welders responsible for making connections in steel construction should be certified by the American Welding Society. All supervisors and inspectors must be able to interpret the standard welding symbols shown in Figure 13-14. The major types of structural welds include fillet welds, groove (or butt or vee) welds, and plug or rivet welds. These are illustrated in Figure 13-15.

In addition to the use of qualified welders, requirements for producing satisfactory electric welds include the proper preparation of the base metal, the use of proper electrodes, and the use of the correct current, voltage, and polarity settings.

There are a number of inspection methods available for determining the quality of welds. Test methods include visual inspection, destructive testing, radiographic inspection, ultrasonic inspection, magnetic-particle inspection, and liquid-penetrant inspection. Visual inspection is the quickest, easiest, and most widely used method

Figure 13-13 Welded steel construction. (Courtesy of American Institute of Steel Construction)

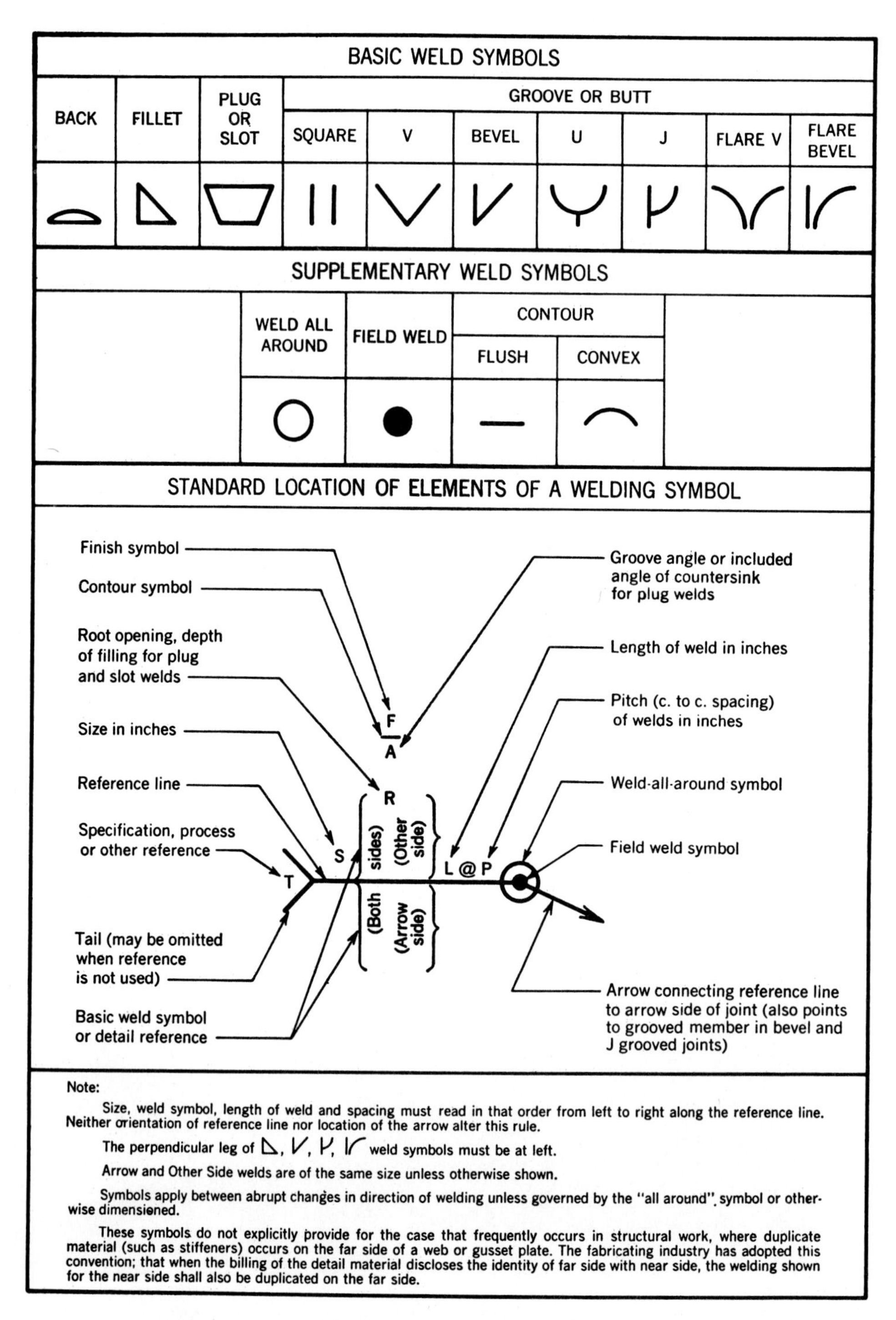

Figure 13-14 Standard welding symbols. (Courtesy of American Institute of Steel Construction)

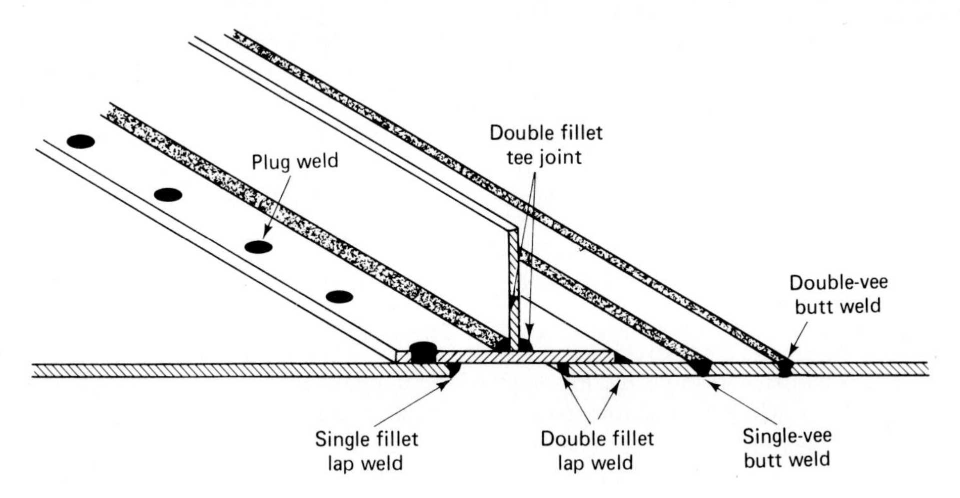

Figure 13-15 Weld types.

of inspection. However, to be effective it requires the use of highly trained and experienced inspection personnel. It is also the least reliable method for ensuring adequate weld strength. Destructive testing is used primarily in welder qualification procedures. Its use may also be necessary to determine the actual strength of welds when nondestructive test methods indicate questionable weld quality. Radiographic inspection involves producing an X-ray picture of the weld. When properly employed, it can detect defects as small as 2% of the joint thickness. Ultrasonic inspection uses high-frequency vibration to detect defects. The nature of the signals that are reflected back from the weld gives an indication of the type, size, and location of any defect. Magnetic-particle inspection utilizes magnetic particles spread on a weld to indicate defects on or near the weld surface. However, it cannot be used on nonmagnetic metals such as aluminum. Liquid-penetrant inspection involves spraying the weld with a liquid penetrant, drying the surface, and then applying a developing fluid which shows the location where penetrant has entered the weld. The method is inexpensive and easy to use but can detect only those flaws that are open to the surface.

13-5 SAFETY

As stated earlier, steel erection is a very hazardous construction task. As a result, a number of safety requirements have been developed and many of these are contained in OSHA safety regulations (reference 4). Some of these requirements have been described earlier in the chapter. Two additional safety areas that deserve comment are the use of protective equipment and the hazards presented by site conditions.

Protective Equipment

OSHA regulations contain a number of requirements for the use of personal protective equipment. Hardhats and gloves are standard requirements for steel erection. Eye protection must be provided for workers engaged in welding, cutting, and chipping operations, as well as those working nearby. Employees working above ground level require protective measures against falls. Temporary floors and scaffolds with guard rails should be provided whenever possible. If these are not feasible, lifelines and safety belts must be used. Where the potential fall exceeds 25 ft (7.6 m) or two stories, safety nets should also be used. When used, safety nets should be placed as close under the work surface as practical and extend at least 8 ft (2.4 m) beyond the sides of the work surface.

Site Hazards

Weather is responsible for many of the hazards at the steel erection site. High and gusty winds may throw workers off balance and cause steel being lifted to swing dangerously. Tag lines must be used for all hoisting operations. Since steel workers will be walking on members shortly after they are lifted, care must be taken to prevent the surfaces of members from becoming slippery. Wet and icy surfaces are obvious hazards. Structural members should be checked to ensure that they are free of hazards such as dirt, oil, loose debris, ice, and wet paint before being hoisted into place.

PROBLEMS

1. Identify the maximum fabrication tolerance of a steel beam in terms of length, depth, width, and straightness.
2. Identify the three principal types of structural steel and their strength.
3. Identify the following steel sections.
   (a) W30 × 124
   (b) C10 × 25
   (c) S24 × 100
   (d) L6 × 6 × ⅝
4. Briefly describe open-web steel joists and their use.
5. Describe the procedure for tightening tension control or tension set bolts.
6. Name four methods of handling and lifting structural steel members into final position.
7. Describe two methods for tightening standard high-strength steel bolts.
8. Identify at least six safety precautions that should be observed in structural steel construction.
9. What is a steel box girder?
10. Write a computer program that will provide an inventory of structural steel required for a building project. Provide for all shapes listed in Table 13-2. Input should include shape and size, length, quantity, and unit weight. Output should include a summary by steel shape as well as the total weight of steel for the project. Using your computer program, provide an example inventory.

REFERENCES

1. *Standard Specifications, Load Tables, and Weight Tables.* Steel Joist Institute, Myrtle Beach, S. C., 1988.
2. *Manual of Steel Construction,* 8th ed. American Institute of Steel Construction, Inc., New York, 1980.
3. OPPENHEIMER, SAMUEL P. *Erecting Structural Steel.* New York: McGraw-Hill, 1960.
4. *OSHA Safety and Health Standards Digest: Construction Industry* (OSHA 2202). U.S. Department of Labor, Washington, D.C., 1990.

14
Masonry Construction

14-1 BRICK MASONRY

Masonry Terms

There are a number of specialized terms used in brick masonry construction. The reader should be familiar with these terms in order to understand the construction practices described in this section and observed in the field. Many of these terms and procedures described in this section are also applicable to concrete masonry construction, which is presented in Section 14-2. Figure 14-1 illustrates the terms applied to the six possible positions in which an individual brick may be placed. The six surfaces of a brick are identified as the face, the end, the side, the cull, and the beds, as shown in Figure 14-2. Brick frequently must be cut to fit into corners and other places where a whole brick cannot be used. Several common shapes are shown in Figure 14-3.

Figure 14-4 illustrates the terms applied to the components of a brick wall. A *course* is a horizontal layer of brick in the plane of the wall. In this illustration the individual bricks in each course, except the top course, are in the *stretcher* position. A *wythe* is a vertical section one brick thick. A *header* is a brick placed with its long axis perpendicular to the direction of the wall. Headers are used to bond two wythes together. The bricks in the top course of Figure 14-4 are in the header position. A *bed joint* is a horizontal layer of mortar (or bed) on which bricks are laid. *Headjoints* are vertical mortar joints between brick ends. A *collar joint* is a vertical joint be-

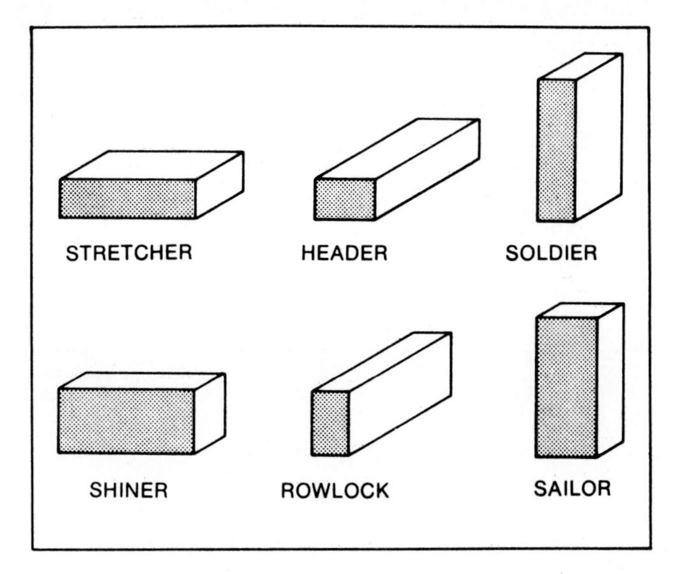

Figure 14-1 Terms applied to brick positions. (Courtesy of Brick Institute of America)

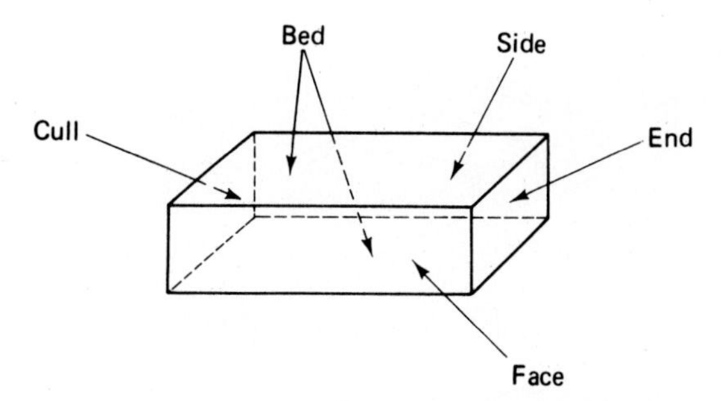

Figure 14-2 Identification of brick surfaces.

tween brick wythes. The usual thickness of mortar joints is ¼ in. (0.6 cm) for glazed brick and tile and either ⅜ in. (1.0 cm) or ½ in. (1.3 cm) for unglazed brick and tile. The exposed surfaces of mortar joints may be finished by troweling, tooling, or raking, as shown in Figure 14-5. A *trowled joint* is formed by cutting off excess mortar with the trowel and then compacting the joint with the tip of the trowel. Troweled joints include the flush joint, the struck joint, and the weather joint. A *tooled joint* is formed by using a special tool to compact and shape the mortar in the joint. The two most common tooled joints are the concave joint and the V-joint. Tooled joints form the most watertight joints. Raked joints are formed by removing a layer of mortar from the joint with a special tool. *Raked joints* are often used for appearance but are difficult to make completely watertight.

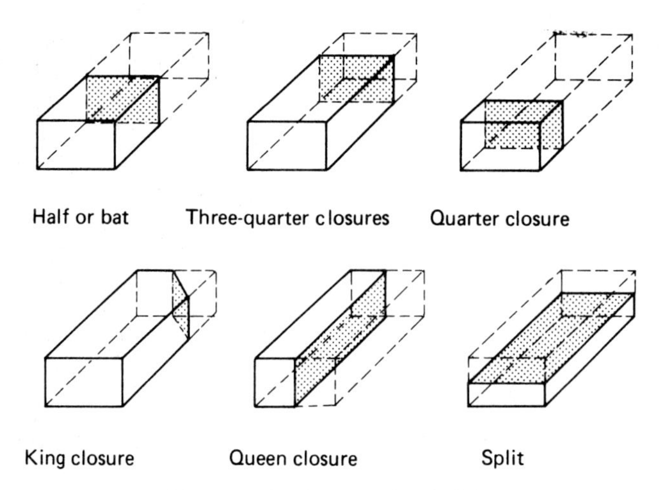

Figure 14-3 Names of cut brick.

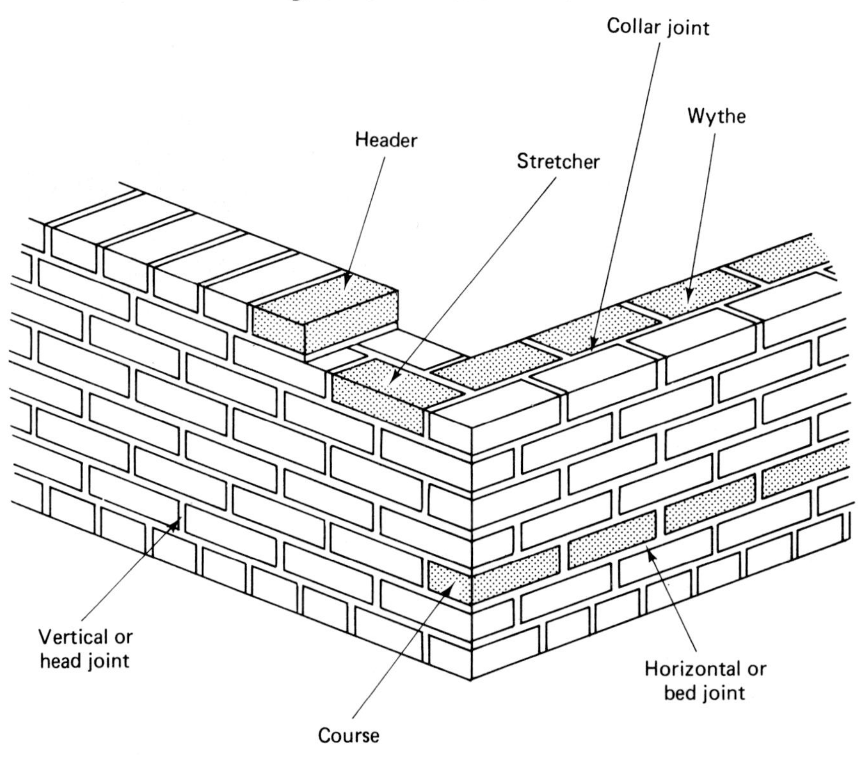

Figure 14-4 Elements of a brick wall.

Materials

Brick is manufactured in a number of sizes and shapes. Typical brick shapes are illustrated in Figure 14-6. The actual size ($W \times H \times L$) of standard nonmodular brick is $3\frac{3}{4} \times 2\frac{1}{4} \times 8$ in. ($9.5 \times 5.7 \times 20.3$ cm). Oversized nonmodular brick is $3\frac{3}{4} \times$

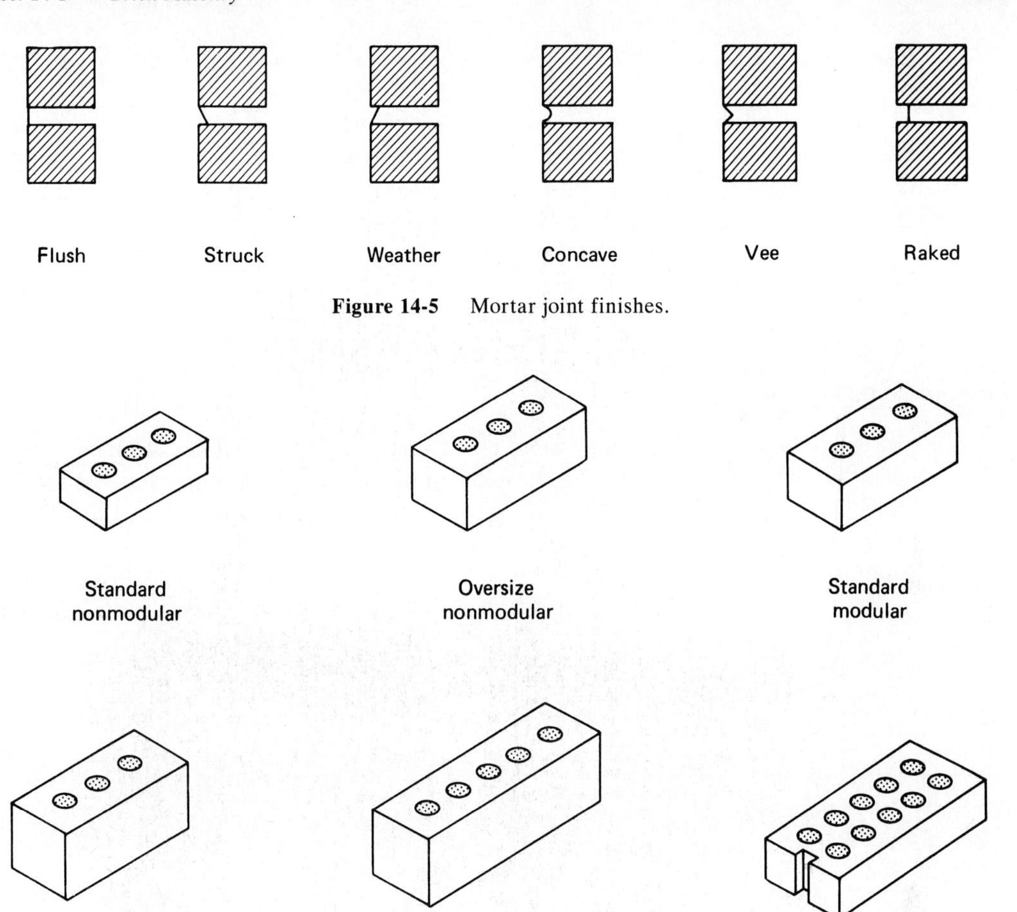

Figure 14-5 Mortar joint finishes.

Figure 14-6 Typical brick shapes.

2¾ × 8 in. (9.5 × 7.0 × 20.3 cm). The other bricks shown are a few of the modular shapes available. The actual sizes of these brick are: Standard Modular, 3⅝ × 2¼ × 7⅝ in. (9.2 × 5.7 × 19.4 cm); Economy, 3⅝ × 3⅝ × 7⅝ in. (9.2 × 9.2 × 19.4 cm); Utility, 3⅝ × 3⅝ × 11⅝ in. (9.2 × 9.2 × 29.5 cm); and SCR, 5⅝ × 2¼ × 11⅝ in. (14.3 × 5.7 × 29.5 cm).

The compressive strength of individual brick produced in the U.S. range from about 2500 psi (17.2 MPa) to over 22,000 psi (151.7 MPa). The overall compressive strength of brick assemblies is a function of both the compressive strength of the individual brick and the type of mortar used. Reference 2 describes procedures for determining the design strength of brick structural units by either performing tests on masonry assemblies or by using assumed strength values based on brick strength and mortar type. Assumed 28-day compressive strength values for masonry range from 530 psi (3.7 MPa) for 2000 psi (13.8 MPa) brick with Type N mortar and no inspection to 4600 psi (31.7 MPa) for 14,000 psi (96.5 MPa) brick with Type M mortar and construction inspection by an architect or engineer.

Mortars for unit masonry are covered by ASTM Standard C270 (Standard Specifications for Mortar for Unit Masonry) and ASTM C476 (Standard Specifications for Mortar and Grout for Reinforced Masonry). The principal mortar types include types M, S, and N. Type M mortar is a high-strength mortar for use whenever high compressive strength and durability are required. Type S mortar is a medium-high strength mortar for general-purpose use. Type N mortar is a medium-strength mortar for general use except that it should not be used below grade in contact with the earth.

Pattern Bonds

Structural bonding of masonry units is accomplished by the adhesion of mortar to masonry and by interlocking the masonry units or by embedding ties in the mortar joints. The manner in which the masonry units are assembled produces a distinctive pattern, referred to as *pattern bond.* The five most common pattern bonds are the running bond, common bond, Flemish bond, English bond, and stack bond, shown in Figure 14-7. *Running bond* uses only stretcher courses with head joints centered

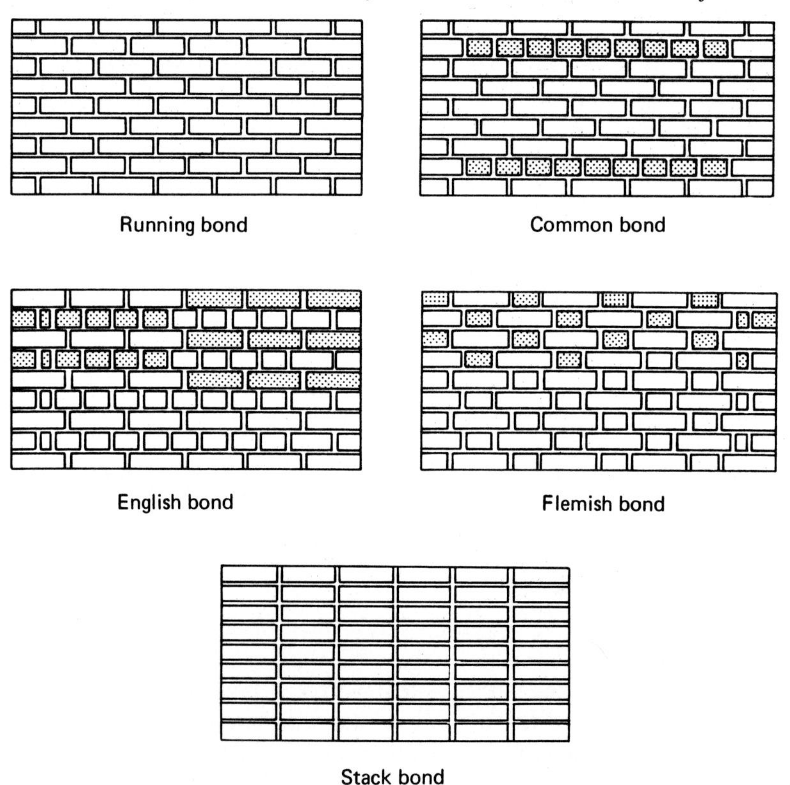

Figure 14-7 Principal brick pattern bonds.

over stretchers in the course below. *Common bond* uses a header course repeated at regular intervals; usually every fifth, sixth, or seventh course. Headers provide structural bonding between wythes. *Flemish bond* alternates stretchers and headers in each course with headers centered over stretchers in the course below. *English bond* is made up of alternate courses of headers and stretchers, with headers centered on stretchers. *Stack bond* provides no interlocking between adjacent masonry units and is used for its architectural effect. Horizontal reinforcement should be used with stack bond to provide lateral bonding.

Hollow Masonry Walls

Masonry *cavity walls* are made up of two masonry wythes separated by an air space 2 in. (5 cm) or more in width and tied together by metal ties. Brick cavity walls combine an exterior wythe of brick with an interior wythe of brick, structural clay tile, or concrete masonry. Cavity walls have a number of advantages over a single solid masonry wall. These advantages include greater resistance to moisture penetration, better thermal and acoustical insulation, and excellent fire resistance. A hollow masonry bonded wall constructed of utility brick, called a *utility wall,* is shown in Figure 14-8. While masonry bonded walls are not as water resistant as are cavity walls, they can resist water penetration when properly constructed. Recommended practice for construction of the utility wall includes bonding every sixth course (with alternating headers and stretchers), installing flashing at the bottom of the wall, and providing weep holes along the bottom exterior brick course on 24-in. (61-cm) centers, using type S mortar, and providing concave tooled joints on the exterior surface. The cavity between wythes may be filled with insulation if desired.

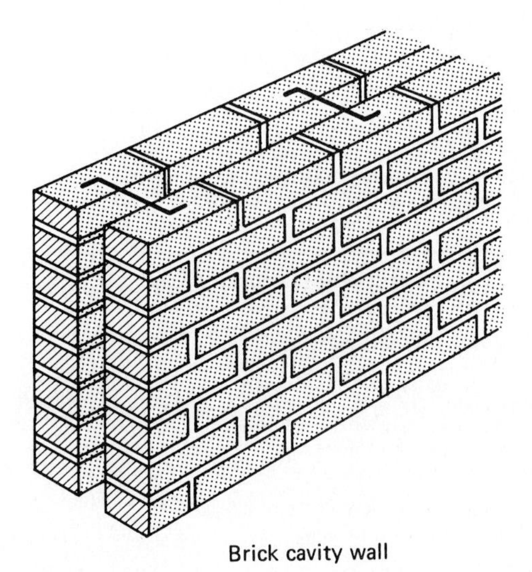

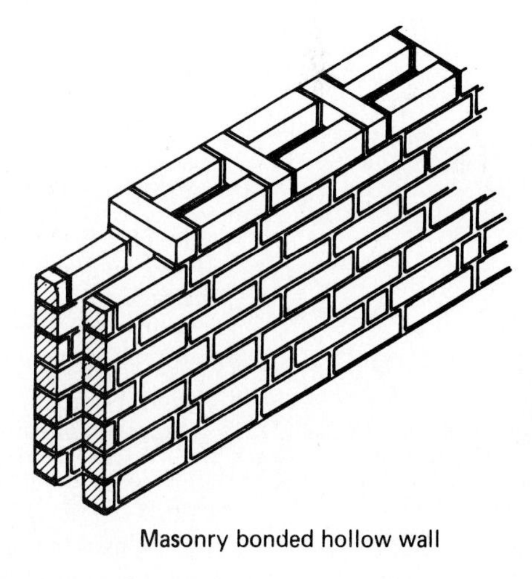

Figure 14-8 Brick cavity and masonry bonded hollow walls.

Reinforced Brick Masonry

The terms *reinforced brick masonry* (or RBM) is applied to brick masonry in which reinforcing steel has been embedded to provide additional strength. Typical reinforced brick masonry wall construction is illustrated in Figure 14-9. Notice the construction is basically the same as that of a cavity wall except that reinforcing steel has been placed in the cavity and the cavity then filled with portland cement grout. Design requirements for reinforced brick masonry are presented in reference 2. A 17-story apartment building constructed with 11-in. (28-cm)-thick reinforced brick masonry bearing walls is shown in Figure 14-10. Prefabricated reinforced brick panels are now being used to provide special shapes in wall construction. Such panels may be rapidly erected in the field, even during inclement weather. The minimum suggested amount of mortar protection for masonry reinforcement is shown in Table 14-1.

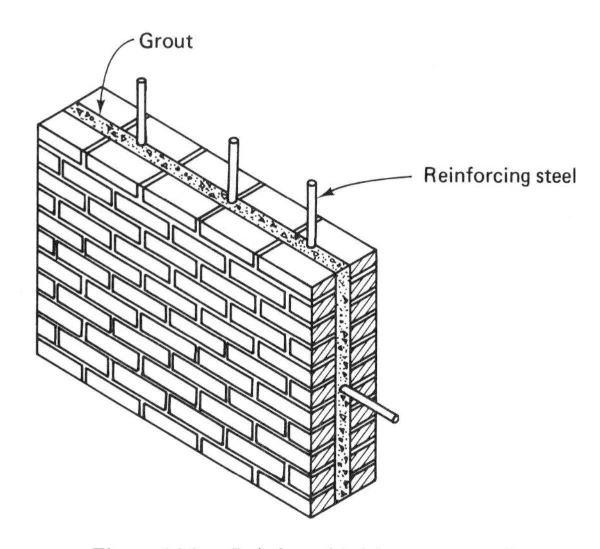

Figure 14-9 Reinforced brick masonry wall.

Table 14-1 Protection for masonry reinforcement

Application	*Minimum Cover (Exposed Face)*
Bottom of footings	3 in. (76 mm)
Columns, beams, or girders not exposed to weather or soil	$1\frac{1}{2}$ in. (38 mm)
Horizontal joint reinforcement	$\frac{5}{8}$ in. (16 mm)
All other	
Not exposed to weather or soil	1 bar diameter but at least $\frac{3}{4}$ in. (19 mm)
Exposed to weather or soil	2 in. (51 mm)

Figure 14-10 Seventeen-story building constructed with reinforced brick masonry bearing walls. (Courtesy of Brick Institute of America)

Bond Beams and Lintels

A *bond beam* is a continuously reinforced horizontal beam of concrete or masonry designed to provide additional strength and to prevent cracking in a masonry wall. Bond beams are frequently placed at foundations and roof levels but may be used at any vertical interval specified by the designer. Support over opening in masonry walls may be provided by lintels or by masonry arches, as shown in Figure 14-11. *Lintels* are short beams of wood, steel, stone, or reinforced brick masonry used to span openings in masonry walls.

Control Joints and Flashings

Expansion or *control joints* in masonry walls are used to permit differential movement of wall sections caused by shrinkage of concrete foundations and floor slabs, temperature and moisture changes, and foundation settlement. Control joints are grooves placed in masonry to control shrinkage cracking. The usual procedure is to separate walls into sections with vertical expansion joints where differential movement may occur. Long, straight walls should be divided into sections. Other expansion joints are placed at window and door openings, at columns and pilasters, at wall offsets, at cross walls, and under shelf angles in multistory buildings. Structural

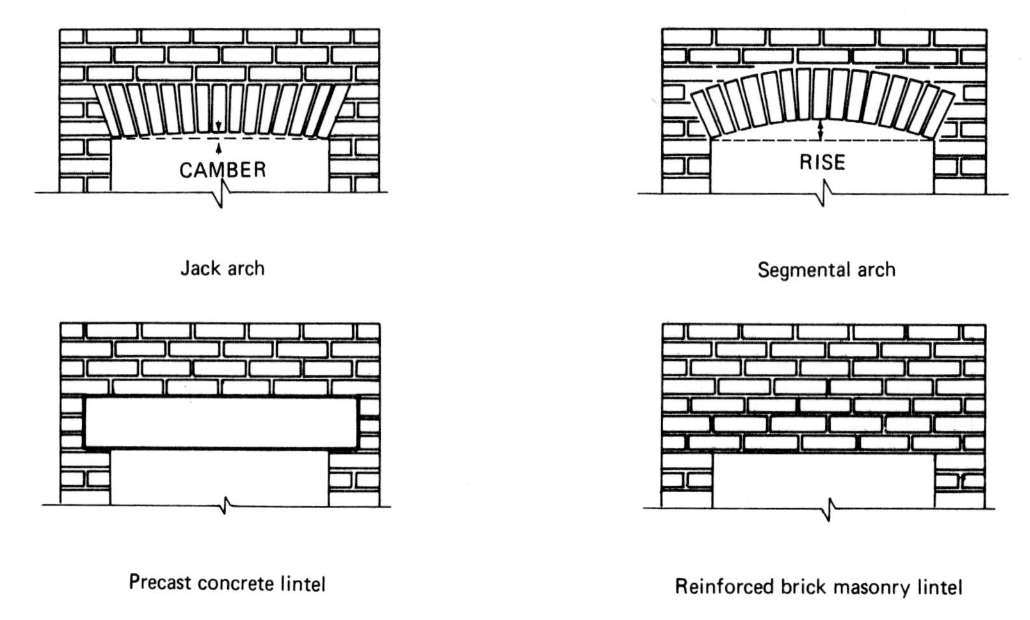

Figure 14-11 Support over openings in brick walls.

bonding across the expansion joint may be provided by interlocking construction or by flexible ties extending across the joint. Some typical expansion joints used for brick walls are shown in Figure 14-12. The exterior of expansion joints must be sealed with a flexible sealant to prevent moisture penetration.

Flashing consists of layers of impervious material used to seal out moisture or to direct any moisture that does penetrate back to the outside. Flashing is used above vertical joints in parapet walls, at the junction of roofs and walls, at window sills and other projections, around chimney openings, and at the base of exterior walls. Typical flashings used with brick masonry are illustrated in Figure 14-13. Flashings used where roofs intersect walls or chimneys are frequently composed of two parts, a base flashing and a counterflashing. The base flashing covers the joint between intersecting surfaces while the counterflashing seals the joint between the base flashing and the vertical surface, as shown in the chimney of Figure 14-13.

14-2 CONCRETE MASONRY

Materials

Concrete masonry units are classified as concrete brick, concrete tile, solid load-bearing concrete block, hollow load-bearing concrete block, and hollow non-load-bearing concrete block. Concrete block that is glazed on one or more surfaces is

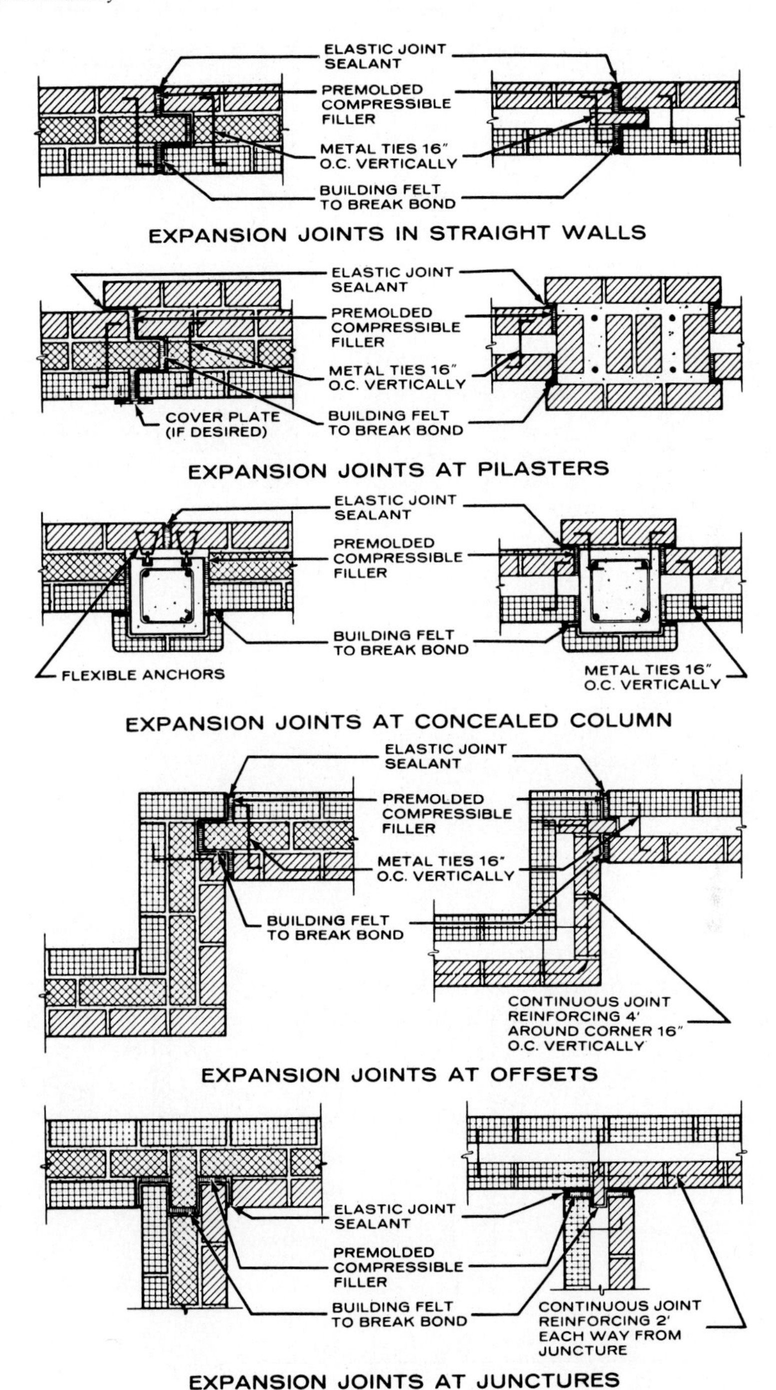

Figure 14-12 Expansion joints in brick masonry. (Courtesy of Brick Institute of America)

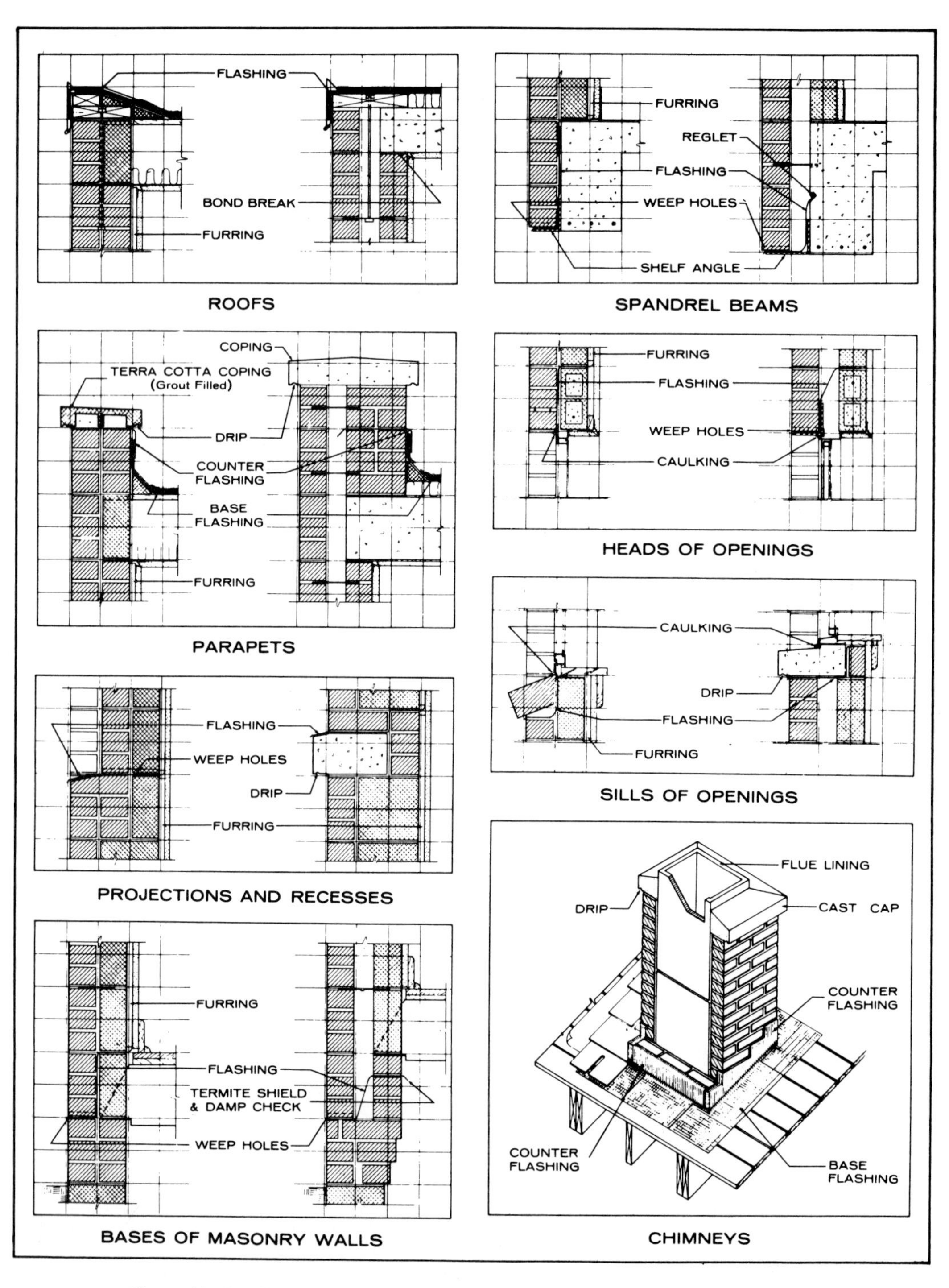

Figure 14-13 Flashing used in brick masonry construction. (Courtesy of Brick Institute of America)

available. Such units are used for their appearance, ease of cleaning, and low cost. *Solid concrete block* must have at least 75% of its cross section made up of concrete. Block having over 25% of its cross-sectional area empty is classified as *hollow block*. The usual hollow concrete block has a core area making up 40 to 50% of its cross section. Typical shapes and sizes of concrete masonry units are shown in Figure 14-14. The most common nominal size of standard block is 8 × 8 × 16 in. (20 × 20 × 40 cm) [actual size 7⅝ × 7⅝ × 15⅝ in. (19.4 × 19.4 × 39.7 cm)]. Half-thick (4-in. or 10-cm) and half-length (8-in. or 20-cm) block are also available. Concrete block is available as either heavyweight or lightweight block, depending on the type of aggregate used. Heavyweight load-bearing block typically weighs 40 to 50 lb (18.1 to 22.7 kg) per unit, while a similar lightweight unit might weigh 25 to 35 lb (11.3 to 15.9 kg). Mortars used for concrete masonry units are the same as those used for brick masonry (Section 14-1). Mortar joints are usually ⅜ in. (0.95 cm) thick. Joints in exterior walls should be tooled for maximum watertightness.

Concrete block may also be laid without mortar joints. Either standard or ground block may be stacked without mortar and then bonded by the application of a special bonding material to the outside surfaces. In this case, the bonding agent provides structural bonding as well as waterproofing the wall. The time required to construct a concrete block wall using this method may be as low as one-half the time required for conventional methods. In addition, the flexural and compressive strength of a surface bonded wall may be greater than that of a conventional block wall. There are also special types of concrete block made with interlocking edges to provide structural bonding as the units are stacked without mortar.

Lintels spanning openings in concrete block usually consist of precast concrete shapes, cast-in-place concrete beams, or reinforced concrete masonry.

Reinforced Concrete Masonry

Reinforced concrete masonry construction is used to provide additional structural strength and to prevent cracking. Figure 14-15 illustrates several methods of reinforcing a one-story concrete block wall. At the top of the wall a concrete bond beam (A) is created by filling U-shaped block (called lintel block or beam block) with reinforced concrete. Vertical reinforcement is provided by placing reinforcing steel in some of the block cores and filling these cores with concrete (B). Additional horizontal reinforcement is obtained from reinforcing steel placed in the mortar joints (D). This type of construction is appropriate for areas of high design loads, such as earthquake and hurricane zones.

Additional applications of horizontal joint reinforcement illustrated in Figure 14-16 include tieing face units to backup units, bonding the two wythes of cavity walls, and reinforcing single wythe walls. Reinforced concrete masonry construction is also used in high-rise building construction. Depending on wall height and design load, some or all of the concrete block cores may be filled with reinforced concrete. A high-rise motel using reinforced concrete masonry walls is shown under construction in Figure 14-17. Details of the placement of reinforcement in a

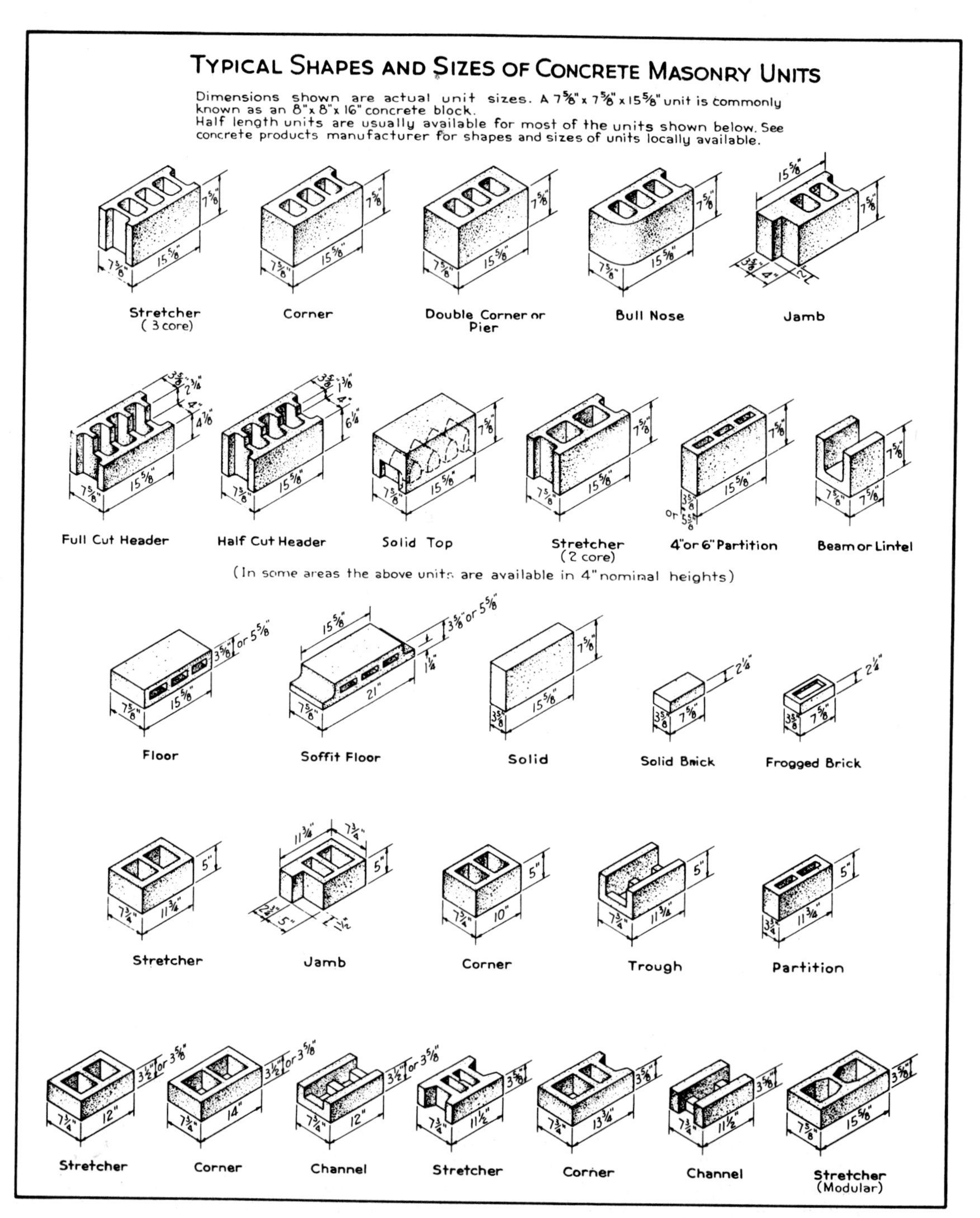

Figure 14-14 Typical concrete masonry units. (Courtesy of Portland Cement Association)

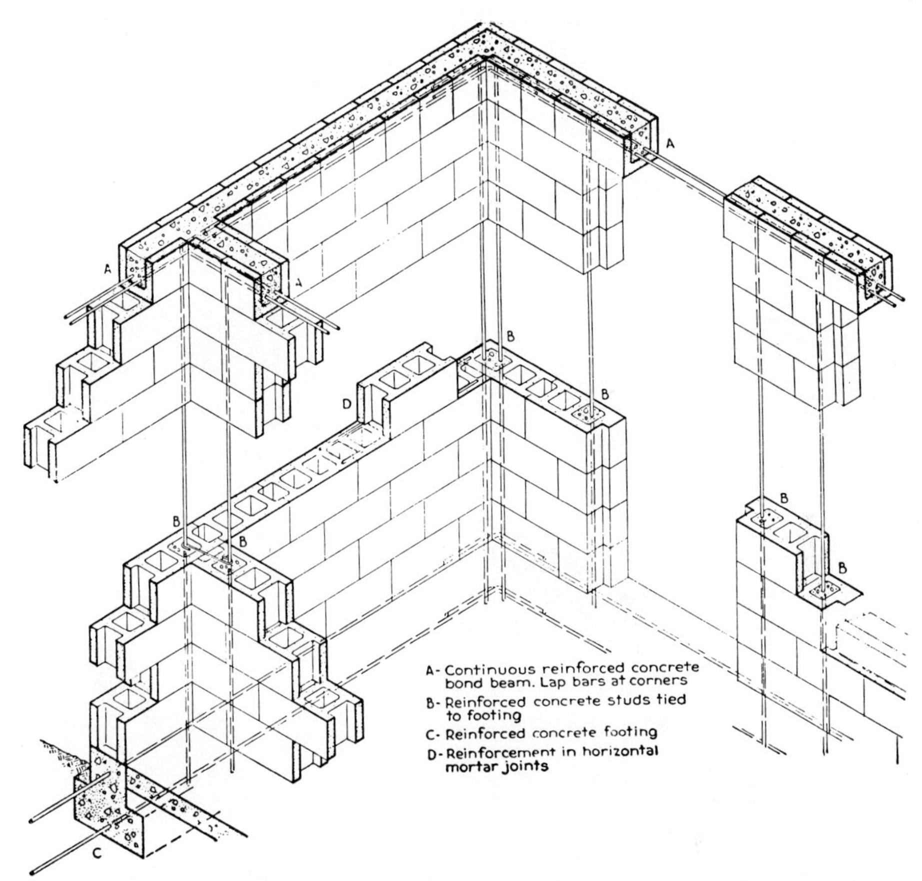

Figure 14-15 Some methods for reinforcing concrete masonry walls. (Courtesy of Portland Cement Association)

reinforced concrete masonry wall are shown in Figure 14-18. Figure 14-19 shows grout being pumped into the block cores of a reinforced concrete masonry wall. The minimum suggested amount of mortar protection for masonry reinforcement is shown in Table 14-1.

Pattern Bonds

The running bond is probably the most common pattern bond used in concrete masonry as it is in brick masonry. However, a number of other pattern bonds have been developed to provide architectural effect. Several of these patterns are illustrated in Figure 14-20. The term *ashlar masonry* is carried over from stone masonry and is now commonly used to identify masonry of any material which uses rectangular units larger than brick laid in a pattern resembling stone.

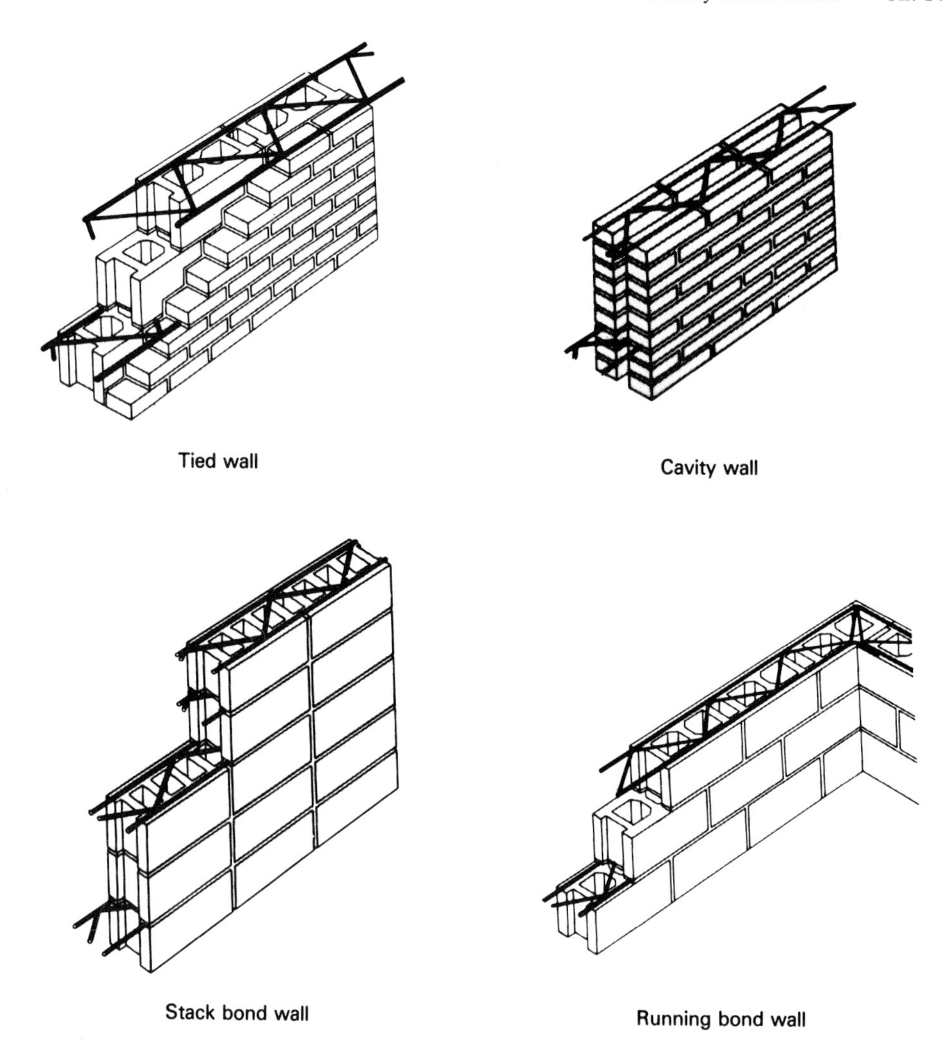

Figure 14-16 Horizontal joint reinforcement. (Courtesy of Dur-O-waL, Inc.)

14-3 OTHER MASONRY MATERIALS

In addition to brick and concrete, masonry units of stone and clay tile are also available. Load-bearing structural clay tile is used in a manner similar to concrete block. However, structural clay tile is seldom used in the United States today, and only a small quantity of glazed structural clay tile is currently being manufactured in this country. Stone and architectural terra-cotta are used primarily as wall veneers. Stone veneer is held in place by ties embedded in the mortar joints or by mechanical

Figure 14-17 Construction of a high-rise motel using reinforced concrete masonry walls. (Courtesy of Portland Cement Association)

Figure 14-18 Placement of reinforcing steel in reinforced concrete masonry wall. (Courtesy of Portland Cement Association)

Figure 14-19 Pumping grout into reinforced concrete masonry wall. (Courtesy of Portland Cement Association)

anchors fastened to the supporting structure, as shown in Figure 14-21. Shaped stone is also used for window sills, lintels, parapet coping (caps), and wall panels. Construction details for use of large architectural stone panels on a multistory building are shown in Figure 14-22.

14-4 CONSTRUCTION PRACTICE

Wind Load on Fresh Masonry

Masonry walls must be designed to safely resist all expected loads, including dead loads, live loads, and wind loads. While the designer must provide a safe structural design, the builder must erect the structure as designed and must also be able to determine the support requirements during construction. Many failures of masonry walls under construction have occurred as the result of inadequate bracing against wind load.

The maximum safe height of an unbraced masonry wall under construction may be calculated by setting the overturning moment produced by wind force equal to the resisting moment produced by the weight of the wall. Referring to Figure 14-23, we will analyze moments about the toe of the wall (A) for a unit length of wall. The design wind load in lb/sq ft (kPa) obtained from the local building code may be used to compute bracing requirements. Alternatively, the maximum anticipated wind velocity can be estimated from local weather records and converted to wind load using Table 14-2. The method of analysis is as follows:

$$\text{Overturning moment } (M_o) = P \cdot \frac{h}{2} = qh \cdot \frac{h}{2} = \frac{qh^2}{2}$$

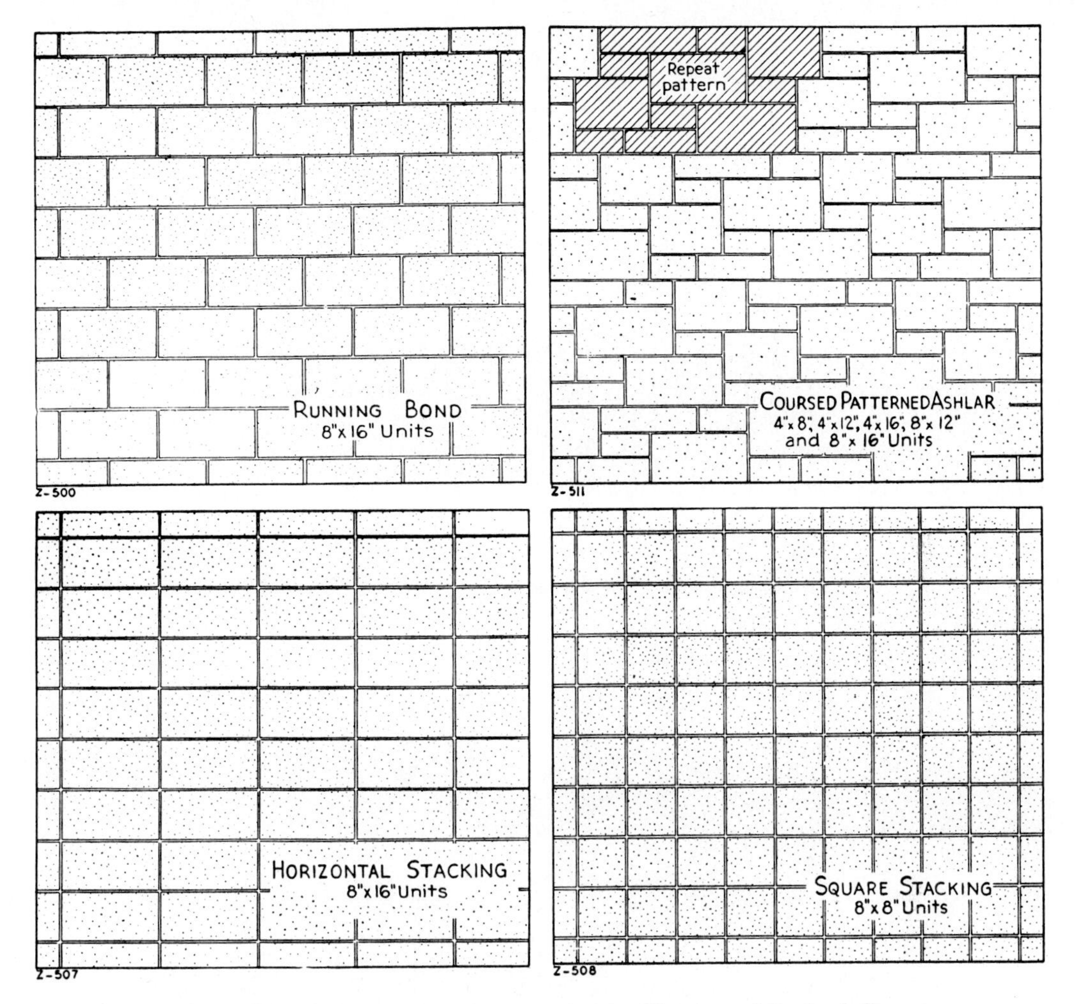

Figure 14-20 Concrete masonry pattern bonds. (Courtesy of Portland Cement Association)

$$\text{Resisting moment } (M_r) = W \cdot \frac{t}{2} = d \cdot h \cdot \frac{t}{2} = \frac{dht}{2}$$

For equilibrium, $\Sigma M_A = 0$

Hence $M_o - M_r = 0$

and $M_o = M_r$

Substitution yields

$$\frac{qh^2}{2} = \frac{dht}{2}$$

$$h_s = \frac{dt}{q} \tag{14-1}$$

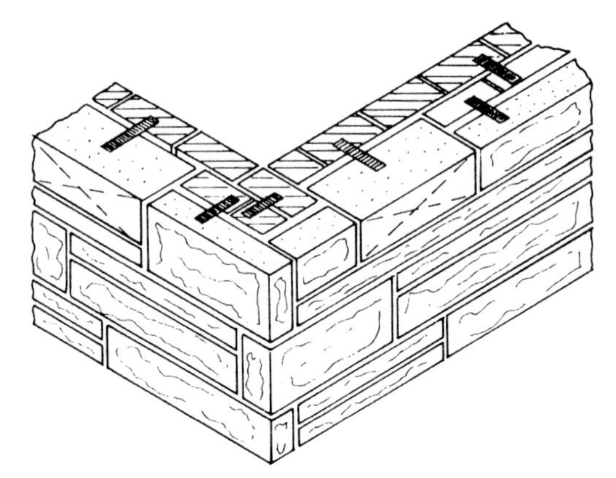

Stone Veneer with Bond Stone Anchored to Brick Backing

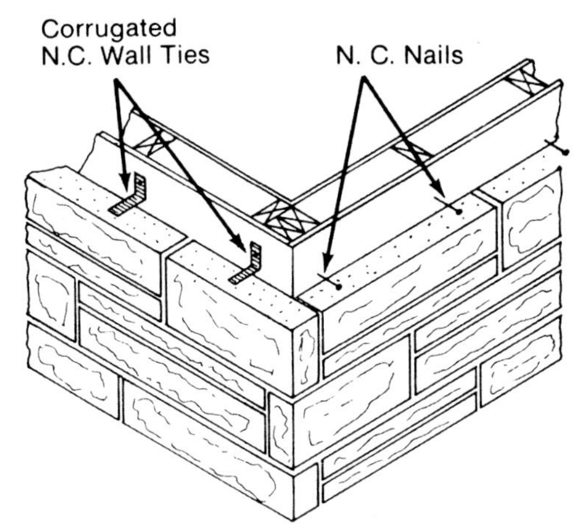

Stone Veneer Anchored to Wood Frame Backing

N.C. = noncorrosive

Figure 14-21 Stone veneer. (Courtesy of Indiana Limestone Institute of America, Inc.)

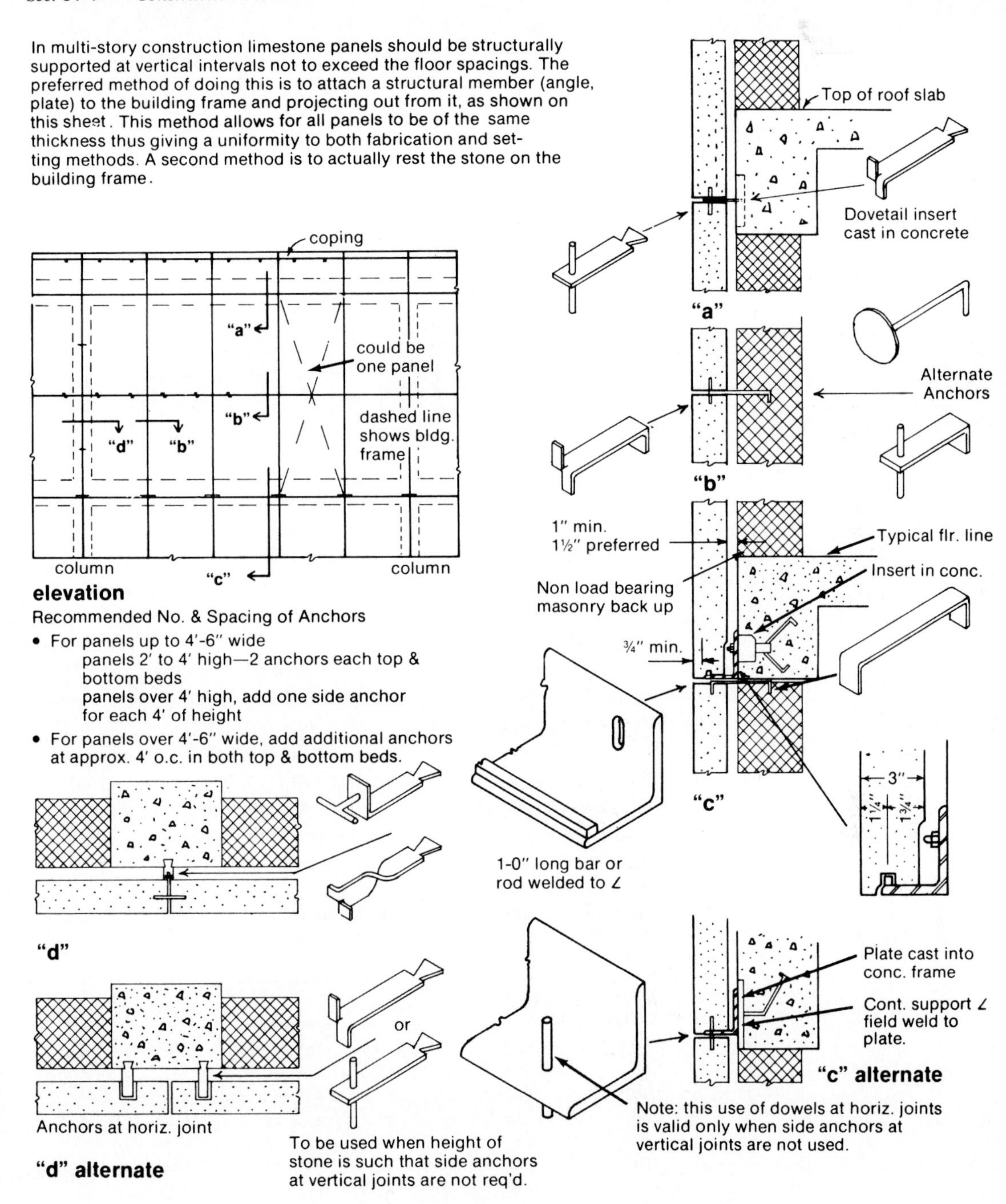

Figure 14-22 Anchorage of stone panels to multistory building. (Courtesy of Indiana Limestone Institute of America, Inc.)

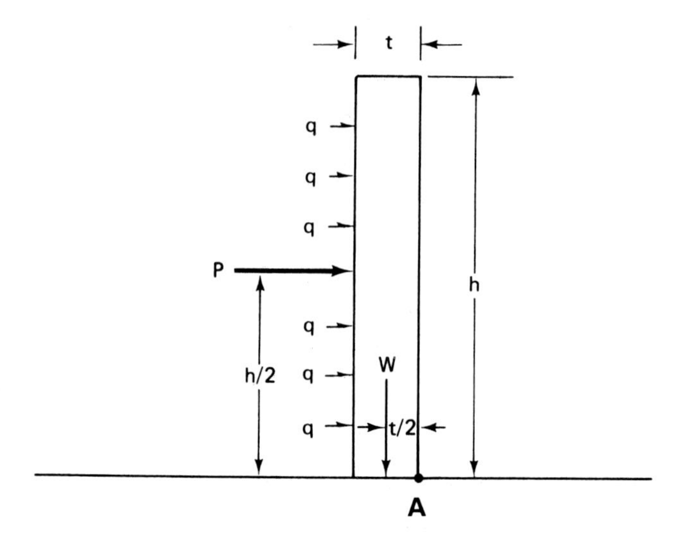

Figure 14-23 Analysis of loads on fresh masonry wall.

where q = wind force (lb/sq ft or kPa)
H = wall height (ft or m)
h_s = safe unbraced height (ft or m)
t = wall thickness (ft or m)
d = weight of wall per unit of surface (lb/sq ft or kN/m^2)
P = resultant wind force (lb or kPa)
W = resultant weight force (lb or kN)

Typical values for the weight of masonry walls per unit of height are given in Table 14-3. Notice that this analysis does not include any specific factor of safety. However, it does neglect all bonding provided by the partially set mortar. In practice, this should provide an adequate factor of safety except in unusual cases.

Table 14-2 Design wind load pressure

Wind Velocity		*Design Wind Load**	
mi/h	*km/h*	*lb/sq ft*	*kPa*
50	80	6	0.29
60	96	7	0.34
70	112	11	0.53
80	128	15	0.72
90	144	20	0.96
100	161	26	1.24
110	177	32	1.53
120	193	39	1.87
130	209	45	2.15

*Effective wind pressure on ordinary structures less than 30 ft (9.1 m) high in flat, open country (ANSI A58.1-1972).

Table 14-3 Typical unit weight for masonry walls

Type of Wall	*Weight per Unit of Wall Surface* *lb/ft²*	*kN/m²*
Heavyweight concrete block		
4-in. (10-cm)	29	1.39
6-in. (15-cm)	44	2.11
8-in. (20-cm)	56	2.68
12-in. (30-cm)	80	3.83
Lightweight concrete block		
4-in. (10-cm)	21	1.01
6-in. (15-cm)	30	1.44
8-in. (20-cm)	36	1.72
12-in. (30-cm)	49	2.35
Brick (solid)		
4-in. (10-cm)	40	1.92
6-in. (15-cm)	60	2.87
8-in. (20-cm)	80	3.83
12-in. (30-cm)	120	5.75

The proper bracing of a concrete masonry wall under construction is illustrated in Figure 14-24. Shores, forms, and braces must not be removed until the mortar has developed sufficient strength to carry all construction loads. Concentrated loads should not be applied to a masonry wall or column until 3 days after construction.

Figure 14-24 Bracing of a concrete masonry wall under construction. (Courtesy of Portland Cement Association)

Example 14-1

PROBLEM Find the maximum safe unsupported height in feet and meters for an 8-in. (20-cm) heavyweight concrete block wall if the maximum expected wind velocity is 50 mi/h (80 km/h).

SOLUTION

$$h_s = \frac{d \times t}{q} \tag{14-1}$$

$d = 56$ lb/ft^2 (2.68 kN/m^2) (Table 14-3)

$t = 8/12$ ft (0.20 m)

$q = 6$ lb/sq ft (0.29 kPa) (Table 14-2)

$$h_s = \frac{(56)\,(8/12)}{6} = 6.2 \text{ ft}$$

$$\left[= \frac{(2.68)\,(0.20)}{0.29} = 1.9 \text{ m} \right]$$

Masonry Materials

Concrete masonry units should be stored and laid in a dry condition. Brick having absorption rates greater than 20 g of water per minute should be wetted before being placed to reduce its absorption rate. However, such brick should be allowed to dry after wetting so that it is in a saturated, surface-dry condition when laid. Mortar should be thoroughly mixed before use. Mortar that has stiffened from evaporation may be retempered by adding additional water and remixing. However, to avoid the possibility of using mortar that has stiffened due to hydration, mortar should be discarded 2½ h after initial mixing.

Placing Masonry and Reinforcement

Masonry units should be placed with joints of the specified width. Brick should be laid with full bed and head joints. In general, concrete block may be laid with either full mortar bedding or face-shell bedding (illustrated in Figure 14-25). However, full mortar bedding should be used for the bottom or starting course of block and for high-load-bearing units. Tooled mortar joints should be carefully compacted with a finishing tool after the mortar has partially stiffened to provide maximum watertightness. Maximum construction tolerances for brick masonry specified by the

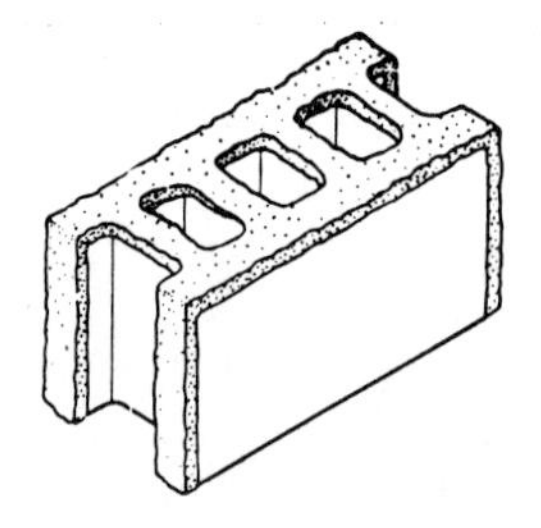

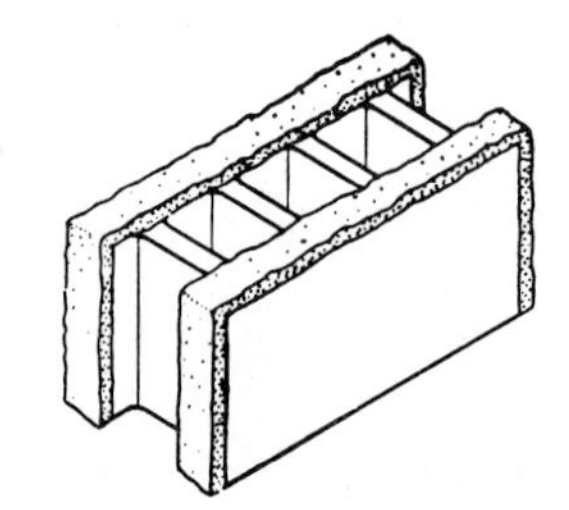

Figure 14-25 Mortar bedding of concrete block. (Courtesy of Portland Cement Association)

Brick Institute of America include vertical or plumb variations of ¼ in. (0.6 cm) in 10 ft (3 m) and ½ in. (1.3 cm) in 40 ft (12 m); horizontal or grade variation of ¼ in. (0.6 cm) in 20 ft (6 m) and ½ in. (1.3 cm) in 40 ft (12 m) or more; variation from the plan position of ½ in. (1.3 cm) in 20 ft (6 m) and ¾ in. (1.9 cm) in 40 ft (12 m) or more; and variation in section thickness of minus ¼ in. (0.6 cm) to +½ in. (1.3 cm). Reinforcing steel must be protected by the minimum thickness of cover described in Chapter 10. Masonry reinforcement not exceeding ¼ in. (0.6 cm) in diameter should have a minimum mortar cover of ⅝ in. (1.6 cm) on exterior faces and ½ in. (1.3 cm) on interior faces. This type of reinforcement should be lapped at least 6 in. (15 cm) at splices to provide adequate bonding at the splice.

Bonding Masonry

Adequate bonding must be provided where masonry walls intersect, between the wythes of cavity walls or multiple-wythe walls, and between units in stack bond construction. Bonding may be provided by masonry bonding units, by corrosion-resistant metal ties, or by truss or ladder-type masonry reinforcement. The size and spacing of bonding specified by the designer must be used. Care must also be exercised to ensure that expansion joints are properly filled with elastic material and kept clean of mortar and other rigid materials.

Weather Protection

Concrete masonry units must be dried to the specified moisture content before use. After drying, they should be stored off the ground and protected from rain. The top of exposed concrete and brick masonry under construction should be protected from rain by covering with a waterproof material whenever work is stopped. Masonry walls that are saturated by rain during construction may require months to completely dry out and will undergo increased shrinkage during drying. Efflorescence, or staining of brick surfaces by dissolved salts, often results when brick walls are saturated during construction.

The precautions required for placing masonry units during cold weather are similar to those described in Chapter 10 for placing concrete. When masonry units are to be placed in air temperatures below 40°F (5°C), the requirements of reference 6 for heating the masonry materials and maintaining an acceptable temperature during placing and curing should be observed. Do not place masonry on a frozen base or bed since proper bond will not be developed between the bed mortar and the frozen surface. If necessary, thaw the supporting surface by careful use of heat. Do not lay wet or frozen masonry units.

PROBLEMS

1. What is the nominal size and typical weight of a standard concrete block?
2. What purpose do expansion or control joints serve in masonry construction?
3. Describe the difference between a masonry cavity wall and a masonry bonded hollow wall.

4. Identify the principal types of masonry mortar and their use.

5. How many 3¾ × 2¼ × 8 in. (9.5 × 5.7 × 20.3 cm) bricks are required for a double wythe wall 8 ft high × 14 ft wide (2.44 × 4.27 m) having one opening 48 × 72 in. (122 × 183 cm) and one opening 32 × 48 in. (81 × 122 cm). Mortar joints are ½ in. (13 mm). Allow 2% for brick waste.

6. What purpose does a bond beam serve in masonry construction?

7. Find the maximum safe unsupported height of a 6-in. (15-cm) lightweight concrete block wall if the design wind load is 15 lb/sq ft (0.72 kPa).

8. Find the maximum safe unsupported height of an 8-in. (20-cm) solid brick wall for a wind velocity of 100 mi/h (161 km/h).

9. What minimum mortar cover is required for a No. 5 rebar from the exterior face of a reinforced brick masonry wall exposed to the weather?

10. Write a computer program to calculate the maximum safe height of an unbraced masonry wall under construction using Equation 14-1. Solve Problem 8 using your computer program.

REFERENCES

1. *Building Code Requirements for Concrete Masonry Structures* (ACI 531). American Concrete Institute, Detroit, Mich., 1981.
2. *Building Code Requirements for Engineered Brick Masonry.* Brick Institute of America, Reston, Va., 1969.
3. *Concrete Masonry Handbook.* Portland Cement Association. Skokie, Ill., 1985.
4. *Principles of Brick Masonry.* Brick Institute of America, Reston, Va., 1989.
5. *Recommended Practice for Engineered Brick Masonry.* Brick Institute of America, Reston, Va., 1975.
6. *Recommended Practices and Guide Specifications for Cold Weather Masonry Construction.* International Masonry Industry All-Weather Council, International Masonry Institute, Washington, D.C., 1984.
7. *Technical Notes on Brick Construction,* Series. Brick Institute of America, Reston, Va., various dates.

PART THREE
CONSTRUCTION MANAGEMENT

15
Planning and Scheduling

15-1
INTRODUCTION

Planning and Scheduling

As you already know, some planning must be done in order to perform any function with a minimum of wasted time and effort. This is true whether the function is getting to work on time or constructing a multimillion dollar building. A schedule is nothing more than a time-phased plan. Schedules are used as guides during the performance of an operation in order to control the pace of activities and to permit completion of the operation at the desired or required time.

Scheduling is utilized for many different phases of the construction process, from master planning through facility construction to facility operation and maintenance. In the construction phase itself, schedules are useful for a number of purposes before starting a project and after completion of the project as well as during the actual conduct of construction work. Some of the principal uses for schedules during each of these phases of construction are listed below.

Before Starting

1. Provides an estimate of the time required for each portion of the project as well as for the total project.
2. Establishes the planned rate of progress.
3. Forms the basis for managers to issue instructions to subordinates.

4. Establishes the planned sequence for the use of manpower, materials, machines, and money.

During Construction

1. Enables the manager to prepare a checklist of key dates, activities, resources, and so on.
2. Provides a means for evaluating the effect of changes and delays.
3. Serves as the basis for evaluating progress.
4. Aids in the coordination of resources.

After Completion of Construction

1. Permits a review and analysis of the project as actually carried out.
2. Provides historical data for improving future planning and estimating.

Scheduling Principles

There are a number of different forms of schedules that may be used, including written schedules, bar graph schedules, network schedules, and others. In this chapter, we will consider only bar graph, network, and linear scheduling methods. Regardless of the scheduling method employed, the following general principles of scheduling should be observed.

1. Establish a logical sequence of operations.
2. Do not exceed the capabilities of the resources that are available.
3. Provide for continuity of operations.
4. Start project controlling (or critical) activities early.

It must be recognized that the accuracy or validity of scheduling depends on the validity of the work quantity and productivity estimates used. The accuracy of an estimate of the time it will take to perform a construction operation is a function of the kind of work involved and prior experience in doing that sort of work. For example, one expects a more accurate estimate of the time required to install a wastewater line in a residential structure than of the time required to install a cooling line in a nuclear power plant. Methods for dealing with the uncertainty associated with activity-time estimates will be discussed later in this chapter.

In addition to valid time estimates for activities, the planner must have a thorough understanding of the nature of the work to be performed and the relationships between the various work items making up the project. One of the major deficiencies of the bar graph schedule described in the following section is the fact that the bar graph fails to show relationships between work items. That is, what activities must be started or completed before other activities can be started or completed?

15-2 BAR GRAPH METHOD

The Bar Graph Schedule

The *bar graph* or *bar chart schedule* is a graphical schedule relating progress of items of work to a time schedule. The bar schedule traces its origin to a chart developed by Henry L. Gantt, a pioneer in the application of scientific management methods to industrial production. These charts, referred to as *Gantt charts,* took several different forms, depending on their application. Because of their origin, all forms of bar graph schedules are sometimes called Gantt charts. In spite of the advent of network planning methods, the bar graph schedule is still the most widely used schedule form found in construction work. Its continued popularity in the face of the significant deficiencies that are described in the next section is undoubtedly due to its very graphic and easily understood format. A simple bar graph schedule for a construction project is shown in Figure 15-1. The major work items, or activities, making up the project are listed on the left side of the schedule with a time scale across the top. The column headed "Hours" indicates the estimated number of labor-hours required for each activity. The column headed "Weight" indicates the portion of the total project effort accounted for by each activity. For example, "Clearing and Stripping" requires 750 labor-hours of work, which represents 4.7% of the 15,900 labor-hours required for the entire project. While a weighting column is not always present on a bar chart, its presence is very useful when calculating cumulative project progress. Activities may be weighted on any desired basis. However, dollar value and labor-hours are most frequently used as a weighting basis.

Notice that two horizontal blocks are provided opposite each activity. The upper block (SCH) represents scheduled progress and the lower block (ACT) is used to record actual progress as work proceeds. For each block, a bar extends from starting to ending times. The numbers above each bar indicate percentage of activity completion at each major time division. Again such a system greatly simplifies calculation of scheduled cumulative progress and its comparison with actual progress. To aid in the evaluation of progress, it is suggested that the actual progress of each activity be inserted at the end of each major time period, as shown in Figure 15-1.

Cumulative Project Progress

Figure 15-2 shows a cumulative progress-versus-time curve for the bar graph of Figure 15-1. The vertical scale represents cumulative project progress in percent and the horizontal scale indicates time. Once the bar graph schedule has been prepared and weighting factors calculated for each activity, scheduled cumulative progress can be calculated and plotted as shown on the figure. Actual cumulative progress is calculated and plotted as work progresses. To construct the scheduled cumulative progress curve, scheduled cumulative progress must be calculated and plotted for a sufficient number of points to enable a smooth curve to be drawn. In Figure 15-2 scheduled cumulative progress for the bar graph schedule of Figure 15-1 has been calculated and plotted at the end of each week. Cumulative progress may be calculated as follows:

$$\text{Cumulative progress} = \sum_{i=1}^{n} (\text{Activity progress})_i \times (\text{Weight})_i \qquad (15\text{-}1)$$

Example calculations for the scheduled cumulative progress for the first 3 weeks of the project whose bar graph schedule appears in Figure 15-1 are given below.

Construction Progress Chart

Project *Construct Runway xyz*

Date *4/10/99*

Operation	Hours	Weight (%)		Week 1	Week 2	Week 3	Week 4	Week 5	Week 6	Week 7	Week 8	Week 9	Week 10
1. Clear and Strip	750	4.7	SCH		20	70							
			ACT			45	90						
2. Drainage	150	0.9	SCH			60							
			ACT			30	100						
3 Subgrade	4200	26.4	SCH			15	40	65	85				
			ACT			5	25						
4. Base Course	4000	25.2	SCH					10	30	60	85		
			ACT										
5. Pave	6800	42.8	SCH						20	45	65	80	95
			ACT										
			SCH										
			ACT										
			SCH										
			ACT										
			SCH										
			ACT										
			SCH										
			ACT										
			SCH										
			ACT										

Figure 15-1 Bar graph schedule for construction project.

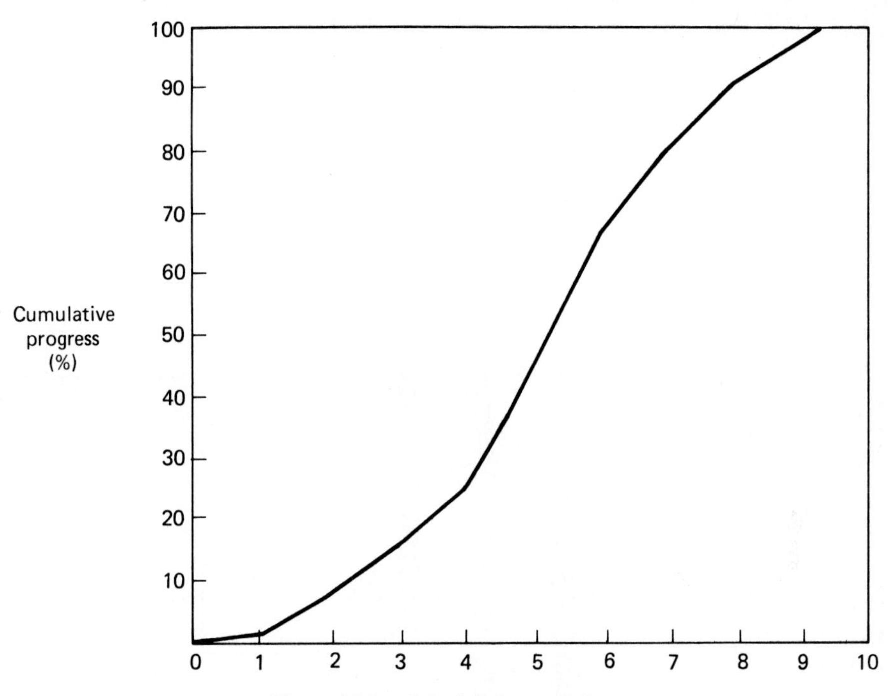

Figure 15-2 Scheduled cumulative progress.

End of First Week

$$\text{Progress} = \underset{[\text{Activity 1}]}{(0.20 \times 4.7)} = 0.9\%$$

End of Second Week

$$\text{Progress} = \underset{[\text{Activity 1}]}{(0.70 \times 4.7)} + \underset{[\text{Activity 2}]}{(0.60 \times 0.9)} + \underset{[\text{Activity 3}]}{(0.15 \times 26.4)} = 7.8\%$$

End of Third Week

$$\text{Progress} = \underset{[\text{Activity 1}]}{(1.00 \times 4.7)} + \underset{[\text{Activity 2}]}{(1.00 \times 0.9)} + \underset{[\text{Activity 3}]}{(0.40 \times 26.4)} = 16.2\%$$

Frequently, cumulative progress curves for a project are superimposed on the project's bar graph schedule as illustrated in Figure 15-3.

The Normal Progress Curve

At this point, let us consider the probable shape of a cumulative progress-versus-time curve. Observation of a large number of projects indicates that the usual shape of the curve is that shown in Figure 15-4. As the curve indicates, progress is slow at

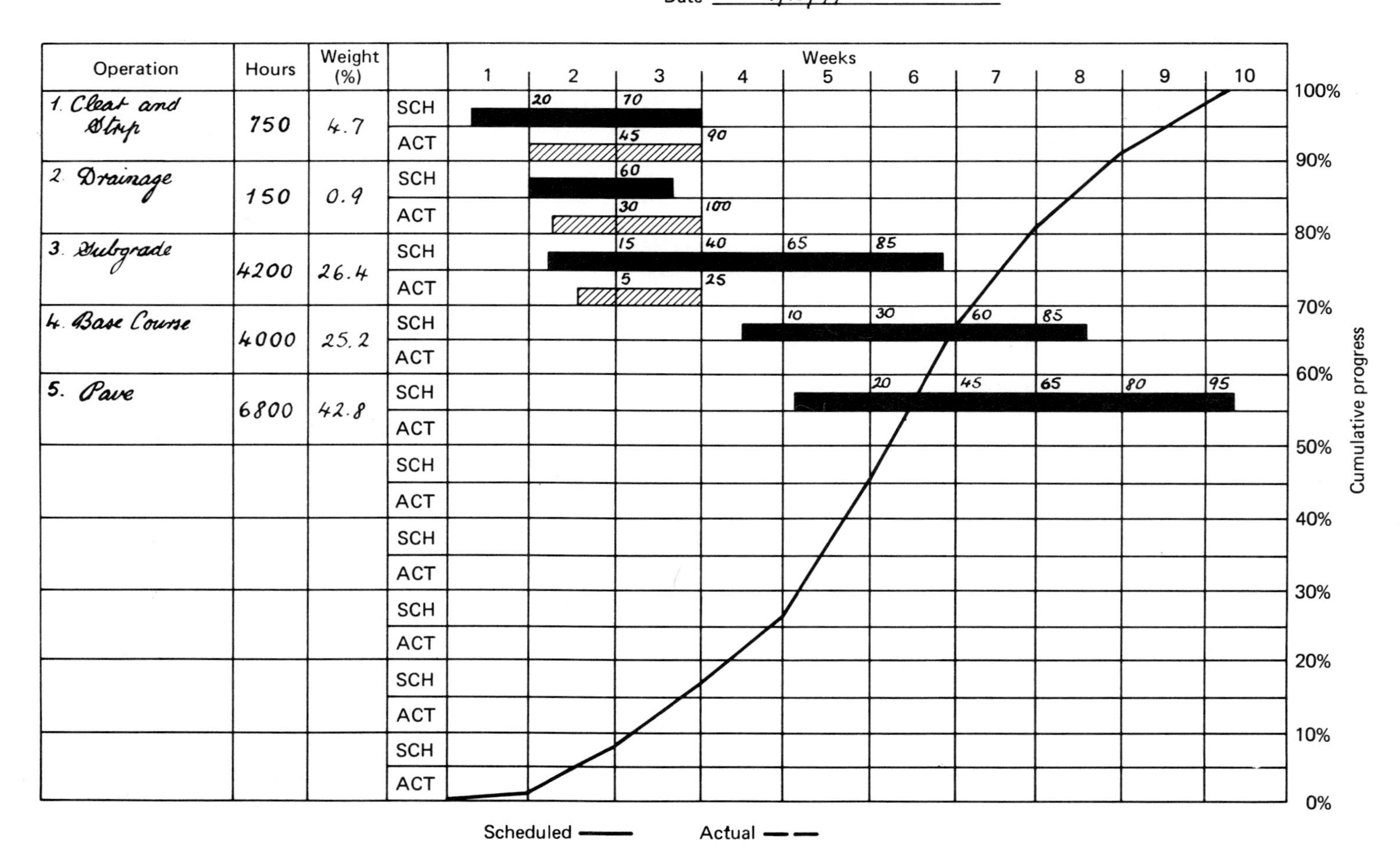

Figure 15-3 Bar graph with cumulative progress curve.

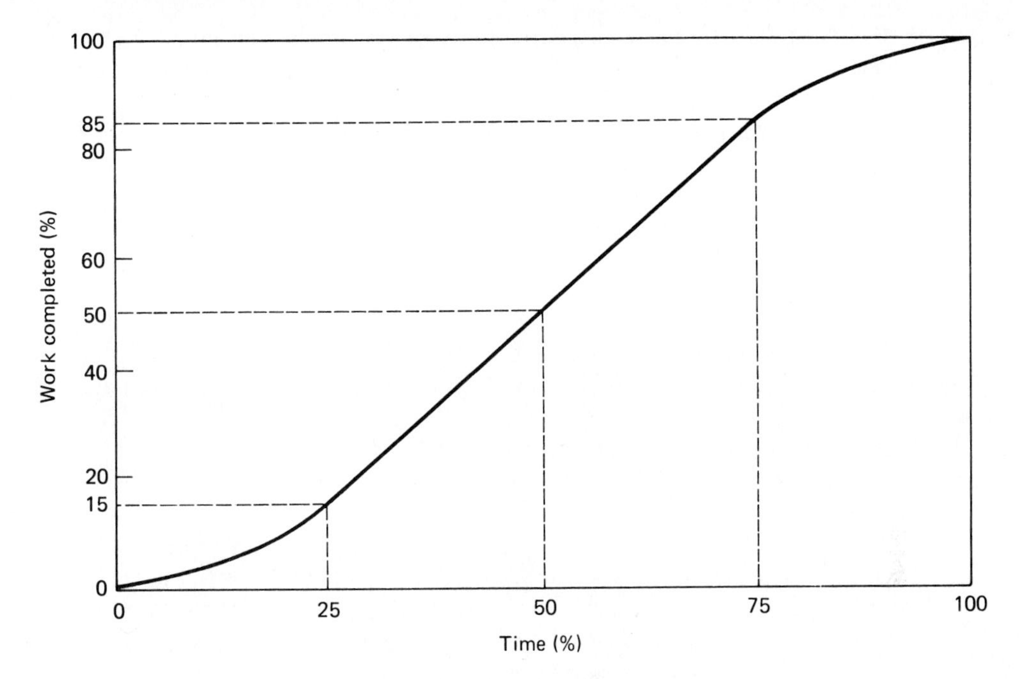

Figure 15-4 Normal progress curve.

the beginning of a project as work is organized and workers become familiar with work assignments and procedures. Thus, only about 15% of the project is completed in the first 25% of project time. After that, progress is made at a rather constant rate until 85% of the work is completed at the end of 75% of project time. Progress again slows as finishing work and project demobilization takes place. The progress curve illustrated in Figure 15-4 is referred to as a *normal progress curve* or *S-curve* and will generally apply to any type of nonrepetitive work. If you find that the shape of a scheduled cumulative progress curve deviates substantially from the curve of Figure 15-4, you should carefully investigate the reason for this deviation. If progress is based on dollar value, the presence of a few high-dollar-value items may cause the curve to assume an abnormal shape. Otherwise, it is likely that a mistake has been made and that the planned rate of progress is unrealistic.

15-3 CPM—THE CRITICAL PATH METHOD

Deficiencies in Bar Graph Schedules

As indicated earlier, a major deficiency of the bar graph schedule is its failure to show relationship between project activities. Thus there is no way to determine from a bar graph schedule whether the person preparing the schedule was, in fact, aware of these relationships. A related weakness of the bar graph schedule is its failure to identify those activities which actually control the project duration. We will refer to such duration-controlling activities as *critical activities.* As a result of its failure to identify activity relationships and critical activities, the bar graph schedule also fails to show the effect of delay or change in one activity on the entire project. Recognizing these weaknesses in bar graph schedules, planners have for a number of years attempted to devise improved planning and scheduling methods.

However, it was not until the development of network planning methods in 1957–1958 that a major improvement in planning and scheduling methods took place. During this period the *Critical Path Method (CPM)* was developed jointly by the DuPont and Remington Rand Companies as a method for planning and scheduling plant maintenance and construction projects utilizing computers. At almost the same time, the Special Projects Office of the U.S. Navy, with Booze, Hamilton, and Allen as consultants, was developing the *Program Evaluation and Review Technique (PERT)* for planning and controlling weapons systems development. Successful application of both CPM and PERT by their developers soon led to widespread use of the techniques on both governmental and industry projects.

Both CPM and PERT use a network diagram to graphically represent the major activities of a project and to show the relationships between activities. The major difference between CPM and PERT is that PERT utilizes probability concepts to deal with the uncertainty associated with activity-time estimates, whereas CPM assigns each activity a single fixed duration.

The Network Diagram

As indicated, a network graphically portrays major project activities and their relationships. There are basically two methods of drawing such networks: the *activity-on-arrow diagram* and the *activity-on-node diagram.* Special forms of activity-on-node diagram, such as *precedence diagrams,* will be discussed later in the chapter. While activity-on-node diagrams have certain advantages, activity-on-arrow format will be utilized to illustrate network construction and time calculations.

In the activity-on-arrow format, each activity is represented by an arrow that has an associated description and expected duration. Each *activity,* as illustrated in Figure 15-5, must start and terminate at an *event* (represented by a circle). Events are numbered for identification purposes and event numbers are also utilized to identify activities on the diagram. That is, activities are identified by citing the event

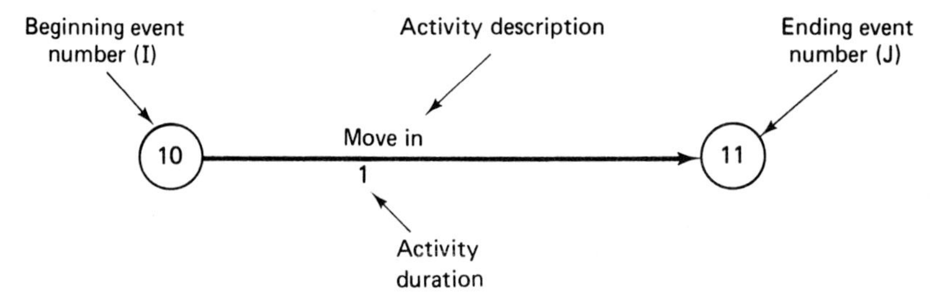

Figure 15-5 Activity-on-arrow notation.

number at the tail of the arrow (I number) followed by the event number at the head of the arrow (J number). Thus activity 10-11 refers to the activity starting at event 10 and ending at event 11, as seen in Figure 15-5. This activity numbering system is referred to as the *I-J numbering system*. An event is simply a point in time and, as used in network diagramming, is assumed to occur instantaneously when all activities leading into the event have been completed. Similarly, all activities leading out of an event *may* start immediately upon the occurrence of an event. Figure 15-6 shows a simple network diagram for a construction project. As mentioned earlier, the diagram graphically indicates the relationship between activities. These relationships are *precedence* (what activities must precede the activity?), *concurrence* (what activities can go on at the same time?), and *succession* (what activities must follow the activity?). In Figure 15-6, activity 1-2 must precede activity 2-5, Activities 1-2 and 1-3 are concurrent, and activity 2-5 succeeds activity 1-2. Activities progress in the direction shown by the arrows. Good diagramming practice requires that diagrams present a clear picture of the project logic and generally flow from left to right. Arrows should not point backward, although they may point straight up or down.

Notice the dashed arrow in Figure 15-6. This is called a *dummy activity* or simply a *dummy.* Dummies are used to impose logic constraints and prevent duplication of activity I-J numbers. They do not represent any work and, hence, always have a duration of zero.

Event-Time Calculations

Once a network diagram has been drawn that represents the required relationships between activities, network time calculations may be made. The first step is to calculate the earliest time at which each event may occur based on an arbitrary starting time of zero. This earliest event occurrence is referred to as *early event time,* com-

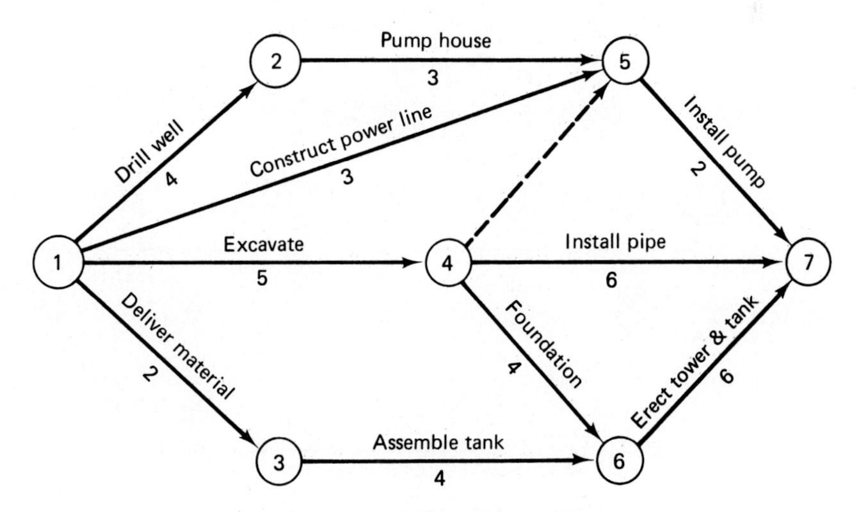

Figure 15-6 Example network diagram.

monly abbreviated *EET.* It is usually placed above the event circle as shown in Figure 15-7. Calculations then proceed from left to right, starting with 0 at the first event. This calculation is referred to as the forward pass through the network. At each event the early event time is found as the early event time of the previous event plus the duration of the activity connecting the two events. Thus the early event time of event 2 is found as the sum of the early event time at event 1 plus the duration of activity 1-2 ($0 + 4 = 4$). When two or more activity arrows meet at an event, the largest value of possible early event times is chosen as the early event time because, by definition, the event cannot occur until all activities leading into the event have been completed. In Figure 15-7, note the activity early completion times of 3, 5, and 7 at event 5, leading to the proper early event-time value of 7. The early event time at the last event is, of course, the minimum time required to complete the project.

When all early event-time values have been calculated and entered on the network, a backward pass is made to compute the latest possible time at which each event may occur without changing the project duration. As a starting point, the *late event time (LET)* of the last event is set equal to the early event time of the event. Starting with the assigned late event time at the last event, work backward through the network, calculating each late event time as the late event time of the previous event minus the duration of the activity connecting the events. The results are illustrated in Figure 15-8. The late event time of event 6 is found as the late event time of event 7 minus the duration of activity 6-7 ($15 - 6 = 9$). When two or more activities meet at an event, the lower of possible times is chosen as the late event time because, by definition, the event must occur before any activity leading out of the event may start. In order for all activities to be completed within the allotted time, the event must occur at the earliest of the possible time values. In Figure 15-8, note the possible late times at event 4 of 5, 9, and 13, leading to a late event time of 5.

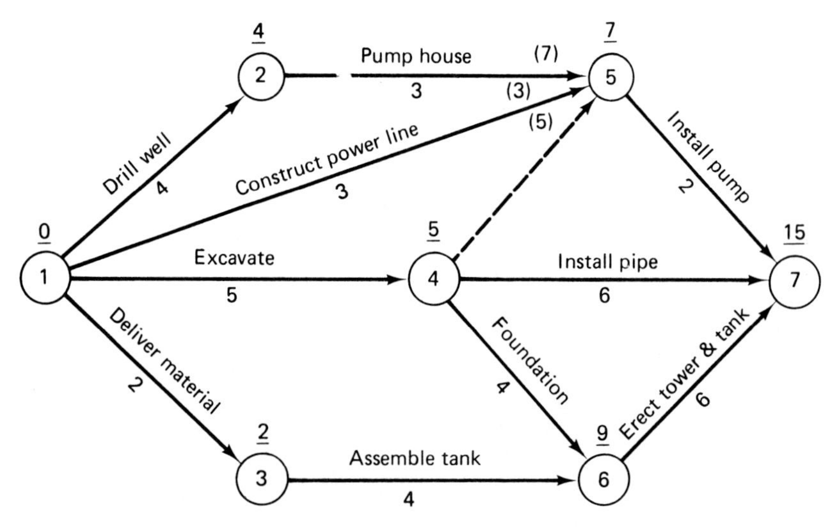

Figure 15-7 Example network—early event times.

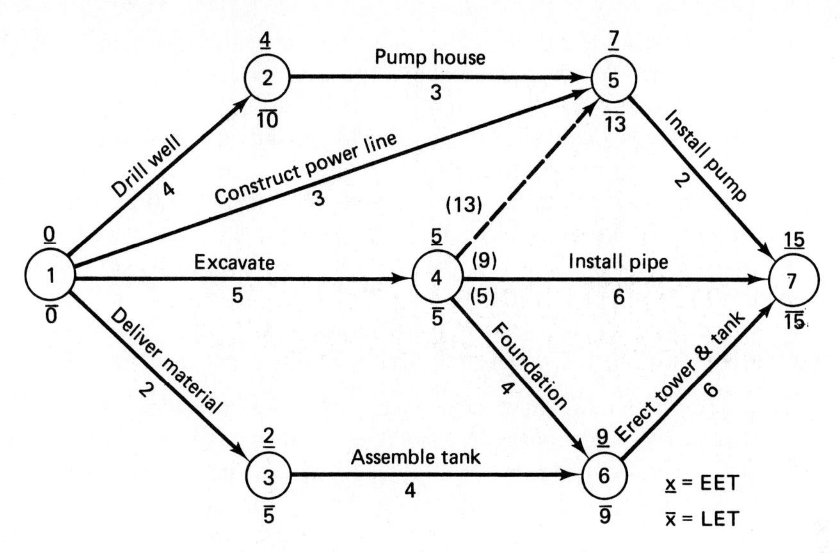

Figure 15-8 Example network—late event times.

The Critical Path

That path through the network which establishes the minimum project duration is referred to as the *critical path.* This path is the series of activities and events that was used to determine the project duration (or early event time of the final event) in the forward pass. However, it is usual to wait until all early and late event times have been calculated to mark the critical path. Notice in Figure 15-9 that the critical path passes through all events whose early event times are equal to their late event times.

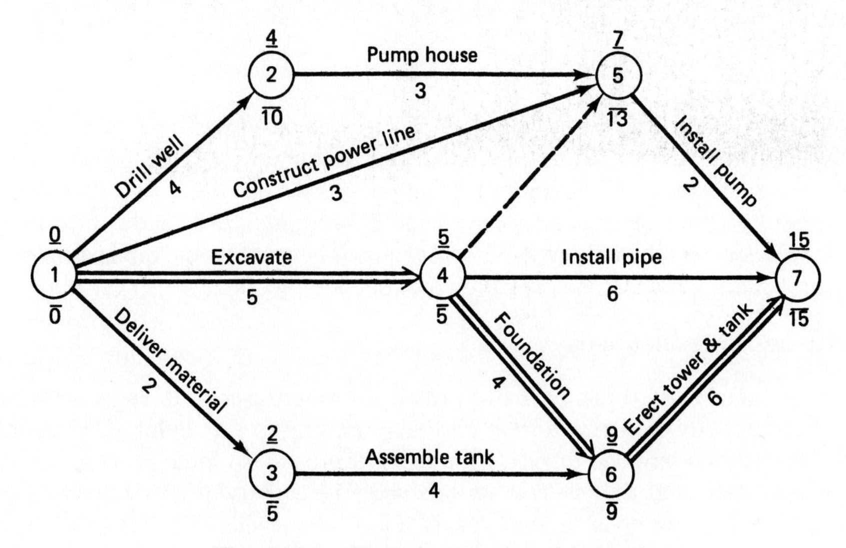

Figure 15-9 Example network—critical path.

Critical activities are those which make up the critical path and, of course, connect critical events. Where parallel activities connect critical events, however, only the activities whose duration equals the difference between event times at the ends of the arrow are critical. Thus in Figure 15-9, activities 4-7, as well as activities 4-6 and 6-7, connect the critical events 4 and 7. However, the time difference between event times at event 7 and event 4 is 10 units, while the duration of activity 4-7 is only 6 units. Hence activity 4-7 is *not* critical. When the critical path has been identified, it should be clearly indicated on the network by color, double arrow (as used in Figure 15-9), or similar means.

Activity Times

Up to this point, we have determined the minimum duration of our project and identified the critical path. The next step is to calculate the earliest and latest starting and finishing time and the total float (scheduling leeway) for each activity based on the event times already calculated. These time values are used as the basis for scheduling and resource allocation. Two of these values, early start and late finish, may be read directly off the network while the remaining values must be calculated. The following relations may be used to determine activity times

$$\text{Early start (ES)} = \text{Early event time of preceding (I) event} \quad (15\text{-}2)$$

$$\text{Early finish (EF)} = \text{Early start} + \text{Activity duration} \quad (15\text{-}3)$$

$$\text{Late finish (LF)} = \text{Late event time of following (J) event} \quad (15\text{-}4)$$

$$\text{Late start (LS)} = \text{Late finish} - \text{Activity duration} \quad (15\text{-}5)$$

$$\text{Total float (TF)} = \text{Late finish} - \text{Early finish} \quad (15\text{-}6)$$

or

$$\text{Total float (TF)} = \text{Late start} - \text{Early start} \quad (15\text{-}7)$$

Activity-time values for the example network are given in Figure 15-10. Note that activity times are not usually calculated for dummy activities. *Float* (*slack* in PERT terminology) is the amount of scheduling leeway available to an activity. While several different types of float have been defined, *total float* is the most useful of these values and is the only type of float that will be used here. Application of float to the scheduling process is covered in the following section.

Activity-on-Node Diagrams

As stated earlier, there are two principal formats used in drawing network diagrams. The activity-on-arrow format has been used up to this point. The second format is the activity-on-node format. This technique uses the same general principles of network logic and time calculations as does the activity-on-arrow technique. However, the activity-on-node network diagram looks somewhat different from the activity-on-arrow diagram because the node (which represented an event in the activity-on-arrow

	Act no.	Description	Duration	Early start	Early finish	Late start	Late finish	Total float
	1-2	Drill well	4	0	4	6	10	6
	1-3	Deliver material	2	0	2	3	5	3
*	1-4	Excavate	5	0	5	0	5	0
	1-5	Power line	3	0	3	10	13	10
	2-5	Pump house	3	4	7	10	13	6
	3-6	Assemble tank	4	2	6	5	9	3
*	4-6	Foundation	4	5	9	5	9	0
	4-7	Install pipe	6	5	11	9	15	4
	5-7	Install pump	2	7	9	13	15	6
*	6-7	Erect tower and tk	6	9	15	9	15	0

* = critical activity

Figure 15-10 Activity-time data for example network.

method) is now used to represent an activity. A simple form of the activity-on-node diagram is the *circle diagram* or *circle notation,* in which each activity is represented by a circle containing the activity description, an identifying number, and the activity duration.

Figure 15-11 illustrates a circle diagram for a five-activity construction project. In the activity-on-node technique, notice that arrows are used to represent logic constraints only. Thus all arrows act in the same manner as do dummies in the activity-on-arrow format. This feature has been found to make activity-on-node diagramming somewhat easier for beginners to understand. The principal disadvantage of activity-on-node diagramming has been the limited availability of computer programs for performing network time calculations. However, there are now a number of such programs available, and the use of the activity-on-node techniques is expected to increase.

Another form of activity-on-node diagram is illustrated in Figure 15-12. Here an enlarged node is used to provide space for entering activity time values directly on the node. This format is particularly well suited to manual network calculation, because activity times may be entered directly on the network as they are calculated. When time calculations are performed in this manner, it is suggested that calculations be performed independently by two individuals and the results compared as an error check.

The third form of activity-on-node diagram is the *precedence diagram.* Because of its special characteristics, it is described in greater detail in the following paragraphs.

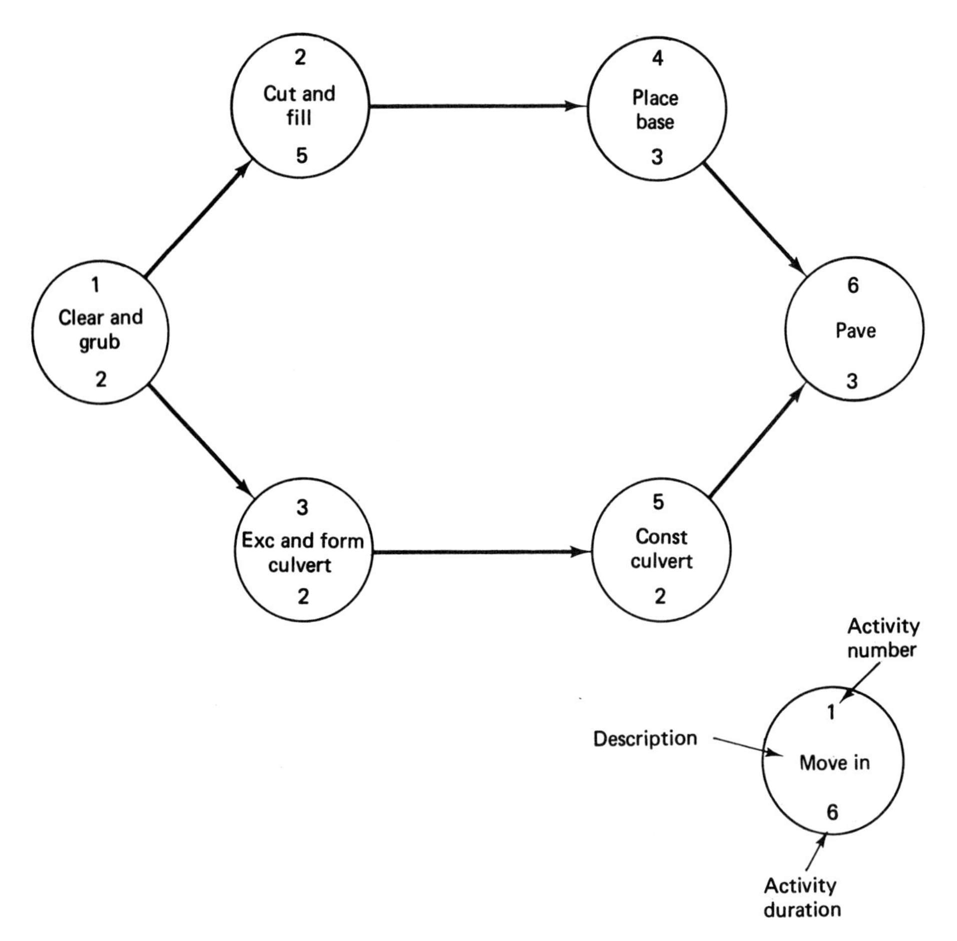

Figure 15-11 Circle diagram for a project.

PRECEDENCE DIAGRAMS

The precedence diagram is an extension of the activity-on-node format that provides for incorporation of lag-time factors as well as permitting additional precedence relationships. The use of lag time is very useful when diagramming construction project relationships, where activities can often start as soon as a portion of a preceding activity is completed. In addition to the usual finish-to-start precedence relationship, this technique permits start-to-start and finish-to-finish relationships. These relationships and the use of lag times are illustrated in Figure 15-13.

To appreciate the value of incorporating lag-time relationships, it is useful to consider how such relationships could be represented in the usual network diagramming techniques. For example, consider the network of Figure 15-12. The planner decides that activity 2, cut and fill, can start when activity 1, clear and grub, is 50% complete (equivalent to 1 day's work). Figure 15-14 illustrates how this situation would be represented in both conventional CPM and in a precedence diagram. To represent this situation in conventional CPM, it is necessary to split activity 1 into

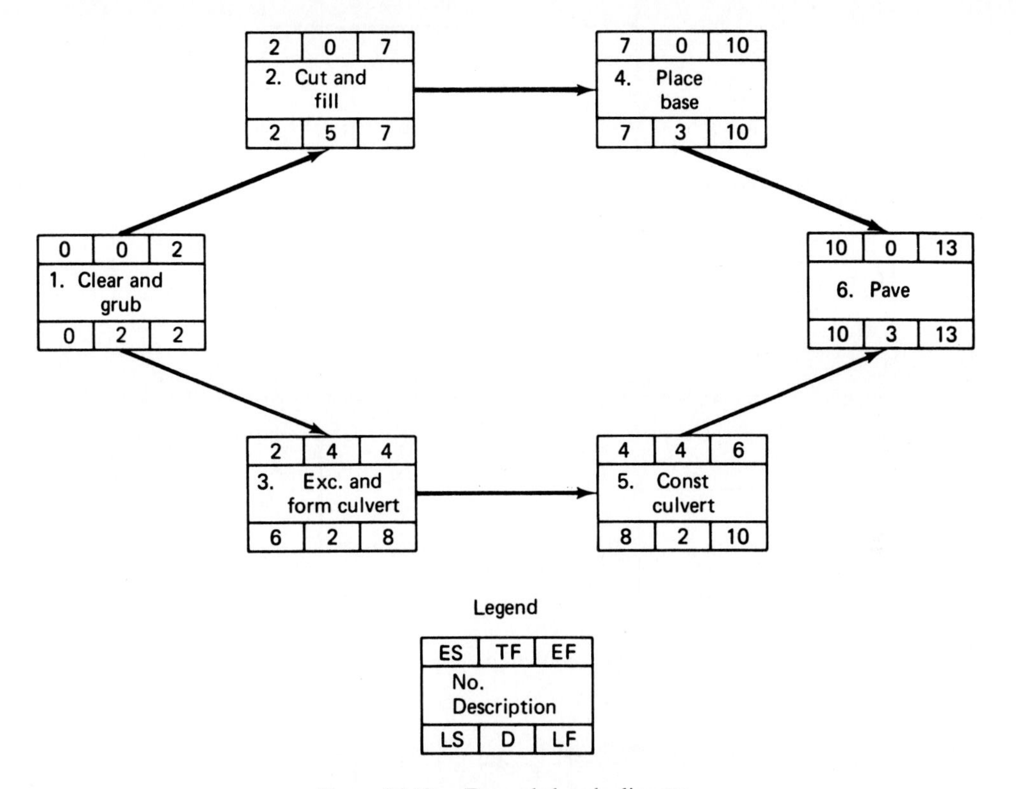

Figure 15-12 Expanded node diagram.

two activities, each having a duration of 1 day (Figure 15-14a). Using precedence diagram procedures, a 1-day lag time is simply inserted in a start-to-start relationship from activity 1 to activity 2 (Figure 15-14b).

The precedence diagram for the example project is shown in Figure 15-15. Notice that this diagram is essentially the same as any other activity-on-node diagram for the project since only finish-to-start relationships are employed and no lag times are used.

Suppose that we now add the following logic constraints to the example project:

1. Erection of the tower and tank cannot begin until 3 days after completion of the foundation.
2. Installation of pump cannot be completed until 1 day after completion of pipe installation.
3. The foundation can start 3 days after start of excavation.

The precedence diagram for the revised project is shown in Figure 15-16. Notice that the early start of activity 8 (day 3) is the early start of activity 4 (day 0) plus a 3-day lag time. The early finish of activity 10 (day 12) is determined by the early finish of activity 7 (day 11) plus a 1-day lag time. The early start of activity 10 is, therefore, day 10 (early finish minus duration). The early start of activity 11 (day 10) is the early finish of activity 8 (day 7) plus a lag time of 3 days. As noted earlier, the

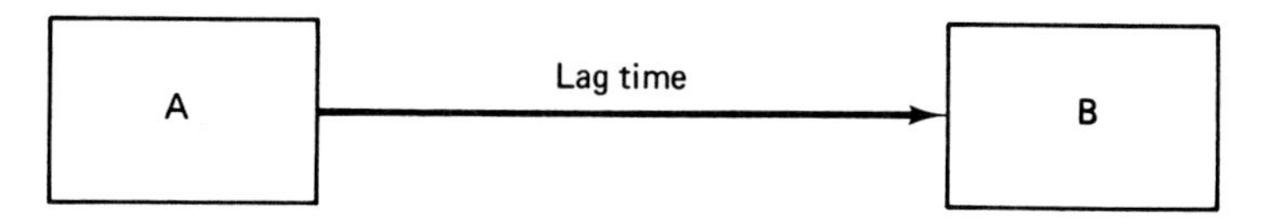

a. Finish-to-start: start of B depends on finish of A plus lag time.

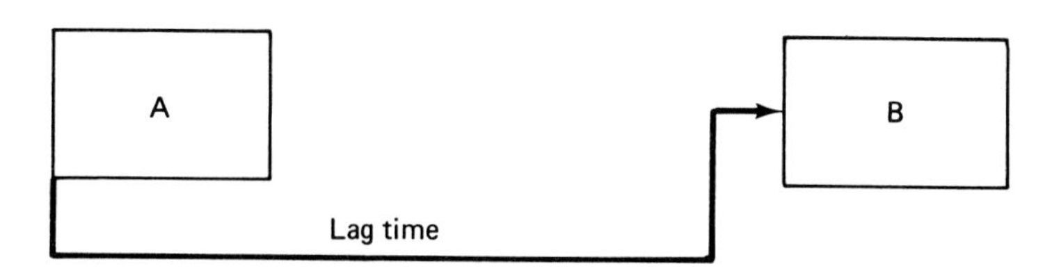

b. Start-to-start: start of B depends on start of A plus lag time.

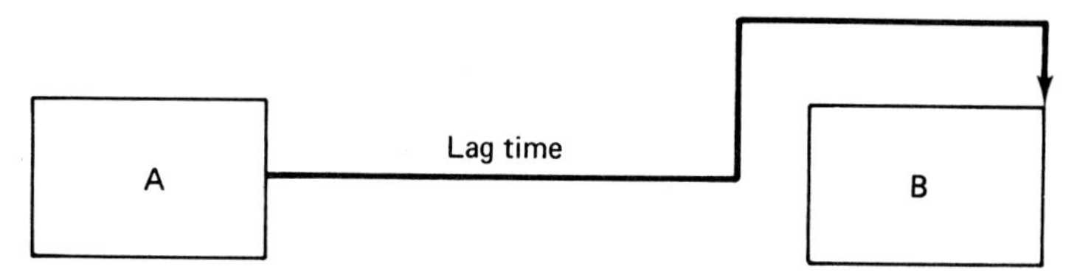

c. Finish-to-finish: finish of B depends on finish of A plus lag time.

Figure 15-13 Precedence diagram relationships.

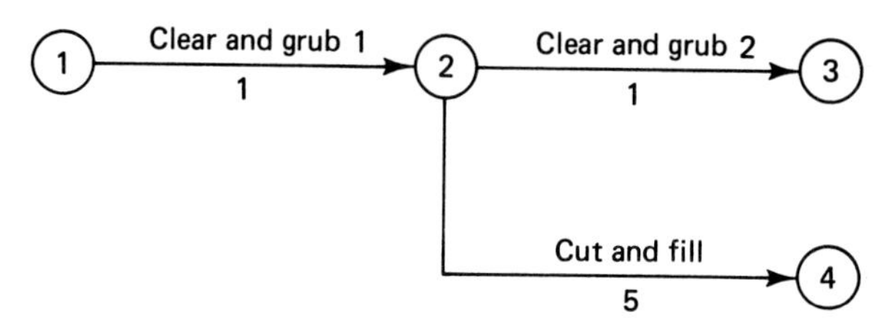

a. Activity-on-arrow diagram

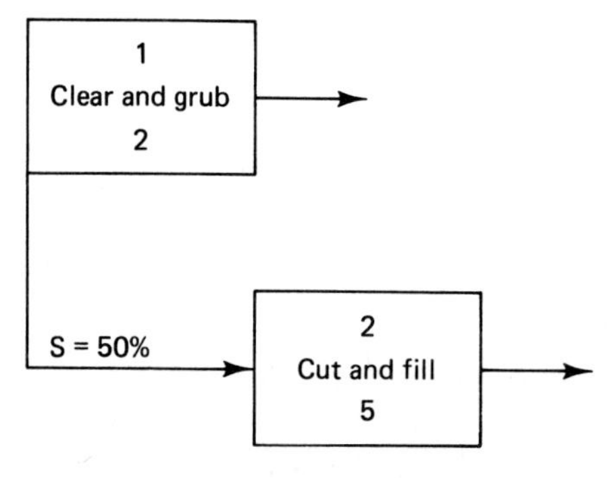

b. Precedence diagram equivalent to (a)

Figure 15-14 Comparison of CPM diagram and precedence diagram.

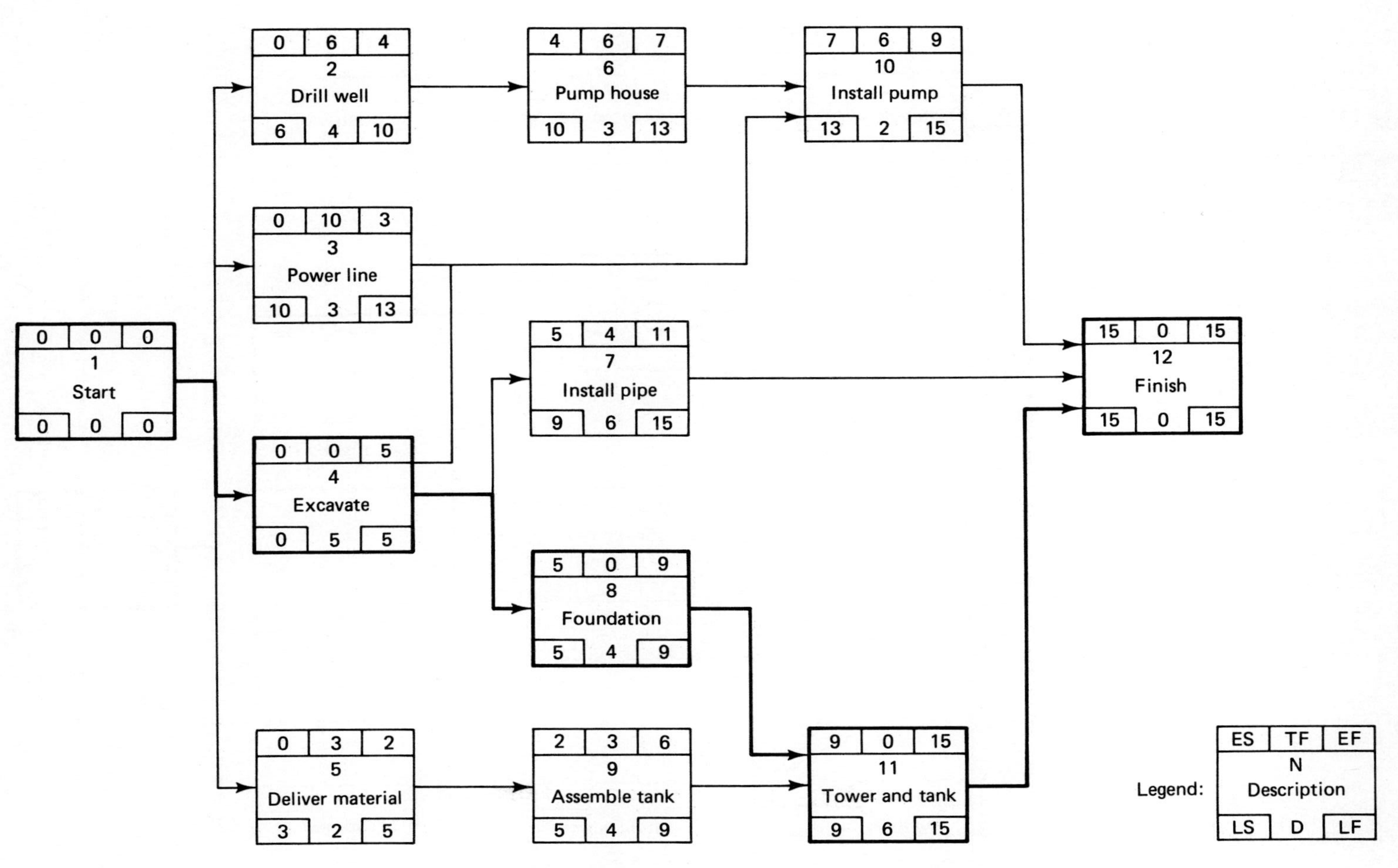

Figure 15-15 Precedence diagram for example project.

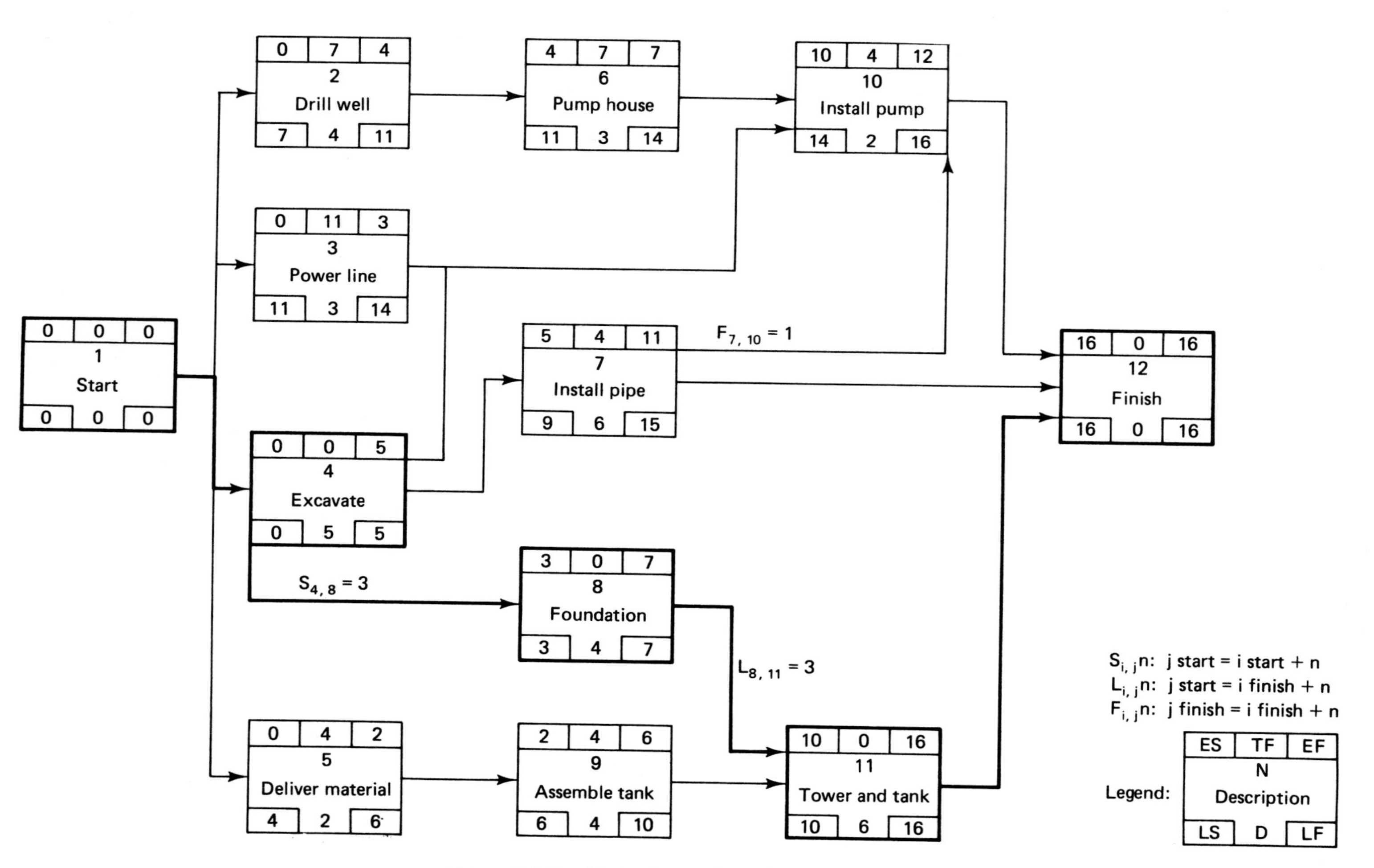

Figure 15-16 Revised example project.

increased flexibility of the precedence diagram comes at the cost of increased computational complexity.

15-4 SCHEDULING AND RESOURCE ASSIGNMENT USING CPM

The Early Start Schedule

The activity times calculated in Section 15-3 form the basis for a project schedule but in themselves do *not* constitute a schedule. For example, in Figure 15-10, activity 1-5, which has a duration of 3 days, has an early start time of 0 and a late finish time of 13 with 10 days of float. Thus activity 1-5 may be scheduled to occur on any 3 days between the beginning of day 1 and the end of day 13 without changing the project duration of 15 days.

When all activities are scheduled to start at the earliest allowable time, such a schedule is referred to as an *early start schedule.* To produce a schedule based on a calculated network, it is suggested that a time line first be drawn between the early start time and late finish time for each activity. Figure 15-17 illustrates this procedure applied to the example network of Figure 15-9. Note that the line starts at the end of the early start time tabulated in Figure 15-10 and extends to the end of the late finish time. For activity 1-2, the time line extends from time 0 (beginning of day 1) through time 10 (end of day 10). Notice also that the time has been filled in solid for activities on the critical path (activities 1-4, 4-6, and 6-7). This serves to warn the scheduler that these activities can only be scheduled at the time indicated unless the project duration is to be changed.

Each activity may now be scheduled at any position desired on the time line. If all activities are started at the beginning of their time line, the early start schedule of Figure 15-18 is produced. Here each workday is indicated by an asterisk while each day of float is represented by the letter F. Float may be used to rearrange the schedule as desired by the scheduler without changing project duration.

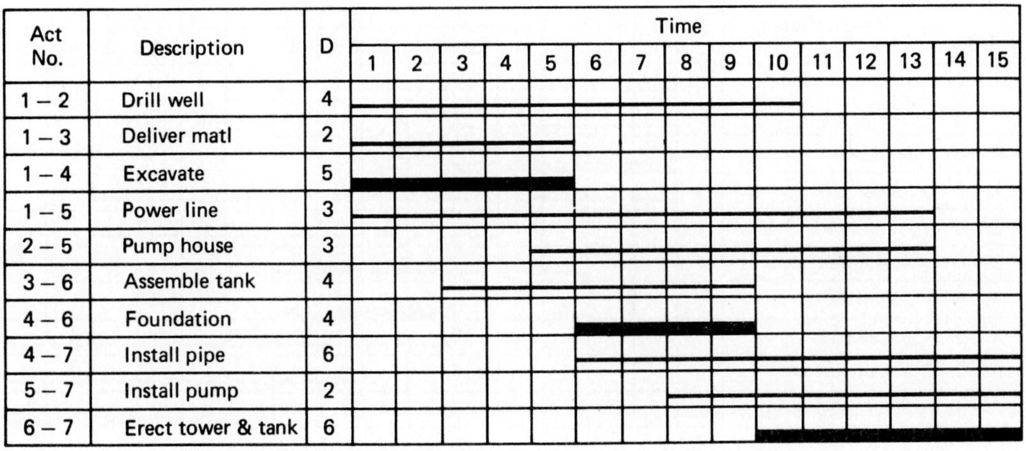

Act No.	Description	D	Time														
			1	2	3	4	5	6	7	8	9	10	11	12	13	14	15
1 – 2	Drill well	4															
1 – 3	Deliver matl	2															
1 – 4	Excavate	5															
1 – 5	Power line	3															
2 – 5	Pump house	3															
3 – 6	Assemble tank	4															
4 – 6	Foundation	4															
4 – 7	Install pipe	6															
5 – 7	Install pump	2															
6 – 7	Erect tower & tank	6															

Figure 15-17 Allowable activity time span for example network.

Some of the uses for float in scheduling are to incorporate preferential logic, to satisfy resource constraints, and to allow weather-sensitive activities to be scheduled when weather conditions are expected to be most favorable. *Preferential logic* is that network logic which is imposed by the planner solely because the planner prefers to conduct the operation in that sequence. In other words, it is not logic imposed by the fundamental nature of the process. An example would be the scheduling of all concreting activities in sequence so that only one concrete crew would be required for the project.

Late Start and Other Schedules

When all activities are started at their latest allowable starting time, a *late start schedule* is produced, as shown in Figure 15-19. Note that all float is used before the activity starts. An obvious disadvantage to the use of such a schedule is that it leaves no time cushion in the event that an activity requires longer than its estimated duration. In practice, the usual schedule is neither an early start nor a late start schedule. Rather, it is an intermediate schedule produced by delaying some activities to permit resource leveling or the incorporation of preferential logic while retaining as much float as possible.

When producing a schedule other than an early start schedule, care must be taken to ensure that no activity is scheduled to start before its predecessor event has occurred. This, of course, would be a violation of network logic. Such an error may be prevented by referring to the network diagram each time an activity is scheduled. However, a simple technique makes use of the activity numbers on the schedule to check logic constraints. Referring to Figures 15-17 and 15-9, we see activity 2-5 may be started as soon as all activities ending with event 2 (J number = 2) have been completed. In this case, this is only activity 1-2. In using this technique, some provision must be made for incorporating the logic constraints imposed by dummies. This may be done by putting a third number in parentheses after the usual activity I-J number.

Act No.	Description	D	Time														
			1	2	3	4	5	6	7	8	9	10	11	12	13	14	15
1 – 2	Drill well	4	*	*	*	*	F	F	F	F	F	F					
1 – 3	Deliver matl	2	*	*	F	F	F										
1 – 4	Excavate	5	*	*	*	*	*										
1 – 5	Power line	3	*	*	*	F	F	F	F	F	F	F	F	F	F		
2 – 5	Pump house	3					*	*	*	F	F	F	F	F	F		
3 – 6	Assemble tank	4			*	*	*	*	F	F	F						
4 – 6	Foundation	4						*	*	*	*						
4 – 7	Install pipe	6						*	*	*	*	*	*	F	F	F	F
5 – 7	Install pump	2								*	*	F	F	F	F	F	F
6 – 7	Erect tower & tank	6										*	*	*	*	*	*

* = work day F = float

Figure 15-18 Early start schedule for example network.

Act no.	Description		Time														
			1	2	3	4	5	6	7	8	9	10	11	12	13	14	15
1 – 2	Drill well	4	F	F	F	F	F	F	*	*	*	*					
1 – 3	Deliver matl	2	F	F	F	*	*										
1 – 4(5)	Excavate	5	*	*	*	*	*										
1 – 5	Power line	3	F	F	F	F	F	F	F	F	F	F	*	*	*		
2 – 5	Pump house	3					F	F	F	F	F	F	*	*	*		
3 – 6	Assemble tank	4			F	F	F	*	*	*	*						
4 – 6	Foundation	4						*	*	*	*						
4 – 7	Install pipe	6						F	F	F	F	*	*	*	*	*	*
5 – 7	Install pump	2								F	F	F	F	F	F	*	*
6 – 7	Erect tower & tank	6										*	*	*	*	*	*

* = work day F = float

Figure 15-19 Late start schedule for example network.

The number in parentheses represents the event number at the end of the dummy. For example, in Figure 15-19, note that activity 1-4 has been identified as activity 1-4 (5). Here the number 5 indicates that activities starting with number 5 (I number = 5) cannot begin until activity 1-4 is completed. Reference to the network diagram of Figure 15-9 shows that this is the correct logic and is the result of the presence of dummy 4-5. Thus the real predecessors of activity 5-7 are activities 1-4, 1-5, and 2-5.

Resource Assignment

In planning the assignment of resources to a project, the planner is usually faced with two major considerations. For each type of resource, these are (1) the maximum number of resources available during each time period, and (2) the desire to eliminate peaks and valleys in resource requirements (i.e., resource leveling).

If the asterisk designating a workday in Figures 15-18 and 15-19 is simply replaced by a number representing the quantity of the resource required for the activity during that time period, it is a simple matter to determine the total quantity of the resource required for each time period for any particular schedule. Thus Figure 15-20 illustrates the number of workers needed on each day for the early start schedule (Figure 15-18) of the example network. The daily labor requirements are rather uneven, varying from 19 workers on the first 2 days to 8 workers on the twelfth day. Unless the contractor has other nearby projects that can utilize the excess labor produced by these fluctuations, labor problems would soon develop. A far better policy would be to attempt to level out the daily labor requirements. This can often be done by simply utilizing float to reschedule activities.

A quick calculation will indicate that the total resource requirement for the example network indicated in Figure 15-20 is 195 worker days. This yields an average requirement of about 13 workers per day. By utilizing float to reschedule activities, the revised schedule of Figure 15-21 may be obtained. The daily requirements of this schedule only vary between 12 and 15 workers. Similar procedures may be ap-

Act No.	Description	D	Time														
			1	2	3	4	5	6	7	8	9	10	11	12	13	14	15
1 – 2	Drill well	4	4	4	4	4											
1 – 3	Deliver matl	2	4	4													
1 – 4(5)	Excavate	5	5	5	5	5	5										
1 – 5	Power line	3	6	6	6												
2 – 5	Pump house	3					2	2	2								
3 – 6	Assemble tank	4			3	3	3	3									
4 – 6	Foundation	4						7	7	7	7						
4 – 7	Install pipe	6						4	4	4	4	4	4				
5 – 7	Install pump	2								5	5						
6 – 7	Erect tower & tank	6										8	8	8	8	8	8
	Total		19	19	18	12	10	16	13	16	16	12	12	8	8	8	8

Resources (Workers) Required by Activity

Activity	Number required
1 – 2	4
1 – 3	4
1 – 4	5
1 – 5	6
2 – 5	2
3 – 6	3
4 – 6	7
4 – 7	4
5 – 7	5
6 – 7	8

Figure 15-20 Resource assignment—early start schedule.

plied when maximum resource limits are established. However, it will often be necessary to extend the project duration to satisfy limited resource constraints. The daily requirements for each resource must be calculated separately, although several resources may be tabulated on the same schedule sheet by utilizing different colors or symbols for each resource. While the manual technique suggested above will be satisfactory for small networks, it is apparent that the procedure would become very cumbersome for large networks. Thus computer programs have been developed for both resource leveling and limited resource problems. Reference 1 identifies a number of such computer programs.

15-5 PRACTICAL CONSIDERATIONS IN NETWORK USE

When to Use Network Methods

The methodology involved in drawing a network diagram forces the planner to consider in some detail and to put down on paper the manner in which the project is to be carried out. In addition, the network diagram is an excellent communications

Act No.	Description	D	Time														
			1	2	3	4	5	6	7	8	9	10	11	12	13	14	15
1 – 2	Drill well	4	4	4	4	4											
1 – 3	Deliver matl	2	4	4													
1 – 4(5)	Excavate	5	5	5	5	5	5										
1 – 5	Power line	3			6	6	6										
2 – 5	Pump house	3					2	2	2								
3 – 6	Assemble tank	4						3	3	3	3						
4 – 6	Foundation	4						7	7	7	7						
4 – 7	Install pipe	6								4	4	4	4	4	4		
5 – 7	Install pump	2														5	5
6 – 7	Erect tower & tank	6										8	8	8	8	8	8
	Total		13	13	15	15	13	12	12	14	14	12	12	12	12	13	13

Figure 15-21 Improved level of resource assignment.

device for transmitting this information to everyone involved in a project. For these reasons, preparation of a network diagram is useful for any project, regardless of size. The size of the network used will, of course, depend on the size and complexity of the project. Rules of thumb on network size and the need for a network diagram proposed by some experts are based on a particular method of operation and do not necessarily apply to your situation. Projects as large as a $3.5 million 25-story building have been successfully managed with a CPM network consisting of less than 90 activities. Where repetitive operations are involved, it may be worthwhile to draw a subnetwork to show each operation in some detail while using only a single activity to represent the operation on the major network.

Even prior to bid submittal, a summary or outline network can be very useful. For example, the network diagram can be used to determine whether the project can reasonably be completed within the time specified on the bid documents. Thus a decision can be made at this point whether a bid should even be submitted. If the decision is to proceed, the network diagram can then be used as a framework for developing the project's cost estimate for the bid. Upon award of the construction contract, a full network should be prepared to the level of detail considered necessary for carrying out the project. Some of the factors to be considered in determining the level of detail to be used include the dollar value, size, complexity, and duration of the project.

Preparing the Network

Regardless of the size of the planning group (which may be as small as the project manager alone) chosen to develop the network, it is important that input be obtained from the field personnel most familiar with the construction techniques to be applied. If specialty subcontractors are not represented in the planning group, it is important that they review the plan prior to its finalization to ensure that they can carry out their work in the manner contemplated.

Manual or Computer Techniques

One of the major factors that has sometimes led to dissatisfaction with network methods has been the excessive or inappropriate use of computers. Manual techniques have much to recommend them, particularly for personnel not well versed in network procedures. The manual preparation and calculation of a network is one of the best ways for a manager to really understand a project and to visualize potential problems and payoffs.

When it is necessary to utilize a network of more than several hundred activities, the use of computers for performing time calculations is advantageous. However, do not let yourself or your subordinates become inundated with unnecessary computer output. Output should be carefully selected to provide all levels of management with only the information they can effectively utilize.

An obvious advantage of the computer is its ability to rapidly update network calculations and to provide reports in any format and quantity desired. However, the preparation of reports too frequently or in an excessive quantity is simply a waste of paper and computer time. While the network diagram at the project site should always be kept current, computer reports should be produced on a more limited basis. For projects of average duration and importance an updating interval of 2 to 4 weeks should be satisfactory. As with all computer operations, the output is only as good as the input, so care must be taken to ensure that data are correctly entered before running a network program.

Advanced Network Techniques

There are a number of more sophisticated network-based management techniques that have been developed. Among these are selection of an optimum (lowest total cost) project duration based on project time-cost relations, minimizing project cost through financial planning and cost control techniques, and resource leveling over multiple projects. Although beyond the scope of this chapter, many of these techniques are described in the end-of-chapter references.

15-6 LINEAR SCHEDULING METHODS

Scheduling Repetitive Projects

Many in the construction industry feel that conventional network methods such as CPM are not well suited to highly repetitive work. Such projects include highways, airfields, pipelines, multiple housing units, and high-rise buildings. Highway projects whose activities progress linearly from one end of the project to the other are particularly difficult to adequately represent in CPM. As a result, linear scheduling techniques are increasingly being employed on such projects.

The *Linear Scheduling Method (LSM)* is similar to the Line of Balance (LOB) scheduling technique developed in the early 1950s for industrial and aerospace projects and is sometimes identified by the same name. The objective of the LOB technique is to ensure that components or subassemblies are available at the time they are required to meet the production schedule of the final assembly. The objective of the LSM technique is to display and prevent interference between repetitive activities that progress linearly from one end of a project to the other. A brief explanation of the Linear Scheduling Method applied to a highway construction project is provided below.

A Linear Scheduling Method Diagram

An LSM diagram of the highway construction project of Figure 15-1 is shown in Figure 15-22. The five activities involved are Clear and Strip, Drainage, Subgrade, Base Course, and Pave. Notice that activities are represented by a line or band representing time versus location.

The height (A) of an activity at a specific time represents the distance over which that activity is being carried out at that instant. Thus, Subgrade work extends from station 13 to station 21 at the start of week 3. The width (B) of an activity indicates the time from start to finish of that activity at a specific location. Hence, at station 40 Subgrade work starts at week 4.4 and continues to week 5.0. The horizontal distance between activities (C) represents the time lag or interval between the finish of one activity and the start of the succeeding activity at a specific location. Thus, the start of Base work at station 10 lags the completion of Subgrade work at that location by 1.4 weeks.

Notice that the Drainage activity follows Clear and Strip but extends only to station 30. That is, no drainage work is required from station 30 to station 50. The Drainage activity could also overlap the Clear and Strip activity if it were determined that the two activities could be carried out concurrently at the same location without interference between the two.

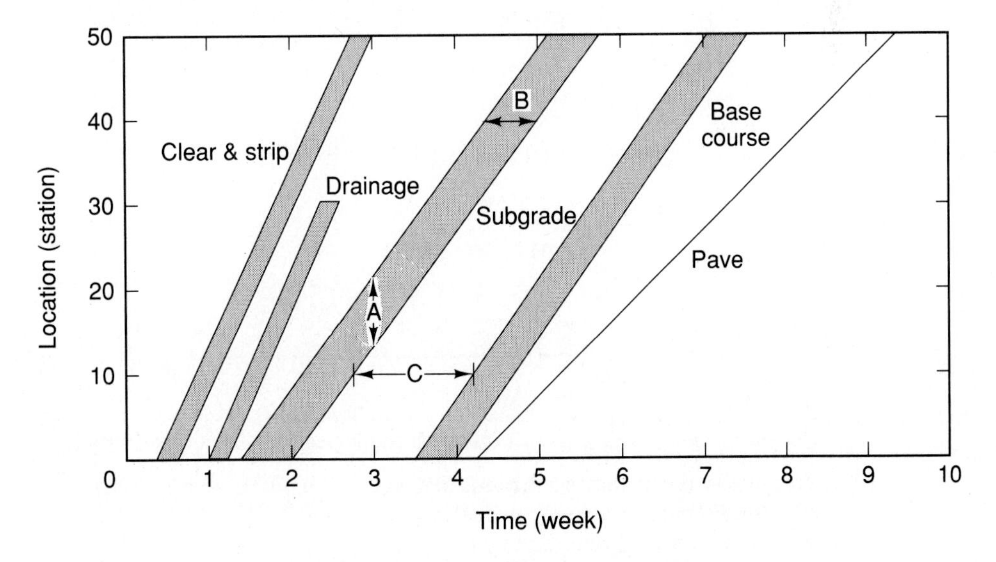

Figure 15-22 LSM diagram for project of Figure 15-1.

PROBLEMS

1. How does the actual progress at the end of the second week in Figure 15-1 compare with the scheduled progress? Express the answer as the percentage of scheduled progress that has actually been attained.

2. Draw an activity-on-arrow network diagram representing the following logical relationships.

Activity	*Depends on Completion of Activity:*
A	—
B	A
C	A
D	—
E	D
F	—
G	B
H	C
I	G and H

3. Draw an activity-on-arrow network diagram representing the following logical relationships.

Activity	*Depends on Completion of Activity:*
A	C and K
B	F
C	F
D	C and I
E	G
F	H
G	H
H	—
I	E and F
J	A and D
K	B

4. Draw a precedence diagram representing the logical relationships of Problem 3.

5. Redraw the accompanying network diagram, adding early and late event times to the diagram. Prepare an activity time tabulation showing early start, late start, early finish, late finish, and total float times.

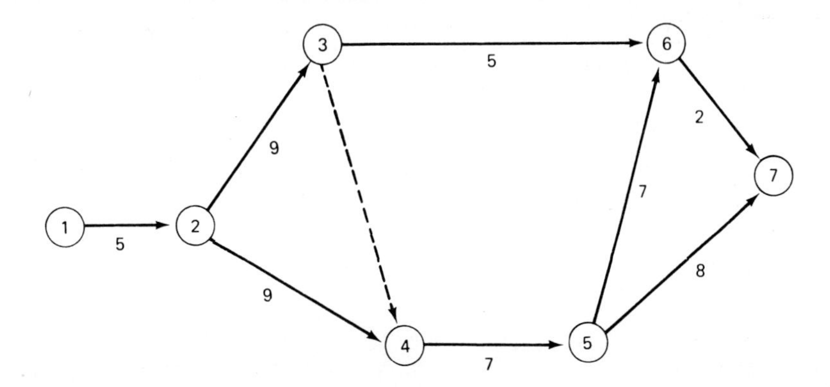

6. Redraw the precedence diagram of Figure 15-15 adding the relationships given below. Enter the early start, late start, early finish, late finish, and total float times on the diagram. Mark the critical path.

Activity Relationships			
Start to Start	*Finish to Start*	*Finish to Finish*	*Lag Time*
8 to 7			3
	7 to 10		2
		8 to 11	4

7. Prepare an early start schedule for the network shown. Indicate the allowable time span for each activity. Activity durations are in days.

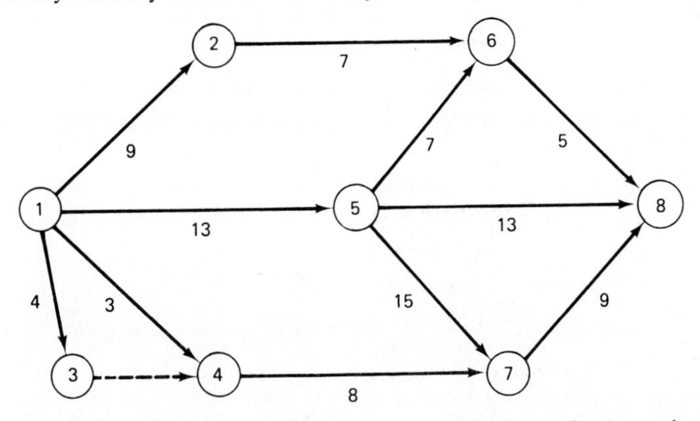

8. For the network of Problem 7, assign resources based on an early start schedule. Indicate the total resource requirement for each time period. Adjust resource assignments so that a maximum of 25 workers are used on the first 10 days and a maximum of 22 workers are needed thereafter. Resource requirements for each activity are as follows:

Activity	Workers Required
1–2	6
1–3	4
1–4	15
1–5	10
2–6	5
4–7	6
5–6	5
5–7	8
5–8	6
6–8	4
7–8	7

9. For the LSM diagram of Figure 15-22, how long does it take to complete the Base activity at a specified location?

10. Utilizing a computer CPM program, solve Problem 5 by use of a computer.

REFERENCES

1. AHUJA, H. N. *Project Management.* New York: Wiley, 1984.
2. ARDITI, DAVID, AND M. ZEKI ALBULAK. "Line-of-Balance Scheduling in Pavement Construction," *ASCE Journal of Construction Engineering and Management,* vol. 112, no. 3 (1986), pp. 411–424.
3. CLOUGH, RICHARD, AND GLENN A. Sears. *Construction Project Management,* 2nd ed. New York: Wiley, 1979.
4. HARRIS, ROBERT B. *Precedence and Arrow Networking Techniques for Construction.* New York: Wiley, 1978.
5. JOHNSTON, D.W. "Linear Scheduling Method for Highway Construction," *ASCE Journal of the Construction Division,* vol. 107, no. C02 (1981), pp. 247–261.
6. MODER, JOSEPH J., AND CECIL R. PHILLIPS. *Project Management with CPM, PERT, and PRECEDENCE Diagramming,* 3rd ed. New York: Van Nostrand Reinhold, 1983.
7. O'BRIEN, JAMES J. *CPM in Construction Management,* 3rd ed. New York: McGraw-Hill, 1984.
8. PINNELL, STEVEN S. "Critical Path Scheduling: An Overview and a Practical Alternative," *Civil Engineering—ASCE* (May 1981).
9. WHITEHOUSE, GARY E. "Critical Path Program for a Microcomputer," *Civil Engineering—ASCE* (May 1981), pp. 54–56.

16
Construction Economics

16-1
INTRODUCTION

As has been noted on a number of occasions, construction contracting is a highly competitive business. Therefore, the financial management of a construction company is equally as important to company success as is its technical management. As a matter of fact, many successful constructors have evolved from a background of business and finance rather than from construction itself. However, there is a little question that a strong technical base supported by business skills and management ability provides the best foundation for success as a construction professional.

A complete discussion of the many facets of construction economics is beyond the scope of this book. Rather, the purpose of this chapter is to introduce the reader to the terminology and basic principles involved in determining the owning and operating costs of construction plant and equipment, analyzing the feasibility of renting or leasing rather than purchasing equipment, and the financial management of construction projects.

16-2
TIME VALUE OF MONEY

Everyone is aware that the amount of money held in a savings account will increase with time if interest payments are allowed to remain on deposit (compound) in the account. The value of a sum of money left on deposit after any period of time may be calculated using Equation 16-1.

$$F = P(1 + i)^n \tag{16-1}$$

where F = value at end of n periods (future value)
P = present value
i = interest rate per period
n = number of periods

The expression $(1 + i)^n$ is often called the *single-payment compound interest factor.* Equation 16-1 can be solved to find the present value (present worth) of some future amount, resulting in Equation 16-2.

$$P = \frac{F}{(1 + i)^n} \tag{16-2}$$

The expression $1/(1 + i)^n$ is called the *single-payment present worth factor.* Expressions have also been developed that yield the value of a series of equal periodic payments at the end of any number of periods (*uniform series compound amount factor*), the present worth of such a series (*uniform series present worth factor*), the periodic payment required to accumulate a desired amount at some future date (*sinking fund factor*), and the annual cost to recover an investment, including the payment of interest, over a given period of time (*capital recovery factor*). Other expressions have been developed to find the present and future worth of gradient (nonuniform) series of payments.

These equations form the basis of a type of economic analysis commonly called *engineering economy.* The methods of engineering economy are widely used to analyze the economic feasibility of proposed projects, to compare alternative investments, and to determine the rate of return on an investment. However, because of their complexity and the difficulty of accounting for the effects of inflation and taxes, these techniques have not been widely used within the construction industry. Construction equipment owning costs, for example, are usually determined by the methods described in the following section rather than by employing engineering economy techniques. A present worth analysis, however, is very helpful when comparing the cost of different alternatives. This is illustrated by the rent–lease–buy analysis described in Section 16-4.

16-3 EQUIPMENT COST

Elements of Equipment Cost

In earlier chapters, we have discussed the proper application of the major items of construction equipment and some methods for estimating equipment's hourly production. We then divided the equipment's hourly cost by its hourly production to obtain the cost per unit of production. However, up to this point we have simply assumed that we knew the hourly cost of operation of the equipment. In this section we consider methods for determining the hourly cost of operation of an item of equipment. Although the procedures explained in this section are those commonly employed in the construction industry, they are not the only possible methods.

In following the procedures of this section, you will note that it is necessary to estimate many factors, such as fuel consumption, tire life, and so on. The best basis for estimating such factors is the use of historical data, preferably that recorded by your construction company operating similar equipment under similar conditions. If such data are not available, consult the equipment manufacturer for recommendations.

Equipment *owning and operating costs* (frequently referred to as *O & O costs*), as the name implies, are composed of owning costs and operating costs. Owning costs are fixed costs that are incurred each year whether the equipment is operated or not. Operating costs, however, are incurred only when the equipment is used.

Owning Costs

Owning costs are made up of the following principal elements:

- Depreciation.
- Investment (or interest) cost.
- Insurance cost.
- Taxes.
- Storage cost.

Methods for calculating each of these items are described below.

DEPRECIATION

Depreciation represents the decline in market value of an item of equipment due to age, wear, deterioration, and obsolescence. In accounting for equipment costs, however, depreciation is used for two separate purposes: (1) evaluating tax liability, and (2) determining the depreciation component of the hourly equipment cost. Note that it is possible (and legal) to use different depreciation schedules for these two purposes. For tax purposes many equipment owners depreciate equipment as rapidly as possible to obtain the maximum reduction in tax liability during the first few years of equipment life. However, the result is simply the shifting of tax liability between tax years, because current tax rules of the U.S. Internal Revenue Service (IRS) treat any gain (amount received in excess of the equipment's depreciated or book value) on the sale of equipment as ordinary income. The depreciation methods explained below are those commonly used in the construction equipment industry. Readers familiar with the subject of engineering economics should recognize that the methods of engineering economy may also be employed. When the methods of engineering economics are used, the depreciation and investment components of equipment owning costs will be calculated together as a single cost factor.

In calculating depreciation, the initial cost of an item of equipment should be the full delivered price, including transportation, taxes, and initial assembly and servicing. For rubber-tired equipment, the value of tires should be subtracted from the amount to be depreciated because tire cost will be computed separately as an element of operating cost. Equipment salvage value should be estimated as realistically as possible based on historical data.

The equipment life used in calculating depreciation should correspond to the equipment's expected economic or useful life. The IRS guideline life for general construction equipment is currently 5 years, so this depreciation period is widely used by the construction industry.

The most commonly used depreciation methods are the straight-line method, the sum-of-the-years'-digits method, the double-declining balance method, and IRS-prescribed methods. Procedures for applying each of these methods are explained below.

Straight-Line Method

The *straight-line method* of depreciation produces a uniform depreciation for each year of equipment life. Annual depreciation is thus calculated as the amount to be depreciated divided by the equipment life in years (Equation 16-3). The amount to be depreciated consists of the equipment's initial cost less salvage value (and less tire cost for rubber-tired equipment).

$$D_n = \frac{\text{Cost} - \text{Salvage} \ (- \text{ tires})}{N} \qquad (16\text{-}3)$$

where N = equipment life (years)
n = year of life (1, 2, 3, etc.)

Example 16-1

PROBLEM Using the straight-line method of depreciation, find the annual depreciation and book value at the end of each year for a track loader having an initial cost of $50,000, a salvage value of $5000, and an expected life of 5 years.

SOLUTION

$$D_{1,2,3,4,5} = \frac{50{,}000 - 5000}{5} = \$9000$$

Year	*Depreciation*	*Book Value (End of Period)*
0	0	$50,000
1	$9,000	41,000
2	9,000	32,000
3	9,000	23,000
4	9,000	14,000
5	9,000	5,000

Sum-of-the-Years'-Digits Method

The *sum-of-the-years'-digits method* of depreciation produces a nonuniform depreciation which is the highest in the first year of life and gradually decreases thereafter. The amount to be depreciated is the same as that used in the straight-line method. The depreciation for a particular year is calculated by multiplying the amount to be depreciated by a depreciation factor (Equation 16-4). The denominator of the depreciation factor is the sum of the years' digits for the depreciation period (or $1 + 2 + 3 + 4 + 5 = 15$ for a 5-year life). The numerator of the depreciation factor is simply the particular year digit taken in *inverse* order (i.e., $5 - 4 - 3 - 2 - 1$). Thus for the first year of a 5-year life, 5 would be used as the numerator. The procedure is illustrated in Example 16-2.

$$D_n = \frac{\text{Year digit}}{\text{Sum of years' digits}} \times \text{Amount to be depreciated} \tag{16-4}$$

Example 16-2

PROBLEM For the loader of Example 16-1, find the annual depreciation and book value at the end of each year using the sum-of-the-years'-digits method.

SOLUTION Using Equation 16-4:

$$D_1 = \frac{5}{15} \times (50{,}000 - 5000) = 15{,}000$$

$$D_2 = \frac{4}{15} \times (50{,}000 - 5000) = 12{,}000$$

$$D_3 = \frac{3}{15} \times (50{,}000 - 5000) = 9{,}000$$

$$D_4 = \frac{2}{15} \times (50{,}000 - 5000) = 6{,}000$$

$$D_5 = \frac{1}{15} \times (50{,}000 - 5000) = 3{,}000$$

Year	*Depreciation*	*Book Value (End of Period)*
0	0	$50,000
1	$15,000	35,000
2	12,000	23,000
3	9,000	14,000
4	6,000	8,000
5	3,000	5,000

Double-Declining-Balance Method

The *double-declining-balance method* of depreciation, like the sum-of-the-years'-digits method, produces its maximum depreciation in the first year of life. However, in using the double-declining-balance method, the depreciation for a particular year is found by multiplying a depreciation factor by the equipment's *book value at the beginning of the year* (Equation 16-5). The annual depreciation factor is found by dividing 2 (or 200%) by the equipment life in years. Thus for a 5-year life, the annual depreciation factor is 0.40 (or 40%). Unlike the other two depreciation methods, the double-declining-balance method does not automatically reduce the equipment's book value to its salvage value at the end of the depreciation period. Since the book value of equipment is not permitted to go below the equipment's salvage value, care must be taken when performing the depreciation calculations to stop depreciation when the salvage value is reached. The correct procedure is as follows:

$$D_n = \frac{2}{N} \times \text{Book value at beginning of year} \tag{16-5}$$

Example 16-3

PROBLEM For the loader of Example 16-1, find the annual depreciation and book value at the end of each year using the double-declining-balance method.

SOLUTION Using Equation 16-5:

$$\text{Annual depreciation factor} = \frac{2.00}{5} = 0.40$$

$$D_1 = 0.40 \times 50{,}000 = 20{,}000$$
$$D_2 = 0.40 \times 30{,}000 = 12{,}000$$
$$D_3 = 0.40 \times 18{,}000 = 7{,}200$$
$$D_4 = 0.40 \times 10{,}800 = 4{,}320$$
$$D_5 = 0.40 \times 6480 = 2{,}592 \quad \text{use } \$1{,}480^*$$

Year	Depreciation	Book Value (End of Period)
0	0	$50,000
1	$20,000	30,000
2	12,000	18,000
3	7,200	10,800
4	4,320	6,480
5	1,480*	5,000

*Because a depreciation of $2592 in the fifth year would reduce the book value to less than $5000, only $1480 ($6480 – $5000) may be taken as depreciation.

IRS-Prescribed Methods

Since the Internal Revenue Service tax rules change frequently, always consult the latest IRS regulations for the current method of calculating depreciation for tax purposes. The Modified Accelerated Cost Recovery System (MACRS) has been adopted by the Internal Revenue Service for the depreciation of most equipment placed in service after 1986. This depreciation method is also referred to as the General Depreciation System (GDS) by the IRS.

Under the MACRS system, depreciation for all property except real property is spread over a 3-year, 5-year, 7-year, or 10-year period. Most vehicles and equipment, including automobiles, trucks, and general construction equipment, are classified as 5-year property. The yearly deduction for depreciation is calculated as a prescribed percentage of initial cost (cost basis) for each year of tax life without considering salvage value. For 5-year property, annual depreciation percentages are 20%, 32%, 19.2%, 11.52%, 11.52%, and 5.76% for years 1 through 6, respectively. Notice that regardless of the month of purchase, only one-half of the normal double-declining-balance depreciation is taken in the year of purchase. The remaining cost basis is spread over a period extending through the year following the recovery life. Thus, depreciation for 5-year property actually extends over a 6-year period. This procedure is referred to by the IRS as using the "half-year convention."

Example 16-4

PROBLEM For the loader of Example 16-1, find the annual depreciation and book value at the end of each year using the MACRS method.

SOLUTION

$$\begin{aligned}
D_1 &= 0.20 \times 50{,}000 = 10{,}000 \\
D_2 &= 0.32 \times 50{,}000 = 16{,}000 \\
D_3 &= 0.192 \times 50{,}000 = 9{,}600 \\
D_4 &= 0.1152 \times 50{,}000 = 5{,}760 \\
D_5 &= 0.1152 \times 50{,}000 = 5{,}760 \\
D_6 &= 0.0576 \times 50{,}000 = 2{,}880
\end{aligned}$$

Year	*Depreciation*	*Book Value (End of Period)*
0	0	$50,000
1	10,000	40,000
2	16,000	24,000
3	9,600	14,400
4	5,760	8,640
5	5,760	2,880
6	2,880	0

INVESTMENT COST

Investment cost (or interest) represents the annual cost (converted to an hourly cost) of the capital invested in a machine. If borrowed funds are utilized, it is simply the interest charge on these funds. However, if the item of equipment is purchased from company assets, an interest rate should be charged equal to the rate of return on company investments. Thus investment cost is computed as the product of an interest rate multiplied by the value of the equipment, then converted to cost per hour. The true investment cost for a specific year of ownership is properly calculated using the average value of the equipment during that year. However, the average hourly investment cost may be more easily calculated using the value of the average investment over the life of the equipment given by Equation 16-6.

$$\text{Average investment} = \frac{\text{Initial cost} + \text{Salvage}}{2} \tag{16-6}$$

The results obtained using Equation 16-6 should be sufficiently accurate for calculating average hourly owning costs over the life of the equipment. However, the reader is cautioned that the investment cost calculated in this manner is not the actual cost for a specific year. It will be too low in the early years of equipment life and too high in later years. Thus this method should not be used for making replacement decisions or for other purposes requiring precise investment cost for a particular year.

INSURANCE, TAX, AND STORAGE

Insurance cost represents the cost of fire, theft, accident, and liability insurance for the equipment. *Tax cost* represents the cost of property tax and licenses for the equipment. *Storage cost* represents the cost of rent and maintenance for equipment storage yards and facilities, the wages of guards and employees involved in handling equipment in and out of storage, and associated direct overhead.

The cost of insurance and taxes for each item of equipment may be known on an annual basis. In this case, these costs are simply divided by the hours of operation during the year to yield the cost per hour for these items. Storage costs are usually obtained on an annual basis for the entire equipment fleet. Insurance and tax cost may also be known on a fleet basis. It is then necessary to prorate these costs to each item. This is usually done by converting total annual cost to a percentage rate by dividing these costs by the total value of the equipment fleet. When this is done, the rate for insurance, tax, and storage may simply be added to the investment cost rate to calculate the annual cost of investment, tax, insurance, and storage.

TOTAL OWNING COST

Total equipment owning cost is found as the sum of depreciation, investment, insurance, tax, and storage. As mentioned earlier, the elements of owning cost are often known on an annual cost basis. However, whether the individual elements of owning cost are calculated on an annual-cost basis or on an hourly basis, total owning cost should be expressed as an hourly cost.

INVESTMENT CREDIT

There are two major tax implications of equipment ownership. The first, depreciation, has already been discussed. The second is investment credit. *Investment credit* is a mechanism that the U.S. government uses to encourage industry to modernize production facilities by providing a tax credit for the purchase of new equipment. When in effect, investment credit provides a direct credit against tax due, not merely a reduction in taxable income. The investment credit last authorized allowed a tax credit of 10% of the investment for the purchase of equipment classified as 5-year property and 6% for equipment classified as 3-year property. However, when investment credit is taken, the cost basis (amount used for calculating cost recovery deductions) must be reduced or a smaller investment credit used. Current IRS tax regulations should always be consulted for up-to-date information, including investment credit procedures.

Operating Costs

Operating costs are incurred only when equipment is operated. Therefore, costs vary with the amount of equipment use and job operating conditions. Operating costs include operators' wages, which are usually added as a separate item after other operating costs have been calculated.

The major elements of operating cost include:

- Fuel cost.
- Service cost.
- Repair cost.
- Tire cost.
- Cost of special items.
- Operators' wages.

FUEL COST

The *hourly cost of fuel* is simply fuel consumption per hour multiplied by the cost per unit of fuel (gallon or liter). Actual measurement of fuel consumption under similar job conditions provides the best estimate of fuel consumption. However, when historical data are not available, fuel consumption may be estimated from manufacturer's data or by the use of Table 16-1. Table 16-1 provides approximate fuel consumption factors in gallons per hour per horsepower for major types of equipment under light, average, and severe load conditions.

SERVICE COST

Service cost represents the cost of oil, hydraulic fluids, grease, and filters as well as the labor required to perform routine maintenance service. Equipment manufacturers publish consumption data or average cost factors for oil, lubricants, and filters

Table 16-1 Fuel consumption factors (gal/h/hp)

	*Load Conditions**		
Type of Equipment	*Low*	*Average*	*Severe*
Clamshell and dragline	0.024	0.030	0.036
Compactor, self-propelled	0.038	0.052	0.060
Crane	0.018	0.024	0.030
Excavator, hoe, or shovel	0.035	0.040	0.048
Loader			
Track	0.030	0.042	0.051
Wheel	0.024	0.036	0.047
Motor grader	0.025	0.035	0.047
Scraper	0.026	0.035	0.044
Tractor			
Crawler	0.028	0.037	0.046
Wheel	0.028	0.038	0.052
Wagon	0.029	0.037	0.046

*Low, light work or considerable idling; average, normal load and operating conditions; severe, heavy work, little idling.

for their equipment under average conditions. Using such consumption data, multiply hourly consumption (adjusted for operating conditions) by cost per unit to obtain the hourly cost of consumable items. Service labor cost may be estimated based on prevailing wage rates and the planned maintenance program.

Since service cost is related to equipment size and severity of operating conditions, a rough estimate of service cost may be made based on the equipment's fuel cost (Table 16-2). For example, using Table 16-2 the hourly service cost of a scraper operated under severe conditions would be estimated at 50% of the hourly fuel cost.

REPAIR COST

Repair cost represents the cost of all equipment repair and maintenance except for tire repair and replacement, routine service, and the replacement of high-wear items, such as ripper teeth. It should be noted that repair cost usually constitutes the largest item of operating expense for construction equipment.

Lifetime repair cost is usually estimated as a percentage of the equipment's initial cost less tires (Table 16-3). It is then necessary to convert lifetime repair cost to an hourly repair cost. This may be done simply by dividing lifetime repair cost by the expected equipment life in hours to yield an average hourly repair cost. Although this method is adequate for lifetime cost estimates, it is *not* valid for a particular year of equipment life. As you might expect, repair costs are typically low for new

Table 16-2 Service cost factors (% of hourly fuel cost)

Operating Conditions	*Service Cost Factor*
Favorable	20
Average	33
Severe	50

Table 16-3 Typical lifetime repair cost (% of initial cost less tires)

	Operating Conditions		
Type of Equipment	*Favorable*	*Average*	*Severe*
Clamshell and dragline	40	60	80
Crane	40	50	60
Excavator, hoe, or shovel	50	70	90
Loader			
Track	85	90	105
Wheel	50	60	75
Motor grader	45	50	55
Scraper	85	90	105
Tractor			
Crawler	85	90	95
Wheel	50	60	75
Truck, off-highway	70	80	90
Wagon	45	50	55

machines and rise as the equipment ages. Thus it is suggested that Equation 16-7 be used to obtain a more accurate estimate of repair cost during a particular year of equipment life.

$$\text{Hourly repair cost} = \frac{\text{Year digit}}{\text{Sum of years' digits}} \times \frac{\text{Lifetime repair cost}}{\text{Hours operated}} \qquad (16\text{-}7)$$

This method of prorating repair costs is essentially the reverse of the sum-of-the-years'-digits method of depreciation explained earlier, because the year digit used in the numerator of the equation is now used in a normal sequence (i.e., 1 for the first year, 2 for the second year, etc.)

Example 16-5

PROBLEM Estimate the hourly repair cost for the first year of operation of a crawler tractor costing $136,000 and having a 5-year life. Assume average operating conditions and 2000 hours of operation during the year.

SOLUTION

$$\text{Lifetime repair cost factor} = 0.90 \quad (\text{Table 16-3})$$

$$\text{Lifetime repair cost} = 0.90 \times \$136{,}000 = \$122{,}400$$

$$\text{Hourly repair cost} = \frac{1}{15} \times \frac{122{,}400}{2000} = \$4.08$$

TIRE COST

Tire cost represents the cost of tire repair and replacement. Among operating costs for rubber-tired equipment, tire cost is usually exceeded only by repair cost. Tire cost is difficult to estimate because of the difficulty in estimating tire life. As always, historical data obtained under similar operating conditions provide the best basis for estimating tire life. However, Table 16-4 may be used as a guide to approximate tire life. Tire repair will add about 15% to tire replacement cost. Thus Equation 16-8 may be used to estimate tire repair and replacement cost.

$$\text{Tire cost} = 1.15 \times \frac{\text{Cost of a set of tires (\$)}}{\text{Expected tire life (h)}} \qquad (16\text{-}8)$$

Table 16-4 Typical tire life (hours)

	Operating Conditions		
Type of Equipment	*Favorable*	*Average*	*Severe*
Dozers and loaders	3,200	2,100	1,300
Motor graders	5,000	3,200	1,900
Scrapers			
Conventional	4,600	3,300	2,500
Twin engine	4,000	3,000	2,300
Push-pull and elevating	3,600	2,700	2,100
Trucks and wagons	3,500	2,100	1,100

SPECIAL ITEMS

The cost of replacing high-wear items such as dozer, grader, and scraper blade cutting edges and end bits, as well as ripper tips, shanks, and shank protectors, should be calculated as a separate item of operating expense. As usual, unit cost is divided by expected life to yield cost per hour.

OPERATOR

The final item making up equipment operating cost is the operator's wage. Care must be taken to include all costs, such as worker's compensation insurance, Social Security taxes, overtime or premium pay, and fringe benefits in the hourly wage figure.

Total Owning and Operating Costs

After owning cost and operating cost have been calculated, these are totaled to yield total owning and operating cost per hour of operation. Although this cost may be used for estimating and for charging equipment costs to projects, notice that it does not include overhead or profit. Hence overhead and profit must be added to obtain an hourly rental rate if the equipment is to be rented to others.

Example 16-6

PROBLEM Calculate the expected hourly owning and operating cost for the second year of operation of the twin-engine scraper described below.

Cost delivered = $152,000
Tire cost = $12,000
Estimated life = 5 years
Salvage value = $16,000
Depreciation method = sum-of-the-years'-digits
Investment (interest) rate = 10%
Tax, insurance, and storage rate = 8%
Operating conditions = average
Rated power = 465 hp
Fuel price = $0.40/gal
Operator's wages = $8.00/h

SOLUTION

Owning Cost

Depreciation cost:

$$D_2 = \frac{4}{15} \times (152{,}000 - 16{,}000 - 12{,}000) = \$33{,}067 \qquad (16\text{-}4)$$

$$\text{Depreciation} = \frac{33{,}067}{2000} = \$16.53/\text{h}$$

Investment, tax, insurance, and storage cost:

$$\text{Cost rate} = \text{Investment} + \text{tax, insurance, and storage} = 10 + 8 = 18\%$$

$$\text{Average investment} = \frac{152{,}000 + 16{,}000}{2} = \$84{,}000 \qquad (16\text{-}6)$$

$$\text{Investment, tax, insurance, and storage} = \frac{84{,}000 \times 0.18}{2000} = \$7.56/\text{h}$$

$$\text{Total owning cost} = 16.53 + 7.56 = \$24.09/\text{h}$$

Operating Cost

Fuel cost:

$$\text{Estimated consumption} = 0.035 \times 465 = 16.3 \text{ gal/h} \quad \text{(Table 16-1)}$$

$$\text{Fuel cost} = 16.3 \times 0.40 = \$6.52/\text{h}$$

Service cost:

$$\text{Service cost} = 0.33 \times 6.52 = \$2.15/\text{h} \quad \text{(Table 16-2)}$$

Repair cost:

$$\text{Lifetime repair cost} = 0.90 \times (152{,}000 - 12{,}000) = \$126{,}000 \quad \text{(Table 16-3)}$$

$$\text{Repair cost} = \frac{2}{15} \times \frac{126{,}000}{2{,}000} = \$8.40/\text{h} \qquad (16\text{-}7)$$

Tire cost:

$$\text{Estimated tire life} = 3000 \text{ h} \quad \text{(Table 16-4)}$$

$$\text{Tire cost} = 1.15 \times \frac{\$12{,}000}{3000} = \$4.60/\text{h}$$

Special item cost: None

$$\text{Operator wages} = \$8.00/\text{h}$$

$$\text{Total operating cost} = 6.52 + 2.15 + 8.40 + 4.60 + 8.00 = \$29.67/\text{h}$$

Total O & O Cost

$$\text{Owning and operating cost} = 24.09 + 29.67 = \$53.76/\text{h}$$

16-4 THE RENT–LEASE–BUY DECISION

The question of whether it is better to purchase a piece of construction equipment rather than renting or leasing the item is difficult to answer. Leasing involves a commitment for a fixed period and may include a purchase option in which a portion of the lease payments is credited toward the purchase price if the option is exercised. Renting is a short-term arrangement subject only to the availability of rental equipment and a minimum rental period (usually 1 day). In recent years there has been a trend toward increased leasing and renting of construction equipment. Some of the reasons for this trend include inflation, the high cost of borrowed funds, and the wide fluctuation in the rate of demand for construction services. There are some construction companies that make it a policy to rent or lease all major items of equipment.

Advantages of equipment ownership include governmental tax incentives (investment credit and depreciation), full control of equipment resources, and availability of equipment when needed. However, leasing and renting require little initial capital (usually none for renting) and equipment costs are fully tax deductible as project expenses. A rational analysis of these alternatives for obtaining equipment is complex and must include cost under the expected conditions as well as equipment availability and productivity. In general, purchasing equipment will result in the lowest hourly equipment cost if the equipment is properly maintained and fully utilized. However, as we noted earlier, equipment owning costs continue whether equipment is being utilized or sitting idle. Therefore, renting is usually least expensive for equipment which has low utilization. Leasing is intermediate between the two and may be the best solution when capital is limited and equipment utilization is high. The lease-with-purchase option may provide an attractive opportunity to purchase the equipment at low cost after lease costs have been paid under a cost-type contract.

One approach to comparing the cost of buying, leasing, and renting an item of equipment is illustrated in Example 16-7. The analysis considers net after-tax cash flow and its present value (present worth). The example is based on a method suggested by David J. Everhart of Caterpillar Inc. In making the calculations for present value, the present worth factors for midyear were used for yearly costs. To ensure that alternatives were compared under equal conditions, maintenance and repair costs were excluded from all calculations, although maintenance and repair is often included in rental rates. Notice that such an analysis depends on the specific tax rules applied (in this case, ACRS depreciation and 10% investment credit).

Under the particular circumstances of Example 16-7, buying is significantly less expensive than either leasing or renting if the equipment is fully utilized for the planned 5 years or 10,000 hours. However, notice that the cost difference is considerably smaller when considered on a present worth basis. Figure 16-1 illustrates the effect on hourly cost (present value) when equipment utilization declines. Since total capital cost is constant over the 5-year period for both leasing and buying, hourly capital cost increases as utilization declines for both of these alternatives. Since the 5-year cost of leasing is fixed, leasing is always more expensive than owning in these circumstances. Since the hourly cost for renting is constant, the hourly cost for renting and buying become equal at 42% utilization or 4200 h of use. As utilization continues to decline, renting becomes even more advantageous.

Example 16-7

PROBLEM Analyze the cost of renting, leasing, and purchasing an item of construction equipment under the conditions described. Evaluate total net after-tax cash flow and its present value.

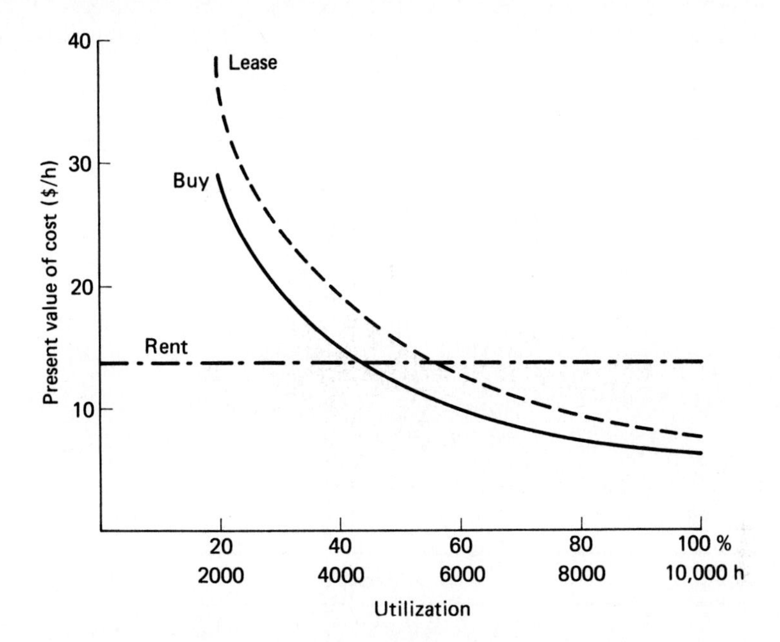

Figure 16-1 Hourly cost of buying, leasing, and renting for Example 16-7.

Basic assumptions:

Company's marginal tax rate = 46%
Company's after-tax rate of return = 8%
Planned equipment use = 2000 h/yr for 5 years

Purchase assumptions:

Equipment cost = $150,000
Estimated resale value after 5 years = $60,000
Cost recovery method = 5-yr ACRS
(yearly depreciation of 15%, 22%, 21%, 21%, and 21%)
Investment credit (10%) = $15,000
Cost basis = $142,500 (cost less $\frac{1}{2}$ investment credit)
Loan period = 36 months
Loan interest rate = 12%
Down payment (20%) = $30,000

Lease assumptions:

Term of lease = 5 years
Lease payment = $2800 per month
Initial payment = 3 months in advance

Rental assumptions:

Rental period = 5 years, month to month
Rental rate = $5150 per month

SOLUTION

Purchase Cost								
	Initial	Year 1	Year 2	Year 3	Year 4	Year 5	Final	Total
Payments	30,000	47,820	47,820	47,820	0	0	0	173,460
Resale	0	0	0	0	0	0	(60,000)	(60,000)
Tax at resale	0	0	0	0	0	0	27,600	27,600
Tax savings—depreciation	0	(9,832)	(14,421)	(13,766)	(13,766)	(13,765)	0	(65,550)
Tax savings—interest	0	(5,738)	(3,688)	(1,366)	0	0	0	(10,792)
Investment credit	0	(15,000)	0	0	0	0	0	(15,000)
Net cost	30,000	17,250	29,711	32,688	(13,766)	(13,765)	(32,400)	49,718
Present value of net cost	30,000	16,611	26,491	26,987	(10,523)	(9,743)	(22,051)	57,772

Lease Cost								
	Initial	Year 1	Year 2	Year 3	Year 4	Year 5	Final	Total
Payments	8,400	33,600	33,600	33,600	33,600	25,200	0	168,000
Tax savings—payments	(3,864)	(15,456)	(15,456)	(15,456)	(15,456)	(11,592)	0	(77,280)
Net cost	4,536	18,144	18,144	18,144	18,144	13,608	0	90,720
Present value of net cost	4,536	17,472	16,178	14,980	13,870	9,632	0	76,668

Rental Cost								
	Initial	Year 1	Year 2	Year 3	Year 4	Year 5	Final	Total
Payments	0	61,800	61,800	61,800	61,800	61,800	0	309,000
Tax savings—payments	0	(28,428)	(28,428)	(28,428)	(28,428)	(28,428)	0	(151,800)
Net cost	0	33,372	33,372	33,372	33,372	33,372	0	166,860
Present value of net cost	0	32,136	29,756	27,552	25,510	23,621	0	138,575

16-5 FINANCIAL MANAGEMENT OF CONSTRUCTION

The high rate of bankruptcy in the construction industry was pointed out in Chapter 1. Statistics compiled by Dun & Bradstreet on construction company failure in the United States indicate that the four major factors of inadequate financing, underestimating costs, inadequate cost accounting, and poor management account for over

80% of all failures. Thus the basis for the statement earlier in this chapter that "the financial management of a construction company is equally as important to company success as is its technical management" is apparent. In this section, we shall consider the basic principles of financial planning and cost control for contruction projects.

Financial Planning

Financial planning for a construction project includes cost estimating prior to bidding or negotiating a contract, forecasting project income and expenditure (or cash flow), and determining the amount of work that a construction firm can safely undertake at one time.

Cost estimating for a project, as the name implies, involves estimating the total cost to carry out a construction project in accordance with the plans and specifications. Costs that must be considered include labor, equipment, materials, subcontracts and services, indirect (or job management) costs, and general overhead (off-site management and administration costs). Cost estimating for bidding purposes is discussed further in Chapter 17.

A finance schedule or cash flow schedule shows the planned rate of project expenditure and project income. It is common practice in the construction industry (as discussed in Chapter 17) for the owner to withhold payment for a percentage of the value of completed work (referred to as "retainage") as a guarantee until acceptance of the entire project. Even when periodic progress payments are made for the value of completed work, such payments (less retainage) are not received until some time after the end of each accounting period. Hence project income will almost always lag behind project expenditure. The difference must be provided in cash from company assets or borrowed funds. The construction industry relies heavily on the use of borrowed funds for this purpose. Therefore, the finance charges associated with the use of such funds, as well as the maximum amount of funds available, are important considerations in the financial planning for a construction project. While a financial schedule may be developed manually from any type of project schedule, the use of CPM methods will facilitate preparation of a financial schedule. The use of CPM procedures also makes it easy to determine the effect on cash flow of different project schedules. Figure 16-2 shows a graph of project cost versus time for three different schedules: an early start schedule, a late start schedule, and a proposed schedule which is between these limits. Figure 16-3 illustrates a financial schedule showing project expenditures, value of completed work, and receipts for a particular project schedule.

Another important consideration in financial planning is the capacity of a firm to undertake additional projects. It has been found that most construction contracts require a minimum working capital of about 10% of the contract value. This working capital is needed to cover the difference between project income and project expenditures described above. The availability of working capital also affects the type of construction contract that might be appropriate for any additional work to be undertaken. When working capital is marginal, any additional work should be limited to low-risk projects such as cost-reimbursable contracts.

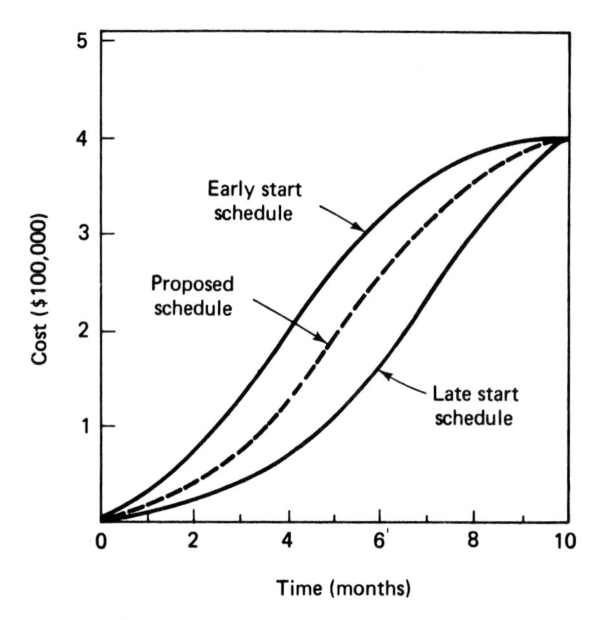

Figure 16-2 Project cost versus time.

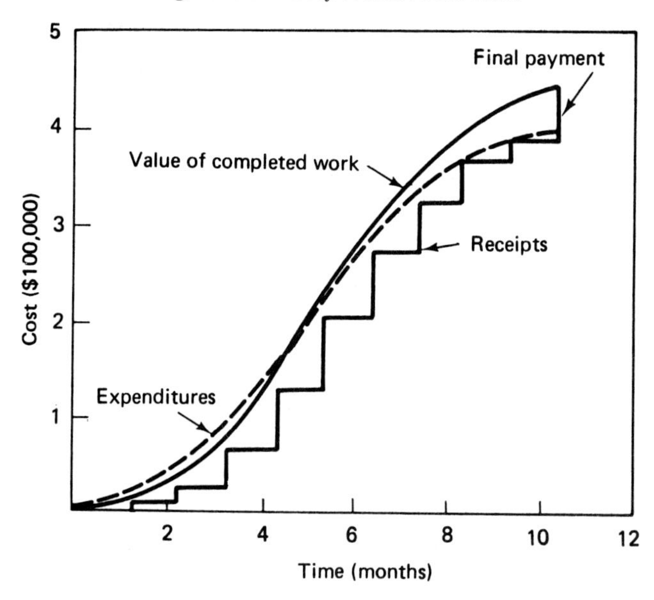

Figure 16-3 Project financial schedule.

Project Cost Control

Project cost control involves the measurement and recording of project costs and progress and a comparison between actual and planned performance. The principal objective of project cost control is to maximize profit while completing the project on time at a satisfactory level of quality. Proper cost control procedures will also result in the accumulation of historical cost data, which are invaluable in estimating and controlling future project costs.

To carry out project cost control it is necessary to have a method for identifying cost and progress by project work element. The use of CPM procedures greatly simplifies this process, because major work items have already been identified as activities when preparing the project network diagram. A cost code system is usually combined with activity numbering to yield a complete system of project cost accounts. It is essential that the coding system permit charging all labor, material, equipment, and subcontract costs to the appropriate work item. Indirect and overhead costs for the project are usually assigned a separate cost code. Record-keeping requirements for foremen and other supervisors should be kept to a minimum consistent with meeting the objectives of accurate and timely reporting. Labor costs may be easily computed if time sheets are coded to identify the activity on which the time is expended. Plant and equipment costs may be similarly computed if a record is kept of the time spent on each activity by each machine. The cost of materials used may be based on priced delivery invoices coded to the appropriate activity. Since subcontract and service work may not be billed at the same interval at which costs are recorded, it may be necessary to apportion such costs to the appropriate activity at each costing interval.

To permit a comparison of project progress versus cost, it is necessary that progress reporting intervals coincide with cost reporting intervals. The interval between reports will depend on the nature and importance of the project. Monthly intervals are commonly used as the basis for requesting progress payments. However, construction management may desire weekly or even daily cost and performance reports for the control of critical construction projects.

A number of systems have been developed for relating project cost and progress and for forecasting time and cost to project completion. One such system, called PERT/Cost, has been developed by the U.S. government and has been extensively used by government agencies for project control. Figure 16-4 illustrates a PERT/Cost report for a project. Note that project progress and cost to date are graphed, together with the value of the work completed to date and projections of final completion time and cost. While PERT/Cost itself has not been widely used in the construction industry, systems have been developed within the construction industry which offer similar capabilities tailored to the construction environment.

PROBLEMS

1. A crawler tractor cost $100,000, has an estimated salvage value of $10,000, and a 5-year life. Using the double-declining-balance method of depreciation, find the tractor's yearly depreciation and book value at the end of each year.

2. Find the annual depreciation and book value at the end of each year for the tractor of Problem 1 using the MACRS method.

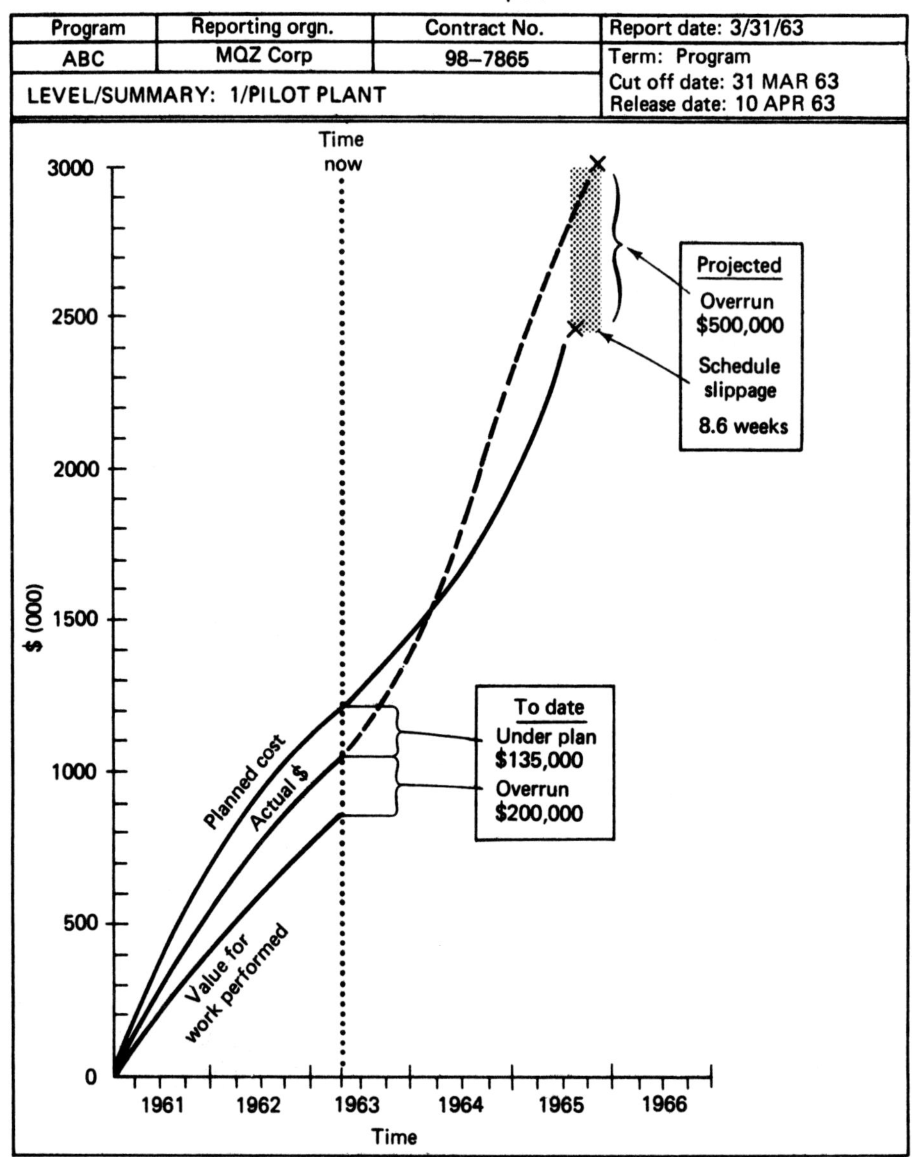

Figure 16-4 PERT/Cost progress and cost report. (PERT Coordinating Group, U.S. government)

3. Find the annual depreciation and book value at the end of each year for the tractor of Problem 1 using the sum-of-the-years'-digits method of depreciation.

4. Find the present value (worth) of $40,000 to be received when a piece of equipment is sold 5 years from now if the rate of return on investment is 7% per year.

5. Calculate the average hourly owning cost for the tractor of Problem 1 if the tractor is operated 2000 hours per year. The interest, tax, and insurance rate is 12% and the rate for storage and miscellaneous costs is 4%.

6. Find the hourly operating cost for the first year of life of the tractor of Problem 1. Use the following additional data:

Rated power = 300 hp
Fuel price = $0.50/gal ($0.132/ℓ)
Load conditions = average
Operating conditions = average
Hours operated = 2000 h/yr
Operator cost = $8.00/h

7. Analyze the cost of renting, leasing, and purchasing a backhoe using the data given below. Evaluate net after-tax cash flow and its present value.
Basic assumptions:

Marginal tax rate = 46%
After-tax rate of return = 7%
Planned equipment use = 2000 h/yr for 5 years
Present value factors (i = 7%):

Year	P/F
1	0.967
2	0.904
3	0.845
4	0.790
5	0.738
End of 5	0.713

Purchase assumptions:

Cost = $100,000
Estimated resale value after 5 years = $40,000
Cost recovery method = 5-yr MACRS
Down payment = $20,000
Loan terms = $80,000, 12%, 36 months = $2657.20/month

Interest payments:

Year	Interest
1	$8333
2	5346
3	1979

Lease assumptions:

Term of lease = 5 years
Lease payments = $1850/month
Initial payment = 3 months in advance

Rental assumptions:

Rental period = 5 years, month to month
Rental rate = $3375/month

8. For the project whose cost data are given below, plot the cumulative project expenditures, value of work, and progress payments received versus time. What is the contractor's maximum negative cash flow, and when does it occur? Progress payments are calculated at the end of each month and received the middle of the following month. Retainage is 10% until project completion. Assume that final project payment, including released retainage, is received the middle of the month following project completion.

	End-of-Month Cumulative Values	
Month	Expenditures	Value of Work
1	$12,000	$8,000
2	24,000	20,000
3	41,000	38,000
4	60,000	58,000
5	85,000	84,000
6	120,000	120,000
7	175,000	180,000
8	230,000	240,000
9	240,000	250,000
10	245,000	260,000
11	246,000	268,000
12	247,000	275,000

9. Determine the probable average cost per hour over the life of the equipment for owning and operating a bottom-dump wagon under the conditions listed below. Use the straight-line method of depreciation.

Operator's wage = \$8.00/h
Operating conditions = average
Delivered price = \$142,000
Cost of a set of tires = \$12,000
Expected life of wagon = 5 years
Expected annual use = 2000 h
Estimated salvage value = \$42,000
Fuel cost = \$0.50/gal (\$0.132/ℓ)
Vehicle horsepower = 440 hp
Rate for interest, tax, insurance, and storage = 15%

10. Write a computer program to calculate cumulative project expenditures, value of work completed, progress payments received, and cash flow over the life of a project. Identify the maximum negative cash flow and its time of occurrence. Input should include estimated expenditures and value of work completed for each time period, percent retainage, and time lag for receipt of progress payments. Solve Problem 8 using your computer program.

11. Write a computer program to determine the probable average hourly owning and operating cost of a piece of construction equipment over the life of the equipment. Use the straight-line method of depreciation. Using your computer program, solve Problem 9.

REFERENCES

1. *Caterpillar Performance Handbook,* 21st ed. Caterpillar Inc., Peoria, Ill., 1990.
2. COLLIER, COURTLAND A., AND WILLIAM B. LEDBETTER. *Engineering Cost Analysis,* 2nd ed. New York: Harper & Row, 1988.
3. *Contractors' Equipment Cost Guide.* Dataquest Incorporated, San Jose, Calif.
4. *Euclid Hauler Handbook,* 15th ed. VME Americas, Inc., Cleveland, Ohio, 1982.
5. JELEN, FREDERIC C., AND JAMES H. BLACK, eds. *Cost and Optimization Engineering,* 3rd ed. New York: McGraw-Hill, 1990.
6. NUNNALLY, S.W. *Managing Construction Equipment.* Englewood Cliffs, N. J.: Prentice Hall, 1977.
7. *Production and Cost Estimating of Material Movement with Earthmoving Equipment.* Terex Corporation, Hudson, Ohio, 1981.
8. THUESEN, G. J., AND W. J. FABRYCKY. *Engineering Economy,* 7th ed. Englewood Cliffs, N. J.: Prentice Hall, 1989.

17
CONTRACT CONSTRUCTION

17-1 INTRODUCTION

The Construction Process

The several organizational and management methods by which construction may be accomplished were described in Chapter 1. Construction by a general contractor employed under a prime construction contract is only one of these methods. However, since this method of obtaining construction services is widely used, it will form the basis for this chapter's discussion of contract construction, including bidding and contract award, construction contracts, plans and specifications, and contract administration.

Construction Contract Law

Construction professionals are not usually lawyers and therefore should not attempt to act as their own lawyers. However, construction professionals must have a thorough understanding of the customary practices and underlying legal principles involving contract construction. Virtually every action taken by a contractor, construction manager, or architect/engineer at a construction site has legal implications. There is simply not time to consult a lawyer every time a decision must be made. Thus construction professionals must understand the contractual consequences of their activities and be able to recognize when legal advice should be secured. Hence the purpose of the discussion of contract law in this chapter is to familiarize the reader with the general principles of construction contract law and practice and to provide a basis for further study. A study of the summaries of court decisions pertaining to construction contract disputes found in many professional magazines will also be very helpful in acquiring further knowledge in this area.

17-2 BIDDING AND CONTRACT AWARD

Bid Preparation

In the United States, as in much of the world, construction contracting is a highly competitive business. To prosper and grow a construction company must achieve a reputation for quality workmanship and timely completion while achieving a reasonable return on its capital investment. Thus profit is an obvious and principal motive for bidding on a construction contract. However, there are a number of other reasons why a contractor may choose to bid on a project. During times of low construction activity, contractors may submit bids with little or no profit margin in order to keep their equipment in operation and prevent the loss of skilled workers and managers. Although such a policy may be successful on a short-term basis, it is apparent that it will lead to financial disaster if long continued. Other reasons for bidding on a project include a desire for prestige and the maintenance of goodwill with regular clients. Projects that receive wide publicity because of their national importance or their unusual nature are often bid at low profit margins for the prestige they confer on the builder. In these cases, the loss of potential profit can be justified by the public recognition gained. Likewise, contractors sometimes bid on relatively undesirable projects in order to maintain a relationship with an owner. In such cases, the profit margin used for the bid would be expected to be high.

Regardless of motivation, once a contractor has decided to bid on a project, he must prepare a detailed cost estimate for the execution of the project. The first step in preparation of a cost estimate is to take off (or extract) the quantities of material required by the plans and specifications. These quantities are then extended (or multiplied by unit cost estimates) to provide a total estimated material cost for the project. Similar estimates are made for labor, equipment, and subcontract costs. The cost of equipment, labor, and material are often referred to as *direct costs.* Next, estimates are made of the administrative and management expenses that will be incurred at the project site. These costs are often referred to as job overhead or *indirect costs.*

After all project costs have been estimated, it is necessary to add an additional amount (or markup) for general overhead and profit. *General overhead* must cover the cost of all company activities not directly associated with individual construction projects. Major items of general overhead include salaries of headquarters personnel (company officials, estimators, clerks, accountants, etc.), rent and utilities, advertising, insurance, office supplies, and interest on borrowed capital. The usual procedure for prorating general overhead expenses to projects is to estimate total annual overhead expense, divide by the expected dollar volume of construction work for the year, and then multiply by the project bid price. The amount to be added for *profits* is, of course, a management decision. Although some projects may be bid with low profit margins for the reasons discussed earlier in this section, in the long run, construction operations must yield a reasonable return on invested capital. Unless the return on capital is greater than the yield of standard commercial investments, the owner would be better off investing in such items than in operating a construction business.

Bidding strategy, or the selection of a specific bid price for fixed-price construction contract, is a mixture of art and science that is beyond the scope of this book. Methods used range from statistical analyses and application of game theory to seat-of-the-pants decisions. The bid price actually submitted by a contractor is usually based on an analysis of the expected competition and the state of the construction market in addition to the contractor's estimate of his cost to execute the project.

Bidding Procedure

The principal steps in the bidding procedure for a fixed-price construction contract include solicitation, bid preparation, bid submission, bid opening, selection of the lowest qualified bid, and contract award. Solicitation may range from an invitation sent to a selected few contractors to public advertisement. Except in special circumstances, U.S. governmental agencies are required to solicit bids by public advertising. To ensure adequate competition, at least three bids should be obtained.

Contractors indicating an interest in bidding should be supplied with at least one complete set of contract documents. A deposit may be required to ensure the return of project plans and specifications furnished to unsuccessful bidders. The time allowed for bid preparation should be based on the size and complexity of the project. Three weeks has been suggested as a reasonable minimum time.

The Associated General Contractors of America, in cooperation with other professional organizations, have developed recommended bidding procedures for both building construction and engineered construction which are designed to ensure fairness to both contractors and owners (references 6 and 8). Among these recommendations are the use of standard bid proposal forms, specifying the order of selection of alternates, and suggested minimum times to be allowed for bid preparation. Alternates are optional items beyond the basic project scope. Since alternates may or may not be selected by the owner, their order of selection will affect the determination of the lowest bid price.

Bid openings are frequently open to the public, and in such cases bid prices are announced as the bids are opened. To facilitate communications between contractors and subcontractors immediately prior to bid submission, the deadline for submission of bids should not occur on a holiday or the day immediately following a holiday.

Contract Award

After opening the bids, they are evaluated by the owner to determine the lowest qualified bid. The *qualification* of a contractor is the determination that the contractor possesses both the technical and financial ability to perform the work required by the contract. The method of qualification used will depend on the owner involved. U.S. government regulations require the *contracting officer* (person empowered to execute contracts binding the government) to make a formal finding that a contractor is qualified to perform before a contract may be awarded.

Another method of bidder qualification is called *prequalification*. Under this procedure only those contractors determined to be capable of performing are invited to submit bids for the project. A more common, although indirect, method of prequalification is to require bonding of the contractor. Bonds used in construction include bid bonds, performance bonds, and payment bonds. A *bid bond* guarantees that a contractor will provide the required performance and payment bonds if awarded the contract. A *performance bond* guarantees completion of the project as described in the contract documents. A *payment bond* guarantees the payment of subcontractors, laborers, and suppliers by the contractor. After identification of the lowest responsible (i.e., one from a qualified bidder) and responsive (i.e., complying with bid requirements) bid, the winning bidder is notified by a letter of acceptance or notice of award. This document brings into force the actual construction contract between the owner and the contractor.

Subcontracts

Subcontracts are contracts between a prime contractor and secondary contractors or suppliers. Subcontracts are widely used in building construction for the installation of electrical, plumbing, and heating and ventilating systems. The contractual arrangements between the prime contractor and his subcontractors are similar to those between the owner and the prime contractor. However, subcontractors are responsible only to the prime contractor (not to the owner) in the performance of their subcontracts. Subcontracts are included in this section only for the purpose of relating them to the bidding process.

Since subcontract costs often make up a major portion of the cost for a project, it is essential that the prime contractor obtain timely and competitive prices for subcontract services. In fairness, the successful prime contractor should execute contracts with those subcontractors whose prices he has used for the preparation of his bid. However, after receiving the contract award, some contractors attempt to obtain lower subcontract prices by negotiating with other subcontractors. This practice is referred to as *bid shopping* and is widely considered an unethical practice which leads to poor subcontractor performance. As a result, bidding procedures often require the bidder to identify subcontractors at the time of bidding and to use only these subcontractors on the project. Some governmental agencies even go so far as to award separate prime contracts for general construction and for each area of specialty work. While protecting the subcontractors, such a procedure greatly complicates project control and coordination.

17-3 CONSTRUCTION CONTRACTS

Contract Elements

The legally essential elements of a construction contract include an offer, an acceptance, and a consideration (payment for services to be provided). The offer is normally a bid or proposal submitted by a contractor to build a certain facility according to the plans, specifications, and conditions set forth by the owner. Acceptance takes the form of a notice of award, as stated earlier. Consideration usually takes the form of cash payment, but it may legally be anything of value.

Contract Types

Contracts may be classified in several ways. Two principal methods of classification are by method of award and by method of pricing. The types of contract by *method of award* are formally advertised contracts and negotiated contracts. The procedure for the solicitation and award of an advertised construction contract was described in the previous section. A *negotiated contract,* as the name implies, is one negotiated between an owner and a construction firm. All terms and conditions of the final contract are those mutually agreed to by the two parties. While federal procurement regulations establish formally advertised competitive bidding as the normal process, negotiated contracts are permitted under special circumstances. Private owners may, of course, award a contract in whatever manner they choose.

The two types of contract by method of pricing are *fixed-price contracts* and *cost-type contracts.* Each of these types has a number of variations. There are two principal forms of fixed-price contracts: firm fixed-price contracts and fixed price with escalation contracts. Other classifications of fixed-price contracts include *lump-sum contracts* and *unit-price contracts.* A *lump-sum contract* provides a specified payment for completion of the work described in the contract documents. Unit-price contracts specify the amount to be paid for each unit of work but not the total contract amount. Such contracts are used when the quantities of work cannot be accurately estimated in advance. The principal disadvantages of unit-price contracts are the requirement for accurately measuring the work actually performed and the fact that the precise contract cost is not known until the project is completed. A combination of lump-sum and unit-price provisions may be used in a single contract.

Fixed price with escalation contracts contain a provision whereby the contract value is adjusted according to a specified price index. Such contracts reduce the risk to the contractor during periods of rapid inflation. Since the alternative during periods of inflation is for the contractor to add a large contingency amount to protect himself, the use of an escalation clause may well result in a lower cost to the owner than would a firm fixed-price contract. In spite of this, fixed-price construction contracts with escalation clauses have not been widely used in the United States.

Cost-type (or cost-plus) contracts are available in a number of forms. Some of these include:

- Cost plus percentage of cost.
- Cost plus fixed fee.
- Cost plus fixed fee with guaranteed maximum cost.
- Cost plus incentive fee.

A *cost plus a percentage of cost contract* pays the contractor a fee that is a percentage of the project's actual cost. This type of contract may not be used by U.S. government agencies because it provides a negative incentive for the contractor to reduce project cost. That is, the higher the project cost, the greater the contractor's fee. The most widely used form of cost reimbursement contract, the *cost plus fixed fee contract,* does not reward the contractor for an increased project cost but still fails to provide any incentive to minimize cost. The *cost plus fixed fee with guaranteed maximum cost contract* adds some of the risk of a fixed-price contract to the cost reimbursement contract because the contractor guarantees that the total contract price will not exceed the specified amount. Hence it is to be expected that the contractor's fee for this type of contract will be increased to compensate for the added risk involved. The *cost plus incentive fee contract* is designed to provide an incentive for reducing project cost. In this type of contract, the contractor's nominal or target fee is adjusted upward or downward in a specified manner according to the final project cost. Thus the contractor is rewarded by an increased fee if he is able to complete the project at a cost lower than the original estimate. All cost-type contracts should clearly define the items of cost for which the contractor will be reimbursed and specify the basis for determining the acceptability of costs.

Contract Documents

A construction contract consists of the following documents:

- Agreement.
- Conditions of the Contract (usually General Conditions and Special Conditions).
- Plans.
- Specifications.

The *agreement* describes the work to be performed, the required completion time, contract sum, provisions for progress payments and final payment, and lists the other documents making up the complete contract. The *general conditions* contain those contract provisions applicable to most construction contracts written by the owner. The *special conditions* contain any additional contract provisions applicable to the specific project. The contents of the *plans* and *specifications* are discussed in Section 17-4.

The Associated General Contractors of America, in cooperation with the American Institute of Architects, the American Society of Civil Engineers, and other professional organizations, has developed standard construction contract provisions and a number of associated forms. The federal government also utilizes standard contract documents. The use of such standard contract forms will minimize the amount of legal review that the contractor must perform before signing a contract. However, even if the contractor is familiar with the standard contract forms being used, care must be taken to fully evaluate all special conditions as well as the plans and specifications. The principal contract clauses and their interpretation are discussed in Section 17-5.

Construction contracts may contain a *value engineering* (VE) *clause.* Value engineering is the analysis of a design with the objective of accomplishing the required function at a lower cost. This objective may also be expressed as eliminating gold plating. When included in a construction contract, a value engineering clause encourages the contractor to propose changes in the project that will reduce project cost without affecting the ability of the facility to perform its intended function. The cost savings resulting from value engineering proposals accepted by the owner are shared between the contractor and owner on the basis specified in the contract. The usual clause prescribes a 50/50 split between the owner and contractor.

Contract Time

The time allowed (expressed as either days allowed or as a required completion date) for completion of a construction project is normally specified in the contract along with the phrase "time is of the essence." If no completion date is specified, a "reasonable time," as interpreted by the courts, is allowed. If the phrase "time is of the essence" is included in a contract and the project is not completed within the specified time, the contractor is liable for any damages (monetary loss) incurred by the owner as the result of late completion. In such a case, the courts will hold the contractor responsible for the actual damages that the owner incurs. A *liquidated damages clause* in the contract may be used to simplify the process of establishing the amount of damages resulting from late completion. Such a clause will specify the amount of damages to be paid by the contractor to the owner for each day of late completion. If challenged in court, the owner must prove that the amount of liquidated damages specified in the contract reasonably represents the owner's actual loss. If the liquidated damages are shown to be reasonable, the courts will sustain their enforcement.

Construction contracts normally contain provisions for time extensions to the contract due to circumstances beyond the control of the contractor, such as owner-directed changes, acts of God (fire, flood, etc.), and strikes. The purpose of such provisions is, of course, to reduce the contractor's risk from events beyond his control. If such provisions were not included, the contractor would have to increase his bid price to cover such risks. It should also be pointed out that the owner is financially responsible to the contractor for any owner-caused delays. The subject of changes and delays is discussed further in Section 17-5.

17-4 PLANS AND SPECIFICATIONS

Plans

Construction plans are drawings that show the location, dimensions, and details of the work to be performed. Taken together with the specifications, they should provide a complete description of the facility to be constructed. Types of contract drawings include site drawings and detailed working drawings. Contract drawings are usually organized and numbered according to specialty, such as structural, electrical, and mechanical.

Specifications

Construction technical specifications provide the detailed requirements for the materials, equipment, and workmanship to be incorporated into the project. Contract drawings and specifications complement each other and must be used together. An item need not be shown on both the plans and specifications to be required. Frequently, the item may be identified on only one of these documents. However, when the provisions of the plans and specifications conflict, the General Conditions of the contract generally provide that the requirements of the specifications will govern. In the absence of such a provision, the courts have commonly held that the requirements of the specifications will govern. The two basic ways in which the requirements for a particular operation may be specified are by method specification or by performance specification. A *method specification* states the precise equipment and procedure to be used in performing a construction operation. A *performance* (or result or end-result) *specification,* on the other hand, specifies only the result to be achieved and leaves to the contractor the choice of equipment and method. Recent years have seen an increase in the use of performance specifications, particularly by governmental agencies. Specification writers should avoid specifying both method and performance requirements for the same operation. When both requirements are used and satisfactory results are not obtained after utilizing the specified method, a dispute based on impossibility of performance will invariably result.

The format most widely used for construction specifications consists of 16 divisions, organized as shown in Table 17-1. This format was developed by the Construction Specifications Institute (CSI) and is usually identified as the CSI format or Uniform System for Building Specifications. Although developed for use on building construction projects, it is also widely used for other types of construction.

Shop Drawings and Samples

Shop drawings are drawings, charts, and other data prepared by a contractor or supplier which describe the detailed characteristics of equipment or show how specific structural elements or items of equipment are to be fabricated and installed. Thus they complement but do not replace the contract drawings. Samples are physical ex-

Table 17-1 Organization of the Uniform System for Building Specifications

Division	*Title*
1	General Requirements
2	Site Work
3	Concrete
4	Masonry
5	Metals
6	Wood and Plastics
7	Thermal and Moisture Protection
8	Doors and Windows
9	Finishes
10	Specialties
11	Equipment
12	Furnishings
13	Special Construction
14	Conveying Systems
15	Mechanical
16	Electrical

amples of materials, equipment, or workmanship which are submitted to the owner for approval prior to their incorporation in a project.

Contract documents should contain the specific requirements for submission of shop drawings and samples. Some suggested provisions include:

- Identification of items requiring samples or shop drawings.
- Procedure for submission of shop drawings, including format, marking, and number and distribution of copies.
- Procedure for submission of samples, including size and number required.
- Eliminating the requirement for shop drawings and samples when standard catalog items are to be used.

17-5 CONTRACT ADMINISTRATION

Progress Reports and Payment

Construction contracts commonly require the contractor to submit a proposed progress schedule to the owner shortly after contract award. Upon approval by the owner or his representative, this schedule forms the basis for judging the contractor's progress toward project completion. The contract may require the contractor to submit his plan and schedule in the CPM format (Chapter 15) and may also require periodic updating of the schedule as work progresses. The owner's representative must continuously evaluate the contractor's progress to keep the owner informed and to provide a basis for the approval of the contractor's requests for progress payment

Failure of the contractor to attain a satisfactory rate of progress may provide the basis for termination of the contract by the owner, as described later in this section.

For projects expected to require more than a few months to complete, it is customary for the owner to make *progress payments* to the contractor. Progress payments are made at the interval specified in the contract, usually monthly or upon completion of certain milestones. Payment is customarily made for the work completed, materials delivered to the work site, and work prefabricated but not yet incorporated into the project. It is customary to withhold a percentage of the value of work completed as a guarantee against defective work and to ensure that the remaining work can be completed within the unpaid amount of the contract. The amount withheld is referred to as *retainage* or *retention.* A retainage of 10% is rather typical.

Changes and Delays

It is rare indeed if a construction project is completed without changes being made. The usual construction contract contains a clause authorizing the owner or his representative to order changes to the project within the general scope of the contract. The document directing such a change is referred to as a *change order.* The contract also provides that an equitable adjustment in time and contract value will be made for such changes. The majority of changes are due to design modifications initiated by the owner or designer. However, change orders may also be used to formalize adjustments to the contract required by site conditions differing from those anticipated at the time of contract award (commonly referred to as "changed conditions").

To minimize disputes, all change orders issued should contain an adjustment in contract time and price which is mutually acceptable to the contractor and owner. However, it is frequently not possible to delay issuing a change order until such an agreement has been reached without delaying the work in progress. As a result, many change orders are issued before an agreement has been reached on the corresponding price and time adjustment. Agreement must therefore be reached later as work progresses or the item will end up as a dispute. In estimating the cost associated with a change or owner-caused delay, the contractor must be careful to evaluate its effect on other project activities. Frequently, it will be found that changes or delay in one activity will necessitate changes in resource allocation or progress on other activities that result in additional project cost. These costs are sometimes referred to as *consequential costs.* To obtain reimbursement of consequential costs, the contractor must be able to document their existence. A CPM network is a valuable aid in identifying and justifying consequential costs.

Delays in the orderly progress of a construction project may result from a multitude of causes. The three general categories of delay include those beyond the control of either the contractor or owner ("acts of God"), those under the control of the owner, and those under the control of the contractor. The general principles established by law and precedent for financial and time adjustments to the contract as a result of such delays are as follows. In the case of fire, flood, earthquake, or other disaster, and strikes, a compensating time extension to the contract will be made. Any financial compensation to the contractor would be provided by the contractor's insurance, not by the owner. If the owner is responsible for the delay (such as by the late delivery of owner-provided equipment), the owner must compensate the contractor for any additional costs incurred as well as provide an appropriate time extension to the contract. If the delay is under the control of the contractor, no compensation or time extension is provided to the contractor. Rather, the contractor is responsible for reimbursing the owner for any damages (actual or liquidated) resulting from the delay.

Acceptance and Final Payment

The acceptance of a completed project is customarily based on a final inspection performed by the owner's representative and conditioned upon the correction of any deficiencies noted. The list of deficiencies to be corrected which is prepared at the final inspection is sometimes referred to as the *punch list of record.* If the facility or a portion thereof is substantially complete, the owner's representative will execute a *certificate of substantial completion* for the work. The contractor may then request and receive a final progress payment for the completed portion of the project. However, sufficient retainage is withheld to ensure the correction of any remaining deficiencies. The certificate of substantial completion should clearly state the responsibilities of the contractor and the owner of maintenance, utility service, and insurance until final acceptance.

Upon correction of all deficiencies on the punch list of record, the contractor should notify the owner's representative of this fact and submit a *request for final payment,* together with any other documents required by the contract (such as releases of liens, an affidavit that all payrolls and bills connected with the project have been paid, consent of surety to final payment, etc.). When inspection confirms the correction of all deficiencies, the owner's representative will issue a final *certificate of payment.* The contract customarily provides a warranty against defective work for some period, usually 1 year. Any deficiencies discovered after preparation of the punch list of record should be handled under the warranty provision of the contract. Final payment and its acceptance by the contractor usually constitute a waiver of all claims by either the owner or contractor except for unsettled liens and claims and deficiencies falling under warranty provisions.

Claims and Disputes

A *claim* is a request by the contractor for a time extension or for additional payment based on the occurrence of an event beyond the contractor's control that has not been covered by a change order. Examples of such events include unexpected site conditions, delays in delivery of owner-provided property, and changes directed by the owner. The usual construction contract empowers the owner's representative (architect/engineer or government contracting officer) to decide on the validity of such claims. However, if the contractor is not satisfied with the decision, the matter becomes a dispute.

Disputes are disagreements between the contractor and owner over some aspect of contract performance. In addition to unsettled claims, disputes may involve such matters as substitution for specified materials, the responsibility for delays in project completion, and the effect of changes ordered by the owner. In recent years there has been an increase in the use of *alternate dispute resolution (ADR)* methods instead of taking the matter to court. When successful, these nonjudicial techniques greatly reduce the time and expense involved in settling disputes. Some ADR techniques include negotiation, mediation, arbitration, nonbinding minitrials, and neutral fact finding. Probably the most common of these techniques are negotiation and arbitration. In 1966, the American Arbitration Association, together with a number of professional organizations involved in construction, established arbitration procedures for the construction industry, known as the *Construction Industry Arbitration Rules.* Under these procedures one or more independent professionals are appointed to resolve the dispute. Hearing procedures are less formal than those of a trial and the arbitrators are not bound by the legal rules of evidence. Because the parties to the dispute must agree to the use of arbitration, no appeal of the arbitration award is usually possible. State laws governing the use of arbitration vary and some states do not recognize the use of a contract clause requiring arbitration of all disputes arising under the contract.

Contract Termination

Although contract termination is usually envisioned as an adversary process, there are a number of nonadversary methods by which a contract may be terminated. Most construction contracts are terminated by satisfactory performance, one method of contract termination. Other nonadversary methods of contract termination include mutual agreement and impossibility of performance.

The principal adversary basis for contract termination is for breach of contract. Either the owner or the contractor may terminate a contract for breach of contract. The basis for termination by the contractor based on breach of contract is usually the failure of the owner to make the specified progress payments or owner-caused delay of the project for an unreasonable period of time. Termination by the owner for breach of contract is most commonly due to failure of the contractor to make reasonable progress on the project or to default by the contractor. When termination is due to breach of contract by the owner, the contractor is generally held to be entitled to payment for all work performed and the expenses of demobilization and cancellation of orders, plus profit. When termination is due to breach of contract by the contractor, the contract commonly permits the owner to take possession of the work site and all on-site equipment and tools owned by the contractor and to complete the project at the contractor's (or surety's) expense.

PROBLEMS

1. Briefly describe the steps that a contractor takes in preparing a bid for a fixed-price construction contract.
2. How does a construction company's bonding capacity (bonding limit established by a bonding company) affect the ability of the construction firm to bid on projects?

3. What is bid shopping, and why is it considered a bad practice?
4. What type of construction contract provides the greatest incentive for a construction contractor to minimize project cost?
5. Explain the purpose and operation of a value engineering clause in a construction contract.
6. Briefly explain the importance of complete and accurate specification preparation for a construction contract.
7. What are shop drawings, and who prepares them?
8. Explain the difference between a claim and a dispute involving a construction contract.
9. What methods are available for settling a contract dispute between a contractor and an owner?
10. Write a computer program that can be used to maintain the current status of all active contracts of a construction firm. Input should include contract number and description, contract amount, date of contract award, date work started, required completion date, current work status (percent complete), projected completion date, amount billed to date, payments received to date, payments due but not received, number and value of contract modifications, and number and value of pending modifications and claims. Provide output in a format that can be used by company management as a summary of contract status. Using your computer program, solve an example problem.

REFERENCES

1. *Avoiding and Resolving Disputes During Construction.* American Society of Civil Engineers, New York, 1991.
2. CLOUGH, RICHARD H. *Construction Contracting,* 5th ed. New York: Wiley, 1986.
3. COLLIER, KEITH. *Managing Construction Contracts.* Reston, Va.: Reston, 1982.
4. DUNHAM, CLARENCE W., AND ROBERT D. YOUNG. *Contracts, Specifications, and Law for Engineers,* 4th ed. New York: McGraw-Hill, 1985.
5. FISK, EDWARD R. *Construction Project Administration,* 3rd ed. New York: Wiley, 1988.
6. *Recommended Guide for Competitive Bidding Procedures and Contract Awards for Building Construction* (AIA Document A501). The Associated General Contractors of America, Inc., Washington, D.C., 1982.
7. ROSEN, HAROLD J. *Construction Specifications Writing,* 2nd ed. New York: Wiley, 1981.
8. SMITH, ROBERT J. *Recommended Bidding Procedures for Construction Projects.* Engineers Joint Contracts Documents Committee, American Society of Civil Engineers, New York, 1987.

18 CONSTRUCTION SAFETY AND HEALTH

18-1 IMPORTANCE OF SAFETY

Construction is inherently a dangerous process that has traditionally had a high accident rate. In the United States, national concern over the frequency and extent of industrial accidents and health hazards led to the passage of the *Occupational Safety and Health Act (OSHA)* of 1970, which established specific safety and health requirements for virtually all industries, including construction. The extent and nature of OSHA requirements is discussed further in Section 18-2. However, the concern over OSHA regulations and penalties has tended to obscure the fact that there are at least two other major reasons for construction management to be seriously concerned about safety. These reasons are humanitarian and financial.

Everyone is understandably distressed when a fellow employee is killed or disabled, so the humanitarian basis for safety is apparent. However, many managers do not fully appreciate the financial consequences of accidents. Worker's compensation insurance premiums, for example, are based on a firm's accident rate. Public liability, property damage, and equipment insurance rates are also affected by accident rates. It has been shown that a construction firm can lose its competitive bidding position simply because of the effect of high insurance premiums resulting from a poor safety record. In addition to the visible cost of accidents represented by insurance and worker's compensation payments, there are other costs, which are difficult to estimate. Such costs associated with an accident include the monetary value of lost project time while the accident is investigated and damages are repaired, the time required to replace critical materials and equipment and to train replacement workers, as well as the effect on those portions of the project not directly involved in the accident.

18-2 OSHA

The U.S. Occupational Safety and Health Act (OSHA) has produced a comprehensive set of safety and health regulations, inspection procedures, and record-keeping requirements. The law also established both civil and criminal penalties for violations of OSHA regulations. Table 18-1 indicates the maximum penalty for major categories of violations. As shown in Table 18-1, civil penalties of $7000 per day may be assessed for failure to correct a cited violation. Under criminal proceedings, a fine of $20,000 and imprisonment for 1 year may be adjudged for a second conviction of a violation resulting in the death of an employee. OSHA officials may also seek a restraining order through a U.S. District Court to stop work or take other action required to alleviate a condition identified as presenting imminent danger of serious injury or death.

Under OSHA regulations employers are required to keep records of all work-related deaths, injuries, and illnesses. It is not necessary to record minor injuries that require only first-aid treatment. However, all injuries involving medical treatment, loss of consciousness, restrictions on work or body motion, or transfer to an-

Table 18-1 Maximum penalties under OSHA

Administrative Proceedings

Violation	*Maximum Penalty*
Willful or repeated	$70,000/violation
Routine or serious	$7,000/violation
Failing to correct cited violation	$7,000/day
Failing to post citation near the place where violation exists	$7,000/violation

Criminal Proceedings

Violation	*Maximum Fine*	*Maximum Imprisonment*
Killing, assaulting, or resisting OSHA officials	$10,000	Life
Willful violation resulting in death of employee, first conviction	10,000	6 months
Willful violation resulting in death of employee, second conviction	20,000	1 year
Falsifying required records	10,000	6 months
Unauthorized advanced notice of inspection	1,000	6 months

other job must be recorded. A special report of serious accidents resulting in one or more deaths or the hospitalization of five or more employees must be made to OSHA officials within 48 hours.

One of the major inequities of OSHA is that only management may be penalized for safety violations. Thus even though an employee willfully violates both OSHA and company safety regulations, only the company and its management can be penalized under OSHA for any safety violation. Therefore, the only way in which management may enforce safety regulations is to discipline or fire workers engaging in unsafe acts.

OSHA safety regulations for construction (reference 2) consist largely of safety standards developed by segments of the construction industry. Requirements for equipment safety include rollover protection (ROPS), seat belts, back-up alarms, improved brake systems, and guards for moving parts. Maximum noise levels are also set for equipment operators and other workers. A number of OSHA safety requirements have been mentioned in earlier chapters in connection with specific construction operations. However, supervisors at all levels must be familiar with all applicable OSHA standards.

It should be pointed out that OSHA safety regulations are considered to be the minimum federal safety standards and that the various states may impose more stringent safety standards for construction within the state. The U.S. Department of Labor has also delegated to certain states the authority and responsibility for enforcing OSHA regulations within those states.

18-3 SAFETY PROGRAMS

All construction firms need a carefully planned and directed safety program to minimize accidents and ensure compliance with OSHA and other safety regulations. However, no safety program will be successful without the active support of top management. Job-site supervisors have traditionally neglected safety in their haste to get the job done on time and within budget. Only when supervisors are convinced by higher management that safety is equally as important as production will the benefits of an effective safety program be achieved. An effective safety program must instill a sense of safety consciousness in every employee.

Although there are many ingredients in a comprehensive safety program, some of the major elements are listed below.

1. A formal safety training program for all new employees. *Note:* OSHA Regulations require every employer to "instruct each employee in the recognition and avoidance of unsafe conditions and the regulations applicable to his work environment. . . ."
2. Periodic refresher training for each worker.
3. A formal supervisory safety training program for all supervisors.
4. A program of regular site visits by safety personnel to review and control job hazards.

5. Provision of adequate personal protective equipment, first-aid equipment, and trained emergency personnel.
6. An established procedure for the emergency evacuation of injured workers.
7. Provisions for maintaining safety records and reporting accidents in compliance with OSHA requirements.

The accident prevention manual of the Associated General Contractors of America, Inc. (reference 6) provides many suggestions for an effective construction safety program.

18-4 SAFETY PROCEDURES

It has been found that most serious construction accidents involve construction equipment operations, trench and embankment failure, falls from elevated positions, collapse of temporary structures and formwork, or the failure of structures under construction. OSHA safety regulations (reference 2) are quite specific in many of these areas, and special management attention should be devoted to the safety of these activities.

Reference 4 provides an excellent discussion, along with examples, of construction failure involving both design and construction practice. The safety manuals published by the Construction Industry Manufacturers Association (reference 3) provide safety rules and suggestions for the safe operation of many types of construction equipment.

A complete list of construction safety rules would fill a large volume. Such a list is clearly beyond the scope of this book. However, the following list of major safety precautions should be helpful as a general guide.

General

Good housekeeping on a project site is both a safety measure and an indicator of good project supervision. Lumber, used formwork, and other material lying around a work area increase the likelihood of falls and puncture wounds.

Equipment Operations

- Require operators and mechanics to use steps and hand holds when mounting equipment.
- Utilize guides or signalmen when the operator's visibility is limited or when there is danger to nearby workers. Backup alarms or guides must be used when equipment operates in reverse.
- Exercise extreme caution and comply with safety regulations when operating near high-voltage lines. In case of contact with a high-voltage line, the operator should attempt to move the equipment enough to break contact. If unsuccessful, the operator should remain on the equipment until the line can be deenergized.
- Make sure that machines are equipped with required safety features and that operators use seat belts when provided.

- Use care when operating equipment on side slopes to prevent overturning.
- When operating cranes, be extremely careful not to exceed safe load limits for the operating radius and boom position. Electronic load indicators are available.
- Do not allow workers to ride on equipment unless proper seating is provided.
- Haul roads must be properly maintained. Items to check include condition of the road surface (holes, slippery surface, excess dust), visibility (curves, obstacles, intersections, and dust), and adequate width for vehicles to pass (unless one-way).
- Park equipment with the brake set, blade or bowl grounded, and ignition key removed at the end of work.
- Equipment used for land clearing must be equipped with overhead and rear canopy protection. Workers engaged in clearing must be protected from the hazards of irritant and toxic plants and instructed in the first-aid treatment for such hazards.
- When hauling heavy or oversized loads on highways, make sure that loads are properly secured and covered if necessary. Slow-moving and over-sized vehicles must use required markings and signals to warn other traffic.
- Take positive action to ensure that equipment under repair cannot be accidentally operated.
- Utilize blocking, cribbing, or other positive support when employees must work under heavy loads supported by cables, jacks, or hydraulic systems.
- Ensure that any guards or safety devices removed during equipment repair are promptly replaced.
- Shut down engines and do not allow smoking during refueling.

Construction Plant

- Set equipment containing hot or flammable fluids on firm foundations to prevent overturning. Clearly mark high-temperature lines and containers to prevent burns. Be especially careful of live steam. Provide fire extinguishers and other required safety equipment.
- Aggregate bins and batching plants should be emptied before performing major repairs.
- When electrical equipment is being repaired, shut off and tag electrical circuits.
- Ensure that wire rope and cable is of the proper size and strength, well maintained, and inspected at least weekly.

Excavations

- The sides of excavations must be properly shored or sloped to the angle of repose to prevent cave-ins. OSHA regulations require that banks over 5 ft (1.5 m) must be shored, cut back to a stable slope, or otherwise protected.
- When employees are required to work in a trench 4 ft (1.2 m) or more in depth, provide an adequate means for emergency exit located within 25 ft (7.6 m) of the worker.

- Avoid the operation of equipment near the top edge of an excavation because this increases the chance of slope failure. The storage of materials near the top edge of an excavation, vibration, and the presence of water also increase the chance of slope failure. When these conditions cannot be avoided, additional measures must be taken to increase slope stability. If workers are required to enter the excavation, no spoil or other material may be stored within 2 ft (0.6 m) of the edge of the excavation.
- Ensure that workers are not allowed under loads being handled by excavators or hoists.
- Watch out for buried lines and containers when excavating. Possible hazards include toxic and flammable gases, electricity, and collapse of side slopes due to sudden release of liquids. If a gas line is ruptured and catches fire, get personnel and flammable material away from the fire and have the gas turned off as quickly as possible. Do not attempt to extinguish the fire because an accumulation of unburned gas poses a greater threat than does a fire.

Construction of Structures

- Properly guard all openings above ground level.
- Provide guard rails, safety lines, safety belts, and/or safety nets for workers on scaffolds or steelwork.
- Ensure that temporary structures are properly designed, constructed, and braced.
- Special caution should be exercised in high-rise concrete construction. Forms must be of adequate strength and properly braced. The rate of pour must be maintained at or below design limits. Shoring and reshoring must be adequately braced and not removed until the concrete has developed the required strength.

Marine or Over-Water Construction

Marine or over-water construction operations present all of the usual construction hazards plus additional hazards posed by the marine environment. These additional hazards include drowning, slippery surfaces, increased tripping and height hazards, as well as weather and wave action. Some of the major safety precautions that should be taken are listed below.

- Unless workers can safely step onto vessels, a ramp or safe walkway must be provided. Access ways must be adequately illuminated, free of obstructions, and located clear of suspended loads.
- Working areas should have nonslip surfaces, be maintained clear of obstructions, and be equipped with adequate handrails.
- Workers on unguarded decks or surfaces over water must wear approved lifejackets or buoyant vests. Life rings and a rescue boat must also be available. Workers more than 25 ft (7.6 m) above a water surface must be protected by safety belts, safety nets, or similar protective equipment.

18-5 ENVIRONMENTAL HEALTH IN CONSTRUCTION

Increased governmental interest in occupational safety has been accompanied by an increased concern for occupational health and environmental controls. The major environmental health problems encountered in construction consist of noise, dust, radiation (ionizing and nonionizing), toxic materials, heat, and cold. These hazards and appropriate control measures are discussed in the following paragraphs.

Noise

OSHA construction safety and health regulations (reference 2) prescribe maximum noise levels to which workers may be exposed. Permissible noise levels are a function of length of exposure and range from 90 dBA (decibels measured on the A-scale of a standard sound meter) for an 8-h exposure to 140 dBA for impulse or impact noise. When a satisfactory noise level cannot be attained by engineering controls, personal ear protection must be provided.

Noise controls have resulted in the increasing use of cab enclosures on construction equipment to protect equipment operators from equipment noise. The use of such enclosures has necessitated improved equipment instrumentation to enable the operator to determine whether the machine is operating properly without depending on the sound of the equipment's operation. Although the use of operator enclosures permits an improved operator environment, it also creates a safety hazard, because it is difficult for workers outside the enclosures to communicate with the equipment operator. As a result, increased attention must be given to the use of guides, backup alarms, and hand signals if accidents are to be avoided.

Dust

In addition to creating a safety hazard due to loss of visibility, dust may be responsible for a number of lung diseases. Silica dust and asbestos dust are particularly dangerous and produce specific lung diseases (asbestosis and silicosis). Asbestos dust has also been found to be a cancer-producing agent. As a result, OSHA safety and health standards limit the concentration of dust to which workers may be exposed. The allowable concentration of asbestos particles is, as you might expect, quite low.

Radiation

Ionizing radiation is produced by X-ray equipment and by radioactive material. Such radiation may be encountered on the construction site when X-raying welds, measuring soil density, or performing nondestructive materials testing. Any use of such equipment must be accomplished by trained personnel in accordance with regulations of the Nuclear Regulatory Commission.

Nonionizing radiation is produced by laser equipment and electronic microwave equipment. Laser equipment is coming into widespread use for surveying and for alignment of pipelines, tunnels, and structural members. Again, only well-trained employees should be permitted to operate such equipment. OSHA regulations limit the exposure of workers to both laser output and microwave power output. Workers must be provided antilaser eye protection when working in areas having a potential exposure to laser light output greater than 5 mW.

Toxic Materials

Construction workers may accidentally encounter toxic materials at any time, particularly on reconstruction projects. However, the most frequent hazards consist of buried utility lines and underground gases. Every effort must be made to locate and properly protect utility lines during excavation operations. The air in a work area should be tested whenever an oxygen deficiency or toxic gas is likely to be encountered. Emergency rescue equipment such as breathing equipment and lifelines should be provided whenever adverse atmospheric (breathing) conditions may be encountered. Specific safety procedures and protective equipment should be provided if hazardous liquids or solids are likely to be encountered.

Heat

Construction workers are often required to work under high-temperature conditions. Fortunately, the human body will acclimate itself to high-temperature conditions within a period of 7 to 10 days. However, serious heat illness may result when workers are not properly acclimated and protected. Medical effects range from fatal heat stroke to minor heat fatigue. It is particularly important to health that the body's water and salt levels be maintained. Heat cramps result when the body's salt level drops too low. Factors that have been found to increase the heat strain experienced by workers include drug consumption, fever from an infection, exposure to low-frequency noise, and exposure to environmental gases such as carbon monoxide.

Methods for reducing heat effect on workers include use of mechanical equipment to reduce physical labor requirements, scheduling hot work for the cooler part of the day, use of sun shields, providing cool rest areas [optimum temperatures about 77 °F (25 °C)], providing a water and salt supply easily accessible to workers, and the use of proper hot-weather clothing.

Cold

Extreme cold-weather conditions, although not encountered as often as heat conditions, pose essentially opposite problems to those of hot-weather operations. The human body will acclimate itself to cold as it will to heat, but the acclimation period for cold is much longer. Medical effects of cold include frostbite, trenchfoot, and general hypothermia (reduction of the core body temperature). General hypothermia is usually fatal when the body core temperature drops below 65 °F (18 °C).

Military operations have demonstrated that human beings can successfully perform in temperatures much lower than those encountered in the continental United States when they are properly clothed, fed, and acclimated. Thus the major requirement for successful cold-weather construction appears to be the provision of adequate clothing and warming areas. The use of bulky cold-weather clothing, however, reduces manual dexterity and may increase the possibility of accidents.

PROBLEMS

1. How might a poor safety record cause a construction firm to become noncompetitive in bidding for projects?
2. Briefly describe at least four major elements of a comprehensive construction safety program.
3. What is the maximum penalty provided under OSHA for a willful violation resulting in the death of an employee?
4. Identify the major safety hazards involved in crane operations, and describe how these hazards might be minimized.
5. Explain the action that should be taken if an excavator ruptures a natural gas line.
6. Identify the major categories of environmental health problems encountered in construction.
7. In what areas of construction operations do most serious accidents occur?
8. Briefly describe the major health hazards involved in construction operations under high-temperature conditions and how these hazards can be minimized.
9. What job site conditions serve to increase the chance of an excavation slope failure?
10. Suggest at least two ways in which a microcomputer might be employed to improve the safety and health performance of a construction company.

REFERENCES

1. *AGC Guidelines for a Basic Safety Program.* The Associated General Contractors of America, Inc., Washington, D.C., 1989.
2. *Code of Federal Regulations,* Title 29, Chapter XVII, Part 1926. U.S. Department of Labor, Washington, D.C.
3. *Construction Machine Operators' Safety Manuals.* Construction Industry Manufacturers Association, Milwaukee, Wis., for the following machines: cold planer, crane, crawler tractor/loader, dumper (off-highway truck), hydraulic excavator, motor grader, portable air compressor, roller compactor, scraper, wheel tractor/loader.
4. FELD, JACOB. *Construction Failure.* New York: Wiley, 1968.
5. LEE, D. H. K. *Heat and Cold Effects and Their Control* (Public Health Monograph No. 72). U.S. Department of Health, Education, and Welfare, Washington, D.C., 1964.
6. *Manual of Accident Prevention in Construction.* The Associated General Contractors of America, Inc., Washington, D.C., 1992.
7. *OSHA Safety and Health Standards Digest: Construction Industry* (OSHA 2202). U.S. Department of Labor, Washington, D.C., 1990.

19
Improving Productivity and Performance

19-1
THE BIG PICTURE

State of the Industry

The serious decline in U.S. construction industry productivity during the 1960s and 1970s led the Business Roundtable to conduct its Construction Industry Cost Effectiveness (CICE) study of the industry described in Chapter 1. This study, completed in 1982, was probably the most comprehensive study ever made of the U.S. construction industry. Although the study found that the U.S. construction industry faced a number of problems in remaining competitive in the international construction market, it concluded that the majority of problems could be overcome by improved management of the construction effort. At the project management level, the study discovered inadequate management performance in a number of areas. These included construction safety, control of the use of overtime, training and education, worker motivation, and failure to adopt modern management systems. Thus the purpose of this chapter is to look at ways in which construction managers can improve construction productivity and performance.

What Is Productivity?

There is serious disagreement about the proper definition of the term "productivity" within the construction industry. As usually employed, the term means the output of construction goods and services per unit of labor input. Obviously, such a definition ignores the contribution of technology and capital investment to the measured productivity. The heavy construction element of the industry has demonstrated that the use of larger and more productive earthmoving equipment can increase productivity and lower unit production costs in the face of generally rising labor and materials costs. The continued rapid growth of technology in the world economy makes it likely that new technology such as robotics and industrialized building processes will have a significant impact on construction productivity in the not too distant future.

Probably a better measure of construction industry performance is cost-effectiveness as used by the Business Roundtable CICE study. However, for the purpose of this chapter we will use the traditional definition of productivity as output per unit input of labor and focus our attention on ways in which construction industry productivity and cost-effectiveness can be increased by improved management.

Tools for Better Management

A number of studies, including the CICE study, have shown that most on-site delays and inefficiencies lie within the control of management. Management is responsible for planning, organizing, and controlling the work. If these management responsibilities were properly carried out, there would be few cases of workers standing idle waiting for job assignment, tools, or instructions. As you see, the scope of management responsibility is great and the techniques for efficiently carrying out these responsibilities are varied and complex. Many books have been written on individual topics and techniques in this area. Thus the purpose of this chapter is simply to introduce the reader to some of these techniques and their potential for improving the management of construction.

One of the major tools for improving construction productivity is *work improvement:* that is, the scientific study and optimization of work methods. Such techniques are also known as *work simplification, motion and time study, work study,* and *methods analysis.* Human factors, often not adequately considered, also play an important part in productivity. Workers' physical capacity, site working conditions, morale, and motivation are important elements in determining the most effective work methods and the resulting productivity for a particular task.

Other techniques available to assist the construction manager in improving construction productivity and cost-effectiveness include network planning methods, economic analyses, safety programs, quantitative management methods, simulation, and the use of computers. Many of these topics have been introduced in previous chapters. Other major topics are discussed in the following sections.

19-2 WORK IMPROVEMENT

What Is It?

Techniques for improving industrial production by scientific study of work methods can be traced back many centuries. However, only in this century did such techniques begin to be widely adopted by manufacturing industries. Frederick W. Taylor and Frank Gilbreth were among the early pioneers of what came to be known as *scientific management* and which today forms the basis for the field of industrial engineering. However, it is interesting to note that Frank Gilbreth began his career as a bricklayer and performed his early studies in that field. Another pioneer, D. J. Hauer, published his book *Modern Management Applied to Construction* in 1918. In spite of these early efforts by Gilbreth, Hauer, and others, work improvement methods were never widely adopted by the construction industry. Today there is renewed interest in these techniques by a construction industry faced with declining productivity and cost-effectiveness.

An important but often overlooked component of work improvement is preplanning, that is, detailed planning of work equipment and procedures prior to the start of work. Physical models as well as traditional work improvement charts and diagrams may be used to advantage in the preplanning process. Models are often used for large and complex projects such as power plants, dams, and petrochemical process plants to check physical dimensions, clearance between components, and general layout. Carried to a greater level of detail, they are very useful in planning concrete placement, blockouts for placing equipment, erection of structural components, and the actual procedure for placing equipment into a structure. Computer graphics and computer-aided design (CAD) can perform similar functions faster and at lower cost than can physical models or other manual techniques.

Traditional work improvement techniques, described in more detail below, include time studies, flow process charts, layout diagrams, and crew balance charts.

Time Studies

Time studies are used to collect time data relating to a construction activity for the purpose of either statistical analysis or of determining the level of work activity. In either case, it is important that the data collected be statistically valid. Hence a random-number procedure is usually employed in selecting the time for making each observation. The number of observations required for statistical validity depends on the type of study being made. For time analysis, the number of observations required depends on sample size, the standard deviation of the sample, and the level of accuracy and confidence desired. For effectiveness ratings, the number of observations required depends on the confidence desired, the acceptable error, and the measured percentage of effectiveness.

Work sampling is the name for a time study conducted for the purpose of determining the level of activity of an operation. A study of a construction equipment operation, for example, may classify work activity into a number of categories, each designated as either active or nonworking. The number of active observations divided by the total number of observations will yield the level of activity. The distribution of observations by category will provide an indication to management of how machine time is being spent.

The types of work sampling performed to determine labor utilization and effectiveness include field ratings and 5-minute ratings. *Field ratings* are used to measure the level of activity of a large work force. At the selected random times, each worker is observed and instantaneously classified as either working or nonworking. The number of working observations divided by the total number of observations yields the level of activity. The *5-minute rating* is used primarily to measure the level of activity of a crew. Each crew member is observed for a minimum of 1 minute (or a minimum of 12 minutes per crew). If the crew member is working more than 50% of the time observed, the observation is recorded as working.

Sampling for labor effectiveness may also divide observations into categories. Some common categories used include effective work, essential contributory work, ineffective work, and nonworking. Analysis of work by category will again assist management in determining how labor time is being utilized and provide clues to increasing labor effectiveness.

Although time studies are traditionally made using stopwatches and data sheets, there is growing use of time-lapse photography and video time-lapse recording for this purpose. The use of time-lapse equipment for conducting work improvement studies on construction projects provides several advantages over stopwatch studies. A permanent record of the activity is provided which can be studied as long as necessary to obtain necessary time data. In addition, a historical record of the activity is obtained which may be useful in training managers and supervisors as well as providing evidence in event of legal disputes. Modified super-8mm cameras and projectors provide a relatively inexpensive method of recording and analyzing time-lapse film. One super-8mm film cartridge will provide approximately 4 hours of recording at a filming interval of one frame per 4 seconds.

Flow Process Charts

A *flow process chart* for a construction operation serves the same purpose as does a flowchart for a computer program. That is, it traces the flow of material or work through a series of processing steps (classified as operations, transportations, inspections, delays, or storages). Depending on the level of detail, it usually indicates the distance and time required for each transportation and the time required for each operation, inspection, or delay. From the chart the manager should be able to visualize the entire process and to tabulate the number of operations, transportations, inspections, delays, and storages involved, and the time required for each category.

In preparing a flow process chart (see Figure 19-1), list in sequence a brief description of each step as it occurs. Trace the work flow by connecting the appropriate symbol in the second column. Enter the transportation distance and the time involved for each step. Figure 19-1 illustrates a flow process chart for the assembly of the roof truss shown in Figure 19-2 employing a crew of two workers and a forklift. The production rate for the process is determined by the time required to perform those steps that cannot be performed concurrently. This time is called the *control factor.* The control factor can be reduced only by speeding up these steps or by devising a method that permits some of its activities to be performed concurrently.

FLOW PROCESS CHART		NUMBER 101	PAGE NO. 1	NO. OF PAGES 1
PROCESS Assemble Truss				
[X] MAN OR [ ] MATERIAL				
CHART BEGINS Parts stack	CHART ENDS Parts stack			
CHARTED BY J. Doe	DATE 7/13			
ORGANIZATION E Z Construction				

SUMMARY

ACTIONS	PRESENT NO.	PRESENT TIME	PROPOSED NO.	PROPOSED TIME	DIFFERENCE NO.	DIFFERENCE TIME
○ OPERATIONS	10	137				
⇨ TRANSPORTATIONS	9	90				
□ INSPECTIONS	0					
D DELAYS	0					
▽ STORAGES	0					
DISTANCE TRAVELLED (Feet)	300					

DETAILS OF [X] PRESENT [ ] PROPOSED METHOD

No.	Details	Operation / Transportation / Inspection / Delay / Storage	Distance in feet	Quantity	Time (sec)	Notes
1	Remove chords from stack	○		2	3	
2	Transport chord to jig	⇨	25	2	10	
3	Position chords in jig	○		2	5	
4	Return to parts stack	⇨	25		6	
5	Remove rafters from stack	○		2	3	
6	Transport rafters to jig	⇨	25	2	10	
7	Position rafters in jig	○		2	5	
8	Return to parts stack	⇨	25		6	
9	Remove diagonals	○		2	3	
10	Transport diagonals	⇨	25	2	10	
11	Position diagonals in jig	○		2	5	
12	Return to parts stack	⇨	25		6	
13	Remove hanger from stack	○		1	3	
14	Transport hanger to jig	⇨	25	1	10	
15	Position hanger in jig	○		1	5	
16	Fasten truss plates	○		12	85	
17	Remove truss from jig	○		1	20	
18	Trans & stack truss	⇨	50	1	15	Using forklift
19	Return to parts stack	⇨	75		17	
20						Cycle time
21						= 227 sec

Analysis columns (blank): WHY? — WHAT? WHERE? WHEN? WHO? HOW?; CHNGE — ELIMINATE, COMBINE, SEQUENCE, PLACE, PERSON, IMPROVE

Figure 19-1 Flow process chart.

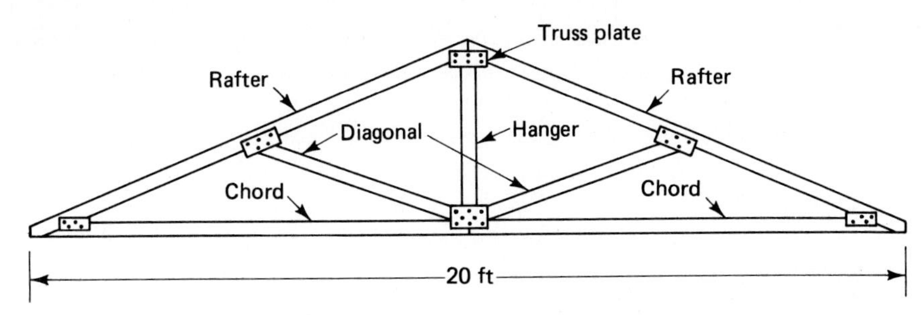

Figure 19-2 Roof truss diagram.

After preparing a flow process chart, it should be analyzed and revised to reduce the number of operations, movements, storages, and delays, as well as the control factor, to a minimum. Challenge each step in the process. Ask yourself: Is it necessary? Is it being done at the proper place and in the most efficient manner? How can it be done faster and safer?

Layout Diagrams

A *layout diagram* is a scaled diagram that shows the location of all physical facilities, machines, and material involved in a process. Since the objective of a work improvement study is to minimize processing time and effort, use a layout diagram to assist in reducing the number of material movements and the distance between operations. A *flow diagram* is similar to a layout diagram but also shows the path followed by the worker or material being recorded on a flow process chart. The flow diagram should indicate the direction of movement and the locations where delays occur. Step numbers on a flow diagram should correspond to the sequence numbers used on the corresponding flow process chart.

It should be apparent that flow process charts, flow diagrams, and layout diagrams must be studied together for maximum benefit and must be consistent with each other. Since layout diagrams and flow diagrams help us to visualize the operation described by a flow process chart, these diagrams should suggest jobs that might be combined, storages that might be eliminated, or transportations that might be shortened. The objective is to position the materials and machines so that the shortest possible path can be used without creating traffic conflicts or safety hazards.

Crew Balance Charts

A *crew balance chart* uses a graphical format to document the activities of each member of a group of workers during one complete cycle of an operation. A vertical bar is drawn to represent the time of each crew member during the cycle. The bar is then divided into time blocks showing the time spent by that crew member on each activity which occurs during the cycle. The usual convention utilizes a color code to indicate the level of activity during each time block. The darker the color, the higher the level of activity. Thus effective work might be shown by a dark block, contributory work by a lighter-colored block, and noneffective work or idle time by a white block. As the name indicates, the crew balance chart enables us easily to compare the level of activity of each worker during an operation cycle. Often its use will suggest ways to reduce crew size or to realign jobs so that work is equalized between crew members.

A crew balance chart for the assembly of the roof truss of Figure 19-2 is illustrated in Figure 19-3. However, note that here the crew size has been increased to four members instead of the two members used in the flow process chart of Figure 19-1.

Crew balance charts are sometimes referred to as *multiman charts.* Charts showing both crew activities and machine utilization are called *man-machine charts* or *multiman-and-machine charts.*

Human Factors

In attempting to improve construction productivity and cost-effectiveness, it is important to remember that people are the essential element in the construction process. Workers who are fatigued, bored, or hostile will never perform at an optimum level of effectiveness. Some major human factors to be considered include environmental conditions, safety conditions, physical effort requirements, work hours, and worker morale and motivation. Safety and health considerations, including work in extreme heat and cold, are discussed in Chapter 18.

There are several considerations involved in assessing the effect of physical exertion on workers. It has been found that the maximum long-term rate of human energy expenditure for the average worker is approximately equal to the energy expended in walking. Attempts at sustained higher levels of effort will only result in physical fatigue and lower performance. Therefore, physical work requirements should be adjusted to match worker capability. For example, Frederick Taylor found in his early studies that the best performance of workers loading material using hand shovels was obtained by matching shovel style, size, and weight to individual worker characteristics. Physical fatigue can be caused by holding an object in a fixed position for an extended period of time as well as by overexertion.

Studies (see reference 9) have shown that worker productivity is seriously reduced by sustained periods of overtime work. In general, a 40-hour workweek appears to be the optimum for U.S. construction workers. When construction workers are put on a scheduled overtime basis, productivity usually drops sharply during the first week, recovers somewhat during the following three weeks, then continues to decline until it finally levels off after about 9 weeks. When first put on overtime, total worker production per week is initially higher than for a standard 40-hour week. However, as productivity continues to decline, the total output for a 50-hour or 60-hour week falls to that of a 40-hour week after about 8 weeks. When the premium cost of overtime is considered, it is apparent that the labor cost per unit of production will always be higher for overtime work than for normal work. As the length of

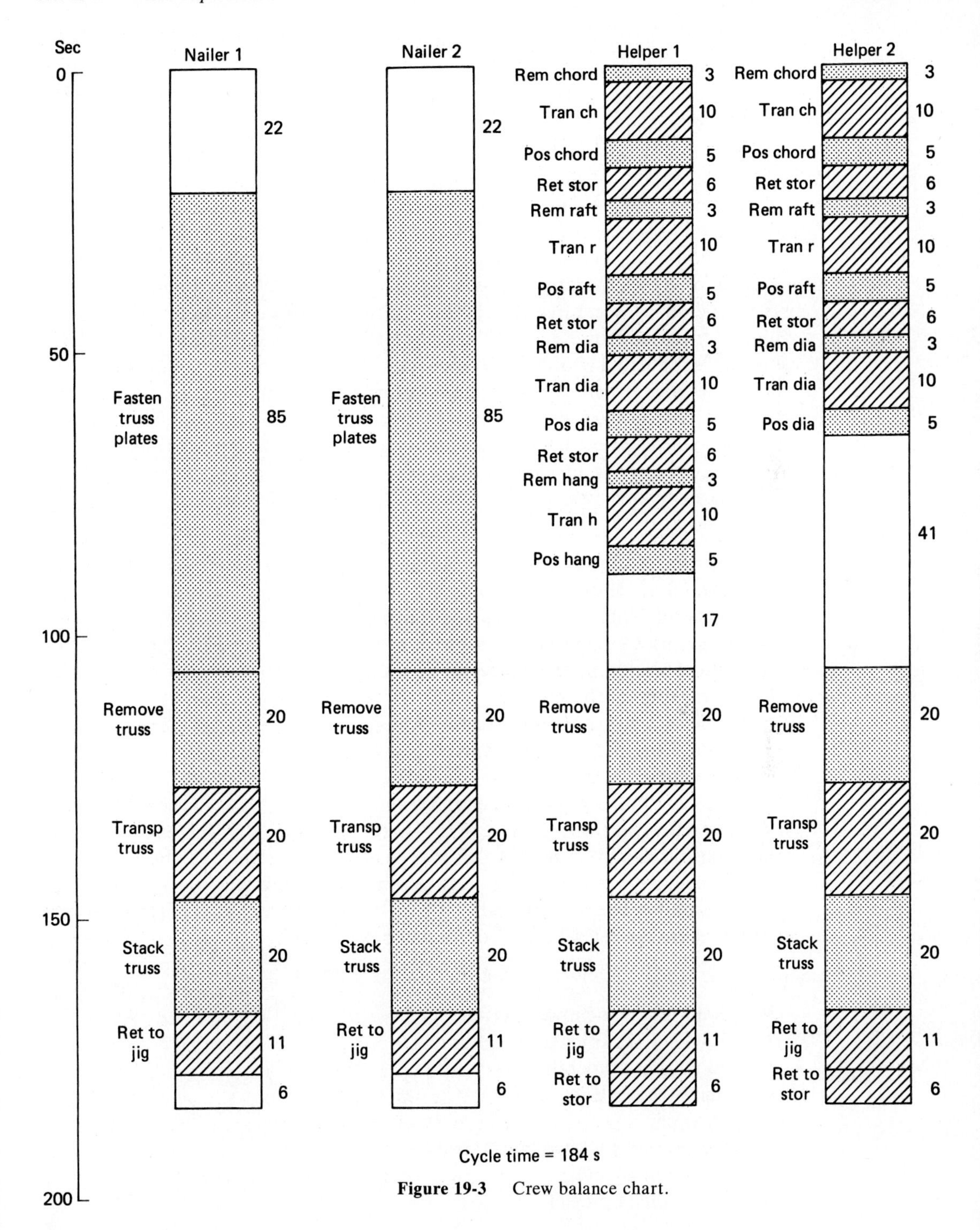

Figure 19-3 Crew balance chart.

the overtime period increases, the cost differential becomes sizable. For example, if the hourly pay rate for overtime work (work beyond 40 hours) is 150% of the standard rate, the labor cost per unit of production for a 60-hour week after 8 weeks would be more than 80% higher than for a 40-hour week.

Worker morale and motivation have also been found to be important factors in construction worker productivity. In studies of 12 large power-plant construction projects, Borcherding and Garner (reference 1) analyzed the factors inhibiting craft productivity on these projects. Of the factors studied, nonavailability of material was the most significant, followed by nonavailability of tools, and the need to redo work. Interfacing of crews, overcrowded work areas, delays for inspection, craft turnover, absenteeism, changes in foremen, and incompetence of foremen were also found to inhibit productivity. However, these factors were much less significant than were the first three factors cited above. The same study identified a number of circumstances that acted as worker motivators or demotivators on these projects. As might be expected, the most productive projects tended to have the highest number of worker motivators and the lowest number of worker demotivators. It appears from the study that the presence of worker demotivators has more effect on productivity than does the presence of worker motivators. Some of the worker demotivators identified by the study included:

- Disrespectful treatment of workers.
- Lack of sense of accomplishment.
- Nonavailability of materials and tools.
- Necessity to redo work.
- Discontinuity in crew makeup.
- Confusion on the project.
- Lack of recognition for accomplishments.
- Failure to utilize worker skills.
- Incompetent personnel.
- Lack of cooperation between crafts.
- Overcrowded work areas.
- Poor inspection programs.
- Inadequate communication between project elements.
- Unsafe working conditions.
- Workers not involved in decision making.

Some of the worker motivators identified in the study include:

- Good relations between crafts.
- Good worker orientation programs.
- Good safety programs.
- Enjoyable work.
- Good pay.
- Recognition for accomplishments.
- Well-defined goals.
- Well-planned projects.

19-3 QUANTITATIVE MANAGEMENT METHODS

The science that uses mathematical methods to solve operational or management problems is called *operations research.* After World War II, operations research techniques began to be employed by industry to provide management with a more logical method for making sound predictions and decisions. Basically, these techniques deal with the allocation of resources to various activities so as to maximize some overall measure of effectiveness. Although a number of mathematical optimization techniques are available, linear programming is by far the most widely used for management purposes. In this section we consider briefly the application of linear programming to construction management.

Linear Programming — Graphical Solution

As the name implies, all relationships considered in linear programming must be linear functions. To apply linear programming, it is necessary to have a set of linear *constraint* (boundary) *equations* and a linear *objective function* which is to be maximized or minimized. We consider first a graphical solution technique that may be employed when only two variables are present. This relatively simple case, which will be illustrated by Example 19-1, should enable us to visualize the nature of the solution procedure. As you recognize, it would be impossible to use a graphical procedure to solve a problem involving more than three variables.

Example 19-1

PROBLEM A project manager for a large earthmoving project is faced with the task of selecting the dozers to be used on a relatively remote project. The project manager is advised by the equipment division manager that both heavy and medium dozers are available for the project. However, only 10 heavy dozers are available. The supply of medium dozers is relatively unlimited. Because of time and transportation limitations, a maximum of 1080 tons of dozers may be transported to the site. The project manager also has the following information on dozer performance and weight.

Dozer	Weight (tons)	Production Index
Heavy	60	2 units/day
Medium	40	1 unit/day

SOLUTION We must first formulate the constraint equations defining the limits of the solution. Obviously, the number of each type of dozer must be zero or greater. (This assumption is implicit in all linear programming solution procedures.) If we let X_1 represent the number of heavy dozers and X_2 represent the number of medium dozers, these constraints become

$$X_1 \geq 0 \tag{1}$$

$$X_2 \geq 0 \tag{2}$$

Another constraint is that the number of heavy dozers cannot exceed 10. Hence

$$X_1 \leq 10 \tag{3}$$

Finally, the maximum weight to be transported is 1080 tons. Hence

$$60X_1 + 40X_2 \leq 1080 \tag{4}$$

After establishing the constraints, we must define the objective function that is to be maximized or minimized. In this case we want to maximize some measure of production of the dozer fleet. Since each heavy dozer will produce twice as much as each medium dozer, the objective function can be expressed as

$$\text{(maximize)} \quad 2X_1 + X_2$$

Summarizing the equations, we have:

Constraints:

$$X_1 \geq 0 \tag{1}$$
$$X_2 \geq 0 \tag{2}$$
$$X_1 \leq 10 \tag{3}$$
$$60X_1 + 40X_2 \leq 1080 \tag{4}$$

Objective function:

$$\text{(maximize)} \quad 2X_1 + X_2$$

The graphical solution procedure is illustrated in Figure 19-4. The feasible region for a solution as defined by the constraint equations is shown in Figure 19-4a. In Figure 19-4b, the objective function has been set equal to 10 and to 20. Notice that the objective function can be represented by a family of lines having a slope of -2. As this line is moved away from the origin, the value of the objective function increases. Since we wish to maximize the value of this function, the optimum value of the objective function will be obtained when this line is as far from the origin as possible while remaining in the feasible region. In this case, the optimum value occurs at the point (10, 12) defined by the intersection of constraints 3 and 4. Following the usual convention for mathematical optimization procedures, we designate optimum values using an asterisk. Hence for our problem

$$X_1^* = 10$$
$$X_2^* = 12$$
$$\text{OF}^* = 32$$

As the reader may recognize, this very simple problem could easily be solved by analytical procedures. However, for more complex problems, linear programming usually provides a much faster and simpler solution procedure than do other techniques.

Computer Solution

Since the graphical solution technique cannot be used for problems involving more than three variables, a more general solution procedure is required. A manual solution algorithm, the simplex method, is available for solving the general linear pro-

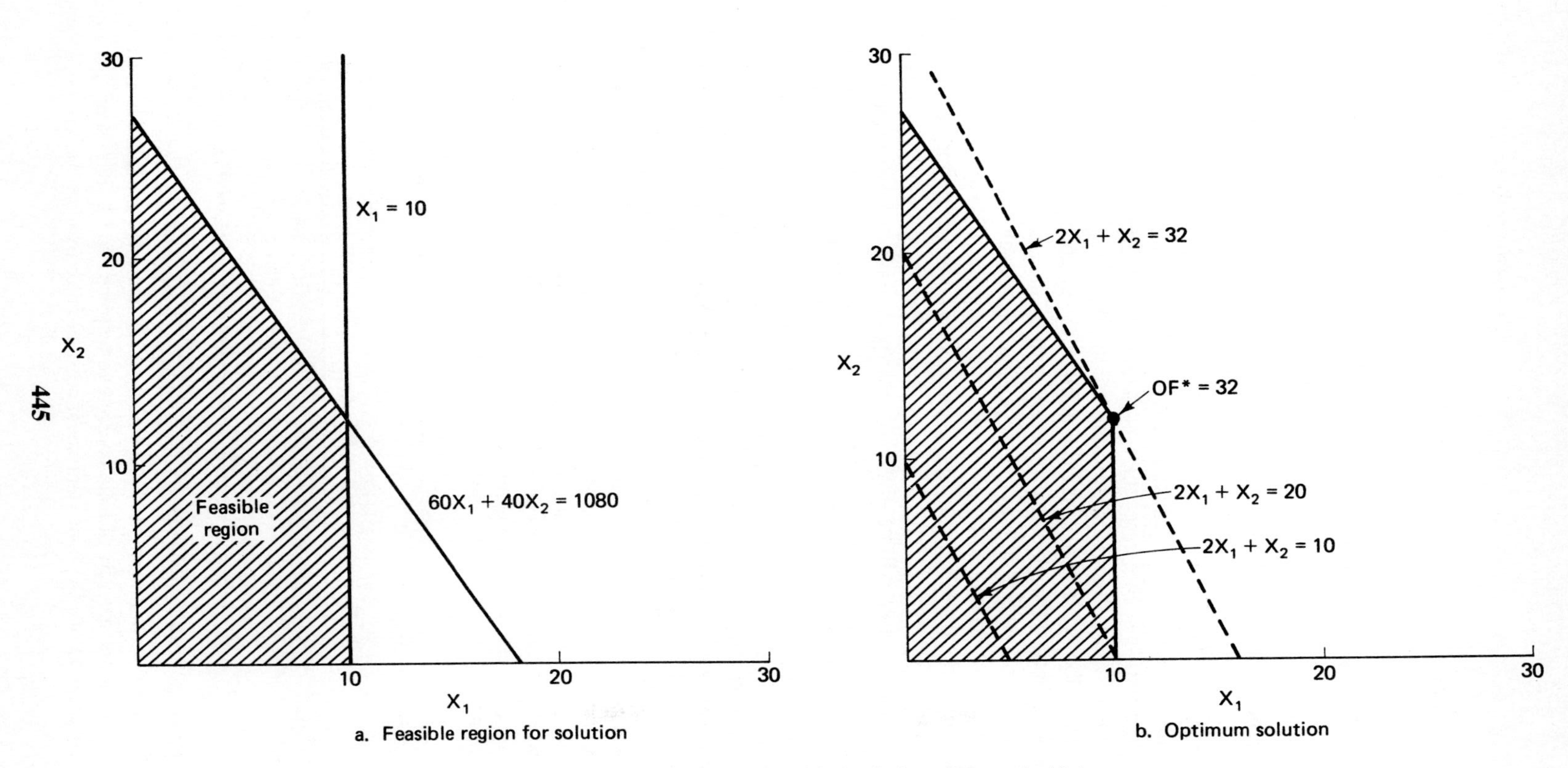

Figure 19-4 Graphical solution of Example 19-1.

gramming problem. However, the procedure is computationally cumbersome and, therefore, of practical value only for the solution of small problems. Moreover, computerized solution techniques are available which can rapidly solve linear programming problems involving thousands of variables and constraints.

Because of the wide availability of microcomputers and linear programming software, this discussion will be confined to the use of computers for solving the general linear programming problem. As we have learned, the essential elements in the formulation of a linear programming problem are a set of constraint equations and an objective function. Care must be taken to ensure that constraint equations are not mutually exclusive, which would result in there being no feasible region for a solution. In such a case, the computer output will advise you that no feasible solution exists. While linear programming computer programs differ somewhat, the user is normally required to enter the number of variables, number of constraint equations, and whether the objective function is to be maximized or minimized. Then, for each constraint equation, enter the coefficient for each variable, the equality relationship ($\leq$, $=$, or $\geq$), and the right-hand-side constant. Finally, the coefficient of each variable in the objective function is entered. The program output will indicate whether a feasible solution exists. If it does, the optimum value of each variable and of the objective function will be given. A sensitivity analysis is often provided which indicates the effect on the objective function resulting from a unit change in the right-hand-side constant of each binding constraint. A computer solution produced by a microcomputer for the problem of Example 19-2 is shown in Figure 19-5.

Example 19-2

PROBLEM A paving contractor is planning his work schedule for the following week. He has a choice of either of two types of concrete, plain concrete or concrete with an additive. The use of additive concrete reduces concrete finishing time but increases the time required for placement. Cost records indicate that the contractor can expect a profit of $4.00 per cubic yard for plain concrete and $3.00 per cubic yard for additive concrete. Naturally, the objective of the contractor is to maximize his profits. However, he does not want to hire additional workers. Labor requirements [in man-hours per cubic yard (mh/cy)] for each type of concrete are given below. Assuming that sufficient demand exists, how many cubic yards of each type of concrete should the contractor place the following week? The contractor works a 40-hour week.

		Labor Required	
Type	Number	Plain	Additive
Foreman	2	0.11 mh/cy	0.21 mh/cy
Laborer	11	0.81 mh/cy	1.01 mh/cy
Finishers	4	0.52 mh/cy	0.22 mh/cy

SOLUTION From the table of labor requirements it is determined that 80 foreman man-hours, 440 laborer man-hours, and 160 finisher man-hours are available each week. We will let X_1 represent the quantity of plain concrete to be placed and X_2 represent the quantity of additive concrete. Hence the constraint equations and objective function are as follows:

```
THE FOLLOWING LINEAR OPTIMIZATION MODEL WILL BE MAXIMIZED

THE OBJECTIVE FUNCTION =      +4.000X 1     +3.000X 2

SUBJECT TO THE FOLLOWING CONSTRAINTS

          +0.11X 1     +0.21X 2   <=   80
          +0.81X 1     +1.01X 2   <=   440
          +0.52X 1     +0.22X 2   <=   160

THE FEASIBLE SOLUTION FOUND AFTER    2    ITERATIONS

    ITERATION          3                   OBJECTIVE =       0
    ITERATION          4                   OBJECTIVE =       1230.77
    ITERATION          5                   OBJECTIVE =       1600

VARIABLES IN THE SOLUTION

    VARIABLES          2                   AMOUNT =          282.353
    VARIABLE           1                   AMOUNT =          188.235

VARIABLES OUT OF THE SOLUTION

BINDING CONSTRAINTS

    CONSTRAINT         3                   SHADOW PRICE = 6
    CONSTRAINT         1                   SHADOW PRICE = 8

SLACK CONSTRAINTS

    CONSTRAINT         2                   SLACK =           2.35294

THE OPTIMUM OBJECTIVE FUNCTION IS           1600
```

Figure 19-5 Computer solution of Example 19-2.

Constraints:

$$0.11X_1 + 0.21X_2 \leq 80$$
$$0.81X_1 + 1.01X_2 \leq 440$$
$$0.52X_1 + 0.22X_2 \leq 160$$

Objective function:

$$\text{(maximize)} \quad 4.00X_1 + 3.00X_2$$

The optimum solution, shown in Figure 19-5, is

$$X_1^* = 188.235 \text{ cu yd (plain concrete)}$$
$$X_2^* = 282.353 \text{ cu yd (additive concrete)}$$
$$\text{OF}^* = \$1600.00 \text{ (profit)}$$

The shadow prices shown for the binding constraints indicate the amount by which the objective function (profit) would be increased if the respective constraint constant were increased by one unit. That is, profit would be increased by $8.00 if the number of foreman man-hours available were increased to 81. Similarly, profit would be increased by $6.00 if the number of finisher man-hours were increased to 161. Note also that there are slightly over 2 excess laborer man-hours available at the optimum solution.

19-4 COMPUTERS AND OTHER TOOLS

Computers in Construction

A number of the end-of-chapter problems in the preceding chapters have illustrated the use of small computers for solving construction engineering and management problems. For the most part, solutions to these problems have been obtained using computer programs written in a traditional computer language such as BASIC, FORTRAN, or Pascal. However, there is a growing library of packaged computer software as well as special-purpose computer programming languages available for small computers. Many of these can be profitably employed by the construction manager. Some of the widely used software packages include word processors, electronic spreadsheets, data base programs, communications programs, graphics programs, and project management programs. Integrated software packages that include several of these programs utilizing a common file structure are also available. Some of the software written specifically for the construction industry includes estimating programs, bidding programs, project management programs, and programs for maintaining cost and performance data for equipment and labor.

Word processors are used for general correspondence as well as for the preparation of memos, reports, training manuals, and procedures manuals. They are particularly useful for the preparation of repetitive documents such as contract specifications, where much of the material is standard (often called "boiler plate"), but are modified somewhat for each specific project. Word processors often have associated spelling checkers which identify words not contained in its standard dictionary. The user can either correct the word, accept it without adding it to the dictionary, or add it to the dictionary. Mailing-list programs are also available for many word processors. These enable the user to prepare form letters and associated address files. Address files can be used to prepare mailing labels or envelopes as well as to merge names, addresses, and other data into form letters.

An *electronic spreadsheet* is a more powerful form of the familiar row-and-column spreadsheet used for tabulating such data as quantity, unit cost, total cost, sales price, and profit. By allowing the user to specify mathematical relationships between cells, results for any input data can be quickly calculated. For example, the value for each row of column 4 may be specified as that obtained by multiplying column 2 by column 3. Many electronic spreadsheet programs also contain built-in functions such as interest calculations, loan amortization, present value, future value, and internal rate of return. The use of such a program will enable the manager quickly to determine the effect produced by any change in the assumed or actual data.

Data base programs are used to organize, maintain, and manipulate a collection of data. Special-purpose data base programs are written for a specific purpose such as inventory control. Although such specialized programs are relatively easy to learn and put into use, they can be used only for the specific purpose for which they were written. Therefore, general-purpose data base programs are much more widely used. General-purpose programs are very flexible but must be customized for each specific application. The two major types of general-purpose data base programs are file managers and relational data bases. File managers are simpler and usually less expensive than are relational data bases. However, they can access only one data file at a time. Relational data bases, on the other hand, are capable of using data from a number of files at the same time. For example, they are capable of integrating material, labor, and equipment cost data from many different cost files to produce total project cost.

Communications programs are used for communications between a small computer and another small computer or a large computer. They are also used to access such information services as electronic mail or electronic bulletin boards. Used with a computer and modem, they allow computer communications over an ordinary telephone line.

Graphics programs greatly speed up and facilitate the preparation of graphic material. They are widely used for preparing charts and other illustrations for reports and presentations. Computer-aided design and drafting programs are becoming widely used for construction design.

Project management programs are usually built around the network planning techniques described in Chapter 15. They often provide for maintaining and forecasting cost and resource data as well as time data. Some programs also contain functions capable of resource leveling.

Advanced Techniques

In addition to the quantitative management techniques and computer software described above, there are several more advanced techniques available to the progressive construction manager. For example, there are various optimization techniques available for the solution of optimization problems involving nonlinear functions. Computer simulation is a powerful tool for analyzing problems not easily solved by analytical methods. The application of simulation techniques to network planning methods has been described in Chapter 15. Construction operations such as earthmoving may be simulated by the use of packaged simulation programs or by writing a simulation program in a simulation language. Reference 6 describes several earthmoving simulation programs written in the GPSS simulation language.

In this day of rapid technological advance, one can never predict the exact impact of new technology on construction. As we have seen, the wide availability of the microcomputer or personal computer has placed a powerful new tool at the disposal of all construction professionals. Computers have already begun to be applied to the control systems of earthmoving equipment.

19-5 ROBOTS IN CONSTRUCTION

Robots, or manipulator machines controlled by computer have been employed on industrial production lines for some years. As the technology has improved, robots have found increasing use in a number of industries, including the automobile manufacturing industry. Advantages of robots over human workers include higher speed, greater accuracy, absence of worker fatigue or boredom, and the ability to work under hazardous conditions without endangering worker health. While robots do displace some production workers, they increase the demand for skilled workers to design, manufacture, program, and maintain the robots.

Despite the advantages that robots can offer, robot manufacturers and construction firms have been slow to apply robots to construction tasks. Many argue that the field environment and unique characteristics of each construction project make the use of robots impractical for construction. Despite these obstacles, progress is being made in the application of robots and automated equipment to construction tasks.

Recent Developments

Considerable research and development work on the use of robots in construction is taking place in universities and construction research facilities, particularly in Japan. Some of the construction tasks to which automation and robotics have been successfully applied are described in references 4 and 8. These include:

BUILDING CONSTRUCTION

- Finishing of concrete floor slabs.
- Fireproofing of structural steel after erection.
- Positioning steel members for steel erection.
- Spray linings for silos and similar structures.
- Surfacing of walls and other building components.

HEAVY CONSTRUCTION

- Automated asphalt and concrete plants.
- Automated excavators.
- Automated tunnel boring machines.
- Concrete demolition in radioactive areas.
- Manufacture of precast concrete beams.

The Future

It can be expected that the use of automation and robots in construction will grow as successful applications are demonstrated and proven. Certainly, tomorrow's construction professional faces an exciting future.

PROBLEMS

1. What were the principal conclusions of the Business Roundtable CICE study of the U.S. construction industry?

2. What is the purpose of work sampling, and how is it performed as applied to construction?

3. Explain the effect of sustained overtime on the labor cost per unit of construction production.

4. Prepare a flow process chart for precutting the chords of the roof truss of Figure 19-2 using a single table-mounted power saw. Notice that two cuts at different angles (a vertical or plumb cut at one end and an angle cut at the other end) must be made on each chord. Two chords are required per truss. The steps in the process to be charted are as follows: a piece of raw material is removed from the storage pile, carried to the saw, positioned on the saw table, one cut is made, and the partially precut piece is removed and placed in a temporary storage pile. After the pieces have all been cut, the saw is reset for the second-cut angle and the process repeated for the second cut. However, after the second cut the piece is placed into a precut storage pile for use in truss assembly. Use the following job planning data in preparing your flow process chart. The subject to be charted is material.

   Hand transport rate, loaded = 2.5 ft/s (0.76 m/s)
   Hand transport rate, unloaded = 4.5 ft/s (1.37 m/s)
   Job efficiency = 40 min/h
   Make saw cut = 2 s
   Position piece at saw = 2 s
   Power saw to precut storage = 25 ft (7.6 m)
   Power saw to temporary storage = 15 ft (4.6 m)
   Raw material stack to saw = 15 ft (4.6 m)
   Remove cut piece from saw = 3 s
   Remove material from stack = 3 s
   Stack material = 3 s

5. (a) The control factor for the process of precutting the rafters of the truss of Figure 19-2 is 12 s based on the cutting rate of one saw. What is the maximum number of precut rafters that can be produced using one saw in a 40-min hour when labor supply is unlimited?
   (b) Using a crew of two workers (one worker carrying material and one saw operator) with one saw the cycle time for precutting rafters is 52 s. Using this crew, how long would it take to precut the rafters for 200 trusses?

6. How could the process of precutting chords described in Problem 4 be made more efficient?

7. What is an electronic spreadsheet computer program?

8. Why have few robots been employed by the construction industry?

9. Solve the problem of Example 19-2 using the graphical solution technique for linear programming.
10. Use a linear programming computer program to solve the problem of Example 19-1.

REFERENCES

1. BORCHERDING, JOHN D., AND DOUGLAS F. GARNER. "Work Force Motivation and Productivity on Large Jobs," *ASCE Journal of the Construction Division,* vol. 107, no. C03 (1981), pp. 443–453.
2. *Construction Labor Motivation,* CICE Report A-2. The Business Roundtable, New York, August 1982.
3. DREWIN, F. J. *Construction Productivity.* New York: Elsevier, 1982.
4. KANGARI, ROOZBEH, AND TETSUJI YOSHIDA. "Prototype Robotics in Construction Industry," *ASCE Journal of Construction Engineering and Management,* vol. 115, no. 2 (1989), pp. 284–301.
5. MUNDEL, MARVIN E. *Motion and Time Study,* 6th ed. Englewood Cliffs, N.J.: Prentice Hall, 1985.
6. NUNNALLY, S.W. *Managing Construction Equipment.* Englewood Cliffs, N.J.: Prentice Hall, 1977.
7. OGLESBY, C. H., H.W. PARKER, AND G. A. HOWELL. *Productivity Improvement in Construction.* New York: McGraw-Hill, 1989.
8. PAULSON, BOYD C., JR. "Automation and Robotics for Construction," *ASCE Journal of Construction Engineering and Management,* vol. 111, no. 3 (1985), pp. 190–207.
9. *Scheduled Overtime Effect on Construction Projects,* CICE Report C-2. The Business Roundtable, New York, November 1980.
10. STARK, ROBERT M., AND ROBERT H. MAYER, JR. *Quantitative Construction Management.* New York: Wiley, 1983.

APPENDIX A

METRIC CONVERSION FACTORS

Multiply English Unit	By	To Obtain Metric Unit
foot	0.3048	meter (m)
square feet	0.0929	square meters (m^2)
square yard	0.8361	square meters (m^2)
cubic feet	0.0283	cubic meters (m^3)
cubic yards	0.7646	cubic meters (m^3)
gallon	3.785	liter (ℓ)
gallons per square yard	4.527	liters per square meter (ℓ/m^2)
inch	2.540	centimeter (cm)
miles	1.609	kilometers (km)
miles per gallon	0.4251	kilometers per liter (km/ℓ)
miles per hour	1.609	kilometers per hour (km/h)
pound	0.4536	kilogram (kg)
	4.448	newton (N)
pounds per cubic foot	0.01602	gram per cubic centimeter (g/cm^3)
	16.02	kilograms per cubic meter (kg/m^3)
pounds per cubic yard	0.5933	kilograms per cubic meter (kg/m^3)
pounds per square inch	0.06895	bar
	6.895	kilopascals (kPa)
pounds per square foot	47.88	newtons per square meter (N/m^2) or Pascals (Pa)
pounds per square yard	0.5425	kilograms per square meter (kg/m^2)
	5.320	newtons per square meter (N/m^2)
ton (2000 lb)	0.9072	metric ton (t)
	8.896	kilonewtons (kN)

APPENDIX B

Construction Industry Organizations

In the preceding chapters, reference has been made to the following construction industry organizations.

American Concrete Institute
22400 West Seven Mile Road, Detroit, MI 48219

American Institute of Steel Construction, Inc.
400 North Michigan Avenue, Chicago, IL 60611

American Plywood Association
P.O. Box 11700, Tacoma, WA 98411

American Society of Civil Engineers
345 East 47th Street, New York, NY 10017-2398

Asphalt Institute
P.O. Box 14052, Lexington, KY 40512-4960

Associated General Contractors of America, Inc.
1957 E Street N.W., Washington, DC 20006

Brick Institute of America
11490 Commerce Park Drive, Suite 300, Reston, VA 22091

Business Roundtable
200 Park Avenue, New York, NY 10166

Concrete Reinforcing Steel Institute
933 North Plum Grove Road, Schaumburg, IL 60195

Construction Industry Manufacturers Association
111 East Wisconsin Avenue, Milwaukee, WI 53202

Contractors Pump Bureau
P.O. Box 5858, Rockville, MD 20855

National Forest Products Association
1619 Massachusetts Avenue, Washington, DC 20036

Portland Cement Association
5420 Old Orchard Road, Skokie, IL 60077-1083

Prestressed Concrete Institute
201 North Wells Street #1410, Chicago, IL 60606

Steel Joist Institute
1205 48th Avenue North, Suite A, Myrtle Beach, SC 29577

Index

C

D

E

F

G

H

I

J

K

L

M

N

O

P

Q

R

S

T

U

V

W

Z